FUTURE COMMUNICATION, INFORMATION AND COMPUTER SCIENCE

PROCEEDINGS OF THE INTERNATIONAL CONFERENCE ON FUTURE COMMUNICATION, INFORMATION AND COMPUTER SCIENCE, BEIJING, CHINA, 22–23 MAY 2014

Future Communication, Information and Computer Science

Editor

Dawei Zheng
International Research Association of Information and Computer Science, Beijing, China

CRC Press
Taylor & Francis Group
Boca Raton London New York Leiden

CRC Press is an imprint of the
Taylor & Francis Group, an **informa** business

A BALKEMA BOOK

CRC Press/Balkema is an imprint of the Taylor & Francis Group, an informa business

© 2015 Taylor & Francis Group, London, UK

Typeset by MPS Limited, Chennai, India
Printed and bound in Great Britain by CPI Group (UK) Ltd, Croydon, CR0 4YY.

Published by: CRC Press/Balkema
                P.O. Box 11320, 2301 EH Leiden, The Netherlands
                e-mail: Pub.NL@taylorandfrancis.com
                www.crcpress.com – www.taylorandfrancis.com

ISBN: 978-1-138-02653-7 (Hardback)
ISBN: 978-1-315-75229-7 (eBook PDF)

Future Communication, Information and Computer Science – Zheng (Ed.)
© 2015 Taylor & Francis Group, London, 978-1-138-02653-7

Table of contents

Preface

We cordially invite you to attend the 2014 International Conference on Future Communication, Information and Computer Science (FCICS 2014), Beijing, China during May 22–23, 2014. The main objective of FCICS 2014 is to provide a platform for researchers, engineers, academicians as well as industrial professionals from all over the world to present their research results and development activities in Future Communication, Information and Computer Science. This conference provides opportunities for the delegates to exchange new ideas and experiences face to face, to establish business or research relations and to find global partners for future collaboration.

FCICS 2014 received over 280 submissions which were all reviewed by at least two reviewers. As a result of our highly selective review process about 109 papers have been retained for inclusion in the FCICS 2014 proceedings, less than 40% of the submitted papers. The program of FCICS 2014 consists of invited sessions, and technical workshops and discussions covering a wide range of topics. This rich program provides all attendees with the opportunity to meet and interact with one another. We hope your experience is a fruitful and long lasting one. With your support and participation, the conference will continue its success for a long time.

The conference is supported by many universities and research institutes. Many professors play an important role in the successful holding of the conference, so we would like to take this opportunity to express our sincere gratitude and highest respects to them. They have worked very hard in reviewing papers and making valuable suggestions for the authors to improve their work. We also would like to express our gratitude to the external reviewers, for providing extra help in the review process, and to the authors for contributing their research results to the conference. Special thanks go to our publisher. At the same time, we also express our sincere thanks for the understanding and support of every author. Owing to time constraints, imperfection is inevitable, and any constructive criticism is welcome.

We hope you will have a technically rewarding experience, and use this occasion to meet old friends and make many new ones. Do not miss the opportunity to explore in Beijing, China. And do not forget to sample the many and diverse attractions in the rest of the China.

We wish all attendees an enjoyable scientific gathering in Beijing, China. We look forward to seeing all of you at next year's conference.

The Conference Organizing Committees
May 22–23, 2014
Beijing, China

Organizing Committees

General co-Chairs

E. Ariwa, *London Metropolitan University, UK*
M. Mansor, *University Malaysia Perlis, Malaysia*

Publication Chair

R. Zheng, *International Research Association of Information and Computer Science, China*

Technical Committees

A. M. Leman, *Universiti Tun Hussein Onn Malaysia, Malaysia*
Y. Bai, *General Research Institute for Nonferrous Metals, China*
L. Bi, *Beijing Institute of Specialized Machinery, China*
G. S. Chen, *Marshall University, USA*
H. AKEB, *ISC Paris Business School, France*
S. Arun Kumar, *VIT University, India*
S. Saravanan, *SASTRA University, India*
H. Yaghoubi, *Iran Maglev Technology (IMT), Iran*
N. Mollaee, *University of Applied Science and Technology, Iran*
G. Z. Chen, *Heriot-Watt University, UK*
Z. Wang, *University of Kansas, USA*

Future Communication, Information and Computer Science – Zheng (Ed.)
© 2015 Taylor & Francis Group, London, 978-1-138-02653-7

The boundary of cell density in mobile communication networks

M. Zhang, P. Butovitsch & Y. Wu
Technology, Ericsson Region Northeast Asia, Beijing, China

ABSTRACT: We investigate the cell density that maximizes capacity in mobile cellular networks. The study is based on system-level simulations. The results show different trends for downlink and uplink performance when continuously adding cells within open areas. For the downlink of WCDMA/HSPA, the system performance deteriorates for inter-site distances below 200 meters. Careful considerations have to be taken if an operator intends to improve network performance by adding cells to their networks in dense urban cities.

Keywords: interference; cellular networks; WCDMA/HSPA; LTE; densification; performance

1 INTRODUCTION

Cell splitting or densification is the most fundamental rule for network capacity expansion in cellular networks, and has been deployed all over the world as mobile communication develops. In the dense urban areas of modern cities, an Inter-Site Distance (ISD) in the range of 100 m is already a reality in certain areas in e.g. Japan or Hong Kong.

With the forthcoming 50 billion connections and explosion of mobile broadband data, heterogeneous networks with a combined approach consisting of improving, densifying and adding cells, are the future way to enhance network capacity and performance [1]. However, in a high interference scenario, with many Line-Of-Sight (LOS) interferers around, will keep adding cells, regardless of the power or form of the cell to be added, always help to improve the network performance? Is there a boundary in mobile communication networks where added interference will out-play the benefits of splitting gain? Or is there an upper limit where we cannot squeeze more capacity/performance gains by adding cells? In this paper, we try to investigate this problem.

Let us describe one example that was observed in a city of north-east Asia: along the street many macro cells are deployed. Around the center of the area, a small cell is planned to be added to the network with the intention to improve the network performance. The macro cells are all deployed on the top of the buildings on the sides of the street, thus many of them have LOS interference with the small cell which forms an open-space area with free space propagation. Another example in real networks are areas around a huge stadium complex or festival venue, e.g., the national stadium in Beijing where tens of cells could be deployed in an area of less than 1 square kilometer to handle the traffic generated by an extremely high density of users during events like the Olympic Games.

The cells mostly have LOS interference from each other in the open area within and around the stadium. Under such circumstances, careful considerations have to be made as to whether adding cell is preferred in order to improve network performance.

In this paper we mainly focus on the High Speed Packet Access (HSPA) technology in the Wideband Code Division Multiple Access (WCDMA) evolution track. On one hand, HSPA is still the main technology for mobile broadband access in terms of subscriptions worldwide and will still be one of the most important mobile broadband access technologies in many years to come [2]. With the largest subscriber base and their increasing packet data usage, the high-interference scenario under investigation is already happening in HSPA systems and will be a reality in many more urban cities in the near future. Therefore an investigation of this problem in HSPA systems has important practical implications. As a comparison, LTE systems have been in commercial operations only for a couple of years. The loads of the LTE systems are still in a low to moderate range. Nonetheless, in Section IV we will also briefly address LTE systems with regard to the problem under investigation. On the other hand, the pilot pollution is a well-known problem for WCDMA systems in both literature [3, 4] and practice of mobile network optimization [5]. However, the boundary of adding cells in the context of pilot interference has not been thoroughly investigated. In [6] we addressed the problem with an analysis of CINR and simulation results on WCDMA systems, and in this paper we further extend the work by extensive simulations on different ISDs and also to the LTE system.

2 METHODOLOGY

In a real network deployment, measurements can be made and network performance indicators can be taken to evaluate a specific deployment. However, such evaluations are highly dependent on implementation specifics such as geographic and building characteristics, user distributions, traffic change, how well the configurations of the network are optimized to specific scenarios, etc. Therefore we want to use a more generalized approach to investigate the problem with system-level simulations.

We chose to simulate a cellular network deployed in an open space area, where the impact of geographic variance and different building models is eliminated. With a common hexagon shaped cellular deployment the surrounding cells will have LOS interference with each other, thus we used a free space propagation model to reflect a worst-case scenario close to the examples mentioned in Section I. It is also noted in [7] that in a very dense urban area where the LOS (Line of Sight) interference dominates, the fading channel model can be approached by an open space model.

The simulated system consists of a large number of cells to fully reflect the interference generated possibly by many layers of surrounding cells. Moreover, the cells are assumed to be identically installed on a pole with omni-directional antennas. Such assumptions were made to simplify irrelevant configuration details like antenna models and to focus the study on the problem of interests. We also use uniformly distributed users with a large user density to simulate a dense urban or CBD scenario. The users will generate a certain amount of data ranging from low to high traffic profile to fully explore the network performance in various load conditions.

3 SIMULATION CONFIGURATIONS

In this section we elaborate on the models and configurations used in the simulation.

3.1 *System model*

As briefly described in the previous section, the system is a cellular network built on open space. There are no buildings and thus no associated penetration loss. Each base station is mounted on a pole of the same height. The antennas are also the same for every base station. In this study we used omni-directional antennas to eliminate variance of the antennas in the study. Therefore each cell is represented by a hexagon in Figure 2. Implementation details of the base stations are not considered in the study either. Nonetheless we will make some comments on the impact of antennas on the results in Section IV.

In the results we simulate 127 sites wrapped-around. In such a large system 7 layers of surrounding cells are generating interference to any cell. To reflect different density of base station deployment, the ISD of the network is changed in a large range. The smallest ISD simulated is 25 m, which could be a typical coverage area of a very low power node, e.g., pico or even femto. The largest ISD simulated is 1 km, which

could be a typical ISD in common urban area, but not a dense urban area.

3.2 *Subscriber traffic model*

The subscriber density is assumed to be 20,000 per square kilometer, which is a high density reflecting dense urban area. The subscribers are distributed uniformly on the ground and every subscriber has the same traffic profile.

Each subscriber consumes a certain amount of bytes per month, which is denoted by X in the following and also serves as the horizontal axis in the result figures. As an input to the simulation which determines the offered traffic of the system, X is varied to a large extent to study the system under various load conditions. In today's commercial networks, monthly data usage varies a lot. Typical monthly data package per user could be several tens of MBytes/month or several hundreds of MBytes/month in many countries. High usage of more than 1 GBytes/month is also a reality for some cities.

The simulation takes snapshots of the system in busy hours. The traffic volume generated by each subscriber on each day is assumed to be the same (no busy day) and on busy hours 6% of the daily traffic volume is consumed. Thus the traffic volume generated by each subscriber on a busy hour is $X/30 \times 6\%$. The ratio between downlink and uplink traffic is assumed to be 2:1 in this paper.

3.3 *Propagation model*

The following equation shows the propagation model, where L is the path loss in dB, f is the frequency of the system under evaluation in gigahertz, and D is the distance between the base station antenna and UE antenna in meters. In this paper we use a common band for the system where the downlink frequency is 2.1 GHz and uplink frequency is 1.9 GHz.

$$L = 32.45 + 20\lg(f) + 20\lg(D) \qquad (1)$$

It is noted that in such a dense deployment scenario where the free space propagation model works, the fading degree of the signal becomes lower and the pilot pollution will be aggravated accordingly.

3.4 *Pilot and control channel overhead model*

To reasonably reflect the interference condition in HSPA systems, we model both downlink and uplink overhead. In the downlink, the overhead consumes a certain amount of power, which includes both pilot and control channels. In the uplink, the overhead consumes a certain percentage of the code trees.

4 RESULTS AND DISCUSSIONS

In this section we present the results and key findings. There are a few cases we have investigated. As a

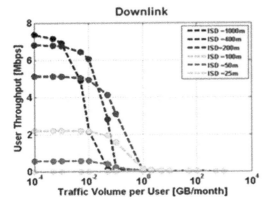

Figure 1. Downlink User Throughput of the baseline case.

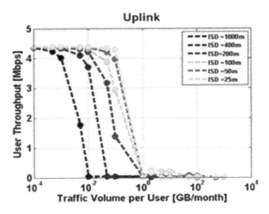

Figure 2. Uplink User Throughput of the Baseline case.

baseline, the base stations are mounted on 20 m height with 20 W maximum output power. The base stations are equipped with a single antenna and the antennas are omni-directional with 10 dBi gain. The UEs are equipped with 2 antennas with both receiving and one transmitting. In downlink, 20% overhead is assumed.

Average user throughput is the key performance indicator we use to evaluate the system performance in this paper. It also represents a typical experience an end user will get in such a system. Nonetheless the trend of the results are of more interests than specific values of the simulations, since specific values of user throughput depend on many factors including user equipment (UE) traffic profile etc.

4.1 Baseline results and findings

Figure 1 plots the mean downlink user throughput of the baseline case versus the monthly traffic volume generated per user. Note that the horizontal axis is not of linear scale. Instead we use an exponential scale to reflect the system performance in a very large load range. Different curves represent different ISDs or site densities.

For a specific curve with certain ISD or site density, the performance in general decreases as the traffic volume or load increases. The user throughput will get to zero after a breakpoint beyond which the system gets too over-loaded and cannot pass through any useful payload to most of the users. The breakpoint for each curve or different site density is somehow different. Of course in commercial networks such a situation would be prevented and the system would not work in the range beyond or even close to the breakpoint. Thus to relate to real networks, the normal working range on the left of the breakpoints shall be of interest.

We see that given a certain offered traffic volume or user data usage, there will be a system with a certain ISD which outperforms systems with different ISDs in terms of mean user throughput. In case of very low traffic, a lower density system performs better, i.e., ISD 1 km outperforms the others. The reason is that

the load of the system is very low under such a traffic condition, and the sparser system consumes less overhead than the denser system, thus the interference generated by the overhead, e.g., by pilot power, is less than with a denser system.

While the traffic volume picks up, the load of the total system and cell load are increased as well. Densification or adding cells is needed to carry the added traffic and load. We see that the optimum ISD changes to a smaller value from 1 km to 400 m to 200 m gradually as traffic increases. Here by "optimum" we mean the system with a certain density the can provide the highest mean user throughput among the curves.

However, when the ISD becomes too small, e.g., 50 m or even 25 m, the system becomes basically "saturated" by the interference generated by overhead. The actual user throughput becomes virtually zero. In such cases the interference generated by overhead e.g., CPICH will deteriorate the signal-to-interference-ratio (SINR) so much that very little or practically no successful data transmission can be made.

Therefore, from Figure 1 we can draw the conclusion that in WCDMA/HSPA systems, adding cells will not always help to improve the system performance in downlink. In some cases adding cells will worsen the performance, and in extreme situations adding cells may even destroy the system performance.

Figure 2 shows the mean uplink user throughput of the baseline case. Different from the downlink results, in the uplink smaller ISD appears not to deteriorate the performance. There are always some gains in user throughput when densification is carried out or cells are added, although the gains seem to diminish gradually after a certain point, e.g., in this case, ISD smaller than 100 m will only give very small or rather negligible gains in terms of user throughput.

The reason behind the difference between downlink and uplink lines is the following: first of all, uplink is more power limited than downlink since the UEs have much less output power than the base stations. Densification or adding cells will get the base station closer to the users thus the signal received at the base station is stronger given a fixed UE output

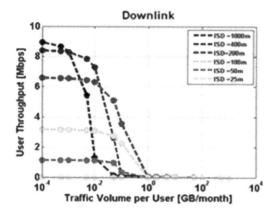

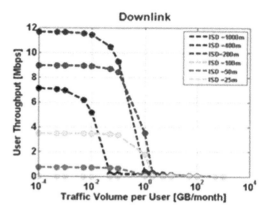

Figure 3. Downlink User Throughput of Case with 10% overhead.

Figure 4. Downlink User Throughput of the Case with 3-sector sites.

power. On the other hand, in downlink the pilot has to be transmitted once the base station is up and running, thus a certain amount of power is consumed in downlink and is always generating interference to data transmission regardless of whether there are active data transmissions in respective base stations; while in uplink there is no such pilot or always-on interference source needed. When a UE doesn't have any active data session, unlike in downlink, it doesn't generate interference to other UEs. Therefore, without a deterioration of SINR due to pilot interference, densification or adding cells will practically share the total offered traffic to more cells, and thus each cell will have less load. The uplink user throughput will be reasonably improved by a lower load per cell.

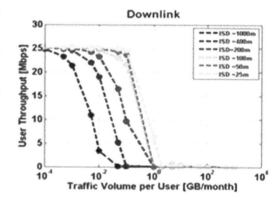

Figure 5. LTE Downlink User Throughput.

4.2 Variances of the baseline

We also simulate the system with variances by changing some of the key parameters to explore the phenomena in a more general context.

Change of output power and base station height: We have changed the base station output power to 5 W and the base station height from 20 m to 5 m. The results show the same trend and conclusion.

Change of overhead percentage: we have changed the overhead percentage from 20% to 10%, which would correspond to e.g., a configuration of smaller CPICH power in reality. The downlink results are shown in Figure 3. As expected, reducing overhead will give a slightly better performance compared to baseline. However, there is no fundamental change in the conclusion.

Finally, we have also simulated a system with 3-sector sites instead of omni sites. The downlink results are shown in Figure 4.

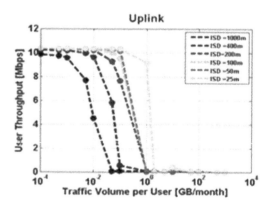

Figure 6. LTE Uplink User Throughput.

4.3 LTE systems

In this sub-section we briefly address the interference saturation problem in LTE systems. As for the simulation of LTE systems the same configurations as in the

baseline results in sub-section 4.1 were used. E.g., the evaluation is also done for LTE systems with 5 MHz bandwidth, with the same number of omni-sites and the same user distribution, traffic profile and so on. The downlink and uplink mean user throughputs are plotted in Figure 5 and Figure 6. For LTE systems, the base stations are equipped with 2 antennas which

4

enable downlink 2×2 MIMO, and the supported modulation schemes are higher than HSPA as is in line with most current live networks, so the absolute values of Figure 5 and Figure 6 are higher than those of Figure 1 and Figure 2 respectively. Nonetheless evolutions of HSPA systems also support MIMO and higher modulation schemes, therefore the absolute values of the figures only reflect the system performance under specific parameter settings of the simulations and the focus of this paper is on the trend of the curves instead of the absolute values.

When we observe the curves in Figure 5, we will see that the trend for LTE systems are different from HSPA systems. The downlink user throughputs are almost always improved by densification. Although the gains of adding cells are diminishing, the performance will not deteriorate. The difference lies in the overhead modeling. In HSPA systems, the pilot and control channel overhead always generate interference to data once powered on, while in LTE systems, the overhead including reference signals consumes only a portion of the resource elements. Since the resource elements are orthogonal with each other, the downlink overhead only generates interference to data transmitted on the same resource elements in surrounding cells. The rest of the resource elements are not affected by overhead interference and are only subject to interference generated by data transmitted on the same resource elements in surrounding cells. As the ISD gets smaller, the traffic load offered to each cell is reduced and thus the chance that same resource elements are used in surrounding cells is getting smaller. In other words, the gain of adding cells plays a dominant role in this case.

The uplink curves in Figure 6 show the same trend as in HSPA systems. The user throughput will not deteriorate with shrinking cells.

5 CONCLUSIONS

In this paper we have investigated the boundary of adding cells in mobile cellular networks through system level simulations. A free-space radio propagation model is used in our study for the following reason: areas characterized by free-space radio propagation, viz. areas where LOS conditions are dominating, become more prevalent when networks are densified and cells become smaller.

The conclusion we can draw from the results is that there is a limit of the cell density for HSPA system beyond which the capacity will be deteriorated by interference. Therefore in terms of downlink user throughput, there will be an optimum point of cell densification or cell size given a certain traffic load in the network. Under the simulated conditions, this optimum ISD is about 200 meters. The uplink performance will, on the other hand, not deteriorate neither in HSPA networks nor LTE networks.

The conclusion has practical implications: when an operator tries to enhance network capacity or performance as traffic increases, adding cells may not be an efficient approach, and can even be destructive to the network performance in certain scenarios. Under such circumstances other solutions have to be taken into consideration to further increase the system performance, e.g., adding carriers or use more advanced technologies.

REFERENCES

[1] Ericsson White Paper "Heterogeneous Networks", http://www.ericsson.com/res/docs/whitepapers/WP-Heterogeneous-Networks.pdf

[2] Ericsson Traffic and Market Report, www.ericsson.com/traffic-market-report

[3] Niemela, J. & Lempiäinen, J. 2004. Mitigation of pilot pollution through base station antenna configuration in WCDMA. VTC2004-Fall: 4270–4274.

[4] Chen, L. 2010. Impact of soft handover and pilot pollution on video telephony in a commercial network. Asia-Pacific Conference on Communications (APCC), 2010.

[5] Zhao, L. 2008. Analysis and optimization of pilot pollution for WCDMA wireless network. Communication Technology, Vol. 41, No. 12, 2008.

[6] Wu, Y. et al. 2014. Capacity Upper Bound for Adding Cells in the Super Dense Cellular Deployment Scenario, VTC2014-Spring.

[7] Rappaport, T.S. 1996. Wireless Communications, Englewood Cliffs, NJ: Prentice-Hall.

Future Communication, Information and Computer Science – Zheng (Ed.)
© 2015 Taylor & Francis Group, London, 978-1-138-02653-7

Factor analysis on regionalized remote sensing satellite data acquisition of territorial resources survey

J.-H. Yang
China Land Surveying and Planning, Beijing, China

P. Huang & G.-B. Ma
Institute of Remote Sensing and Digital Earth, Chinese Academy of Sciences, Beijing, China

L.-S. Shi & Y. Zhan
China Land Surveying and Planning, Beijing, China

Z.-X. Wang
Institute of Remote Sensing and Digital Earth, Chinese Academy of Sciences, Beijing, China

ABSTRACT: This paper analyzed the influential factors of regionalized remote sensing satellite data acquisition of territorial resources survey, elaborated the impact of the factors on regionalized remote sensing satellite data acquisition of territorial resources survey, and clarified the key influential factors of data acquisition. The analytical results showed that regionalized remote sensing satellite data acquisition of territorial resources survey is influenced by various factors, which are related to each other to some extent. Therefore, it is a comprehensive and complex problem. This paper provided the design and implementation of the plan on regionalized remote sensing satellite data acquisition of territorial resources survey with reference and guidance.

Keywords: satellite; sensor; influential factors

1 INTRODUCTION

Along with the rapid development of remote sensing technologies, satellite remote sensing technology has become the core technology and major method for territorial resources survey. Bringing the advantages of remote sensing technologies and striving to promote the application of remote sensing technologies are an important content for strengthening modern scientific and technological support [1]. Featuring high resolution, large breadth, high precision, short acquisition cycle, replication and continuous observation, etc., remote sensing satellite data is the most important data source of territorial resources survey. The data received from the SPOT-5 and SPOT-6 satellites of France, QUICKBIRD, WORLDVIEW, GEOEYE satellites of the United States and ZY-3 satellite of China have been widely applied to territorial resources survey. In the process of remote sensing satellite data acquisition from a target area for territorial resources survey, factors such as features of satellite orbit, capability of sensor, timing of data acquisition, weather and topography will exert an influence on the effect of data acquisition.

In this paper, the influential factors of the results of remote sensing satellite regionalized data acquisition are analyzed and summarized from the perspectives of satellite capacity factors and business Requirement Factors, providing the design and implementation of the plan on regionalized remote sensing satellite data acquisition of territorial resources survey with references.

2 ANALYSIS ON SATELLITE CAPACITY FACTORS

Satellite capacity factors mainly refer to the impact of satellite imaging capability constraints on the result of regionalized remote sensing satellite data acquisition, mainly including sensor capacity factors, satellite orbit factors and constraints on imaging.

2.1 *Sensor capacity factors*

Sensor is the key technology for earth observation [2]. The sensor capacity factor is one of the most essential factors that influence regionalized remote sensing satellite data acquisition of territorial resources survey. Currently, most remote sensing satellites related to territorial resources survey are optical sensor satellites,

Table 1. The factors that need to be considered for different imaging modalities.

Imaging modality	FOV of Sensor	Swing angle of sensor	Swing step of sensor	Swing velocity of sensor
Sub-satellite point scan imaging	√			
Pitch imaging	√	√Pitch angle	√Pitch step	√Pitch velocity
Roll imaging	√	√roll angle	√roll step	√roll velocity
Combination of roll and pitch imaging	√	√Pitch angle √roll angle	√Pitch step √roll step	√Pitch velocity √roll velocity

which is also the focus for the analysis on sensor capacity factors in this paper.

For an optical sensor, its sensor capacity factors mainly include FOV, swing angle, swing step and swing velocity, etc.

1. FOV of the sensor

The FOV of sensor gives an impact on the breadth of the sensor observation area: the greater the FOV, the wider the breadth of the imaging area of the sensor.

2. Swing angle of the sensor

The Swing angle of the sensor consists of the roll angle and the pitch angle. The roll angle impacts the imaging capability of the sensor in the direction vertical to the orbit, while the pitch angle impacts the imaging capability of the sensor along the orbit direction. The impact of the sensor swing angle depends mainly on the roll angle: the larger the roll angle, the stronger the access and re-access capabilities of the sensor. Pitch is mainly used in the same orbit to acquire stereopair and acquire multiple imaging strips from the same orbit in combination with roll, so as to meet the imaging requirements of the area with a relatively large east-west span.

3. Swing step of the sensor

It consists of roll step and pitch step. Roll step impacts the division of sensor imaging strips, while pitch step has no the impact on the imaging strip.

4. Swing velocity of the sensor

It consists of roll velocity and pitch velocity and mainly impacts the time required for the sensor to swing to the designated position or switch over to the roll angle.

Currently, there are the following 4 imaging modalities for common optical sensors:

1) Sub-satellite point scan imaging

Without any swing capability, the sensor scans the earth by completely following the orbit direction of the satellite. Examples are the TM sensor of the LANDSAT-5 satellite and ETM+ sensor of the LANDSAT-7 satellite of the United States.

2) Pitch imaging

Despite the similarity to the sub-satellite point scan imaging sensor, the sensor is capable of acquiring stereopair from the same orbit and its data acquisition capability along the orbit direction is improved.

3) Roll imaging

The roll direction of the sensor is vertical to the direction of the satellite orbit and thus greatly enhances the data acquisition capability of the satellites including the SPOT-5 satellite sensor of France as well as QUICKBIRD and GEOEYE satellite sensors of the United States.

4) Combination of roll and pitch imaging

With both roll and pitch capabilities, the modality significantly enhances the data acquisition capability of the satellite and acquires different imaging strips of the target area from the same satellite orbit, such as the French SPOT-6 satellite sensor.

The factors that need to be considered for different imaging modalities are shown in table 1.

2.2 Satellite orbit factors

Satellite orbit factors mainly consist of semi-major axis (satellite orbit altitude), eccentricity, orbit inclination, RAAN, argument of perigee and the time that the satellite passes the perigee, etc. [3], of which the satellite orbit altitude has a great impact on the imaging capability of the satellite. Under the condition of equal imaging capability of satellite sensors, the higher the satellite orbit altitude, the wider the imaging breadth of the satellite sensor, the lower the corresponding ground spatial resolution, the stronger the access and re-access capabilities; otherwise, the lower the satellite orbit altitude, the narrower the imaging breadth of satellite sensor, the higher the corresponding ground spatial resolution, and the lower the access and re-access capabilities.

Therefore, sensor capacity factors are not only related to the capability of the satellite sensor itself, but also to satellite orbit factors.

2.3 Constraints on imaging

Constraints on imaging include swing frequency in a certain period, the maximum number and length of an imaging strip in a single orbit, the maximum length of a single continuous imaging, multiple continuous imaging interval, solar altitude, etc.

1. Swing frequency in a certain period

It is mainly limited by the capability of the satellite platform, and the roll frequency of the satellite must not exceed the required maximum roll frequency within a certain period.

2. Maximum number and length of the imaging strip in a single orbit

It is a key constraint on the satellite with pitch imaging capability when enabling the pitch imaging modality.

3. Maximum length of single continuous imaging and multiple continuous imaging interval

The maximum length of the satellite single continuous imaging represents the capability of satellite single imaging, while multiple continuous imaging interval represents the distance between the front and rear imaging strips. Its main impact is that, when the maximum imaging length is less than the north-south span of the target area, it is necessary to design the plan on satellite regionalized data acquisition on the basis of the maximum length of satellite single imaging and the interval to the next imaging.

4. Requirements on solar altitude

To ensure high quality of imaging data, solar altitude must be greater than a certain threshold value. Generally, the solar altitude must not be less than 0 degrees.

3 ANALYSIS ON BUSINESS REQUIREMENT FACTORS

Business requirement factors mainly refer to the impact of actual business application requirements on the regionalized data acquisition of a remote sensing satellite. Business requirement factors mainly include target area conditions, requirements on acquisition time, data overlapping ratio, data preference, acquisition efficiency, cloud cover, snow cover, etc.

3.1 *Target area conditions*

The requirements on the design and implementation of the plan on satellite regionalized data acquisition vary with the size, shape and topographic features of the target area.

1. Regional size and shape

The impact of regional size and shape mainly consists of the following 2 aspects. On the one hand, if the area is relatively large (the north-south length of the area is greater than the maximum length of satellite single continuous imaging), it is necessary to select an imaging strip on the basis of the maximum length of satellite single continuous imaging and multiple continuous imaging interval at the time of design and implementation of the plan on satellite regionalized data acquisition. On the other hand, if the north-south length is short, it is necessary to select an imaging strip on the basis of the pitch and roll angles of the satellite at the time of design and implementation of the plan on satellite regionalized data acquisition.

2. Topographic features

If the target area has topographic features of plain, hill, mountain, plateau, etc., in order to ensure that data quality can satisfy the business requirements, when designing the plan on satellite regionalized data acquisition, the requirements on satellite sensor roll angle vary with the area with different topographic features.

At plain area, the requirement of the swing angle is relatively low and a larger sensor swing angle may be accepted; while at mountainous area and plateau, the requirement of the swing angle is relatively high, that is, the swing angle must be less than a certain threshold value, otherwise, the data will be severely deformed and thus cannot be used.

3.2 *Requirements on acquisition time*

Generally, a territorial resources survey is required to be conducted between July and December each year. Exceptionally, it may be extended to January and February of the next year at some areas. In the implementation process of the plan on satellite regionalized data acquisition, data acquisition time shall vary with area due to different weather features. Basically, the acquisition shall not be carried out in cloudy, rainy or snow cover conditions, so as to enhance data acquisition efficiency and ensure data quality.

3.3 *Data overlapping ratio*

Data overlapping ratio refers to overlapping between different imaging strips. According to the imaging plan, the overlapping between imaging strips shall meet the requirement on minimum overlapping ratio. When designing and implementing the plan on satellite regionalized data acquisition, the choice of an imaging strip shall made in a way that ensures that the ratio of overlapping between strips is greater than or equal to the minimum overlapping ratio, otherwise data will be affected.

3.4 *Data preference*

Data preference refers to the users' certain preference for the acquired data. For example, the users may require that the data acquired by a sensor of a satellite cover a certain proportion to the entire target area, that data from several satellites are available and that the data of a certain satellite are given priority.

3.5 *Acquisition efficiency*

When designing and implementing the plan on satellite regionalized data acquisition, some acquired data are likely to cover a very small target area, in this case, the acquisition of such data may be neglected or processed later by setting the acquisition efficiency.

3.6 *Cloud cover*

For an optical imaging satellite, its imaging effect is directly affected by the cloud cover in the atmosphere [4]. Cloud cover has a great impact on the data acquisition quality of a visible light sensor.

Cloud cover is classified into 2 types: the first type is accurate short-term forecast information including the time and the weather of the region during the period (rainy, sunny, overcast and cloudy). It is mainly used for temporary adjustment and change in the implementation process of the plan on satellite regionalized data acquisition. The other type is forecast information

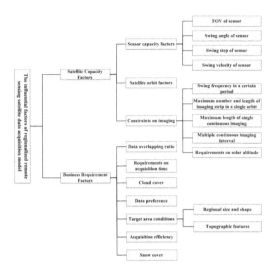

Figure 1. The main influential factors of regionalized remote sensing satellite data acquisition of territorial resources survey.

on the future cloud cover generated through statistical analysis on long-term historical information on cloud cover, including the time and average cloud cover during the period. It is mainly used for design, evaluation and optimization of the plan on satellite regionalized data acquisition.

4 SUMMARY

To sum up, the main influential factors of regionalized remote sensing satellite data acquisition of territorial resources survey are shown in figure 1.

This paper mainly discussed the influential factors of regionalized data acquisition of remote sensing satellite, which are classified into satellite capacity factors and business requirement factors. Moreover, it carried out in-depth analysis on the influential factors and found that sensor capacity factors, target area conditions and cloud cover have particular impact on regionalized data acquisition of territorial resources survey remote sensing satellite. The analytical results showed that regionalized remote sensing satellite data acquisition of territorial resources survey is influenced by many factors, which are related to each other to some extent, and it is a comprehensive and complex problem. This paper provided the design and implementation of the plan on regionalized remote sensing satellite data acquisition of territorial resources survey with reference and guidance.

REFERENCES

[1] Jiang Li-ming, Yang Wu nian, Fang Shi bo. The Application of the Remote Sensing Technology in a New Round of the Survey for the Land and Resources. Tong Qing-xi. The Memoir of China remote sensing advance bravely for twenty years. 2001.
[2] Li Liang-xu, Wu Peng-fei. Ji Jin-hu. An Overview of Remote Sensing Satellite's Sensors at Present. Journal of Xinjiang meteorological. 2003, 26(5):19–22, 26.
[3] Xiao Ya-lun. The spacecraft flight dynamics principle. China Astronautic Publishing House, Beijing, 1995, 43–45.
[4] Shen Ru-song, Song Gui-bao, Lu Wei-min, Peng Shao-xiong. Capabilities of Identifying Targets of Imaging Reconnaissance Satellites. Journal of System Simulation. 2006, 8, 18(2): 34–37.

Future Communication, Information and Computer Science – Zheng (Ed.)
© 2015 Taylor & Francis Group, London, 978-1-138-02653-7

Extending the Deffuant model by incorporating the influence factor

J.P. Cao, H. Wang & F.C. Qiao
Information System and Management School, National University of Defense Technology, Changsha, China

ABSTRACT: Influence is a critical factor for the formation of public opinions. However, it hasn't been considered in classical continuous opinion dynamics. We extended the interacting rules by stressing different influence of agents in Deffuant dynamic systems. Through Monte Calo simulation of our model on Watts-Strogtz (WS) small world, Barab'asi-Albert (BA) network and an ego-network collected from Sina Weibo (weibo.com, a Twitter equivalent in China). We found that if we assign different values to represent individual influence respectively, the time to reach the stationary state gets shorter, but the uncertainty to reach complete consensus gets higher, thus individuals who have bigger influence will lead bias in the final configuration of opinions. Our model is closer to the reality; the conclusions of this paper will enrich our understandings of continuous opinion dynamics.

Keywords: Deffuant model; influence factor; simulation

1 INTRODUCTION

Agreement is a critical aspect for human daily life. There are many circumstances where people need to make a consensus of opinion according to some rules, for example, the election or important proposals. Agreement makes a position stronger, and amplifies its impact on society.

Inspired by the Axelrod model [1], Deffuant et al. [2–4] proposed a continuous opinion dynamics model. The Deffuant model simulates the processes of how these randomly distributed opinions evolve into the final stationary state. In the model, an individual opinion is a stochastic variable of continuous distribution (any value within interval [0,1]), the agent is interacting with others according to some rules. The basic rule is bounded confidence, which defines that only when the absolute difference of agents' opinions is within ε, can they interact with each other and influence their opinions. In most cases, the final configuration of opinions is stationary. The opinion cluster could be one (consensus), two (polarization) or more (fragmentation).

Following the proposition of Deffuant dynamics, researchers studied its properties in different ways. Ben-Naim et al. [5] analytically solved the system opinion distribution when any agent could interact with everyone else; however, since the model randomly selects the interacting peers, Deffuant dynamic isn't analytically solvable. Most studies on Deffuant dynamics simulated the interaction processes to get results. Fortunato studied the complete consensus threshold of ε based on different networks topology,

finding that the threshold of ε is 1/2, independent of the network topology [6,7]. This conclusion is a general property of Deffuant dynamics. In most cases, the model will finally get complete consensus when the model parameter ε is over 1/2; if $\varepsilon < 1/2$, there exists more than one opinion clusters. Stauffer and Meyer-Ortmanns studied the different final opinion configuration of the Deffuant model on BA network. They found the final surviving opinions are proportional to the population of nodes for fixed ε [8]. Laguua et al found parameter μ is the important factor of evolution time step, however, when μ is small enough, the final cluster configuration will be related to μ [9,10].

Deffuant opinion dynamics have a good explanation of describing and simulating the opinion evolution process. However, it ignored the differences of social relationship, just changing the opinions by the same extent which was proportional to the differences of opinions after one interaction. Though it is easy for computing, it isn't true in the real society. Communication of opinions is an influential promulgation in essence rather than exchanges of position. Due to different levels of individual influence, the interaction rules of different agents in Deffuant dynamic systems must be rewritten. In reality, opinion leaders impact the common plain people much more than the counterpart course [11,12]. Opinion leaders are usually stubborn while the followers seem persuadable. Inspired by this idea, we emphasized the influence factor on the modified Deffuant model, using logarithm of the node's degree to represent its impact, setting the move of opinions related to mutual influence (**Sec. 2**). We simulated the evolution of opinion system strictly for enough time on WS small networks, BA scare free networks and an ego-network. The results show that if the degree distribution of the network follows scare free law, the evolution of opinions speeds up, time-steps to reach the final state reduce rapidly, the threshold of getting complete consensus becomes higher; at the same time, there is bias from the opinion center, but it is independent of the initial distribution of opinions (**Sec. 3**).

2 MODEL

Traditional Deffuant model is a bounded confidence continuous opinion dynamics model. It can be illustrated as: if Agent i has opinion x, it will only interact with its peers whose opinions value between $|x - \varepsilon,$ $x + \varepsilon|$. If the differences are over ε, they won't interact with each other. Within the interval, the opinion interaction rule is as following:

$$x_i(t+1) = x_i(t) + \mu[x_j(t) - x_i(t)]$$
$$x_j(t+1) = x_j(t) + \mu[x_i(t) - x_j(t)] \tag{1}$$

It is easy to find that, if x_i and x_j interact with each other, the aggregation of the two opinions will not change. That is $x_i(t + 1) + x_j(t + 1) = x_i(t) + x_j(t)$. The key reason for the equivalence of aggregation is that in Eq. (1) μ is the same in both functions. It can be interpreted that in the Deffuant model, the interaction pair of agents influence each other by the same extent. However, in reality, individuals interact with others by different extent. E.g. an authority in a certain field must influence a fresher greater than the counterpart influence process. Thus Agent i influence j stronger than j to i (or reverse). Therefore the interaction rule of Agent i and j is: $\mu_{i \to j} > \mu_{j \to i}$ or $\mu_{i \to j} < \mu_{j \to i}$. So Eq. (1) should be exactly written as follows:

$$x_i(t+1) = x_i(t) + \mu_{j \to i}[x_j(t) - x_i(t)]$$
$$x_j(t+1) = x_j(t) + \mu_{i \to j}[x_i(t) - x_j(t)] \tag{2}$$

where $\mu_{i \to j} > \mu_{j \to i}$ or $\mu_{i \to j} < \mu_{j \to i}$. In this paper we qualified the influence of each node.

Now the question is how to qualify the influence of each node? Of course we can assume the influence of system nodes as stochastic distribution, assign a stochastic number to each node. However, does the influence of everybody follow the U-distribution in real society? The answer is definitely no. Considering the influence of people is neither equal nor stochastic, it is rather likely to follow the "Matthew effect", which is a small number of people having bigger influence and the majority having thinner influence. This property is similar to the degree distribution of complex networks in reality. So we proposed a nature logarithm of node's degree to represent the influence of the node. As to the Deffuant model, to avoid the circumstances

of denominator equals to 0, we defined the influence of Agent j to i is

$$\mu_{j \to i} = \frac{\log(d_j + 1)}{\log(d_i + 1) + \log(d_j + 1)}$$

And i to j :

$$\mu_{j \to i} = \frac{\log(d_j + 1)}{\log(d_i + 1) + \log(d_j + 1)} \tag{3}$$

Here d_i, d_j represent the degree distribution of node i and j respectively. It is clear that the interaction pair will still reach the same final opinion after one interacting-step. However, the symmetry of the classical model is broken. The bigger influential nodes (nodes with a higher degree) will attract other agents to their own opinion values. So those nodes with a higher degree become stubborn while lower degree nodes become persuadable. How will this effect the Deffuant model? Are the former conclusions still useful here? To answer these questions, we simulated the modifying model on the topology of WS small world network, BA scare free network and a real ego-network drawn from Sina Weibo. The following sections will introduce the contrasting experiments and results of the Deffuant model in three aspects.

3 EXPERIMENTS AND RESULTS

3.1 The structure of experiment networks

Our opinion dynamic experiment was on WS small world networks, BA networks and a real social network of 1383 nodes. The basic parameters of 3 types of topology networks are listed in Table 1. We collected the social network from Sina Weibo, which is an ego-network with a famous professor at its center, containing all of his fans and the fans, relationships among him. Since we studied the opinion dynamics on an indirect social network, we compulsorily symmetrized the network (if there exists an edge, we consider the edge as an indirect edge). After the symmetry, the average path length of the ego-network is 1.98, and its degree distribution is shown in **Fig. 1**:

Obviously, the network has small-world effect, and **Fig. 1** shows that its degree distribution followed scare-free distribution, but it isn't strict. Linear fitting

Table 1. Parameter of 3 types of topology networks.

	N	Number of neighbors	Rewriting probability	Exponent	Average path length
WS network	2500	40	0.3	–	2.76
	10000	10	0.3	–	4.83
BA network	2500	–	–	−2.20	2.70
	10000	–	–	−2.20	3.70
Ego-network	1383	–	–	−1.61	1.98

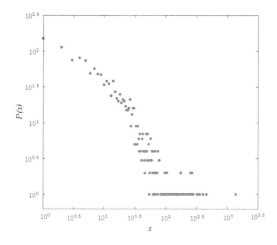

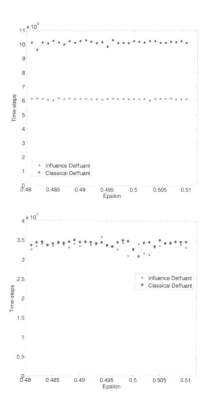

Figure 1. The distribution function of degree distribution of the ego-network with $N = 1,383$ nodes on a double logarithmic scale. The dataset was crawled on May 6th, 2012 from Sina Weibo.

shows that $\alpha = -1.069$ and the sum of squared error is 8.019.

3.2 Opinion evolution results

1) Time–step of evolution

Laguua et al. studied the Deffuant model and concluded that parameter μ mainly effects the time step. Generally speaking, when $\mu \in (0, 1/2]$, the evolution time descends with the increase of μ. This conclusion can be easily understood, when μ is getting bigger, the opinions after interaction are getting nearer to the meaning of the previous two opinion values, thus resulting in that every step is getting nearer to the final opinion, therefore the evolution time is getting shorter. When it comes to influencing the Deffuant model, the situation is quite different from the traditional model. Our experiment showed that, the evolution time-steps of WS small world is nearly the same, but the time-steps of BA network and the ego network which follows scare free law is only about 60%–70% to the traditional model.

Take the BA network (N = 2500, $\varepsilon = 0.505$) as an example, the classical model will compute about 2.1e5 time step getting static, and with influence factor model only need 1.3e5 time step (about 62.2% of the classical model). The evolution is quicker than the classical model. **Fig. 2–4** compare the average evolution time steps on three types of networks to reach the final complete consensus.

2) Evolution bias

In the classical Deffuant model, agents' evolving into the opinion space center forms the final configuration of the model. If the final configuration is complete consensus, all agents' opinions are 0.5, the center of the opinion space, which indicates that all agents are neutral. However, our model broke the symmetry of

Figure 2. Average time-steps of non-complete consensus on WS small-world (2500,10000).

Deffuant dynamics. Does this break effect the final configuration of opinions? To find out whether influence leads to a bias in definite direction (right or left), we tested this by comparing the complete consensus opinion values of the two models. The statistics of the final complete opinions on three types of topology networks suggest that the complete consensus is independent of the initial opinion distribution, and the bias direction is random, independent of the topology of the networks. The difference between the opinion space center and the final complete consensus opinion of influence Deffuant is about 0.2%–0.4%. Remind that the bias represents left/right leaning, if the dynamic system is indicated by the geometric position of the seat of a deputy in the Parliament, the bias is indicated by the left/right leaning party disturbing the balance and leading in the Parliament. However, the simulation showed that the evolution bias is not always in one direction, according to our simulation, the bias equivalently falls in two directions. Our results are shown in **Fig. 5**.

3) Evolution results

As [11] pointed out, the evolution of the system is due to the instability of opinions. In the classical model, all the opinions went smooth towards the space center forming attitude group. Once an attitude group formed and with considering difference to an other attitude group, only agents in the group can interact with each other, which will lead to the final consensus of the

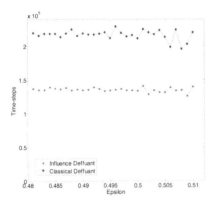

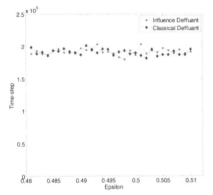

Figure 3. Average time-steps of non-complete consensus on BA scare free network (2500,10000).

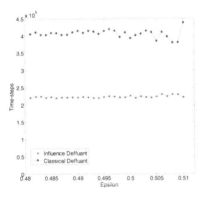

Figure 4. Average time-steps of non-complete consensus on Ego-network.

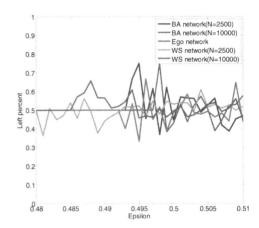

Figure 5. The proportion of final complete opinion under 0.5 of influence Deffuant dynamic system.

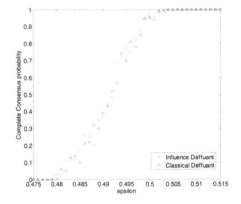

Figure 6. Complete consensus probability of WS small world network (N = 2500,10000).

group individuals. So the final state of the system can be described by a continuous Dirac δ function. Generally speaking, the number of groups and their scales depend on the threshold of ε. When it comes to Influence Deffuant model, how will the attitude numbers and scale change?

We focused on the process of how the opinions get complete consensus. Take the 2500 nodes BA network as an example (**Fig. 7**). From the contrast experiments we can see that, the probability of the classical Deffuant model to form a complete consensus group is

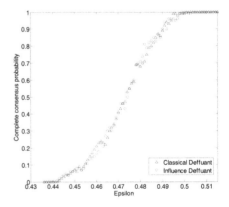

0.01 when N $= 2500$, $\varepsilon = 0.48$, and if $\varepsilon = 0.5$, the probability of getting final complete consensus is near 1. Influence Deffuant model has a similar trend of evolution, however, when $\varepsilon = 0.491$, the first complete consensus group emerged, and when $\varepsilon = 0.52$, most of the cases got complete consensus. In conclusion, the influence Deffuant model requires more strictly forming complete consensus. So, different μ not only

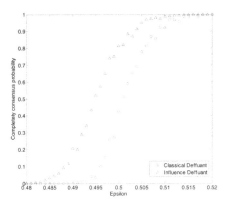

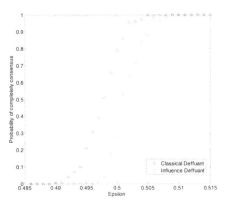

Figure 7. Complete consensus probability of BA Network (N = 2500, 10000).

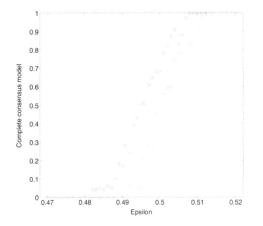

Figure 8. Complete consensus probability of ego-network from Sina Weibo.

impacts on the evolution speed, but also effects the final opinion distribution.

This phenomenon can also be seen on the ego-network (**Fig. 8**), for its degree distribution is similar to BA scare free network. However, such conclusion cannot be applied to WS small world network (**Fig. 6**). For the nodes degree differences are small, and the effect is trivial.

4 CONCLUSION AND DISCUSSION

The experiments showed that since the influence objectively exists, the evolution time-steps decreased, yet the requirement of getting complete consensus is stricter on BA network and those networks which have scare-free effect. However, the forks in static state are usually of two groups. We also found that, when the system evolves into static state, opinion bias state exists; statistics showed that the bias doesn't have a definite direction, and the bias was independent of initial opinion state; it was generated in the evolution process.

Our model takes different influence of agents into consideration, it is closer to reality than traditional Deffuant dynamics, and it shows that people's opinions tend to get static much faster under such assumption. The complete consensus state of influence Deffuant model does not mean neutralization of the system, it has little bias form opinion space center. The partial direction is much more important to real systems, for it defines the final opinion trend. Such partial direction is independent of the initial state, but it depends on the concrete evolution process. Such a phenomenon enlightened us that if we want to intervene on the final opinion state, we don't need to care about the initial opinion distribution; we just need to operate on the system in the process.

We added influence factor into the classical Deffuant model, and studied the system evolution state. Such improvement revealed to us some interesting properties which haven't been found in previous studies. The next step is to consider the direction of the edge, and to test the model on more social networks.

REFERENCES

[1] Axelrod, R., The complexity of cooperation: agent-based models of competition and collaboration 1997, J. Conflict Resolut. 41(2), 203.

[2] Deffuant, G., D. Neau, F. Amblard, and G. Weisbuch, Mixing beliefs among interacting agents, 2000, Advances of Complex. Systems. 3(1–4), 87.

[3] G. Weisbuch, G. Deffuant, F. Amblard, and J.-P.Nadal, Complexity 7, 2002;

[4] G. Deffuant, F. Amblard, G. Weisbuch and T. Faure, Journal of Artificial Societies and Social Simulations 5, issue 4, paper 1 (jasss.soc.surrey.ac.uk) (2002).

[5] Ben-Naim, E., P. Krapivsky, and S. Redner, Bifurcations and Patterns in Compromise Processes, 2003, Physica D 183, 190.

[6] Fortunato, S., Universality of the Threshold for Complete Consensus for the Opinion Dynamics of Deffuant et al. International Journal of Modern Physics. 2004. C 15, 1301.

[7] Lorenz, J., and D. Urbig, 2007, Adv. Compl. Syst. 10(2), 251.

[8] D. Stauffer and H. Meyer-Ortmanns. Simulation of Consensus Model of Deffuant et al on a

Barab'asi-Albert Network, 2004, Int. J. Mod. Phys. C 15(2), 241.

[9] M. F. Laguna, Guillermo Abramson and Dami'an H. Zanette, Minorities in a model for opinion formation.

[10] Porfiri et al 2007 M. Porfiri, M., E. M. Bollt, and D. J. Stilwell, Decline of minorities in stubborn societies, European Physics Journal. 2007, B 57, 481.

[11] Noah E. Friedkin and Eugene C. Johnsen, Social Influence Networks and Opinion Change, Advances in Group Processes, 1999, Volume 16, P.1–29.

[12] Duncan J. Watts Peter Sheridan Dodds, Influentials, Networks, and Public Opinion Formation, Journal of Consumer Reserch, 2007.

[13] Claudio Castellano, Santo Fortunato and Vittorio Loreto, Statistical physics of social dynamics, 2009.

Future Communication, Information and Computer Science – Zheng (Ed.)
© *2015 Taylor & Francis Group, London, 978-1-138-02653-7*

On nonabelian p-group containing a cyclic normal subgroup

L. Zeng

Research Centre of Zunyi Normal College, Zunyi, GuiZhou, China

ABSTRACT: Group theory is generally applied in computer science and many other application sciences. In this article, as tools, we research the modular, dihedral, semidihedral, and quaternion groups first, we have then proved several lemmas about cyclic group, centralizes of group, Sylow p-subgroup and so on. In the end, we have proved that if G is a nonabelian p-group containing a cyclic normal subgroup U of order p^n with certain properties, then G must have a very definite character, this character is very important in the application of groups.

Keywords: nonabelian p-group; cyclic normal subgroup; module; surjective homomorphism; semidihedral

1 INTRODUCTION

Group theory is generally applied in computer science and many other application sciences. In this article, as tools, we research the modular, dihedral, semidihedral, and quaternion groups first, we have then proved several lemmas about cyclic group, centralizes of group, Sylow p-subgroup and so on. In the end, we have proved that if G is a nonabelian p-group containing a cyclic normal subgroup U of order p^n with certain properties, then G must have a very definite character; this character is very important in the application of groups.

Let G be a nonabelian p-group containing a cyclic normal subgroup U of order p^n with $C_G(U) = U$ [1–5], then, what clear characteristic does G have? In order to research this problem, modular, dihedral, semidihedral, and quaternion groups [6–10], are discussed by the author. We have proved several lemmas, first. Then in the end, we finally proved the following:

Let G be a nonabelian p-group [11–16] containing a cyclic normal subgroup U of order p^n with $C_G(U) = U$. Then either

(1) $G \cong D_{2^{n+1}}, Q_{2^{n+1}}$, or $SD_{2^{n+1}}$, or
(2) $M = C_G(\Theta^1(U)) \cong Mod_{p^{n+1}}$ and $E_{p^2} \cong \Omega_1(M)$ *char* G.

2 THE PROOF OF LEMMAS; PREPARATION

Lemma 1 Let G be a finite group and p a prime. Then:

(1) If $G/Z(G)$ is cyclic, then G is abelian [1].
(2) If $|G| = p^2$ then $G \cong Z_{p^2}$ or E_{p^2}.

The proof of this lemma can be found in [3].

Lemma 2 Let G be a group, $x, y \in G$, and assume $z = [x, y]$ centralizes [1] x and y. Then

(1) $[x^n, y^m] = z^{nm}$ for all $n, m \in Z$.
(2) $(yx)^n = z^{n(n-1)/2} y^n x^n$ for all $0 \le n \in Z$.

Proof. Without loss $G = <x, y>$, so $z \in Z(G)$. $z = [x, y]$ so $x^y = xz$. Then, for $n \in Z$, $(x^n)^y = (x^y)^n = (xz)^n = x^n z^n$ as $z \in Z(G)$. Thus $[x^n, y] = z^n$. Similarly $[x, y^m] = z^m$, so $[x^n, y^m] = [x, y^m]^n = z^{mn}$, and (1) holds.

Part (2) is established by induction on n. Namely $(yx)^{n+1} = (yx)^n yx = z^{n(n-1)/2} y^n x^n yx$, while by (1) $x^n y = yx^n z^n$, so that the result holds.

Lemma 3 Let $S_i = S(A_i, G_i, \pi_i)$, $i = 1, 2$, be semidirect products [17–19]. Then there exists an isomorphism [20] $\phi: S_1 \to S_2$ with $A_1 \sigma_{A_1} \phi = A_2 \sigma_{A_2}$ and $G_1 \sigma_{G_1} \phi = G_2 \sigma_{G_2}$ if and only if π_1 is quasiequivalent to π_2 in the category of groups and homomorphisms.

The proof of this lemma can be found in [20].

It is not difficult to see that semidirect products $S_1 = S(A, G, \pi_1)$ and $S_2 = S(A, G, \pi_2)$ can be isomorphic without π_1 being quasiequivalent to π_2.

Lemma 4 Let G be a finite group, p a prime, and X a p-subgroup of G. Then

(1) Either $X \in \text{Syl}_p(G)$ or X is properly contained in a Sylow p-subgroup of $N_G(X)$.
(2) If G is a p-group and X is a maximal subgroup of G, then $X \lhd G$ and $|G : X| = p$.

The proof this lemma can be found in [20]. Observe that, as a consequence of Lemma 4, a p-group G is elementary abelian if and only if $\Phi(G) = 1$.

Recall that if n is a positive integer then $\Omega_n(G)$ is the subgroup of G generated by all elements of order at most p^n.

Lemma 5 Let $G = <x>$ be cyclic of order $q = p^n > 1$ and let $A = Aut(G)$. Then

(1) The map $a \to m(a)$ is an isomorphism of A with the group $U(q)$ of units of the integers modulo a, where

$m(a)$ is defined by $xa = x^{m(a)}$ for $a \in A$. In particular A is abelian of order $\phi(q) = p^{n-1}(p-1)$.

(2) The subgroup of A of order $p-1$ is cyclic and faithful [10] on $\Omega_1(G)$.

(3) If p is odd then a Sylow p-group of A is cyclic and generated by the element b with $m(b) = p + 1$. In particular the subgroup of A of order p is generated by the element b_0 with $m(b_0) = p^{n-1} + 1$.

(4) If $q = 2$ then $A = 1$ while if $q = 4$ then $A = <c> \cong Z_2$, where $m(c) = -1$, $m(b) = 5$.

(5) If $p = 2$ and $q > 4$ then $A = <b> \times <c>$ where b is of order 2^{n-2} with $m(b) = 5$, and c is of order 2 with $m(c) = -1$. The involution b_0 in $<b>$ satisfies $m(b_0) = 2^{n-1} + 1$ and $m(cb_0) = 2^{n-1} - 1$.

Proof. It is very clear that the proof of part (1), so we omit, and observe also that $\alpha : a \mapsto m(a) \mod p$ is a surjective homomorphism of A onto $U(p)$ with kernel $C_A(\Omega_1(G))$. So, as $|U(p)|_{p'} = p - 1 = |U(q)|_{p'}$, the subgroup of A of order $p - 1$ is isomorphic to $U(p)$ and faithful on $\Omega_1(G)$, while $\ker(\alpha) = \{a \in A : m(a) \equiv 1 \mod p\} \in Syl_p(A)$, Next $U(p)$ is the multiplicative group [11] of the field of order p, and hence cyclic, so (2) holds. Thus we may take $q > p$. Evidently if $m(c) = -1$ then c is of order 2. So, as $|A| = 2$ if $q = 4$, (4) holds. Thus we can assume $n > 1$, and $n > 2$ if $p = 2$. Choose b as in (3) or (5). Then $b \in \ker(\alpha)$, so $b^{p^{n-1}} = 1$ Thus if p is odd it remains to show $b^{p^{n-2}} = b_0$ and if $p = 2$ show $b^{p^{n-3}} = b_0$. Observe:

$$(kp^m + 1)^p \equiv (1 + kp^{m+1} + k^2 p^{2m+1}(p-1)/2) \mod p^{m+2}$$
$$\equiv 1 + kp^{m+1}$$

if $m > 1$ or p is odd, hence as $m(b) = 1 + s$ with $s = p$ if p is odd and $s = 4$ if $p = 2$, it follows that $m(b^{p^{n-2}}) = 1 + p^{n-1} = m(b_0)$ if p is odd, while $m(b^{2^{n-3}}) = 2^{n-1} + 1 = m(b_0)$ if $p = 2$. So the proof is complete.

Next the definition of four extremal classes of p-groups. The modular p-group Mod_{p^n} of order p^n is the split extension of a cyclic group $X = <x>$ of order p^{n-1} by a subgroup $Y = <y>$ of order p with $x^y = x^{p^{n-2}+1}$. Mod_{p^n} is defined only when $n \geq 3$, where, by Lemma 5 and Lemma 3, Mod_{p^n} is well defined and determined up to isomorphism. Similar comments hold for the other classes. If $p = 2$ and $n \geq 2$ the dihedral group D_{2^n} extension of X by Y with $x^y = x^{-1}$ and if $n \geq 4$ the semidihedral group SD_{2^n} is the split extension with $x^y = x^{2^{n-2}-1}$.

The fourth class is a class of nonsplit extensions. Let G be the split extension of $X = <x>$ of order $2^{n-1} \geq 4$ by $Y = <y>$ of order 4 with $x^y = x^{-1}$. Notice $\langle x^{2^{n-3}}, y^2 \rangle = Z(G)$. Define the quaternion group Q_{2^n} of order 2^n to be the group $G/\langle x^{2^{n-2}} y^2 \rangle$.

The modular, dihedral, semidihedral, and quaternion groups are discussed in Lemma 6 and Lemma 7, Observe $Mod_8 = D_8$.

Lemma 6 Let $G \cong Mod_{p^n}$, $n \geq 3$, with $Z_{p^{n-1}} \cong X = <x> \triangleleft G$, y of order p in G-X and $x^y = x^{p-1}$,

Then (1) G is of class 2 with $Z(G) = \Phi(G) = <x^p> \cong Z_{p^{n-2}}$.

(2) $G^{(1)} = \langle x^{p^{n-2}} \rangle \cong Z_p$.

(3) $\Omega_m(G) = \langle x^{p^{n-m}}, y \rangle \cong Z_{p^m} \times Z_p$ for $0 < m < n - 1$, unless $p^n = 8$.

Lemma 7 Let $G \cong D_{2^n}, n \geq 2, Q_{2^n}, n \geq 3$, or SD_{2^n}, $n \geq 4$ Let $Z_{2^{n-1}} \cong X = <x> \triangleleft G$ and $y \in G - X$ with y an involution if G is dihedral or semidihedral and y of order 4 if G is quaternion. Then

(1) $G^{(1)} = \Phi(G) = <x^2> \cong Z_{2^{n-2}}$

(2) Either G is dihedral of order 4 or $Z(G) = \langle x^{2^{n-2}} \rangle$ is of order 2.

(3) G is of class $n - 1$.

(4) X is the unique cyclic subgroup of G of index p, unless G is dihedral of order 4 or quaternion of order 8.

(5) G-X is the union of two conjugacy classes [8] of G with representatives [3] y and yx. Each member of G-X is an involution if G is dihedral, each is of order 4 if G is quaternion, while if G is semidihedral then y is of order 2 and xy of order 4.

(6) G has two maximal subgroups distinct from X. If G is dihedral of order at least 8, both are dihedral. If G is quaternion of order at least 16, both are quaternion. If G is semidihedral then one is dihedral and the other quaternion.

(7) Quaternion groups have a unique involution.
Both the proofs of Lemma 6 and Lemma 7 are very easy, so we omit them.

Lemma 8 Let G be a nonabelian group of order p^n I with a cyclic subgroup of index p. Then $G \cong Mod_{p^n}$, D_{2^n}, SD_{2^n}, or Q_{2^n}.

Proof. Notice that, as G is nonabelian, $n \geq 3$ by Lemma 1. Let $X = <x>$ be of index p in G, By Lemma 4, $X \triangleleft G$. As X is abelian but G is not, $X = C_G(X)$ by Lemma 1, So $y \in G - X$ acts nontrivially on X. As $y^p \in X$, y induces an automorphism of X of order p. By Lemma 5, $Aut(X)$ has a unique subgroup of order p unless $p = 2$ and $n \geq 4$, where $Aut(X)$ has three involutions, in the first case by Lemma 5, $x^y = xz$ for some z of order p in X. In the remaining case $p = 2$ and $x^y = x^{-1} z^\varepsilon$, where $\varepsilon = 1$ or 0 and z is the involution in X.

Now if the extension splits we may choose y of order p and by definition

$$G \cong Mod_{p^n}, \quad D_{2^n}, \quad SD_{2^n}, \text{ or } Q_{2^n}.$$

So assume the extension does not split. Observe $C_X(y) = <x^p>$ if $x^y = xz$, while $C_X(y) = <z>$ otherwise. Also $y^p \in C_X(y)$. As G does not split over X, $<y, C_X(y)>$ does not split over $C_X(y)$. So as $<y, C_X(y)>$ is abelian, it is cyclic. Thus $C_X(y) = <y^p>$. Hence we may take $y^p = x^p$ if $x^y = xz$ and $y^2 = z$ otherwise.

Suppose $x^y = xz$. Then $z = [x, y]$ centralizes x and y, so, by Lemma 2, $(yx^{-1})^p = y^p x^{-p} z^{p(p-1)/2} = z^{p(p-1)/2}$, while $z^{p(p-1)/2} = 1$ unless $p = 2$, So, as G does not split, $p = 2$. Here $z = x^{2^{n-2}}$ and if $n \geq 4$ then, setting $i = 2^{n-3} - 1$, $(yx^i)^2 = 1$. If $n = 3$ then $x^y = x^{-1}$, which we handle below.

18

So $p=2$, $x^y = x^{-1}z^\varepsilon$, and $y^2 = z$. If $\varepsilon = 0$, then by definition $G \cong Q_{2^n}$, so $\varepsilon = 1$. Then, as $z \in Z(G)$, $(yx)^2 = y^2x^yx = zx^{-1}zx = 1$, so the extension does indeed split.

3 FINAL THEOREM AND ITS PROOF

Theorem Let G be a nonabelian p-group containing a cyclic normal subgroup U of order p^n with $C_G(U) = U$. Then either

(1) $G \cong D_{2^{n+1}}, Q_{2^{n+1}}$, or $SD_{2^{n+1}}$, or
(2) $M = C_G(\Theta^1(U)) \cong Mod_{p^{n+1}}$ and $E_{p^2} \cong \Omega_1(M)$ char G.

Proof. Let $G^* = G/U$. As $U = C_G(U) G^* = Aut_G(U) \leq Aut(U)$. As G is nonabelian, $G^* \neq 1$ and $n \geq 2$. If G^* is of order p then the lemma holds by Lemma 8 and Lemma 6, so assume $|G^*| > p$. Then by Lemma 5 there exists $y^* \in G^*$ of order p with $u^y = u^{p^{n-1}+1}$, where $U = <u>$. Let $M = <y, U>$. By Lemma 8, $M \cong Mod_{p^{n+1}}$, and by Lemma 6, $E = \Omega_1(M) \cong E_{p^2}$. It remains to show E char G. By Lemma 5, G^* is abelian and either G^* is cyclic, or $p = 2$ and there exists $g^* \in G^*$ with $u^g = u^{-1}$. In the first case $\Omega_1(G^*) = M^*$, so $E = \Omega_1(M) = \Omega_1(G)$ char G. In the second $\Theta^1(U) = <u^2> = <[u, g]>$ and as G^* is abelian, $G^{(1)} \leq U$. Hence $G^{(1)} = \Theta^1(U)$ or U, and in either case $\Theta^1(U)$ char $G^{(1)}$, so $\Theta^1(U)$ char G.

Therefore $E = \Omega_1(C_G(\Theta^1(U)))$ char G.

ACKNOWLEDGMENTS

This work was supported by Natural Science Foundation (13116339) of China; Natural Science Foundation ([2014]2069) of Science and Technology Department of Guizhou; Natural Science Foundation ([2014]712) of Education Department of Guizhou; Hundred of talents project (2012) of Zunyi Normal College.

REFERENCES

[1] Baumslag G (1963). utomorphism groups of residually finite groups. *J London Math Soc*, 38: 117–118.
[2] Gilbert N D, Howie J, Metaftsis V, Raptis E (2000). Tree actions of automorphism groups. *J Group Theory*, 3: 213–223.
[3] Zeng L (2011). Application of group theory to Lorentz transformation. Proceedings of 2011 Asia-Pacific Youth Conference on Communication (2011APYCC), April, 4–6, 2011 (Hangzhou) pp. 437–440.
[4] Mackey, G. W. On induced representations of groups. Amer. J. Math. 73, 576–592. MR13, 106.
[5] Zeng L (2010). Reducible property of a finitely generated module. Advances in Natural Science. 2(3), 270–276.
[6] Robinson, D. J. S (2003). A Course in the Theory of Groups (Second Edition). Springer, New York, 8–16; 31–42.
[7] Zeng L (2011). Application of group theory to Lorentz transformation. Proceedings of 2011 Asia-Pacific Youth Conference on Communication (2011APYCC), April, 4–6, 2011 (Hangzhou) pp. 437–440.
[8] Zeng L (2011). On the classification of the finite subgroups of $SO(3)$. Proceedings of International Conference on Engineering and Business Management (EBM2011), March, 22–24, 2011 (Wuhan), 2289–2292(ISTP).
[9] Zeng L (2010). On the construction of normal subgroups. Studies in Mathematical Sciences. 1(1), 52–60.
[10] Zeng L (2011). On the algebraic integers in cyclotomic fields. Proceedings of International Conference on Engineering and Business Management (EBM2011), March, 22–24, 2011 (Wuhan), 2293–2296(ISTP).
[11] Zeng L (2011). Equivalence on finitely generated $R[G]$ module. Proceedings of 2011 Asia-Pacific Youth Conference on Communication (2011APYCC), April, 4–6, 2011 (Hangzhou) pp. 434–436.
[12] Rotman, J J (2004). Advanced Modern Algebra. Pearson Education Asia Ltd. Prentice Hall, 634–645; 96–115.
[13] Zeng L (2011). Two Theorems about Nilpotent Subgroup. Applied Mathematics, May 2011:2(5), pp. 562–564.
[14] Allenby R B J T, Kim G, Tang C Y (2001). Residual finiteness of outer automorphism groups of certain pinched 1-relator groups. *J Algebra*, 246(2): 849–858.
[15] Wise D (2003). A residually finite version of Rips's constructions. *Bull London Math Soc*, 35: 23–29.
[16] Zeng L (2010). Reducible property of a finitely generated module. Advances in Natural Science. 2(3), 270–276.
[17] Gilbert N D, Howie J, Metaftsis V, Raptis E (2000). Tree actions of automorphism groups. *J Group Theory*, 3: 213–223.
[18] Mackey, G. W. On induced representations of groups. Amer. J. Math. 73, 576–592. MR13, 106.
[19] Kim G, McCarron J, Tang C Y (1995). Adjoining roots to conjugacy separable groups. *J Algebra*, 176(2): 327–345.
[20] Allenby R B J T, Kim G, Tang C Y (2005). Residualniteness of outer automorphism groups of finitely generated non-triangle Fuchsian groups. *Internat J Algebra Comput*, 15(1): 59–72.

Future Communication, Information and Computer Science – Zheng (Ed.)
© 2015 Taylor & Francis Group, London, 978-1-138-02653-7

Computer simulation of sheet metal digital plastic forming process

J. Liu
School of Mechanical and Electrical Engineering, Shenzhen Polytechnic, Shenzhen, China

ABSTRACT: This paper introduced the fundamentals of the sheet metal digital forming process. Sheet metal digital forming process is a new dieless forming process. This process resolves the intricate three-dimensional geometry information of the workpiece into a series of two-dimensional data, which can be used by an NC system to control a forming tool to make a curvilinear movement over the raw sheet metal layer by layer until the desired component is formed. An FEM model of the digital forming process is established, and a typical process is analyzed to instruct the parameters' selection and the optimization of the forming tracks.

Keywords: sheet metal forming; digital forming; FEM simulation

1 INTRODUCTION

The twenty-first century is a new era with the characteristic of a knowledge-based economy and information-based society. The manufacturing industry is facing an unprecedented severe challenge. How to develop new products with high speed, low cost and good quality is the key factor for the sustentation and development of a manufacturer. Therefore, new methods and technologies are greatly demanded in this field to meet the market's need of small in quantity and various in products. Nowadays, sheet metal forming process has a wide application in the manufacturing industry, but the traditional process cannot meet this need for its long manufacturing cycle and the high cost of dies. The sheet metal digital process technique provides an advanced method of rapidly designing and manufacturing.

Not until recent years was much research work done on the dieless sheet forming process. Up to now, the typical dieless forming methods include spin forming, hammer incremental forming, multi-point forming and sheet metal digital plastic forming method [1].

As one of the most frequently used dieless methods, spin forming can only be applied to symmetric workpiece. Multi-point forming replaces traditional dies with variant curved face composed of digital-controlled hydraulic units to complete the three-dimensional, curved surface forming. But its application to complicated component is still problematic.

2 SHEET METAL DIGITAL FORMING PROCESS

Sheet metal forming process originates in the principle of layered manufacturing in rapid prototyping technology. This rapid and flexible process achieves the unification of design and manufacturing by resolving the complicated geometry information into a series of two–dimensional layers and then carrying out plastic forming process on these layers. According to the geometry information of the workpiece, the molding tool carries out plastic process part by part by making a relative three-dimensional curvilinear movement under the guide of the special three-axis digital control apparatus until the sheet attains the required form. The processing principle is illustrated in Fig. 1. At first, the raw sheet is put on the general mould core and is fastened on all sides by a ring, which can move along the guide pillars. This apparatus is fixed on a three-axis linked digital forming machine, after that, a molding tool presses the sheet into a mould core and moves along the contour line under the control of the machine until the required shape is formed.

It has been proved that [2] compared to the traditional single cycle stretch forming process, digital forming process has higher extensibility, fineness and smoothness of the workpiece. Besides, it can be used for machining complicated curved surfaces, which cannot be done by the traditional method.

3 PROCESS CONTROL OF THE SHEET METAL DIGITAL FORMING

The whole process of this digital forming is illustrated in Fig. 2. Firstly, the 3D picture of the workpiece is made with the aid of such software as UG, PRO/E. Then, the 3D information is sliced into layers with the slicing software. Finally, a processing track file is generated, which is later translated into a series of NC codes to direct the numeric-control forming machine to form the required metal prototype.

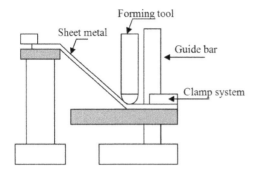

Figure 1. Sheet metal digital forming schematic diagram.

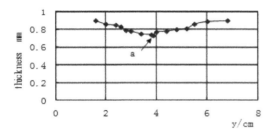

Figure 2. Slab thickness distribution graph.

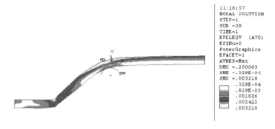

Figure 3. Elastic strain distribution graph.

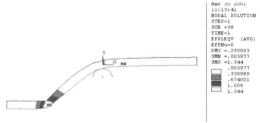

Figure 4. Plastic strain distribution graph.

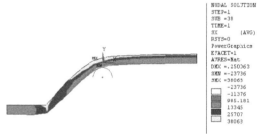

Figure 5. Stress distribution graph.

During the machining process, the computer controls the tool's movement in x-y direction and the gradual movement in z direction. Within the layer of the same height, the molding tool makes a curvilinear movement on the x-y plane. When a layer is finished, the tool moves down a small distance of h along the z-axis and continues to process the next layer till all layers are formed.

4 COMPUTER SIMULATION OF THE FORMING PROCESS

As the sheet metal layered plastic forming process is rather complicated and the parameters and tracks are difficult to be selected by trial process, how to choose feasible parameters is the key factor for an ideal workpiece. The simulation of sheet forming process provides a better solution. Unlike the traditional prototype of single cycle pulling forming, the layered plastic forming prototype simulates the layer forming process on the basis of the material property. To analyze this model, FEM is used to figure out the distribution of deformation and strain, slab thickness distribution, the change in forming force and the prediction of possible defects during the process.

We establish the FEM simulation model. The sheet is in tension in two directions and in compression in one direction, and the strain in slab thickness direction is very small relative to axial and radial direction, so we simplify the 3-dimensional deforming state into 2-dimensional deforming state for calculation.

The simulation result is showed in Fig. 2 to 5. Fig. 2 shows the slap thickness distribution. It can be seen from this figure that the formation is achieved by thinning of the deformation area. Fig. 3 and 4 shows the elastic and plastic strain distribution. It can be seen from these 2 figures that the elastic strain is so small that it can be ignored. The main strain is plastic strain, which mainly appears near the tool head and less deformation at other parts. The maximum deformation lies in the contact point of the tool head and the sheet material, which is the thinnest and is the easiest to evoke tensile rupture. Fig. 5 shows the stress distribution. There is tensile stress outside the bended area, especially the area next to the newly deformed layer. That is to say, the newly deformed area will superpose an extra tension on the existing area, which can reduce the transverse distribution of the stress gradient. As a result, the rebound amplitude of the component will be reduced. Therefore, this forming method can restrain the rebound from the principle basis. This will ensure the dimensional accuracy.

5 CONCLUSION

Sheet metal digital forming technique is a flexible process with no need for special dies, which can greatly decrease the manufacturing cycle and the cost of production by eliminating such time-consuming and capital-wasting steps for the die design, die manufacture, testing and modifying. This technique is suitable for various types, small quantity products and trial manufacture of new products. It can greatly improve the development of sheet metal forming process.

Sheet metal digital forming method is different from traditional solid forming method. Factors affecting the quality of workpiece are very complicated. FEM can prioritize the forming parameter and estimate the forming result and diagnose the defects.

An FEM model of the digital forming process is established, and a typical process is analyzed. The simulation result reveals the mechanics of the formation process, indicates the regularities of distribution of stress and strain, which is close to the practical situation of sheet metal digital forming. The result is instructive to the selection of process parameters' and the track optimization.

REFERENCES

[1] S. J. Yoon and D. Y. Yang, Investigation into a new incremental forming process using an adjustable punch set for the manufacture of a doubly curved sheet metal, Proceedings of the Institution of Mechanical Engineers Part B, Journal of Engineering Manufacture, Vol. 215, Issue 7, pp. 991–1004, 2001.

[2] Z. H. Du, C. K. Chua, Y. S. Chua, K. G. Loh-Lee and S. T. Lim, Rapid Sheet Metal Manufacturing. Part 1: Indirect Rapid Tooling, International Journal of Advanced Manufacturing Technology, Vol. 19, pp. 411–417, 2002.

[3] C. M. Cheah, C. K. Chua, C. W. Lee, S. T. Lim, K. H. Eu and L. T. Lin, Rapid Sheet Metal Manufacturing. Part 2: Direct Rapid Tooling, International Journal of Advanced Manufacturing Technology, Vol. 19, pp. 510–515, 2002.

[4] Trent Maki. Dieless NC Forming and Sheet Fluid Forming, IMTMA Conference, Bombay, India, 11/2002.

[5] Hiroyuki Amino, Yan Lu, et al., Dieless NC Forming, Prototype of Automotive Service Parts [A], Proceedings of the Second International Conference on Rapid Prototyping & Manufacturing, Beijing, 2002:179–185.

[6] S. Matsubara, "A computer numerically controlled dieless incremental forming of a sheet metal", Proceedings of the Institution of Mechanical Engineers, Part B Journal of Engineering Manufacture, Vol. 215, Issue 7, pp. 959–966, 2001.

[7] A. Kochan, Dieless forming, Assembly Automation, Vol. 21, Issue 4, pp. 321–322, 2001.

[8] H. Iseki, T. Shioura and K. Sato, Practical Development of Process-Molding Machine with Small Punching Tool, Advanced Technology of Plasticity 1996, Proc. of the 5th ICTP, Oct. 7–10, 1996.

Future Communication, Information and Computer Science – Zheng (Ed.)
© 2015 Taylor & Francis Group, London, 978-1-138-02653-7

An improved power control algorithm in cognitive radio

B. Wang, C. Zhang & S. Li
Harbin Institute of Technology, Harbin, Heilongjiang, China

ABSTRACT: This paper studies the power control algorithm based on game theory in cognitive radio, proposes a better cost function for power control process. An optimized function related to SINR (signal to interference ratio) is also designed to simplify the algorithm, which balanced the punishment of far users and total communication quality. The comparison of simulation shows that the improved algorithm not only reduces the transit power and increases the total utility, but also gives adequate consideration to fairness and guarantees the communication quality of the farthest users, making it better than other algorithms.

Keywords: cognitive radio; power control; game theory; Nash equilibrium

1 POWER CONTROL ALGORITHMS BASED ON GAME THEORY

In wireless communication systems, users use spectrum for communication in order to ensure the communication quality as far as possible. Generally, this purpose can be achieved easily by raising the transit power of secondary users. However, the arbitrary raise of transit power will definitely lead to the increasing of interference towards other users. In order to solve the questions of power control, many scholars introduce the Game Theory because of its advantage in processing the distributed optimal problems especially the Non-cooperation Game Theory, which is better suited for the solving of power control problems [1].

The NPG algorithm in non-cooperation power control model put forward by Goodman, applies the Game Theory to power control at the earliest [2]. The algorithm finally makes the SNR of all users reach a same level, however the total utility of the system is low. Goodman et al. then propose NPGP algorithm [3], which is a power control algorithm based on cost function. The transit power would be restricted to a low level by sacrificing the SNR of remote subscribers and the improvement of the system's total utility is obtained to some extent. Nevertheless, for the reason that the SNR can't be guaranteed, this algorithm doesn't give consideration to fairness.

With the purpose of partly taking the fairness into account, reference [4] purposes NPGP-LP algorithm which is a power control algorithm based on link gains. The main improvement is in the cost function, taking the various link qualities of different users into consideration by adding a gain factor of the user link. It shows consideration for users whose link qualities are bad and thus balances the fairness partly, but the total utility and the performance of power control extent are sacrificed.

In the NPGP algorithm proposed by Goodman, there is one more cost function than the NPG algorithm. It is a linear function of transit power aiming at punishing the users whose transit power is higher. Nevertheless, this kind of algorithm whose cost function is simply based on the linear function of power obviously can't give enough consideration to users who are distant from a node, and the utility value is also very low. Given this, some scholars purpose the non-linear cost function [5]. The simulation proves that its equity of users is better than that of the NPGP algorithm and the total utility value also increases. There are many studies aiming at improving the cost function. Some literatures propose an algorithm whose cost function dynamically changes as the variation of network congestion situations [6],[7].

All kinds of studies above make the improvements for the cost function, but the utility function itself has flaws. In cognitive radio network, secondary users access the spectrum randomly, so the modulation mode is complex and changeable according to the variousness of environment. In the previous utility function, all users adopt a fixed modulation mode and obviously it doesn't match the need of practical applications.

2 THE IMPROVED POWER CONTROL ALGORITHM NPG-NEW

The present algorithm still has room for improvement such as power control and utility function. With respect to power control, better cost functions could be designed for more effective power control; with respect to utility function, functions correlate with SINR could be designed to simplify the algorithm. Although the NPGP-NL algorithm is capable to well control the power, its punishment for distant users seems typically heavy. Based on this, this paper proposes the new improved NPG-NEW algorithm. At first, the expression of final utility function is given as follows:

$$u_i(p_i, p_{-i}) = \frac{LR}{Mp_i} f(\gamma_i) - c_i(p_i, p_{-i})$$

$$= \frac{LR}{Mp_i} \cdot \frac{1}{1+\exp(10-\gamma_i)} - \lambda_1 h_i \cdot \ln(1+p_i) - \lambda_2 \cdot (\tau - \gamma_i)^2 \quad (1)$$

In formula (1), the utility function still stands for the number of effective bits transferred by unit Joule, R is the data transfer rate, M is the number of whole bits transferred by unit frame and L is the number of data bits transferred by unit frame. The transmit power is expressed as p_i, stands for the expected SNR of secondary user, λ_1 and λ_2 represent the constant parameter, γ_i is the SINR of the ith user:

$$\gamma_i = G \frac{h_i p_i}{\sum_{j \neq i} h_j p_j + \sigma^2} \quad (2)$$

Here h_i stands for the link gain, σ^2 is the background noise, G is the system gain. The denominator of formula (2) still has p_j, which means the utility function of the ith user is closely bound up with the transit power of jth user. $f(\gamma_i)$ is the valid function that has been improved here, and is irrelevant to the modulation mode. Meanwhile, the calculation is simplified:

$$f(\gamma_i) = \frac{1}{1+\exp(10-\gamma_i)} \quad (3)$$

Similarly, in the game process, the process of each user maximizing its self-interest could be shown as follows:

$$\max_{p_i \in P_i} u_i(p_i, P_{-i}), \forall i \in N \quad (4)$$

It can be proved that:

$$\frac{\partial^2 u_i(p_i, p_{-i})}{\partial p_i p_j} \geq 0 \quad (5)$$

So this game model has reached the state of "Nash equilibrium".

3 THE CONTRASTIVE ANALYSIS OF ALGORITHM VIA SIMULATION

At first some values of parameters are defined. The spectrum spreading gain of system $G = 100$, the number of frame bits $M = 40$ bit, the number of data bits for unit frame $L = 32$ bit, the rate $R = 50$ Kbps, the noise $\sigma^2 = 5 \times 10^{-14}$ W, the maximum value of p_i is 2 W, the link gain $h_i = d^{-4}\omega$, $\omega = 9.7 \times 10^{-3}$. The expected SINR of cognitive user is set as 12 dB, the distances between users and node are $d=[460,580,690,810,9$ -20,1080]m. Then

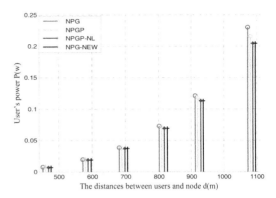

Figure 1. The comparison figure of transit power.

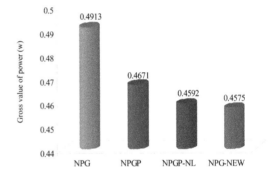

Figure 2. The comparison figure of power gross value.

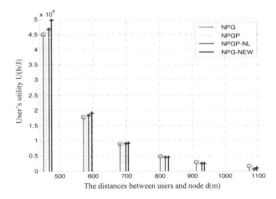

Figure 3. The comparison figure of overall utility.

this paper will simulate the NPG-NEW algorithm and the mentioned other three algorithms simultaneously, and contrastively analyze the results.

Figure 1 shows that the transit power of NPG-NEW algorithm increases with the addition of distance between user and node. The advantage of NPG-NEW algorithm could be seen obviously from the comparison of power value. There is no doubt that with respect to power control, the NPG-NEW algorithm performs best. Figure 3 shows that the utility function value of NPG-NEW algorithm decreases with the addition of distance between

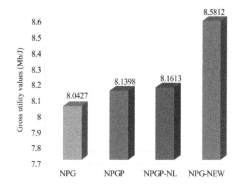

Figure 4.   The comparison figure of gross utility values.

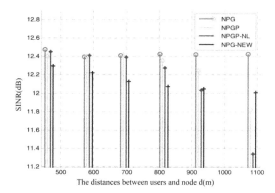

Figure 5.   The comparison figure of user SINRs.

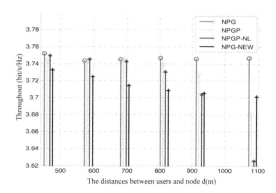

Figure 6.   The comparison figure of user throughputs.

user and node. That is because the design of the utility function is related to SINR, however the link factor in SINR is the fourth power of distance.

The comparison in figure 3 shows that the utility value of NPG-NEW algorithm is much higher than other algorithms for users who near the node. But for users far from a node, the utility value of NPG-NEW algorithm is lower than the NPG and the NPGP algorithms, higher than the NPGP-NL algorithm. It means that the second item in cost function works, the punishment for the farthest user is appropriately reduced.

The advantage of NPG-NEW algorithm, which is not only to make the power well controlled but also increases the total utility of the system, can be seen obviously from the gross utility values in figure 4. Compared to the NPGP-NL algorithm, they both achieve the addition of total utility by sacrificing parts of far users. Simultaneously, the punishment of the farthest users should be taken into consideration, avoiding the situation that the farthest users lose the chances to access the spectrum due to the worse transmission quality.

Figure 5 shows that users' SINRs of the NPG algorithm are almost the same and users who have high transit power are not punished. Nevertheless, as the distance increases, NPGP-NL and NPGP algorithms give the fairness into consideration by rapidly

decreasing the remote users' SINRs. Between them, the NPG-NEW algorithm both considers the fairness and ensures the accesses of remote users by avoiding the SINRs to become too low to communicate. It can be seen that although the NPG-NEW algorithm punishes remote users, the working SINRs of all users are maintained in an acceptable scope, namely assuring all the cognitive users could reach their expected SINRs. From figure 6 we can conclude that the NPG-NEW algorithm which has the middle throughout, both considers the fairness and gives consideration to the normal communication qualities of all users, which is optimal.

The simulations show that compared to the other three algorithms, the NPG-NEW algorithm can effectively control the transit power, while at the same time the total utility value of system is better than others. Because the punishment is at the cost of sacrificing the benefits of remote users, for users who are far from a node the utility values are lower than NPG algorithm. However, it is better than NPG-NL algorithm with the moderate punishment quantity. It can be intuitively seen that at the respects of user SINR and throughput, the NPG-NEW algorithm considers the punishments of the much too high transit powers of remote users and punishes them appropriately, so that its simulation figures of user SINR throughout decreases appropriately, not like the sharp declines of NPGP-NL and NPGP algorithms.

4   CONCLUSION

This paper designs the NPG-NEW algorithm aiming at the following aspects: the utility function which is irrelevant to modulation mode, the non-linear cost function related to link quality and the cost function related to the user's expected SINR. The simulations verified the improvement of system performance brought by the NPG-NEW algorithm. It effectively reduces the transit power and increases the system's total utility, and also takes the fairness into account, guaranteeing that the distant users could reach the SINR set previously.

REFERENCES

[1] Drew Fudenberg & Jean Tirole. Game Theory. Translated by Tao huang, Kai Guo, Peng Gong. Beijing: China Renmin University Press, 2002.

[2] Goodman D & Mandayam N. Power Control for Wireless Data. IEEE Personal Communication Mag, 2000.

[3] Saraydar C, Mandayam N B & Goodman D J. Efficient Power Control via Pricing in Wireless Data Networks. IEEE Trans Commun, 2002.

[4] Shibiao He, Zhiguang Su & Xiandan Ao. A Novel Non-cooperative Power Control Game Algorithm for Cognitive Radio.Telecommunication Engineering, 2008.

[5] Wang, Yilin Cui & Tao Peng. Non-cooperative Power Control Game with Exponential Pricing for Cognitive Radio. Proceedings of IEEE VTC, 2007.

[6] Li shiyin, Liu Mengyun & Liu Qiong. Dynamic Power Control Algorithm and Simulation in Cognitive Radio System. IEEE Communications Surveys & Tutorials, 2005.

[7] Shibiao He, Lunzhi Hu & Xinchun Zhang. Power Control Game Algorithm Based On Interference Temperature in Cognitive Radio. Journal of Chongqing University of Posts and Telecommunications (Nature Science Edition), 2010.

Future Communication, Information and Computer Science – Zheng (Ed.)
© *2015 Taylor & Francis Group, London, 978-1-138-02653-7*

Exploration on teaching computer network course in application-oriented universities

X.H. Xiong
College of Computer & Information, Shanghai Second Polytechnic University, Shanghai, China

A.B. Ning
School of Management, University of Shanghai for Science and Technology, Shanghai, China

ABSTRACT: Computer network course is a core curriculum for computer undergraduates. While many students feel tired to learn it and teachers also find it is hard to teach computer network in traditional teaching means and ways. Combined with the characteristics of computer network course and the students in application-oriented universities, a teaching reform that seeks to innovate teaching content both in theoretical teaching and practical teaching is proposed in this paper. By arranging teaching contents reasonably, reforming teaching methods and means, planning the experiment content scientifically, students can be more competent in their job market by improving their comprehension about computer network and enhancing their practical ability.

Keywords: computer network; teaching reform; application-oriented; teaching methods

1 INTRODUCTION

Computer networks and Internet have become the most critical infrastructure of today's connected world. Computer networks are widely used in almost every area, such as business transaction, education, human communication, scientific research, and even national security. The widespread of networks owes to the rapid development of network technology and the growing of network users. And the demands of talents who master the knowledge of computer networks are increasing sharply. As a result, there is a pressing need to equip students in computer science with a solid foundation in the field. Now a computer network course has become a core curriculum for the computer undergraduate. While at the same time, the computer graduates are not welcomed in the job market mainly for lacking a solid theory foundation and hands-on ability. There are several problems existing in the teaching of computer network course, such as obsolete teaching content, disjoint of theory and practice or preference of theoretical teaching to practical teaching. In application-oriented universities, not like those leading technology universities, the talents' training objectives are to cultivate applied talents who are popular in the job market. And the need to explore the characteristics of computer network course and to try to find a new way to motivate the students to grasp the theory of computer network and enhance the hands-on ability is much more imperative.

This paper considers a foundational course on computer networks for junior students majoring in computer science or computer engineering. In the next section, the difficulties connected to teaching and learning computer networks are presented. In section 3, we describe the reform scheme in theoretical teaching. In section 4 the reform scheme in practical teaching is discussed. Finally, I conclude with some future work in section 5.

2 IDENTIFYING LEARNING AND TEACHING DIFFICULTIES

A computer network course is characterized by strong theoretical property and fast speed of the renewal of knowledge. The course requires students to know the basic principles as well as why they are basics, which requires teaching contents selected by teachers not only including current mainstream technologies, but also reflecting the latest technological developments. Most existing textbooks focus on telling the computer network architecture, protocols, principles (Wu, 2007; Xie, 2008; Kurose, 2009). And at the same time the practical applications of these protocols and principles are less to touch. The disjoint of theoretical teaching and practical teaching, much attention is paid on theory and less on actual operations. Thus students may wrongly conceive that the main task of a computer network course is to memorize the content of protocols and principles and no hands-on ability to solve the problems in real networks. With the rapid development of computer networks, new technology and new standards emerge gradually, new applications continue to develop and expand. The rapid advances in the field also make learning computer networks hard. There are several factors making the teaching and learning of a computer networks difficult. Firstly, the principles underlying computer networks are very complex. Though the layered model helps students to understand and manage the complexity, it is still inadequate to get the whole picture of computer networks. Secondly, to many students, those concepts and protocols are very abstract and boring. Most of these contents are taught by a teacher in the classroom. Students cannot see typical networking equipment and visualize packets and protocols for themselves during lectures. Thirdly, unlike computer programming courses, it is hard to provide students with a hand-on practical experience. Fourthly, many terminologies and acronyms are introduced and used in this field. Some of them are very similar and some of them are not used consistently. Fifthly, due to the common lack of practical experience, some of the networking problems are difficult for students to comprehend or appreciate.

Now in the classrooms of a computer network course, teachers still use traditional teaching models and simple teaching methods and means. The teacher is the center of the lecture and interaction between teacher and student is lacking. Teachers could not fully mobilize the students' enthusiasm. Together with the course's own characteristics, students generally reflect that a computer network course is much more difficult than other courses of computer science.

In application-oriented universities, curriculum design should highlight the application, while taking into account the basic knowledge of computer networks. The framework of curriculum knowledge should not only ensure the integrity of computer network knowledge but also highlight the practicality and keeping up the latest technology on computer networks.

All 4th semester students of the bachelor program at the college of computer and information at the Shanghai second polytechnic have to take part in the course "Computer networks and communication technology". The main contents of the course are concepts and standards of computer networks, hosts and inter-communications. It consists of two hours of weekly lectures, accompanied and supplemented by a three weeks practical course. How to fully mobilize the enthusiasm to learn computer network and avoid the difficulties intrinsically in the course itself is what we want to discuss here.

3 REFORM SCHEME ON THEORETICAL TEACHING

3.1 *Building better teaching contents*

Teaching contents will directly affect the quality of teaching. Considering the characteristic of many teaching contents and few teaching hours, focus should be laid on the basic principles and concepts of computer networks. On the other hand, teaching contents should be updated constantly to follow up the new developments in computer networking. Teachers should simplify those obsolete technologies and increase currently more popular, mature and practical network technologies in time (Li, 2013).

A computer network course has a lot of learning points and those points are very dispersive. Using the TCP/IP architecture as the main line, the theoretical teaching contents can be divided into four parts: basic knowledge of computer networks, local area network technology, network interconnection technology and network applications.

1) *Basic knowledge of computer networks*: This part introduces data communication technology and network architecture.

2) *Local area network technology*: It includes the two lowest layers, namely the physical layer and the data link layer and their common protocols and principles. LAN technology focuses on Ethernet technology and switching and virtual local area network technology.

3) *Network interconnection technology*: It focuses on the wide area network technology. It includes the protocols in IP layer, Transport layer and Application layer. It introduces the position and function of IP layer and network devices and common routing algorithms. In the transport layer, it introduces the services provided by the transport layer and its protocols. In the application layer, it introduces the position and function of this layer, and the application structure, C/S mode, and some common protocols and application including FTP, DNS, email, etc.

4) *Network applications*: It highlights on new network technologies and new network services and new network systems.

3.2 *Reform on teaching methods and means*

Teaching is a bilateral activity of the teacher teaching and students studying. Thus teaching methods must be the combination of teaching methods and learning methods (LiuA & LiuB, 2013). In traditional view, the essence of learning is looked at as a teaching method, highlighting the teaching of "education" side of teaching in classroom teaching. However, with the increasing tendency of modem research, much attention should be paid to "study method of teaching".

1) Combination of using the leading problem and heuristics teaching method

At the beginning of each chapter, teachers can raise a problem and then inspire students to think of some questions. How to solve these problems? Which mature technologies can be used to solve the problem? At the end of each chapter, teachers guide students to see whether the proposed issues have been solved or not.

2) Adopting analogy teaching method

Principles and theories in computer networks are very abstract and hard to understand. Using an analogy teaching method can make students master abstract knowledge visually and vividly. For example, when explaining the difference between circuit switching and message switching, the former is like a phone call and the latter is like a mailing package in a postal office.

3) Adopting case-based teaching method

Case-based teaching method also calls an example approach or case approach (Li, 2008). The definition of case-based teaching is that according to the needs of the teaching objectives and content students learn researching and training capacity through cases under the guidance of teachers. It is good to stimulate students' interests in learning, cultivating creativity and the ability of analyzing and solving problems. In this method, selecting teaching cases is of great importance. Case-based method can help students to understand the principle and where and how to apply these principles. It is a good solution to closely co-coordinating between the theory and practice work.

4) Adopting on-site teaching method

Taking advantages of good research cooperation, organize the students to visit and research the network center of their own universities and other enterprises, to understand more advanced network technology. Invite network engineers to explain the server applications, network topology, network equipment and network technology used, network management methods and knowledge and skills of other aspects, and to deepen their understanding of theoretical teaching content.

5) Applying network studying

It's necessary to apply network teaching and integrate the network education into the traditional education during the stage of undergraduate education. Aiming to solve the problems that some students' learning consciousness is not high but meanwhile the new network technologies and new network products continuously appear, let students use the Internet to consult commonly used network products, network application software, network management software and web development tools, to understand its performance, relevant technology and applications.

Teachers can also put forward some questions which are closer to reality, enable students to solve those questions by accessing information through the Internet. In this way, not only the enthusiasm of students can be greatly mobilized, but also the understanding of students to the network knowledge can be enhanced, the good habits of students to active learning can be trained and finally the ability of students using what they have learned to solve practical problems can be improved.

4 REFORM SCHEME ON PRACTICAL TEACHING

In some universities, computer network course is using the traditional lecture-style teaching methods. And a relatively small number of hours of the course are on practical teaching. They don't pay attention to improve the students' practical ability. As a result,

Table 1. Experiment contents of practical teaching.

	Experiment Name	Type
1	Make twisted-pair cable and know how to use the common network tools	verification
2	Design a Peer-to-peer network	design
3	Analysis and usage of common network command	verification
4	IP address and subnet partition	design
5	Management and configuration of several different Windows 2003 servers	comprehensive
6	Network protocol monitoring with network monitoring software sniffer	comprehensive
7	Basic management and configuration of switch	design
8	Basic management and configuration of router	design
9	Plan and design of a campus network	research

students only get some theoretical knowledge without practical ability. Practical teaching is an indispensable part of a computer network course. Only being combined with practical teaching, abstract definitions and principles can be understood thoroughly and easily. It is a good way to make students learn actively and promote theoretical teaching, too. During the practical teaching, students can have a good place to practice hand-on ability, cultivate engineering experience and promote overall development. How to provide high-quality practical teaching for students becomes an enormous challenge in computer network courses.

4.1 Building better practical teaching contents

The design of experiment contents can be divided into four categories: verification experiments, design experiments, comprehensive experiments and research type experiments. Verification experiments are used to master basic knowledge skills. Designing experiments focus on inspiring the students' creativity under the guidance of teachers. Comprehensive experiments are used to strengthen the students' ability of operating and their comprehensive analysis ability. Research type experiments are used to develop the students' ability to explore new skills and methods. Given the actual situations, such as dormitory network, office network, net bar network, etc., students are asked to present a whole network solution. After the design of a whole network solution, students will know how to apply what they know into a practical situation.

Through the analysis of the labor market and the students' interest, the experiments contents for practical teaching are listed in table 1.

4.2 Reform practical teaching methods

1) Using various simulation tools and virtual machines

Real network environment usually includes hundreds of routers and other network devices. And almost all of the universities' labs could not meet this requirement. Using simulation tools, such as Packet Tracer, simulator of Cisco, can provide an experimental environment in low-cost and repeatable and easy to control (Al-Holou, 2000). Beginners can understand network technology more specifically, a protocol, routing, and packet forwarding, etc. VM ware or Virtual PC software are good to construct a virtual network experimental environment. Students can carry out the innovative experiments under the virtual network surroundings, which is rid of the constraints of the original experimental conditions.

2) Using personal network

Though simulation tools and virtual machines can provide students an experiment environment, students still need to be allowed to interact with the kernel of a real computer network. But in most universities, it is impossible to provide every student such an opportunity to interact with the kernel of a computer network, in terms of both equipment (PCs, hubs, NICs, etc.) and space (Chang, 2004). The only solution to provide students with the opportunity to interact with the kernel of computer network and without resource constraint is to bring the network laboratory to where the students are. We can initiate a class project to let students experiment with a personal computer network with family network background. The project consists of several phases. The first one is to find all the resources to form IP private networks. Students are encouraged to read ahead and understand some of the basic network concepts, such as IP subnets and NAT. The second phase is to observe how network protocols work by conducting experiments on the personal computer network.

Under such a teaching pattern, students are allowed to interact with the kernel of an actual computer network. The environment is the place where students are and it guarantees every student's full participation of the teaching process. The whole process stimulates their thirst for active participation in practical teaching.

5 CONCLUSION

Computer network course is a course combining theory and practice. How to develop good theoretical and practical teaching to grasp the theoretical knowledge and enhance the hands-on ability are the common problems of most universities, especially of application-oriented universities. Combined with our teaching experience, we give some ideas on teaching reform of computer network courses in application-oriented universities. Considering the characteristic of the course itself and the problems existing in current teaching, the contents of theoretical teaching and practical teaching should be arranged reasonably. Effective teaching methods and means should be used to motivate the enthusiasm of students. The teaching process should bring teachers and students together and needs more interaction between them. Through this process, students can have their own thought and understanding of the course. The purpose is not only to enable students to master the basic principles and knowledge of computer network but also to have further insight in the future knowledge of computer network and the competence in solving problems in real-life computer network.

ACKNOWLEDGEMENT

This work is partially supported by a Fine Course Construction Project from Shanghai Second Polytechnic University.

REFERENCES

[1] Al-Holou N., Booth K. & Yaprak E. 2000.Using computer network simulation tools as supplements to computer network curriculum. In *Proc. 30th Annual FIE*:13–16. Kansas City MO, USA.
[2] Chang Rocky K.C. 2004. Teaching Computer Networking with the Help of Personal Computer Networks. in *Proceedings of the 9th annual SIGCSE conference on Innovation and technology in computer science education*, Leeds, United Kingdom.
[3] Kurose J.F. & Ross K.W. 2009. *Computer networking: a top-down approach*. Beijing: Higher Education Press (in Chinese).
[4] Li W. & Li Y.C. 2008. Case studies to explore and practice teaching model. *Computer Education*. 14(3):87–89 (in Chinese).
[5] Liu Q.S. 2013. Exploration of teaching reform in computer network technology based on action-oriented teaching method. *Higher Education Forum*, 169(11):82–83, 110 (in Chinese).
[6] Liu X.C. & Liu R.J. 2013. Research on hierarchical and modular computer network course. *Journal of Suzhou University*. 28(1):106–108 (in Chinese).
[7] Wu G.Y. 2007. *Computer Network*. Beijing: Tsinghua University Press (in Chinese).
[8] Xie X.R. 2008. *The computer network (Fifth edition)*. Beijing: publishing house of electronics industry (in Chinese).

Future Communication, Information and Computer Science – Zheng (Ed.)
© 2015 Taylor & Francis Group, London, 978-1-138-02653-7

Development of a search engine with negative keywords in Chinese and Japanese: Based on a computational model of inductive reasoning

Y.J. Zhang, A. Terai & M. Nakagawa
Tokyo Institute of Technology, Tokyo, Japan

ABSTRACT: In order to search for information more effectively, a search engine was constructed not only with the keywords related to the information we intended to search for, but also with negative keywords, which play an important role in deleting the information that we do not need. Two search engines were constructed in Chinese and Japanese using the computational model of inductive reasoning based on the statistical analysis of each language's data. The results of the search engine show the effectiveness of negative keywords.

Keywords: search engine; negative keyword; inductive reasoning; computational model

1 INTRODUCTION

In recent years, web search engines have become popular around the world and are used frequently in our daily life. When you input keywords in a search engine, the search results are generally presented referring to the keywords. However, some unexpected results are also outputted that we do not need. We expect to search for information more effectively. Therefore, in this research, a search engine was constructed not only with the keywords related to the information we intended to search for, but also with negative keywords, which play an important role in deleting the information that we do not need. The search engine was developed based on a computational model of inductive reasoning in Chinese and Japanese. Inductive reasoning is an important thinking process not only for sciences but also in our daily life. The search engine developed in this study is based on one kind of inductive reasoning argument (e.g., Rips, 1975; Osherson, Smith, Wilkie, Lopez, and Shafir, 1990), such as:

The person likes steak.
The person doesn't like noodles.
The person likes hamburger.

In this type of argument, its strength (the likelihood of the conclusion below the line given the premises above the line) depends mainly on the entities in each sentence (e.g., "steak" (the positive entity), "noodles" (the negative entity), "hamburger" (the conclusion)) since these sentences share the same basic predicate (e.g., "The person likes~." and "The person doesn't like ~."). In the following section, we explain the computational model of this type of inductive reasoning used for the search engine, according to Sakamoto's model (Sakamoto & Nakagawa, 2008).

2 MODEL

We have already constructed the model based on Gaussian kernel functions, which produces the likelihood of a conclusion N_i^c, denoted as $v(N_i^c)$ represented as follows:

$$v(N_i^c) = a\text{SIM}_+(N_i^c) + b\text{SIM}_-(N_i^c) \tag{1}$$

$$\text{SIM}_+(N_i^c) = \sum_i^{n^+} e^{-\beta d_{ij}} \tag{2}$$

$$\text{SIM}_-(N_i^c) = \sum_i^{n^-} e^{-\beta d_{ij}} \tag{3}$$

$$d_{i,i}^+ = \sqrt{\sum_k^m (P(c_k|N_i^c) - P(c_k|N_i^+))^2} \tag{4}$$

$$d_{i,i}^- = \sqrt{\sum_k^m (P(c_k|N_i^c) - P(c_k|N_i^-))^2} \tag{5}$$

where $d_{i,i}^+$ and $d_{i,i}^-$ are word-distance functions based on the latent classes (denoted as c_k). $d_{i,i}^+$ represents the distance between the conclusion entity N_i^c and the positive premise entity N_i^+, while $d_{i,i}^-$ represents the distance between the conclusion entity N_i^c and the negative premise entity N_i^-.

$P(c_k|N_i^c)$, $P(c_k|N_i^+)$, $P(c_k|N_i^-)$ represents the conditional probability of C_k, given N_i^c, N_i^+, N_i^-, respectively, estimated from large scale language data in Chinese and Japanese (ref. section 3). Each word distance function constructs Gaussian kernel functions, such as $\text{SIM}_+(N_i^c)$ and $\text{SIM}_-(N_i^c)$, when combined with nonlinear exponential functions and a parameter β. In this study, these Gaussian kernel functions are regarded as nonlinear similarity functions that reflect the retrieval assumption. $\text{SIM}_+(N_i^c)$ represents the similarities between the conclusion entity N_i^c and the positive premise entities, while $\text{SIM}_-(N_i^c)$ denotes the similarities between N_i^c and the negative premise entities. In the present model, parameter a means the strength of the positive entity effect, while b indicates the strength of the negative entity effect.

Moreover, the estimated values from the proposed model for the multiplier a are positive ($a > 0$), while the estimated values for the multiplier b are negative ($b < 0$) in both conditions. This is also consistent with the model's assumption that argument ratings are based on the similarities of conclusion entities to positive premise entities and the dissimilarities to negative premise entities.

3 PROBABILISTIC LANGUAGE KNOWLEDGE STRUCTURE BASED ON A STATISTICAL ANALYSIS OF JAPANESE AND CHINESE CORPUS

In this study, a large Japanese corpus and Chinese corpus are utilized in the construction of a feature space to represent premise and conclusion entities. Based on the clustering analysis, the conditional probabilities of feature words given particular nouns are computed. Conditional probability is assumed to represent the strengths of relationships between nouns (entities) and their features. The method of clustering is similar in structure to popular methods within natural language processing, such as Pereira's method and PLSI (Hofmann, T., 1999; Kameya, Y., & Sato, T., 2005; Pereira, F., Tishby, N., and Lee, L., 1993). The method assumes that the co-occurrence probability of a term "N_i" and a term "A_j", $P(n_i, a_j)$, are represented as Equation (6):

$$P(n_i, a_i) = \sum_k P(n_i|c_k) P(a_i|c) P(c_k) \qquad (6)$$

where $P(n_i|c_k)$ is the conditional probability of term n_i, given the latent semantic class c_k. Each of the probabilistic parameters in the model, $P(c_k)$, $P(n_i|c_k)$, and $P(a_j|c_k)$ are estimated as values that maximize the likelihood of co-occurrence data measured from a corpus using the EM algorithm (in appendix). In this study, the term "n_i" represents a noun (entity), and the term "a_j" represents a feature word, such as a verb or an adjective.

From the estimated parameters $P(c_k)$, $P(n_i|c_k)$, it is also possible to compute the membership distribution $P(c_k|n_i)$ as follows:

$$P(c_k | n_i) = \frac{P(n_i | c_k) P(c_k)}{P(n_i)} = \frac{P(n_i | c_k) P(c_k)}{\sum_k P(n_i | c_k) P(c_k)} \qquad (7)$$

In this study, the latent class c_k is assumed to be a semantic category that can be described in terms of a typicality gradient (Rosch, E, 1973). In fact, most of the estimated latent classes were identified as meaningful categories in both languages. This conditional probability $P(c_k|n_i)$ is assumed to represent the strengths of relationships between a semantic category

Table 1. An example of a search engine in Japanese with only positive keywords.

	Japanese	English translate
positive keywords	ラーメン	Noodles
	すし	Sushi

Japanese	English translate	$v(N_i^c)$
すし	Sushi	1.702785
ラーメン	Chinese noodles	1.702785
うどん	Japanese noodles	1.454505
クッキー	Cookies	1.419219
ケーキ	Cake	1.413693
おでん	Oden	1.389711
スナック菓子	Snacks	1.388149
大根	Radish	1.382787
せんべい	Crackers	1.376058
キャベツ	Cabbage	1.375948
ハンバーガー	Hamburger	1.372329
アイスクリーム	Ice cream	1.370965
柿	Persimmon	1.366833
野菜	Vegetables	1.36331
キムチ	Kimchi	1.352709
麺	Noodles	1.352536
芋	Potato	1.350056

and an entity. When a certain category has a high conditional probability given a particular noun, it is natural that the entity denoted by the noun has the feature indicated by the category. Thus, by considering each c as a feature dimension, entities can be represented in the feature space constructed from the corpus-analysis results.

4 SEARCH ENGINE

Two search engines were constructed in Chinese and Japanese using the above mentioned computational model based on the statistical analysis of each language's data. The example results of those search engines are shown in the following tables. The specific words included in the first search only with positive keywords are deleted effectively with the negative keywords in the second search, in both cases, Chinese and Japanese.

In every table, words of search results are listed by the value of $v(N_i^c)$, and 17 words from the top ranking are chosen in order to explain the change of results clearly.

In table 2 of the Japanese results, with negative keywords "cake" and "sweets", some words in table 1 like "cake", "cookies", "snacks", "crackers" which are similar to negative keywords are deleted and words like "steak", "pizza", "hamburger" rise to the upper ranks in the list.

Table 2. An example of a search engine in Japanese with positive keywords and negative keywords.

	Japanese	English translate
positive keywords	ラーメン	Noodles
	すし	Sushi
negative keywords	ケーキ	Cake
	お菓子	Sweets

Japanese	English translate	$v(N_i^c)$
すし	Sushi	0.367101
ラーメン	Chinese noodles	0.317022
御飯	Rice	0.135892
雑煮	Rice cake soup	0.120407
焼き肉	Roasted meat	0.119438
うどん	Japanese noodles	0.110882
ステーキ	Steak	0.108299
ピザ	Pizza	0.10046
枝豆	Green soybean	0.093021
ハンバーガ	Hamburger	0.089984
かき氷	Shaved ice	0.087882
ホットドッ	Hot dog	0.08586
うなぎ	Eel	0.083932
焼きもち	Toasted rice cake	0.082843
冷や飯	Cold rice	0.081874
定食	Set meal	0.080871
道草	loiter on the way	0.076467

Table 4. An example of a search engine in Chinese with positive keywords and negative keywords.

	Chinese	English translate
positive keywords	米饭	Rice
	方便面	Instant noodles
negative keywords	蛋糕	Cake
	点心	Chinese dessert

Chinese	English translate	$v(N_i^c)$
米饭	Rice	0.210474
方便面	Instant noodles	0.1816
老酒	Rice wine	0.106877
酒饭	Food and drink	0.08542
中饭	Lunch	0.079402
海鲜	Seafood	0.076044
杂粮	Grains	0.073913
早餐	Breakfast	0.073494
早茶	Morning tea	0.072245
火锅	Chafing dish	0.068802
茶点	Refreshments	0.067948
哑巴亏	Something cannot say	0.066343
蔬菜	Vegetables	0.065137
茶水	Tea	0.064652
饭食	Meals	0.064168
商品粮	Commodity grain	0.063677
早饭	Breakfast	0.063009

Table 3. An example of a search engine in Chinese with only positive keywords.

	Chinese	English translate
positive keywords	米饭	Rice
	方便面	Instant noodles

Chinese	English translate	$v(N_i^c)$
方便面	Instant noodles	1.681904
米饭	Rice	1.681904
红烧肉	Pork	1.560319
面条	Noodles	1.551782
点心	Chinese dessert	1.538836
零食	Snacks	1.498588
盒饭	Bento	1.497101
烧酒	Liquor	1.493163
豆汁	Bean drink	1.489434
豆浆	Soybean milk	1.486998
糖	Candy	1.486236
肉汤	Broth	1.482628
饼	Pie	1.481544
老酒	Rice wine	1.480374
茶点	Refreshments	1.480311
馄饨	Ravioli	1.478663
橘子	Orange	1.477142

In table 4 of the Chinese results, with negative keywords "cake" and "Chinese dessert", some words in table 3 like "Chinese dessert", "snacks", "Bean drink" which are similar to negative keywords are deleted and words like "Lunch", "Grains", "Chafing dish" rise to the upper ranks in the list. In a word, the negative keywords effectively delete some words we do not need and increase the ranking of other words that we need. From the result examples shown in those tables, it is clear how effective the negative keywords are in the search engine for deleting unnecessary information.

5 CONCLUSION

In this study, search engines in Japanese and Chinese were constructed not only with the keywords related to the information which we intended to search for, but also with negative keywords, which play an important role in deleting the information that we do not need. Those search engines were developed based on a computational model of inductive reasoning in Chinese and Japanese.

The results of the search engines show the effectiveness of negative keywords.

In future research, we need practical evaluation of the search engines by users in Chinese and Japanese.

REFERENCES

[1] Hofmann, T. (1999). Probabilistic latent semantic index-ing. *Proceedings of the 22nd International Conference on Research and Development in Information Retrieval: SIGIR' 99.* 50–57.

[2] Kameya, Y., & Sato, T. (2005). Computation of probabilistic relationship between concepts and their attributes using a statistical analysis of Japanese cor-pora. *Proceedings of Symposium on Large-scale Knowl-edge Resources: LKR2005.*

[3] Osherson, D. N., Smith, E. E., Wilkie, O., López, A., & Shafir, E. (1990). Category based induction. Psycholog-ical Review, 97, 185–200.

[4] Pereira, F., Tishby, N., & Lee, L. (1993). Distribu-tional clustering of English words. *Proceedings of the 31st Meeting of the Association for Computational Linguistics.* 183–190.

[5] Rips, L. J. (1975). Inductive judgment about netural cat-egories. Journal of Verbal Learning and Verbal Behavior, 14, 665–681.

[6] Rosch, E. (1973). On the internal structure of perceptual and semantic categories. In T. E. Moore (Ed.), *Cog-nitive Development and the Acquisition of Language* (pp. 111–144). New York: Academic Press.

[7] Sakamoto, K., Terai, A., & Nakagawa, M. (2007). Com-putational models of inductive reasoning using a statis-tical analysis of a Japanese corpus. Cognitive Systems Research, 8, 282–299.

[8] Sakamoto, K.. & Nakagawa, M. (2008). A Compu-tational Model of Risk-Context-Dependent Inductive Reasoning Based on a Support Vector Machine. T. Toku-naga and A. Ortega (Eds.): LKR2008, LNAI 4938, Springer-Verlag Berlin Heidelberg, pp. 295–309

[9] Zhang, Y. J., Terai, A., Dong, Y., Wang, Y., & Nakagawa, M. (2013). The comparison between inductive rea-soning of Chinese and Japanese: Using computational models based on the statistical analysis of language data. Journal of Cognitive Studies in Japanese, 20(4), 439–469.

APPENDIX: EM ALGORITHM

(Kameya, Y., & Sato, T, 2005; Sakamoto, K., Terai, A., Nakagawa, M, 2007)

The EM algorithm attempts to find the parameters h that maximize a posteriori distribution $P(\theta/D)$:

$$P(\theta/D) = \frac{P(D/\theta)P(\theta)}{\int P(D/\theta)P(\theta)d\theta} \propto P(D/\theta)P(\theta)$$

where θ is the vector for the parameters $\sum_k P(n_i|c_k)$, $P(a_i|c_k)$ and $P(c_k)$, and D is the frequency data. We assume here that $P(\theta)$ follows the Dirichlet distribu-tion:

$$P(\theta) = \gamma \prod_k \left\{ P(c_k)^\alpha \prod_i P(n_i/c_k)^{\alpha'} \prod_j P(a_j/c_k)\alpha'' \right\}$$

where γ a normalizing constant, and α, α' and α'' are the hyper parameters of the Dirichlet distribution. From these settings, we obtain an EM algorithm as follows:

In the algorithm, the parameters are updated until the posteriori distribution $P(\theta/D)$ converges. In the E-step, we compute three types of expectations, $E[c_h]$, $E[n_i|c_h]$, and $E[a_j|c_h]$, using current parameters. $E[c_h]$ is the expected count of the class c_h occurring, while $E[n_i|c_h]$ and $E[a_j|c_h]$ are the expected counts of words n_i and a_j, respectively, occurring when the class c_h occurs. In M-step, we update the parameters using these expected counts. V^t is the total number of occur-rences, that is, $V^t = \sum_i \sum_j F(n_i, a_i)$, while V^c is the number of classes, V^n is the number of nouns, V^a is the number of adjectives of predicates and F is the number of co-occurrences of n and a.

E(xpectation)-step:

$$P(c_h|n_i,a_j) = \frac{P(n_i|c_k)P(a_j|c_k)P(c_k)}{\sum_k P(n_i|c_k)P(a_j|c_k)P(c_k)},$$

$$E[c_h] = \sum_{i,j} F(n_i|a_j)P(c_h|n_i,a_j),$$

$$E[n_i|c_h] = \sum_j F(n_i|a_j)P(c_h|n_i,a_j),$$

$$E[a_j|c_h] = \sum_i F(n_i|a_j)P(c_h|n_i,a_j),$$

Where c_h is the hth class, n_i is the ith noun, and a_j is the jth adjective or predicate.

M(axmization)-step:

$$P(c_h) = \frac{E[c_h] + \alpha}{V^t + \alpha V^c}$$

$$P(n_i|c_h) = \frac{E[n_i|c_h] + \alpha'}{E[c_h] + \alpha'V^n}$$

$$P(a_j|c_h) = \frac{E[a_i|c_h] + \alpha''}{E[c_h] + \alpha''V^a}$$

Future Communication, Information and Computer Science – Zheng (Ed.)
© 2015 Taylor & Francis Group, London, 978-1-138-02653-7

Simulation of water flow using Graphical Processing Units (GPUs)

M.N. Rud & O.M. Zamyatina
Tomsk Polytechnic University, Tomsk, Russian Federation

ABSTRACT: In this paper we describe a method to simulate water flow over varying bottom topography. As physical model we use shallow water equations, that represent fluid flow in terms of its velocity and height. We explain how to implement a numerical scheme of solution using parallel computations on graphical processing units with the help of graphics API OpenGL and NVIDIA CUDA technology and compare the obtained results with implementation on a central processor.

Keywords: shallow water equations; parallel computing; CUDA; OpenGL.

1 INTRODUCTION

One of the most complicated and challenging problems of realistically looking terrain modelling is water simulation. It is an area where simulation technologies, physical processing and computer graphics come together. Below are some popular methods to perform this task:

1) Waves simulation. It simulates visual effects of water surface, but not the water flow physics. This approach is suitable for large unbounded surfaces simulation, such as oceans, when it is not necessary for water to interact with the surrounding terrain.

2) Smooth particle hydrodynamics. In this method water is represented by a set of particles; the approach perfectly fits for simulating spray, splashing and runnels. It is computationally expensive and not suitable for simulating rivers, lakes and oceans, because the complexity of computations increases with the amount of water in the domain.

3) Height field. This model represents water surface as grid of cells; each of these cells has its own height and velocity. The advantage of this approach is a transition from 3D-volume to 2D-surface, and, as a consequence, reduction of computational complexity. However, there is no possibility to simulate breaking waves, because at each point of surface there is only one height value.

The above methods have some advantages and disadvantages; it is necessary for a researcher to make an appropriate choice of the method according to the task and hardware resources.

2 TASK

Our task is to develop a water model to be used on Interactive Sandbox, an installation for real-time terrain generating [4]. On this installation we scan sand surface with a Microsoft Kinect sensor and transfer the obtained height map to PC, where each point is colored according to its height. Finally, this image is projected onto the sand surface with a multimedia projector (Figure 1):

A good-looking and physically correct fluid model will add realism and functionality to the installation, because we will be able to simulate real nature phenomena, such as floods, dam breaks, ice melting and volcanic eruptions.

For this purpose we need a water model, which satisfies the conditions below:

1) model must provide physically logical behavior of water;
2) water must interact with surrounding terrain and adopt to its changes in real-time;
3) implementation must be as computationally inexpensive as possible;
4) model must allow regulating viscosity of fluid.

3 FLUID MODEL

3.1 *Shallow water equations*

Shallow water equations (also called Saint Venant equations) are a set of hyperbolic partial differential equations describing the dynamics of a thin fluid surface in terms of its height and flow (see Figure 1):

Here h is the water depth, hu is the discharge along the x-axis, hv is the discharge along the y-axis, g is the gravitational constant, and B is the bathymetry.

The usage of shallow water equations has some limitations. As mentioned above we cannot simulate breaking waves and splashing particles, however, this method is perfect for modelling water domains whose surface area is much greater than its height. With the

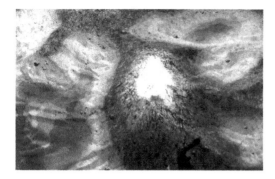

Figure 1. Interactive sandbox.

$$\begin{bmatrix} w \\ hu \\ hv \end{bmatrix}_t + \begin{bmatrix} hu \\ hu^2 + \dfrac{1}{2}gh^2 \\ huv \end{bmatrix}_x + \begin{bmatrix} hu \\ hu^2 + \dfrac{1}{2}gh^2 \\ huv \end{bmatrix}_y = \begin{bmatrix} 0 \\ -ghB_x \\ -ghB_y \end{bmatrix}$$

Figure 2. Shallow water equations.

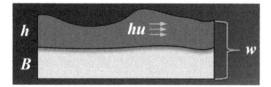

Figure 3. Variables for shallow water equations in one dimension.

help of shallow water equations, we can simulate not only small lakes, rivers, puddles, pools and ponds but also large surfaces of water.

3.2 Numerical scheme

We are interested in an accurate and robust numerical scheme for the Saint-Venant system. A good numerical method for the system should accurately capture both the steady states and their small perturbations (quasi-steady flows). From a practical point of view, one of the most important steady-state solutions is a stationary one (lake at rest):

$u=0, \; h+B=Const.$

Methods having this property are called well balanced. In addition, the method should handle dry ($h = 0$) or near dry zones (positivity preserving property). In these cases, due to numerical oscillations, h can become negative and all computations can simply break down.

Such a method which is both well-balanced and positivity preserving was suggested by A. Kurganov and G. Petrova in 2007 [1]. The system of differential equations is discretized on a regular Cartesian grid using the so-called central upwind scheme:

$$\frac{d}{dt}\overline{\mathbf{U}}_j(t) = -\frac{\overline{\mathbf{H}}_{j+1/2}(t) - \overline{\mathbf{H}}_{j-1/2}(t)}{\Delta x} + \overline{\mathbf{S}}_j(t)$$

Figure 4. Central upwind scheme of discretization in one dimension.

Here $U_j(t)$ is a two-component vector (three-component vector in case of 2D) of conserved quantities w and hu where w is a sum of water height and bathymetry height at the point and hu is the discharge along the x-axis. $S_j(t)$ is an appropriate discretization of the component characterizing bottom topography, and $H_{j+1/2}(t)$ is a numerical flux along the x-axis.

After obtaining all necessary values, the equation (Figure 3) can be solved using a second order stability preserving Runge-Kutta method:

$$U_j^* = U_j^n + \Delta t(R(U_j^n));$$

$$U_j^{n+1} = \frac{U_j^n}{2} + \frac{1}{2}\cdot(U_j^* + \Delta t(R(U_j^*)));\;'$$

where $R(U_j)$ is the right part of equation in Figure 3.

There are stages of performing the entire task:

1) ensuring, that we have bottom topography values and solution values on the previous time step;
2) using these data to obtain values for the equation in Figure 2 (algorithm of computations is described in [1]);
3) computing time step for the Runge-Kutta method. Time step depends on a maximum fluid velocity in the domain [2];
4) solving equation (Figure 3) using second-order Runge-Kutta method;
5) visualizing fluid surface.

4 IMPLEMENTATION

4.1 Implementation using GPU

Unfortunately, implementation of the Kurganov-Petrova [1] scheme on CPU does not give us satisfactory results. Intel Core i5 processor can perform computations only for 50 vertices in each direction with an acceptable frame rate (30 frames per second). This is not enough for two reasons: firstly, the surface is not smooth and does not look real enough, and, secondly, Kinect sensor that is used in Interactive sandbox produces the height map with a resolution of at least 320×240 points, so we need to obtain the appropriate resolution in water calculations.

GPUs have in recent years developed from being hardware accelerators of computer graphics into high-performance computational engines. They are now used not only for visualization, but also for performing general-purpose parallel computations.

Simulating height field water is an example of a highly parallel task. We use computations on GPU to implement the Kurganov-Petrova scheme for all vertices in the grid in parallel.

The first method of the algorithm implementation on GPU is using graphics API OpenGL both for computation and visualization. We send initial data about all vertices (bottom topography obtained from Kinect sensor and solution on previous timestep) to GPU memory by creating a special texture with an appropriate resolution. Such a texture contains information about the whole computational domain. Then we use programmable shader (*shader* – program, that is performed on GPU) to do necessary computations. Obtained results, which contain water surface heights, proceed to visualization stage.

We also implemented the algorithm using technology of general-purpose computing on GPU that is called CUDA (developed by NVIDIA Corporation). This technology uses all benefits of NVIDIA GPUs in process of computation, so we can achieve the best performance. We transfer data about bottom topography to GPU memory and use CUDA function (which is generally called *kernel*) to perform calculations. Obtained results are transferred to visualization function, which is implemented with OpenGL. To avoid transferring data from GPU to CPU and vice versa, we use OpenGL and CUDA interoperability features that allow us to operate with data within GPU memory [4].

4.2 Visualization

To visualize a water surface we use some well-known techniques, which, nevertheless, provide very realistic results.

1) To compute lighting we use a standard ADS-model (which divides the light into three components – ambient, diffuse and specular) [3].
2) We use skybox texture to simulate the effect of real environment around the scene.
3) For water to look realistic we use reflections from skybox and from the terrain part which is above the water surface. To perform this we render the scene with mirrored terrain into the texture and then apply this texture to the water.
4) As in real life, in our implementation water transparency depends on its height. Transparency factor is computed with the formula below:

$$T = 1 - e^{-\alpha \cdot h},$$

where h is height of water surface and α is coefficient, which must be chosen experimentally (we use value 15.0);

5 RESULTS

We obtained a stable and fast implementation of water flow. Fluid interacts with terrain and instantly adopts to its changes. Effects, described in the previous section and inherent to water (refraction, reflection, transparency) provide a realistically looking fluid surface (see Figure 4). In developed software, we can

Figure 4.   Final result of visualization.

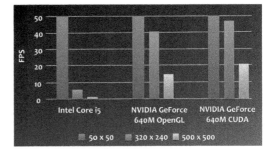

Figure 5.   Performance results.

use bottom topography from both Kinect sensor and previously generated height map.

We tested three grid resolutions: 50×50, 320×240, and 500×500. We are most interested in resolution 320×240, which coincides with Kinect sensor resolution. Using GPUs we obtained almost a tenfold performance increase in comparison with CPU in case of high grid resolutions:

The developed user interface allows the user to add water in certain points with a mouse, change terrain, move the lighting source, and tune water visualization to make it more or less transparent.

Using GPUs we obtained almost a tenfold performance increase in comparison with CPU.

6 CONCLUSION

Our implementation of shallow water equations using GPU provides satisfactory performance results and current visualization techniques provide a realistically looking appearance of the water surface. It allows us to integrate this model into Interactive Sandbox and simulate real nature phenomena with it.

Future work will be performance improving, combining particle system with height field water and using photorealistic lighting techniques, such as ray tracing.

REFERENCES

[1] A. Kurganov, G. Petrova 2007. A second-order well-balanced positivity preserving central-upwind scheme for the Saint-Venant system. *Communications in Mathematical Sciences:* vol. 5, pp. 133–160.

[2] A. R. Brodtcorb, M. L. Setra, M. Altinakar 2010. Efficient shallow water simulations on GPUs: Implementation, visualization, verification, and validation.

[3] D. Shreiner, M. Woo, J. Neider, T. Davis 2007. OpenGL Programming Guide: The Official Guide to Learning OpenGL, 6th ed. *Addison-Wesley*.

[4] Shane Cook 2013. CUDA Programming, *Elsevier Inc.*

Future Communication, Information and Computer Science – Zheng (Ed.)
© 2015 Taylor & Francis Group, London, 978-1-138-02653-7

A novel performance balancing strategy for low latency anonymous communication system

Z.L. Zhuo, Z.J. Zhong, W.N. Niu & X.S. Zhang
University of Electronic Science and Technology of China, Chengdu, China

J.C. Li & L. Liang
National Research Center for Information Security Technology, China

ABSTRACT: Tor is one of the low-latency anonymous communication systems. Its low-latency property is simply achieved by probabilistically choosing nodes with higher bandwidth. However, as our paper shows, this path selection algorithm could put a high pressure on these routers, thus affecting the balanced performance of the whole network. In this paper, we investigate how Tor's path selection algorithm could impact router's traffic overhead. In addition, we propose an approach by updating available bandwidth instead of utilizing the same bandwidth all the time; simulations are conducted to show that our approach could help to reduce the chosen chances of high resources routers by an average of 60~80%, so as to achieve a more balanced network.

Keywords: Tor; path selection algorithm; performance balancing

1 INTRODUCTION

The Tor network [4] which initiated on 20th September 2002 has been deployed world-widely and has gained more than thousands of volunteers to participate. It is an overlay network that is designed to secure a user's privacy, confidential data transmission and communication relationships.

As its name suggests, Tor utilizes the onion routing (Tor) algorithm [5]. It works simply as follows which can be seen from Figure 1.

Once the circuit is established, clients could use it to communicate with the other end anonymously. This kind of anonymity is provided by layered encryption which prevents any single relay from noticing more than just the next hop and previous hop link in the same path. Also, as the Tor network has gained an increasing number of users to join, the anonymity set has become larger, thus making it harder for an adversary to correlate and trace the communication relationship.

The Tor network has been revised many times since its first launch. In the beginning, clients selected the router uniformly at random. Then in 2004, Tor clients follow a new algorithm to choose the nodes, mainly for the load balancing reasons, as well as to achieve the goal of high quality services. As a result, a relay which claims to offer higher bandwidth is selected more often by clients. Another key change is the new design called *entry guard*: all Tor clients would choose a list of relays for their first hop when they first run the program. By default, the list length is 3. For security purposes, the list would not change for about 30~60 days [6].

As the number of clients becomes larger, a higher bandwidth router would be more likely to reach its bandwidth limit. On the other hand, lower bandwidth nodes could probably not even stand a chance to be chosen as a relay.

In this paper, we identify and investigate the Tor's path selection algorithm, and we present a novel approach to balance the router's workload. In this revised version of node selection algorithm, we update the newly introduced available bandwidth of the chosen node each time when a client gets a connection to it, instead of using the same bandwidth calculated by Tor clients. Our result has shown, as the simulation times reach 10000, choosing the same node from a given set, the chances reduce about 68.9%, 63.3%, 82.5% respectively. In other words, we could prevent a single high bandwidth router from being chosen multiple times with equal high probabilities to each client.

The description of original Tor's path selection algorithm and our algorithm are detailed in section 3. The simulation setup and result analysis are presented in section 4. Finally, our conclusion to this study is given in section 5.

2 RELATED WORK

There are two kinds of modern anonymity systems [4], one has higher anonymity but at the cost of

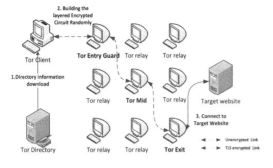

Figure 1. Simplified steps for clients to communicate through Tor network.

higher latency, e.g. Babel [7] and Mixminion [3], the other has lower latency but relative weaker anonymity. For instance, Tor, Anonymizer [1], I2P [2]. Since the earlier anonymity design does not address traffic bottlenecks, new Tor implementation has a congestion control mechanism to solve this problem. This congestion control mechanism uses a simple bucket system, however it does not reduce target router's working burdens. Can Tang et al. proposed an improved algorithm for Tor circuit scheduling [9], they introduce a new way to select the circuit based on exponentially weight moving average (EWMA), thus making most activities over Tor more efficient. Their method, however, incurred non-negligible calculation overhead to single Tor router. According to Snader et al. [8], they suggest that bandwidth estimates used for making routing decisions should be measured by opportunistically sampling actual throughput, rather than nodes reporting their own capacity. This proposal, however, could not guarantee clients to get the optimal choice.

3 MODEL BUILDING

3.1 Tor's routing selection analysis

We analyze the routing selection that Tor uses to choose which nodes to include in a circuit. We choose to analyze the Tor newest version which is 0.2.4.21, released in 2014.

Tor authorities assign flags to different kinds of routers contained in a consensus file, indicating their properties. This flag allocating algorithm is predefined by a directory server. Every node could get one flag if it satisfies corresponding the requirement.

Subject related flags are "Exit", "Guard". Exit flag means the router is more useful for building general-purpose exit circuits than for relay circuits, while the "Guard" flag means the router is suitable for use as an entry guard. Other known flags are documented in dir-spec file [10]. We have to notice that entry guard node can only be selected from routers with "Guard" flag, and exit can only be chosen from nodes with "Exit" flag. A considerable amount of routers may have both "Guard" and "Exit" flags, which suggests that this node could be chosen as entry, middle or exit

relay. While other routers neither have a "Guard" nor an "Exit" flag, we denote these routers as Set-N.

When a Tor client first runs the program, it will download the necessary routers' information in order to build the circuit. At that moment, the Tor client would select 3 guard nodes proportional to its weighted bandwidth from all the nodes that carry a "Guard" flag. Here we denote this set as Set-G. We have to notice that the client would only choose a new entry guard node when one node in that list is unreachable for a certain period of time. And the algorithm always chooses the node by bandwidth weight, only when the node selection function returns null then the Tor client would choose the node only by bandwidth.

Non-Entry node selection algorithm is used when selecting other types of nodes, namely middle node and exit node. We denote these two candidate sets as Set-M, Set-E respectively. Non-Entry node selection algorithm and Entry node selection algorithm use the same algorithm, except that the candidate set for the client to choose from is different and the weight factors are not the same.

The router selection algorithm is described and analyzed as follows.

Step 1. Filtering the candidate nodes, so that no two nodes are in the same family and no two nodes are in the same /16.

Step 2. Calculate the total bandwidth T of the candidate node list.

Step 3. A random number *rand* is generated within the range (1,T).

Step 4. Iterate through the list, add the bandwidth weight of each node to C, return the ith node such that $C \geqslant rand$.

We start to analyze this algorithm from theoretic proving. Suppose that $BW_1, BW_2, \ldots BW_n$ are the bandwidth weights in the filtered candidate list. It is pretty obvious that $T = BW_1 + BW_2 + \cdots + BW_n$, $rand \in (1, T)$. Denote that the first $i - 1$ bandwidth weight sum $S_{i-1} = BW_1 + BW_2 + \cdots + BW_{i-1}$ and $S_i = BW_1 + BW_2 + \cdots + BW_i$. The probability that this algorithm chooses the ith node in the candidate list can be deduced from Equation 1:

$$P(I) = P(S_{i-1} < rand\_bw \leq S_i) = \frac{BW_i}{\sum_{j=1}^{n} BW_j} \tag{1}$$

Here we notice that in Tor network the maximum believable bandwidth is $8*10^4$ Kb/s. And if the advertised bandwidth is bigger than that the bandwidth is then configured to *max-believe-bw*. So the maximum probability one relay can get is about 0.6% for entry guard, 0.6% for middle relay and 1.8% for exit node. The detailed deductions of these estimations can be seen from the result in section 4.

3.2 Performance balancing algorithm

When the client number becomes large enough, a considerable number of clients would end up in

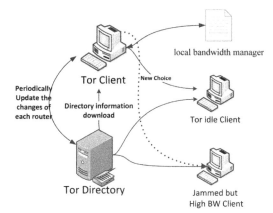

local bandwidth manager

Tor Client — New Choice

Periodically Update the changes of each router

Directory information download

Tor idle Client

Tor Directory

Jammed but High BW Client

Figure 2. The basic steps of Tor network performance balancing algorithm.

Table 1. Number of routers in different categories.

Flag	Number
Guard only	1540
Exit only	522
Set-N	2612
Guard & Exit	447

Table 2. Total Bandwidth of different categories.

Flag	Total Bandwidth(Kb/s)
Guard only	8071518
Exit only	596792
Set-N	1248674
Guard & Exit	4705589

choosing the same entry guard, middle relay or exit node with equal high probability. As a consequence, routers' working loads were not balanced. Due to the unchanged high probability, the busy router would become busier while other potential routers' bandwidths are left unused. We introduce a new parameter called available bandwidth to describe a router's ability to gain more connections to it.

The basic performance balancing algorithm works as follows:

Step 1. Tor client downloads the routers' information list L from the directory services.

Step 2. Tor client selects the nodes according to the list L, instead of using the unchanged value of the bandwidth claimed by the router, our new algorithm uses the available bandwidth which can be periodically updated from the directory services. In addition, our method implements a local bandwidth manager, which updates the selected router's bandwidth to lower the chance of getting the same relay every time.

Step 3. The Tor relay periodically reports its available bandwidth upon reaching certain bandwidth threshold to the directory.

Step 4. Other clients get the updated version of the consensus file for a certain time span, to avoid choosing the same high bandwidth router each time.

The basic procedure of this algorithm can be seen in Figure 2.

The above figure shows how the Tor client using a new strategy to avoid selecting the high bandwidth but already jammed relay (dotted line), instead chooses the relatively idle one.

4 SIMULATION RESULT

4.1 Simulation setup

The simulation of this algorithm is conducted to show that the algorithm can indeed release the burden of certain high resources routers.

The simulation is conducted based on real statistics of the Tor network, which could be downloaded from the Tor metric project [11], which could help researchers to gather real data on the Tor network. Hence the router selecting algorithm could be replayed as real as possible. We assign each router the necessary information according to the consensus file to build the circuit. We simulate the building process for N times, and we record the number of times a certain high bandwidth router has been chosen with and without our algorithm.

4.2 Preliminary result

The number of routers with different flags can be seen in Table 1. *Guard only* means routers with only the "Guard" flag. *Exit only* means routers with only the "Exit" flag, *Guard & Exit* suggests both "Guard" and "Exit" were set in the flag. The relation between those flags and *Set-G, Set-E, Set-M* which we mentioned in section 2, can be clearly described by the following equations.

$$Set_G = Guard_{only} + Guard \& Exit$$
$$Set_E = Exit_{only} + Guard \& Exit$$
$$Set_M = Guard_{only} + Exit_{only} + Guard \& Exit + Set_N$$

The maximum estimation at the end of section 2 could be calculated by the following equations:

$$MAX_{guard} = \frac{MaxBelivableBW}{Guard_{only}BW + GuardBW \& ExitBW \times \alpha}$$

$$MAX_{Exit} = \frac{MaxBelivableBW}{Exit_{only}BW + GuardBW \& ExitBW \times \varepsilon}$$

Where α, ε are the bandwidth weights which can be calculated using the rules described in *dir-spec* file [10]. In our simulation, those weights are listed in Table 3. The total bandwidth of different categories can be seen in Table 2.

Table 3. Specific Bandwidth weight assigned to different categories of routers in different positions.

	Guard only	Exit only	Normal	Guard & Exit
Guard	1.0	0.0	0.0	0.98
Mid	0.81	0.98	1.0	0.80
Exit	0.0	1.0	0.0	0.81

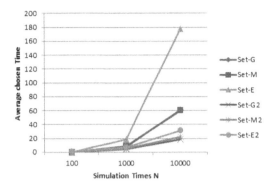

Figure 3. The average number of times a specific highest bandwidth router has been chosen with and without our performance balancing algorithm.

We reasonably assume that the highest bandwidth router also has a high working load. The average times of the one specific highest bandwidth ($8*10^4$ Kb/s) router has been chosen with and without our algorithm can be seen from figure 3. The legends Set-G, Set-M, Set-E denote the result that without our algorithm, while *Set-G2, Set-M2, Set-E2* show the result after adopting our algorithm.

4.3 Result analysis

When comparing the two simulation results shown in the last subsection, we could clearly see the advertised high resources router gets selected more times as a relay as the simulation times grows.

However, when adopting our performance balancing algorithm, the chosen times of the highest advertised router reduce, thus making other routers have a chance to be selected. More specifically, as the simulation times reach 10000, choosing the same node from *Set-G, Set-M, Set-E*, the number of times reduce about 68.9%, 63.3%, 82.5% respectively.

In this way, the Tor network could have a more balanced workload, instead of having most of the routing tasks executed on the same high resources routers.

After simulation, we also evaluate the simulation results with our theoretical proving in section 3. As the results have shown, our results satisfied theoretical proving to a large extent. That is, when the router has a maximum believable bandwidth, the probability it got chosen, approximately satisfied Equation 1.

5 CONCLUSIONS

In this study, we investigate how Tor's path selection algorithm could impact router's traffic overhead. We analyze the path selection algorithm from theoretical proving. As the results have shown, our results satisfied theoretical proving to a large extent.

And we introduce a node selection algorithm simply by updating the parameter, namely available bandwidth each time when a client gets a connection to it, instead of using the same bandwidth. Therefore, we could prevent a single high bandwidth router from being chosen for a substantial amount of times with equal high probabilities.

Simulations are conducted based on real network statistics to show that our approach could contribute to reduce the chosen chances of high resources routers. More specifically, as the simulation times reach 10000, choosing the same node from *Set-G, Set-M, Set-E*, the number of times reduce about 68.9%, 63.3%, 82.5% respectively, thus making other potential routers in the network to get an optimistic chance to be selected.

ACKNOWLEDGEMENT

This work is supported by 863 Science Foundation under Grant 2013AA014702.

REFERENCES

[1] Boyan, J. 1997. The anonymizer. *Computer-Mediated Communication (CMC) Magazine.*
[2] Zantout B. & Haraty R. 2011. I2P data communication system. *ICN 2011, Tenth Int. Conf.* pp. 401–409.
[3] Danezis, G., Dingledine, R. & Mathewson, N. 2003. Mixminion: Design of a type III anonymous remailer protocol. *In Security and Privacy, Proceedings. Symposium on IEEE.* pp. 2–15.
[4] Dingledine, R., Mathewson, N. & Syverson, P. 2004. Tor: The second-generation onion router. *Naval Research Lab Washington DC.*
[5] Dingledine, R. & Mathewson, N. 2013. Tor protocol specification. Technical report, *The Tor Project.*
[6] Elahi, T., Bauer, K., AlSabah, M., Dingledine, R. & Goldberg, I. 2012. Changing of the guards: A framework for understanding and improving entry guard selection in Tor. *In Proceedings of the 2012 ACM workshop on Privacy in the electronic society. ACM.* pp. 43–54.
[7] Gulcu, C. & Tsudik, G. 1996. Mixing E-mail with Babel. *In Network and Distributed System Security, 1996. Proceedings of the Symposium on IEEE.* pp. 2–16.
[8] Snader, R., & Borisov, N. 2008. A Tune-up for Tor: Improving Security and Performance in the Tor Network. *In NDSS* Vol. 8, pp. 127.
[9] Tang, C. & Goldberg, I. 2010. An Improved Algorithm for Tor Circuit Scheduling *Proceedings of the 17th ACM Conference on Computer and Communications Security*, ACM, 329–339.
[10] Tor directory protocol, version 3. Technical report, https://gitweb.torproject.org/torspec.git/blob/HEAD:/dir-spec.txt
[11] Tor Metrics Project https://metrics.torproject.org/data.html

Future Communication, Information and Computer Science – Zheng (Ed.)
© *2015 Taylor & Francis Group, London, 978-1-138-02653-7*

Optimization of relay selection in DF cooperative systems with outdated CSI

L. Fei, J. Zhang & Q. Gao
School of Electronic and Information Engineering, Beihang University, Beijing, P. R. China
National Key Laboratory of CNS/ATM, Beihang University, Beijing, P. R. China

X.-H. Peng
Electronic Engineering, School of Engineering & Applied Science, Aston University, Birmingham, UK

ABSTRACT: Relay selection has been considered as an effective method to improve the performance of cooperative communication. However, the Channel State Information (CSI) used in relay selection can be outdated, yielding severe performance degradation of cooperative communication systems. In this paper, we investigate the relay selection under outdated CSI in a Decode-and-Forward (DF) cooperative system to improve its outage performance. We formulize an optimization problem, where the set of relays that forwards data is optimized to minimize the probability of outage conditioned on the outdated CSI of all the decodable relays' links. We then propose a novel multiple-relay selection strategy based on the solution of the optimization problem. Simulation results show that the proposed relay selection strategy achieves large improvement of outage performance compared with the existing relay selection strategies combating outdated CSI given in the literature.

Keywords: Cooperative communication; outage performance; outdated CSI; relay selection

1   INTRODUCTION

Cooperative communication has been shown to be a promising approach to combat wireless impairments by exploiting spatial diversity without the need of multiple antennas at one node (Laneman et al. 2004, Hong et al. 2007). In a cooperative communication system, intermediate nodes are utilized as relays to forward data from source to destination over independent wireless channels. Recently, it is shown that relay selection that instructs a subset of relays to forward data is an effective method to improve the performance of cooperative communication (Bletsas et al. 2006, Bletsas et al. 2007, Hong et al. 2007).

In the practical system, the relay selection is performed before data transmission, and relays are normally selected based on the instantaneous channel state information (CSI) at the time. Due to the time interval between relay selection and data transmission and the channel fluctuation over time, the CSI used in relay selection could differ from that actually experienced by the data transmitted, i.e., the CSI used in relay selection could be outdated. Thus the selected relays may not actually be the best for data transmission. It is found that the outdated CSI results in a large degradation of the outage probability, diversity order, channel capacity, and symbol error rate (Vicario et al. 2009a, Seyfi et al. 2011a, b, Zhong et al. 2013).

To improve the outage performance of cooperative communications under outdated CSI, some relay selection strategies have been proposed (Vicario et al. 2009b, Li et al. 2011, Chen et al. 2012, Fei et al. 2013). Vicario et al. (2009b) and Fei et al. (2013) propose that the relay with the maximal predicted channel strength is selected to forward data. A minimum mean square error (MMSE) estimator and a maximum *a posteriori* (MAP) estimator are used respectively by Vicario et al. (2009b) and Fei et al. (2013) to predict the channel strength during data transmission based on the outdated CSI. Li et al. (2011) propose a relay selection strategy where the relay causing outage with a minimal probability conditioned on its outdated CSI is selected to participate in data forwarding. The relay selection strategies above focus on single-relay selection. Chen et al. (2012) investigate the outage performance when multiple relays are selected to forward data based on the outdated CSI.

These reported researches can be classified into two categories. The first type of researches (Vicario et al. 2009b, Li et al. 2011, Fei et al. 2013) focuses on proposing relay selection metrics (e.g. the predicted channel strength), which are more appropriate under outdated CSI than the commonly used instantaneous signal-to-noise ratio (SNR), but only single relay is selected to forward data. The single-relay selection (SRS) strategy can achieve excellent outage performance under ideal CSI (assuming that the CSI used in the selection procedure is the same as that experienced by the data transmitted) (Bletsas et al. 2007), but may suffer large performance loss under outdated CSI even with a careful design. The second type of researches (Chen et al. 2012) adopts multiple-relay selection (MRS), but the relays are selected based on the instantaneous SNR and no optimization method is applied for relay selection.

In this paper, we study how to optimize the relay selection under outdated CSI in a decode-and-forward (DF) cooperative communication system to improve the outage performance. We consider an optimization problem, where the set of relays that participate in data forwarding is optimized to achieve the minimal probability of outage conditioned on the outdated CSI of all the decodable relays' links. Multiple relays are allowed to forward data, as long as they can successfully decode the data transmitted from the source. By solving the optimization problem, we propose a novel multiple-relay selection strategy which performs much better than the existing relay selection strategies that deal with outdated CSI.

2 SYSTEM MODEL

Consider a DF cooperative system consisting of one source (S), one destination (D), and K half-duplex relays, where each node is equipped with only one antenna. The relays in the system assist the source to transmit data to the destination. The direct link between the source and the destination does not exist in this system, due to high shadowing (Bletsas et al. 2007, Seyfi et al. 2011b). Suppose that a total transmitting power P_{tot} is used for data transmission in the system. The source consumes a fraction of the total transmitting power (denoted as P_S) to transmit its data, while the rest of the transmitting power (denoted as P_R) is allocated to the relays in the system to forward data. We assume that the transmitting power P_R is evenly distributed among the relays selected to forward data.

2.1 Cooperative scheme

The cooperation period is divided into two time slots. In time slot 1, the source broadcasts data and each relay in the system listens. After receiving the broadcasted data, all relays decode the data and check whether the data is decoded correctly. The relays that have decoded the data correctly (or called decodable relays) are the candidates for data forwarding. Some of the decodable relays are selected to forward data based on the channel states between relays and the destination. In time slot 2, the selected relays forward the received data to the destination simultaneously over the orthogonal channels as adopted by Bletsas et al. (2007) and Hong et al. (2007). The destination combines the signals from the relays using a maximal ratio combiner (MRC) and decodes the received data.

It is assumed that a central unit (e.g. the destination) in the system performs the relay selection as was done by Seyfi et al. (2011a, b). This unit collects all information regarding the channel states between the decodable relays and the destination and feeds the

selection result (i.e. the indices of the relays selected to forward data) back to the relays. The instantaneous channel states of relay-destination links used in the selection can differ from their actual values during time slot 2 since the CSI collection and the feedback of the selection result introduce delay and the channel states fluctuate over time. In other words, the instantaneous CSIs of relay-destination links used in the relay selection are outdated compared with that during time slot 2.

2.2 Channel model

The channel responses of the source-relay link S-k during time slot 1 and the relay-destination link k-D during time slot 2 are denoted as h_{Sk} and $h_{kD}, k = 1, 2, \ldots, K$, respectively. They are assumed to be independent Rayleigh fading distributions, and hence are modeled as zero-mean circularly symmetric complex Gaussian random variables with variances σ_{Sk}^2 and σ_{kD}^2 respectively. Correspondingly, the squared channel strengths of link S-k and link k-D are given by $g_{Sk} = |h_{Sk}|^2$ and $g_{kD} = |h_{kD}|^2$ respectively, which are exponentially distributed and have the average values $\bar{g}_{Sk} = \sigma_{Sk}^2$ and $\bar{g}_{kD} = \sigma_{kD}^2$ respectively. The variance σ_{AB}^2 of link A-B is modeled as $\sigma_{AB}^2 = (\lambda/4\pi d_0)^2 (d_0/d_{AB})^\alpha$, where A-$B \in \{S\text{-}k, k\text{-}D | k = 1, 2, \ldots, K\}$, d_{AB} is the link distance, λ is the carrier wavelength, d_0 is a reference distance, and α is the path-loss exponent. The noise in each link is the additive white Gaussian noise (AWGN) with zero mean and the same variance σ^2.

The channel response of link k-D used in relay selection is an outdated version of the actual one h_{kD} during time slot 2 and is denoted as $\tilde{h}_{kD} \cdot h_{kD}$ and $\tilde{h}_{kD}$ can be related by a first-order autoregressive process as used by Li et al. (2011). Thus h_{kD} conditioned on $\tilde{h}_{kD}$ follows a Gaussian distribution, which is given by (Vicario et al. 2009b, Li et al. 2011, Fei et al. 2013)

$$h_{kD} | \tilde{h}_{kD} \sim \mathcal{CN} \left(\rho_k \tilde{h}_{kD}, \left(1 - \rho_k^2\right) \sigma_{kD}^2 \right) \tag{1}$$

where ρ_k is the correlation coefficient between $\tilde{h}_{kD}$ and h_{kD} and is determined by the time interval T_k between them according to Jakes' model as $\rho_k = J_0(2\pi f_k T_k)$ where f_k is the maximum Doppler frequency shift and $J_0(\cdot)$ denotes the zero-order Bessel function of the first kind. Correspondingly, the squared channel strength of link k-D $\tilde{g}_{kD}$ obtained from $\tilde{h}_{kD}(\tilde{g}_{kD} = |\tilde{h}_{kD}|^2)$ is an outdated version of the actual one g_{kD} during time slot 2. g_{kD} conditioned on $\tilde{g}_{kD}$ follows a non-central chi-square distribution with 2 degrees of freedom, whose probability density function (PDF) is given by (Vicario et al. 2009a)

$$f_{g_{kD}|\tilde{g}_{kD}}(x|y) = \frac{\exp\left(-\dfrac{x + \rho_k^2 y}{\sigma_{kD}^2 \left(1 - \rho_k^2\right)}\right)}{\sigma_{kD}^2 \left(1 - \rho_k^2\right)} I_0\left(\frac{2\sqrt{\rho_k^2 xy}}{\sigma_{kD}^2 \left(1 - \rho_k^2\right)}\right) \tag{2}$$

where $I_0(\cdot)$ denotes the zero-order modified Bessel function of the first kind.

2.3 Outage probability

The transmission outage is declared when the received SNR at the destination is below a predetermined threshold γ_{th} (Laneman et al. 2004). Since only the decodable relays can participate in data forwarding, the outage probability p_{out} is given by

$$p_{out} = \sum_{m=0}^{K} \sum_{\mathcal{D}_m} \Pr\{\text{outage}|\mathcal{D}_m\} \Pr\{\mathcal{D}_m\} \qquad (3)$$

where $\mathcal{D}_m \subseteq \{1, 2, \ldots, K\}$ denotes a set with m decodable relays, called decoding set, $\Pr\{\text{outage}|\mathcal{D}_m\}$ is the probability of outage given the decoding set, and $\Pr\{\mathcal{D}_m\}$ is the probability of that decoding set. The second summation is over all different decoding sets with exactly m decodable relays.

For $m \geq 1$, i.e. at least one relay has correctly decoded the data, $\Pr\{\text{outage}|\mathcal{D}_m\}$ is given by

$$\Pr\{\text{outage}|\mathcal{D}_m\} = \Pr\{\gamma_{\mathcal{D}_m} < \gamma_{th}\} \qquad (4)$$

where $\gamma_{\mathcal{D}_m} = \sum_{k \in \mathcal{D}_m} g_{kD} P_k / \sigma^2$ is the received SNR at the destination and $\gamma_{th} = 2^{2r} - 1$ with the required spectral efficiency r, and $\Pr\{\mathcal{D}_m\}$ is given by

$$\Pr\{\mathcal{D}_m\} = \prod_{i \in \mathcal{D}_m} \Pr\left\{\frac{g_{Si} P_S}{\sigma^2} \geq \gamma_{th}\right\} \prod_{i \notin \mathcal{D}_m} \Pr\left\{\frac{g_{Si} P_S}{\sigma^2} < \gamma_{th}\right\} \qquad (5)$$

For $m = 0$, i.e. no decodable relay, we have

$$\Pr\{\text{outage}|\mathcal{D}_0\} = 1 \qquad (6)$$

and

$$\Pr\{\mathcal{D}_0\} = \prod_{i=1}^{K} \Pr\left\{\frac{g_{Si} P_S}{\sigma^2} < \gamma_{th}\right\} \qquad (7)$$

Thus the outage probability p_{out} can be rewritten as

$$p_{out} = \sum_{m=1}^{K} \sum_{\mathcal{D}_m} \Pr\{\gamma_{\mathcal{D}_m} < \gamma_{th}\} \Pr\{\mathcal{D}_m\} + \Pr\{\mathcal{D}_0\} \qquad (8)$$

3 RELAY SELECTION STRATEGY

In this section, we propose a multiple-relay selection strategy to improve the outage performance of cooperative systems under outdated CSI. From Equation 8, we can see that the overall outage probability of the system p_{out} is determined by the outage probability under each given decoding set $\Pr\{\gamma_{\mathcal{D}_m} < \gamma_{th}\}$. Intuitively, the outage probability for each decoding set can be reduced largely by selecting the relays that minimize the conditional outage probability on the CSI, used in the relay selection, of the links between the decodable relays and the destination. Thus we optimize the set of relays that participate in data forwarding to minimize the conditional outage probability, and then propose a multiple-relay selection strategy based on the solution of the optimization problem.

3.1 Conditional outage probability calculation

The conditional outage probability p_{out}^{C} given a decoding set $\mathcal{D}_m (m \geq 1)$ is given by

$$p_{out}^{C} = \int_0^{\gamma_{th}} f_{\gamma_{\mathcal{D}_m}|\tilde{g}_{n_1 D}, \tilde{g}_{n_2 D}, \cdots, \tilde{g}_{n_m D}}\left(x|y_1, y_2, \cdots, y_m\right) dx \qquad (9)$$

where n_q denotes the index of the qth decodable relay ($q = 1, 2, \ldots, m$ and $n_q \in \mathcal{D}_m$) for facilitating the expression of p_{out}^{C}, and $f_{\gamma_{\mathcal{D}_m}|\tilde{g}_{n_1 D}, \tilde{g}_{n_2 D}, \ldots, \tilde{g}_{n_m D}}(x|y_1, y_2, \ldots, y_m)$ is the conditioned PDF of $\gamma_{\mathcal{D}_m}$ conditioned on $\tilde{g}_{n_q D}, q = 1, 2, \ldots, m$. Then the received SNR at the destination $\gamma_{\mathcal{D}_m}$ can be rewritten as

$$\gamma_{\mathcal{D}_m} = \sum_{q=1}^{m} \gamma_{n_q} \qquad (10)$$

γ_{n_q} represents the received SNR at the destination from the qth decodable relay and is given by

$$\gamma_{n_q} = g_{n_q D} P_{n_q} / \sigma^2 \qquad (11)$$

where P_{n_q} is the transmitting power of the qth decodable relay and it is given by

$$P_{n_q} = \beta_{n_q} P_R / m_S \qquad (12)$$

$\beta_{n_q} = 1$ or 0 denotes whether the relay is selected to forward data or not and $m_S = \sum_{n_q \in \mathcal{D}_m} \beta_{n_q}$ is the number of relays selected to forward data, which is at least equal to one. Since all the channels in the system are assumed to be independent, $f_{\gamma_{\mathcal{D}_m}|\tilde{g}_{n_1 D}, \tilde{g}_{n_2 D}, \ldots, \tilde{g}_{n_m D}}(x|y_1, y_2, \ldots, y_m)$ is given by

$$
\begin{aligned}
&f_{\gamma_{\mathcal{D}_m}|\tilde{g}_{n_1 D}, \tilde{g}_{n_2 D}, \cdots, \tilde{g}_{n_m D}}\left(x|y_1, y_2, \cdots, y_m\right) \\
&= f_{\gamma_{n_1}|\tilde{g}_{n_1 D}}\left(x_1|y_1\right) \otimes f_{\gamma_{n_2}|\tilde{g}_{n_2 D}}\left(x_2|y_2\right) \\
&\otimes \cdots \otimes f_{\gamma_{n_m}|\tilde{g}_{n_m D}}\left(x_m|y_m\right)
\end{aligned} \qquad (13)
$$

where $f_{\gamma_{n_q}|\tilde{g}_{n_q D}}(x_q|y_q)$ is the PDF of γ_{n_q} conditioned on $\tilde{g}_{n_q D}$ and $\otimes$ stands for the convolution.

The convolution of multiple PDFs has high computational complexity. To simplify the calculation, we adopt the approach of moment-generating function (MGF). The MGF of γ_{n_q} conditioned on $\tilde{g}_{n_q D}$ is defined as

$$M_{\gamma_{n_q}|\tilde{g}_{n_q D}}\left(s|y_q\right) = \int_0^{+\infty} f_{\gamma_{n_q}|\tilde{g}_{n_q D}}\left(x_q|y_q\right) \exp(s x_q) dx_q \qquad (14)$$

From (11), we have

$$f_{\gamma_{n_q}|\tilde{g}_{n_q D}}\left(x_q|y_q\right) = \begin{cases} \dfrac{\sigma^2}{P_{n_q}} f_{g_{n_q D}|\tilde{g}_{n_q D}}\left(\dfrac{\sigma^2}{P_{n_q}} x_q \middle| y_q\right), & P_{n_q} > 0 \\[2ex] \delta(x_q), & P_{n_q} = 0 \end{cases} \qquad (15)$$

47

where $\delta(\cdot)$ denotes the Dirac function. $P_{n_q} = 0$ means that the qth decodable relay is not selected to forward data (i.e. $\beta_{n_q} = 0$). To make the convolution of PDFs in Equation 13 remaining valid when $P_{n_q} = 0$, the PDF $f_{\gamma_{n_q}|\tilde{g}_{n_q D}}(x_q|y_q)$ is set to be the Dirac function in this case. Substituting Equation 2 to Equation 15, according to the definition of MGF and using Equation 6.614.3 given by Gradshteyn & Ryzhik (2007), after some algebraic manipulations we obtain $M_{\gamma_{n_q}|\tilde{g}_{n_q D}}(s|y_q)$ as

$$M_{\gamma_{n_q}|\tilde{g}_{n_q D}}\left(s|y_q\right) = \frac{1}{1 - \mu_q s} \exp\left(\frac{v_q y_q s}{1 - \mu_q s}\right) \qquad (16)$$

where $\mu_q = P_{n_q}\sigma_{n_q D}^2(1 - \rho_{n_q}^2)/\sigma^2$ and $v_q = P_{n_q}\rho_{n_q}^2/\sigma^2$. Then we obtain the MGF of $\gamma_{\mathcal{D}_m}$ conditioned on $\tilde{g}_{n_q D}$, $q = 1, 2, \ldots, m$, as

$$M_{\gamma_{\mathcal{D}_m}|\tilde{g}_{n_1 D}, \tilde{g}_{n_2 D}, \cdots, \tilde{g}_{n_m D}}\left(s|y_1, y_2, \cdots, y_m\right) = \prod_{q=1}^{m} M_{\gamma_{n_q}|\tilde{g}_{n_q D}}\left(s|y_q\right) (17)$$

According to the definition of MGF, replacing the s in Equation 17 with $-s$ we obtain the Laplace transform of the conditioned PDF $f_{\gamma_{\mathcal{D}_m}|\tilde{g}_{n_1 D}, \tilde{g}_{n_2 D}, \ldots, \tilde{g}_{n_m D}}(x|y_1, y_2, \ldots, y_m)$. Then, using the inverse Laplace transform, the conditional outage probability p_{out}^C is obtained as

$$p_{out}^C = \frac{1}{2\pi j}\int_{a-j\infty}^{a+j\infty}\frac{1}{s}\prod_{q=1}^{m} M_{\gamma_{n_q}|\tilde{g}_{n_q D}}\left(-s|y_q\right)\exp(s\gamma_{th})ds \qquad (18)$$

where a is an arbitrary real number larger than the real part of all singularities of $1/s\prod_{q=1}^{m} M_{\gamma_{n_q}|\tilde{g}_{n_q D}}(-s|y_q)$. The integral in Equation 18 is conducted along the vertical line $\text{Re}(s) = a$ in the complex plane and can be calculated through a numerical method.

3.2 Conditional outage probability minimization

Based on the derivation above, the conditional outage probability p_{out}^C is a function of $\beta_{n_q}, n_q \in \mathcal{D}_m$. The minimization problem of the conditional outage probability can be formulated as

$$\min_{\beta_{n_q}, n_q \in \mathcal{D}_m} \quad p_{out}^C$$
$$s.t. \quad \beta_{n_q} \in \{0, 1\}, n_q \in \mathcal{D}_m \qquad (19)$$
$$\sum_{n_q \in \mathcal{D}_m} \beta_{n_q} \geq 1$$

This is an optimization problem of selecting relays for data forwarding. The optimal solution $\beta_{n_q}^*, n_q \in \mathcal{D}_m$, indicates which relays should be selected to forward data. The multiple-relay selection strategy can be obtained by solving this problem.

The optimization problem is a nonlinear 0-1 programming problem. As there are m decodable relays and each relay has two choices, there are $2^m - 1$ possibilities (at least one relay is selected to forward data).

This problem can always be solved by the strategy of exhaustive search. However, the computational complexity of this exhaustive strategy is exponential in m. For cooperative systems with a large number of relays, the exponential complexity is undesirable. Thus we need a relay selection strategy with lower complexity (e.g. linear in m).

3.3 Proposed multiple-relay selection strategy

Using the idea of relay ordering proposed by Jing & Jafarkhani (2009), a multiple-relay selection strategy with linear complexity can be obtained. First, the decodable relays are sorted according to a relay ordering function which is related to the channel states of the links in the system. Without loss of generality, assume that the relays are sorted as $n_1, n_2, \ldots, n_m$. Each relay in the ordering has a higher priority to the relays after it for the forwarding action. In other words, if relay n_i does not participate in data forwarding, relay $n_j(j > i)$ should not forward data. Then, to solve the problem in Equation 19, we only need to find the minimum among $p_{out,\{n_1\}}^C, p_{out,\{n_1,n_2\}}^C, \ldots,$ $p_{out,\{n_1,n_2,\ldots,n_m\}}^C$, which is equivalently the conditional outage probability given the relays in the sets $\{n_1\}, \{n_1, n_2\}, \ldots, \{n_1, n_2, \ldots, n_m\}$, respectively. Correspondingly, the optimized set of the relays selected to forward data is obtained.

The determination of the relay ordering function is critical for this strategy. For any relay ordering function, the relay strategy can be obtained but its outage performance may be far away from the optimal one. On the other hand, the relay ordering function which induces a relay strategy approaching the optimal outage performance more closely may have more complex form and higher computational complexity. Thus we need to propose a relay ordering function which has a simple form and meanwhile induces a relay selection strategy with good outage performance. Considering the simplicity of MMSE estimation, we set the relay ordering function as the predicted squared channel strength based on the MMSE estimation, which is given by

$$F(\tilde{g}_{n_q D}, \bar{g}_{n_q D}, \rho_{n_q}) = \rho_{n_q}^2 \tilde{g}_{n_q D} + \left(1 - \rho_{n_q}^2\right)\bar{g}_{n_q D} \qquad (20)$$

Then we propose the multiple-relay selection strategy as follows:

- Step 1) The decodable relays are sorted in descending order according to the function in Equation 20. Assume that the relays are sorted as $n_1, n_2, \ldots, n_m$.
- Step 2) Among the sets $\{n_1\}, \{n_1, n_2\}, \ldots, \{n_1, n_2, \ldots, n_m\}$, the set of relays that achieves the lowest conditional outage probability is selected to forward data, while other relays keep silent.

ρ_{n_q} and $\bar{g}_{n_q D}$ used in this strategy can be obtained with light overhead since they are the statistical information and relatively invariable. The simulation result will show that the proposed relay selection strategy achieves almost the same outage performance as the exhaustive search strategy and performs much better than other relay selection strategies for outdated CSI.

4 SIMULATION RESULTS

In this section, we simulate the proposed relay selection strategy, and compare it to other relay strategies reported in the literature. We assume that five relays are distributed in the cooperative system ($K = 5$) and set $P_S = P_R = P_{tot}/2$ since the equal-power allocation between the source and the relays is a natural choice (Bletsas et al. 2007). Three different simulation scenarios are considered:

- Scenario 1: Both the distances of all the links and the correlation coefficients of all relay-destination links are assumed to be equal. We set $d_{Sk} = d_{kD} = 100$ m and $\rho_k = 0.5$ for $k = 1, 2, \ldots, 5$.
- Scenario 2: The distances of all the links are not identical, while the correlation coefficients of allrelay-destination links are assumed to be equal. We set that the relays are distributed randomly in a 200 m × 200 m square area. The source and the destination are located at the middle of the left-hand side borderline and the right-hand side borderline of the square, respectively. $\rho_k = 0.5$ for $k = 1, 2, \ldots, 5$.
- Scenario 3: The distances of all the links are not identical, nor are the correlation coefficients of all relay-destination links. The nodes are distributed as described in Scenario 2. ρ_k is generated with uniform distribution between 0 and 1.

Other system parameters, in line with that adopted by Bletsas et al. (2007) and Vicario et al. (2009a, b), are summarized in Table 1.

Table 1. System parameters.

Parameters	r	σ^2	λ	d_0	α
	bps/Hz	dBm	m	m	
Values	1	−90	0.125	10	3

Figure 1 shows the outage performance comparison between the proposed strategy and the exhaustive search strategy. It can be seen that the outage probability of the proposed strategy is very close to that of the exhaustive search strategy for different scenarios and different total transmitting power levels. Since the proposed strategy has lower complexity than the exhaustive search strategy, it is straightforward to adopt the proposed strategy instead of the exhaustive search strategy to select relay.

Figure 2 shows the outage performance comparison among the proposed relay selection strategy, opportunistic relay selection (ORS) strategy under ideal CSI, and the existing relay strategies combating outdated CSI in the literature: the single-relay selection strategy based on the predicted channel strength, called PREDICTION strategy (Vicario et al. 2009b), the single-relay selection strategy based on the minimal conditional outage probability, named MOP strategy (Li et al. 2011), and the multiple-relay selection strategy based on outdated CSI, named N+NT-ORS strategy (Chen et al. 2012). It is shown that our proposed strategy performs much better than other relay selection strategies for outdated CSI (especially in large transmitting power region) as the set of selected

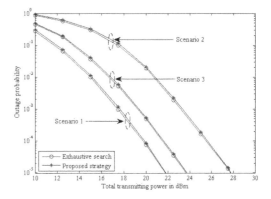

Figure 1. Outage performance comparison between the proposed multiple-relay selection strategy and the exhaustive search strategy.

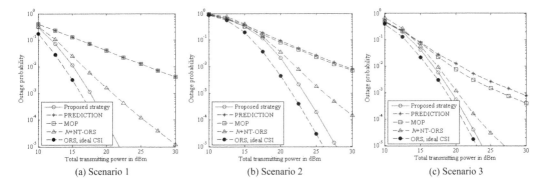

(a) Scenario 1   (b) Scenario 2   (c) Scenario 3

Figure 2. Outage performance comparison among the proposed multiple-relay selection strategy, the relay selection strategies combating outdated CSI in the literature, and ORS under ideal CSI in different scenarios.

relays is optimized to achieve the minimum of the conditional outage probability on the outdated CSI. ORS under ideal CSI achieves the optimal outage performance of the cooperative system (Bletsas et al. 2007). In comparison with it, all of the strategies for outdated CSI suffer degradation of outage performance. How to design a relay selection strategy to further approach the optimal performance is left to our future research.

5 CONCLUSIONS

In this paper, we investigate the optimization of relay selection under outdated CSI in a DF cooperative system to improve its outage performance. We formulize an optimization problem, where the set of relays that forward data is optimized to minimize the conditional outage probability on the outdated CSI of all the decodable relays' links. We then propose a multiple-relay selection strategy by solving the optimization problem. Simulation shows that the proposed strategy largely outperforms the existing relay selection strategies for outdated CSI. Thus our proposed strategy can be used to reduce the impact of outdated CSI on the outage performance effectively in cooperative systems.

REFERENCES

[1] Bletsas, A., Khitsi, A., Reed, D.P. & Lippman, A. 2006. A simple cooperative diversity method based on network path selection. *IEEE Journal on Selected Areas in Communications* 24(3): 659–672.

[2] Bletsas, A., Shin, H. & Win, M.Z. 2007. Cooperative communications with outage-optimal opportunistic relaying. *IEEE Transactions on Wireless Communications* 6(9): 3450–3460.

[3] Chen, M., Liu, T.C.-K. & Dong, X. 2012. Opportunistic multiple relay selection with outdated channel state information. *IEEE Transactions on Vehicular Technology* 61(3): 1333–1345.

[4] Fei, L., Gao, Q., Zhang, J. & Xu, Q. 2013. Relay selection with outdated channel state information in cooperative communication systems. *IET Communications* 7(14): 1557–1565.

[5] Gradshteyn, I.S. & Ryzhik, I.M. 2007. *Table of integrals, series, and products*. New York: Academic Press.

[6] Hong, Y.-W., Huang, W.-J., Chiu, F.-H. & Kuo, C.-C.J. 2007. Cooperative communications in resource-constrained wireless networks. *IEEE Signal Processing Magazine* 24(3): 47–57.

[7] Jing, Y. & Jafarkhani, H. 2009. Single and multiple relay selection schemes and their achievable diversity orders. *IEEE Transactions on Wireless Communications* 8(3): 1414–1423.

[8] Laneman, J.N., Tse, D. & Wornell, G.W. 2004. Cooperative diversity in wireless networks: efficient protocols and outage behavior. *IEEE Transactions on Information Theory* 50(12): 3062–3080.

[9] Li, Y., Yin, Q., Xu, W. & Wang, H.-M. 2011. On the design of relay selection strategies in regenerative cooperative networks with outdated CSI. *IEEE Transactions on Wireless Communications* 10(9): 3086–3097.

[10] Seyfi, M., Muhaidat, S. & Liang, J. 2011a. Average capacity performance of opportunistic relay selection with outdated CSI. *IET Communications* 5(16): 2339–2344.

[11] Seyfi, M., Muhaidat, S., Liang, J. & Dianati, M. 2011b. Effect of feedback delay on the performance of cooperative networks with relay selection. *IEEE Transactions on Wireless Communications* 10(12): 4161–4171.

[12] Vicario, J., Bel, A., Lopez-Salcedo, J. & Seco, G. 2009a. Opportunistic relay selection with outdated CSI: outage probability and diversity analysis. *IEEE Transactions on Wireless Communications* 8(6): 2872–2876.

[13] Vicario, J.L., Bel, A., Morell, A., and Seco-Granados, G. 2009b. A robust relay selection strategy for cooperative systems with outdated CSI. In *IEEE Vehicular Technology Conference: Proc., Barcelona, Spain, 26-29 Apr. 2009*.

[14] Zhong, B., Zhang, Z., Zhang, X., Li, Y. & Long, K. 2013. Impact of partial relay selection on the capacity of communications systems with outdated CSI and adaptive transmission techniques. *EURASIP Journal on Wireless Communications and Networking* 24: 1–13.

Future Communication, Information and Computer Science – Zheng (Ed.)
© *2015 Taylor & Francis Group, London, 978-1-138-02653-7*

A novel hybrid approach for path explosion in concolic execution

W.N. Niu, Z.L. Zhuo, G.W. Yang & X.S. Zhang
University of Electronic Science and Technology of China, Chengdu, China

J.C. Li & L. Liang
National Research Center for Information Security Technology, Beijing, China

ABSTRACT: Concolic execution can alleviate imprecision in symbolic execution using concrete values. This technique is promising for test cases generation and vulnerabilities digging automatically in practice. Since the number of paths grows exponentially with the increase in scale and complexity of the tested programs, exploring all feasible paths is prohibitively expensive. Path explosion is the main challenge against the widespread application of concolic execution. This paper proposes a novel approach that interleaves path merging, heuristics search with constraints simplification to reduce execution time. We have conducted preliminary experiments on 10 programs which belong to GNU COREUTILS. Results show that our hybrid approach can outperform every technique by an order of magnitude in terms of execution time and increase coverage by average of 9.2%, 9.1%, and 8.6%.

Keywords: Concolic execution; path explosion; path merging; heuristics search; constraints optimization

1 INTRODUCTION

Concolic execution performs symbolic execution dynamically, while the program is executed on some concrete input values. Because complex data structures can be reasoned precisely as well as symbolic constraints generated along the path can be simplified using the corresponding concrete values. Concolic execution outperforms pure symbolic execution in program analysis. This technique is promising for high-coverage test cases generation and deep vulnerabilities discovery [1].

Previous work has shown that concolic execution can test small programs effectively. However, the possible number of paths explored is so large that concolic execution ends up exploring small parts of paths in a reasonable time. Thus path explosion remains a major hurdle in applying concolic execution to large programs.

A nature approach for path explosion in concolic execution is devising strategies that could quickly cover a significant portion of the program's path space [2] We propose a novel approach that interleaves static analysis with dynamic analysis. In the hybrid approach, we choose paths to merge based on the control flow graph (CFG) at first. Then we adopt a heuristic search to negate path constraints collected

by concolic performing programs. Previous work has confirmed that heuristics search helps to improve the code coverage more than classical search algorithms. At last, we simplify path constraints negated by heuristics search.

We have implemented the novel hybrid approach in our prototype to generate test cases, and experimentally validate the approach on ten programs over 10 K total lines of code. Our experiments, results demonstrate that the hybrid approach reduces execution time by an average of 12X over path merging, 11X over heuristics search, and 10 X over constraints simplification. And code coverage increases 9.2%, 9.1%, 8.6% on average respectively.

Our contributions are two-fold. First, we present a novel hybrid approach that utilizes static analysis to merge paths as well as constraints simplification to reduce overhead. Second, we successfully apply the hybrid approach to ten programs which belong to GNU COREUTILS. And results on our prototype demonstrate that our hybrid approach outperforms any one of these three strategies.

The rest of the paper is organized as follows. Section 2 summarizes the existing methods of path explosion in program analysis. The presentation of our hybrid approach (in Section 3) follows. The experiments and analyses are presented in Section 4. Finally, a conclusion and a future research trend are given in Section 5.

This work is supported by 863 Science Foundation under Grant 2013AA014702

2 RELATED WORK

Path explosion can be alleviated by heuristic search, summary [2], input space reduction [3], etc. Here, we discuss some interesting solutions to path explosion in detail.

2.1 *Path search algorithms*

Path search algorithm is used to drive programs along different paths. Moreover, optimal search strategies can reduce test time. Thus, some scholars proposed methods to alleviate path explosion from path search perspective.

Albeit simple, traditional depth-first search has a shortcoming. When testing programs with non-terminating loops that relate to symbolic conditions, there is no new path that can be explored if the maximum depth is not set. Setting a maximum depth can prevent infinite loops. However, this method may miss some code especially when the maximum depth is set too small.

Random search avoids starvation since execution paths with large numbers of branches are not predominated by execution chances with the random selection. Random test cases generation and random constraints negation belong to random search. The former does not preserve paths that have explored information, which results in different test cases performing the same path. The latter approach can reduce redundancy, but high code coverage cannot be guaranteed.

Generational search [4] maximizes the number of new test cases generated form each execution to maximize code coverage. But there are two shortages one is slowness, because generational search runs programs using all candidate test cases when selecting a new test case; the other is that scores in test suits have changed after each symbolic execution.

There are some other heuristic methods used to ease path explosion problem, for example meta-strategy adopted by PEX [5] can avoid getting stuck in a particular area of the program by a fixed search order. But design of heuristics is inherently prone to be over-fitting and designing robust heuristics is a tedious time-consuming task.

2.2 *Summary and input space reduction*

A Summary (includes function summary and loop summary) [6] is able to reduce the number of paths, which needs to be explored. However, the application of a summary has three drawbacks. First, it is time-consuming because a summary is usually calculated by hand; second, it is difficult to get a summary of actual functions, because actual functions often have a complex internal logic and a large number of possible paths; third, a loop summary cannot cover all paths of loops.

Symbolic grammar, length abstraction and input partition belong to input space reduction [7]. Symbolic grammar can address tested programs which have highly-structured inputs. However, this technique

```
Require: P, I, threshold
//P is a executable program
//I is the input of the tested program
//threshold is the final excepted value of coverage
newP:=Merge_path(P)
//newP is a new program after merging
pc:=[]
//pc is path constraint
setInput=[]
//setInput are the input sets of program
newInput:=I
coverage:=0
while(coverage<=threshold):
    setInput.add(newInput)
    pc:=DSE(newP,newInput)
    newPC:=Heuristic _search (pc)
    Simplize(newPC)
    newInput:=Solve(newPC)
    Calc(coverage)
return setInput
```

Figure 1. Hybrid approach.

is time-consuming because grammar is constructed manually. Length abstraction is able to alleviate path explosion since it narrows input space. However, it has the significant shortcoming that length abstraction produces false negatives. Input partition can reduce the number of inputs dramatically. But this technique performs well only when the input space of tested programs can be divided into disjoint sub-space.

3 HYBRID APPROACH

3.1 *Framework of hybrid approach*

We propose a hybrid approach to ease path explosion by combining static analysis and dynamic symbols.

At first, we obtain the merged control flow graph (CFG) of programs under analysis by means of static analysis.

Secondly, we perform programs using concolic execution and we add input I to the input sets.

Thirdly, we negate path constraint obtained in step 2 according to the heuristic search algorithms.

Fourthly, we optimize the constraint sets through constraints simplification, and use the latest constraint solver to solve these simplified constraints.

Finally, we calculate basic block coverage. If the coverage does not reach the presented threshold, the algorithm repeats step 2 to step 5.

The hybrid approach is shown in Figure 1.

3.2 *Details of hybrid approach*

The hybrid approach generates test cases of checked programs through combining path merging, heuristics search and constraints simplification. In the following, we describe these three techniques in detail.

Path merging:

An important optimization is to reduce the number of paths explored by merging multiple ones

```
1. int abs (int x){
2.      int X;
3.      if (x<0)
4.          X=-x;
5.      else X=x;
6. }
```

Figure 2. Sample example.

together [8]. We reduce the number of explored paths by applying an aggressive variant of phi-node folding to merge paths of tested programs in this paper.

At first, we find merging nodes that are typically placed at branch conditions. Then, we use phi-node folding technique to merge paths. Phi-node folding statically merges paths to be explored using ite expression. Here, ite(C,p,q) denotes the if-then-else operator that evaluates to p if C is true, and to q otherwise.

Consider the example program in Figure 2, a simple function that gets absolute value of input. The execution paths split at line 3 on the condition C that x is less than zero. Line 6 is then reached by the two paths $(5, C, [X = -x])$ and $(5, \neg C, [X=x])$. These two can be merged into a single $(5, true, [X=ite(C,-x, x)])$.

Heuristic search:

The intuition of heuristic search is using the heuristics to select negated expression in order to get maximum path coverage or branch coverage in test cases generation. So heuristics search algorithms can cover statements which are difficult to reach using traditional path search algorithms. In general, several search algorithms combined with each other can achieve much better results. Thus, in this paper, we use the two search strategies: generational search and random search in a round robin fashion in order to achieve a higher code coverage rate.

Realization idea of generational search can be illustrated as follows:

Step 1: Negating all the branches on the collected path constraints after each symbol execution. Then solving negated path constraints to get new test cases.

Step 2: Running programs with the new inputs and assigning the execution scores, which are equal to the number of new basic blocks explored. Then we put new test cases with scores into the test suits.

Step 3: Selecting a test case with maximum score from the test suits. Then, running programs with that input and deleting this test case from the test suits.

Constraints simplification:

Although current constraint solvers' functions are powerful, simplifying the path constraint can still to reduce execution time. In this paper, we use constraint independence optimization, constraint set simplification and loop counts symbolization to optimize path constraints.

Constraint independence optimization consists of dividing the entire constraints set into several independent constraints subsets and solving them separately [9]. As concolic execution runs programs, a

number of constraints involving the same variables are added to path conditions. In the paper, we use common sub-constraint elimination and fast unsatisfiability check to simply constraint sets. Fast unsatisfiability check works as follows: if any constraint conflicts with other constraints of the new path constraint on the semantic, results are returned immediately without invoking SMT solvers. Loop counts symbolization uses new symbolic variables to represent the number of times of each loop in the program.

3.3 *Analysis*

Path emerging that comes from the hardware domain can decrease the number of paths to be explored effectively. Experiment results demonstrate that path merging may significantly speed-up symbolic execution, an order of magnitude for example for low level software [10]. Heuristics search algorithms can cover statements which are hard-to reach using traditional path search algorithms. In the process of dynamic symbolic execution, a majority of resource consumption is caused by constraint solving. Simplifying constraint expression or reducing the frequency of solving can improve the efficiency of concolic execution.

In conclusion, the hybrid approach can increase path coverage rate. Moreover, a combination of static and dynamic analysis can help automated test generation to scale on large-scale programs.

4 EXPERIMENTS AND EVALUATION

Experiments were run on 2.5 GHZ Core2 Duo servers with 4 GB of RAM and running Ubuntu 10.0.4. We have implemented our hybrid approach on the prototype of concolic execution developed by ourselves to generate test cases. We experimentally evaluated the effectiveness of our hybrid approach by running our prototype on the 96 GNU COREUTILS.

Our prototype uses IDA pro to obtain CFG and call graph of the program to be tested under test. Using Valgrind 3.4.1 to instrument tested programs and collect path constraints. Then the prototype adopts heuristics search to negate collected path constraints. At last, the prototype transforms path constraints for standard STP format, and adopts the latest STP to solve path constraints.

For each tested benchmark, we compare the performance of the hybrid approach with path merging, heuristic search, and constraint simplification respectively. Figure 3 shows the individual results from the hybrid approach and other three methods on the ten coreutils benchmarks.

In the ten coreutils benchmarks, we see that the hybrid approach is dramatically faster than the other three methods. But the performance of the two tested programs is no longer improved dramatically. Because rm is simple and ls has loop a which is not dealt with by hybrid approach.

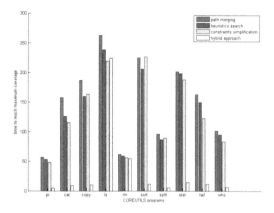

Figure 3. Individual time to get the maximum coverage using these four approaches respectively.

Table 1. Individual results from using hybrid approach and the other three methods on the ten coreutils benchmarks. X represent relative speedup, cov_{inc} is coverage increase.

Programs	Path merging X	cov_{inc}	Heuristic search X	cov_{inc}	Constraints simplification X	cov_{inc}
pr	11.2	12.7	10.4	18.5	9.4	12.3
cat	15.7	11.6	12.4	10.8	11.2	11.1
copy	16.7	12.3	14.2	11.7	14.5	10.6
ls	0.2	2.3	0.1	1.8	0.02	2.1
rm	0.1	0	0.07	0	0.03	0
sort	18.9	10.4	7.3	11.2	9.2	10.8
split	17	10.7	15.3	10.1	15.9	8.3
stat	13.1	15.2	12.9	13.7	12.1	12.5
tail	13.2	12.5	12.1	10.3	9.7	13.4
who	15.3	4.5	14.2	3.2	12.4	5.3
average	**12.1**	**9.2**	**10.9**	**9.1**	**10.4**	**8.6**

Table 1 shows that the execution time has been reduced over 10 X on average. In the best-case scenario, the hybrid reduces execution time about 20 X. In addition, our hybrid approach can increase code coverage over 8% on average. However, code coverage of rm is not increased, because this tested program is so simple that it has achieved full coverage.

Overall, our hybrid approach outperforms any one of these three methods. Our proposed approach needs less time achieving the same code coverage.

5 CONCLUSION

This paper presents a hybrid approach to reduce execution time. This novel technique uses path merging to reduce the number of paths explored, adopts constraints simplification to reduce solving time. Moreover, our proposed approach employs heuristics search to explore new code. This technique combines static analysis with dynamic analysis. Our experimental results demonstrated that the hybrid approach discussed in this article could reduce execution time more than any one of these three techniques. Moreover, a combination of static and dynamic analyses achieves a higher coverage rate in automated test generation.

Hybrid approach is able to alleviate path explosion, but it does not deal with loops and recursive calls effectively. We are now developing an extension of our hybrid approach applicable to programs with loops and recursive calls. Future work includes further experiments with latest-hybrid approach, and its evaluation in comparison to other methods of path explosion. Another research direction is to find a suitable combination of our method for the black box fuzzing in order to increase path coverage rate.

REFERENCES

[1] Bugrara, S. & Engler, D. 2013, June. Redundant State Detection for Dynamic Symbolic Execution. *In Presented as part of the 2013 USENIX Annual Technical Conference*, USENIX, pp. 199–211.

[2] Godefroid, P., Nori, A.V., Rajamani, S.K. & Tetali, S.D. 2010, January. Compositional may-must program analysis: Unleashing the power of alternation. *In ACM Sigplan Notices*, ACM, Vol. 45, No. 1, pp. 43–56.

[3] Majumdar, R. & Xu, R.G. 2007, November. Directed test generation using symbolic grammars. *In Proceedings of the twenty-second IEEE/ACM international conference on Automated software engineering*, ACM, pp. 134–143.

[4] Godefroid, P., Levin, M.Y. & Molnar, D.A. 2008, February. Automated whitebox fuzz testing. *In Proceedings of the Network and Distributed System Security Symposium, 2006*, Vol. 8, pp. 151–166.

[5] Tillmann, N. & De Halleux, J. 2008. Pex–white box test generation for. net. *In Tests and Proofs*, Springer Berlin Heidelberg, pp. 134–153.

[6] Godefroid, P. & Luchaup, D. (2011, July). Automatic partial loop summarization in dynamic test generation. *In Proceedings of the 2011 International Symposium on Software Testing and Analysis*, ACM, pp. 23–33.

[7] Chen, T., Zhang, X.S., Guo, S.Z., Li, H.Y. & Wu, Y. 2013. State of the art: Dynamic symbolic execution for automated test generation. *Future Generation Computer Systems*, Vol. 27, No. 7, pp. 1758–1773.

[8] Collingbourne, P., Cadar, C. & Kelly, P.H. 2011, April. Symbolic crosschecking of floating-point and SIMD code. *In Proceedings of the sixth conference on Computer systems*, ACM, pp. 315–328.

[9] Cadar, C., Dunbar, D. & Engler, D.R. (2008, December). KLEE: Unassisted and Automatic Generation of High-Coverage Tests for Complex Systems Programs. *In OSDI*, Vol. 8, pp. 209–224.

[10] Cadar, C. & Sen, K. (2013). Symbolic execution for software testing: three decades later. *Communications of the ACM*, Vol. 56, No. 2, pp. 82–90.

Future Communication, Information and Computer Science – Zheng (Ed.)
© 2015 Taylor & Francis Group, London, 978-1-138-02653-7

The method of establishing fault dictionary of tolerance circuit based on Fuzzy C-mean clustering algorithm

J. Yin, J.J. Hou & L. Huang
School of Electronic and Information Engineering, Beijing Jiaotong University, Beijing, China

ABSTRACT: This paper proposed an establishment method of fault dictionary of tolerance circuit. It is hard to establish fault dictionary and diagnose faults when there are some tolerance components in the circuit which are allowed to deviate from its nominal values to some extent. First of all we sketched Fuzzy C-mean clustering algorithm. Then we can prove this algorithm valid by introducing a practical example of establishing fault dictionary of tolerance circuit. To testify this, we should distinguish between parametric faults and normal states. Also we obtained the clustering analysis samples by simulating every state. After that, we divided the states into different categories by Fuzzy C-mean clustering algorithm and determined centers of clustering of all states. The example shows that this algorithm can distinguish between normal and fault state and needs only a litter of samples.

Keywords: analog circuit; fault dictionary; clustering analysis; FCM; tolerance

1 INTRODUCTION

The fault diagnosis is an important branch in the research field of circuits. Despite current research achievements on the technique of fault diagnosis in circuits, the analog circuits grow slowly compared with the digital circuits in fault diagnosis (Huang et al. 2013), on account of the circuit response of analog circuits and the tolerance of component parameters. More than that, there are non-linear phenomena and feedback circuits in it. You will find it hard to distinguish whether the circuit is in a normal state of tolerance or failure state (Yang et al. 2011), especially when there are some tolerance components (Mohammadi et al. 2002) in circuits which are allowed to deviate from its nominal values to some extent. The fault diagnosis of analog circuits can be divided into catastrophic faults and parametric faults. The fault diagnosis of parametric faults is more difficult than that of catastrophic faults, because components in a circuit are not completely invalid, but just deviate from its nominal values.

A statistical technique commonly used in determining a profile is cluster analysis (Peng et al. 2012). It divides a set into several categories, and every element in the same category is similar. The method has no need of training samples, for it can obtain every clustering center (YanFei et al. 2011) by iteration. Fuzzy C-mean (FCM) (Lu et al. 2013) clustering algorithm is one of the widely used clustering algorithms based on fuzzy theory. This is a fuzzy partition of similarity, that the elements divided into one category are similar.

In this paper, we proposed an establishment method of fault dictionary (Song et al. 2009) in tolerance circuits. Apart from its practical value at present, the approach for fault dictionary is a classic technique in the fault diagnosis of analog circuits. The precision of fault dictionary in tolerance circuits depends on the quality of fault dictionary. The second part sketched Fuzzy C-mean clustering algorithm. The third part introduced a practical example of establishing fault dictionary in tolerance circuits. In this part, we divided the circuit into normal and fault condition clustering analysis using techniques such as clustering analysis, then obtained centers of clustering of all states.

2 FUZZY C-MEAN CLUSTERING ANALYSIS

Clustering analysis is an analysis process which divided a collection of elements into several categories. The classification is based on similarity, in other words, the elements divided into one category are similar, but elements in different categories are dissimilar.

Fuzzy C-mean clustering algorithm has been widely applied in this field.

$X = |x_i, i = 1, 2, \ldots, n| = \{x_1, x_2, \ldots, x_n\}$ represents the set which is made up of n elements. We divided set X into C types by using Fuzzy C-mean (FCM) clustering algorithm, and obtained centers of clustering of all states, that is

$$C = | c_j, j = 1, 2, \cdots, c | = \{ c_1, c_2, \cdots, c_c \} .$$

μ_{ij} stands for the membership of element x_i to clustering center C_j. μ_{ij} has to satisfy $\mu_{ij} \in [0, 1]$ and $\sum_{j=1} \mu_{ij} = 1$. The membership can be any value between 0 and 1, but the sum of the membership of element x_i to clustering center C_j must be 1. The formula for calculating membership is

$$\mu_{ij} = \frac{1}{\sum_{k=1}^{c} \left(\dfrac{d_{ij}}{d_{ik}} \right)^{\frac{2}{m-1}}} = \frac{1}{\sum_{k=1}^{c} \left(\dfrac{\|x_i - c_j\|}{\|x_i - x_k\|} \right)^{\frac{2}{m-1}}} \quad (1)$$

m stands for weighted index in equation 1, $m > 1$, usually $1.1 \le m \le 5$. d_{ij} stands for the Euclidean distance from element x_i to clustering center C_j: $d_{ij} = \|x_i - c_j\|$. The formula for calculating clustering center is:

$$c_j = \frac{\sum_{i=1}^{n} \mu_{ij}^{m} x_i}{\sum_{i=1}^{n} \mu_{ij}^{m}} \quad (2)$$

Fuzzy C-mean clustering algorithm gains optimal value of least square error based on the optimization of objective function:

$$J(U, C) = \sum_{j=1}^{c} \sum_{i=1}^{n} \mu_{ij}^{m} d_{ij}^{2} \quad (3)$$

The steps of clustering analysis by FCM are:

Step 1. Give the parameters of the objective function: cluster number c, weighted index m, maximum iterations l and threshold value ε.

Step 2. Initialize membership degrees matrix and clustering center;

Step 3. Calculate centers of clustering by equation 2;

Step 4. Calculate objective function by equation 3;

Step 5. Calculate membership degrees matrix U again and then return to step 3.

This method has a clear train of thought and satisfactory convergence for it has no need of training samples, so it can be used in efficient dealing with a huge quantity of experiment data. The MATLAB provides Fuzzy C-mean algorithm. It can use the default parameters of function, so a computer can do this for us. The advantages of FCM make it popular in the pattern recognition field.

3 PARAMETRIC FAULT DIAGNOSIS CASES OF TOLERANCE CIRCUITS

3.1 Tolerance circuits and fault mode

Figure 1 is a common power voltage stabilizing circuit. LM317 is a three-end adjustable integrate device and output reference voltage U_{ref} is 1.25 V. The diode D1 and D2 are used to protect LM317. The capacitors

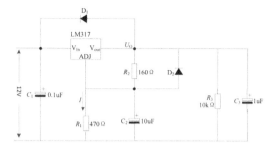

Figure 1. Analog circuit with tolerance.

Table 1. Fault sets.

number	circuit state
1	Normal state, the value of resistance is 5% larger or smaller than standard value.
2	Parametric fault of that the resistance R_1 is larger, R_1 is 120% to 150% times the value of standard resistance.
3	Parametric fault of that the resistance R_1 is smaller, R_1 is 60% to 80% times the value of standard resistance.
4	Parametric fault of that the resistance R_2 is larger, R_2 is 120% to 150% times the value of standard resistance.
5	Parametric fault of that the resistance R_2 is smaller, R_2 is 60% to 80% times the value of standard resistance.

C_1, C_2, C_3 promote the rejection rate of the voltage stabilizer. The output voltage is 5 V when the voltage regulator circuit in Figure 1 works well. The expression of output voltage of power circuit U_o is written as

$$U_o = U_{Ref} \cdot \left(1 + \frac{R_1}{R_2} \right) \quad (4)$$

There are two measured parameters that can be used as diagnostic parameters in Figure 1, such as the output voltage of voltage regulator circuit U_o and the current I through the resistance R_1. We should note that in general we choose voltages as diagnostic parameters because of the complexity of electric current in fault diagnosis of analog circuits, but in this case we cannot do it like that. When the value of resistance R_1 is large and R_2 is small, or R_1 is small and R_2 is large, the faults are overlapped, so we cannot acquire the right diagnosis. Then we should choose the output voltage of voltage regulator circuit U_o and the current I through the resistance R_1.

The five states of tolerance circuit in Figure 1 are shown in Table 1. These states contain a normal state of tolerance and four kinds of parametric faults. The fault diagnosis of parametric faults is more difficult than that of catastrophic faults, because components in circuits are not completely invalid, but just deviate from its nominal values.

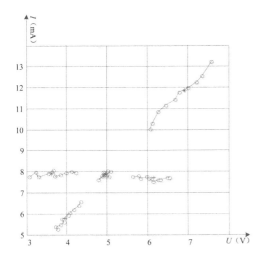

Figure 2. Clustering analysis of 5 states in tolerance circuit.

Table 2. Cluster centers of 5 states in tolerance circuit.

Circuit state	Clustering Center of U_O	Clustering Center of I
1. normal	4.964 V	7.869 mA
2. $R_1 \uparrow$	6.139 V	7.706 mA
3. $R_1 \downarrow$	3.638 V	7.901 mA
4. $R_2 \uparrow$	3.970 V	5.787 mA
5. $R_2 \downarrow$	6.899 V	11.85 mA

3.2 FCM clustering analysis

MULTISIM software can be used to simulate a circuit experiment. In Figure 1, we chose ten samples of every state. Then clustering analysis was carried out by using FCM function of MATLAB software. The parameters in equation 1–3 are as follows: the cluster number $c = 5$, weighted index $m = 2$, maximum iterations $l = 100$ and threshold value $\varepsilon = 10^{-5}$. Figure 2 shows the result of clustering analysis to the circuit in Figure 1 by using FCM algorithm.

In Figure 2, abscissa stands for output voltage of voltage regulator circuit Uo and ordinate represents current I which flows through the resistance R_1. The ten dark spots in the middle are in normal state, and the others are fault samples, such as the dark spots in the right represent that R_1 is larger, the dark spots in the left represent that R_1 is smaller, the dark spots in the left below represent that R_2 is larger, and the dark spots in the right below represent that R_2 is smaller.

These clustering analysis results in Figure 2 are clear and visualized, so we can deal with clustering analysis for five states of tolerance circuit using FCM Clustering Algorithm. And the success rate is almost one hundred percent. As you can see in Table 2, every clustering center is typical and can be used as fault dictionary for fault diagnosis later.

4 CONCLUSION

This method does not need training samples, so it can be used in efficient classifying tolerance circuit by normal and fault states. There is a FCM function in MATLAB software. The engineers can take advantage of it in dealing with the fault diagnosis of circuits. We used techniques such as clustering analysis to obtain the clustering center of all states. The clustering center is a typical value for circuit. Approach for fault dictionary is a highly practical skill in fault diagnosis of analog circuits. In this paper, we can establish fault dictionary for fault diagnosis conveniently and authoritatively.

ACKNOWLEDGEMENTS

This work was financially supported by the National Natural Science Foundation of China (61201363) and the Fundamental Research Funds for the Central Universities (2013JBM016).

REFERENCES

[1] Mohammadi K. et al. 2002. Fault diagnosis of analog circuits with tolerances by using RBF and BP neural networks. Student Conference on. IEEE 317–321.
[2] Huang, L. et al. 2013. Grey entropy relation algorithm of choosing the optimum diagnostic nodes in analogue circuits. *Chinese Journal of Electronics*, 22(3): 615–620.
[3] Lu J. et al. 2013. An analog online clustering circuit in 130 nm CMOS Solid-State Circuits Conference (A-SSCC), IEEE Asian 177–180.
[4] Peng M. et al. 2012. Analog fault diagnosis using decision fusion. International Conference on. IEEE 17–20.
[5] Song G. et al. 2009. Multi-classifier fusion approach based on data clustering for analog circuits fault diagnosis 8th International Conference on. IEEE 1217–1220.
[6] YanFei L. et al. 2011. Application of immune clustering algorithm to analog circuit fault diagnosis. International Conference on. IEEE 1743–1746.
[7] Yang C. et al. 2011. Methods of handling the tolerance and test-point selection problem for analog-circuit fault diagnosis. *Instrumentation and Measurement, IEEE Transactions on*, 60(1): 176–185.

Future Communication, Information and Computer Science – Zheng (Ed.)
© *2015 Taylor & Francis Group, London, 978-1-138-02653-7*

An innovative software testing course based on CDIO

C. Shan, R. Ma, X.L. Zhao & J.F. Xue
School of Software, Beijing Institute of Technology, Beijing, China

ABSTRACT: Software testing is a required course of the software engineering major in Beijing Institute of Technology. In this paper an innovative software testing course based on CDIO is studied and described. This paper covers the objectives, design, organization, and assessment of the software testing course offered to senior undergraduate students in BIT. Project Based Learning (PBL) and Experiential Learning Method (ELM) were used in the course design, and a kind of Peer Evaluation approach was also applied in the course assessment. The work of this paper is a partial achievement of BIT teaching reform project. This course based on CDIO Initiative has been offered twice in BIT, and currently is being taught in the spring 2014 semester.

Keywords: CDIO; software testing; Project Based Learning (PBL); Experiential Learning Method (ELM); peer evaluation

1 INTRODUCTION

Having started in 2000, CDIO Initiative is the result of an international collaboration with initial research funds from the Knut and Alice Wallenberg Foundation for exploring ways to reform the current engineering education (Koo et al. 2008). CDIO stands for Conceive-Develop-Implement-Operate and CDIO Initiative is a program to develop engineering education to meet the requirements for a modern engineer and to emphasize the ability of a student to be able to engineer (Ikonen et al. 2009).

We live in a changing world. Our society becomes more and more dependent on computers and software. The pace of change is steadily increasing, with technology as one important driving factor. To manage the pace and the complexity when developing sophisticated large-scale software intensive systems we need software engineers with the appropriate knowledge, skills and attitudes (Fredrik & Inger 2014).

School of software, Beijing Institute of Technology is one of 37 National Model Software Schools in China. CDIO was introduced to Beijing Institute of Technology in 2008, and then from 2009~2011 a new CDIO syllabus had been built in the software engineering discipline. Software testing as a required course in this CDIO syllabus must be changed. An innovative course based on CDIO was required to design and organization. This paper just introduces the partial achievements of a BIT teaching reform project.

2 RELATED WORK

The CDIO approach is student-centered. A student uses the curriculum as an applying example to build up her or his own knowledge bases. Education process focuses on competency development. It is hence an open system. In concept, traditionally the faculty is acting as a giver and the students as receivers. The CDIO approach would transform the faculty as a facilitator and the students as explorers. The process of the present explorations through the CDIO curricular reform combines personal, inter-personal and system competencies as well as the technological knowledge and reasoning into integrated practices (Koo et al. 2008).

Project Based Learning (PBL) initially promoted the idea of "learning by doing" by John Dewey. Markham described PBL as: "PBL integrates knowing and doing. Students learn knowledge and elements of the core curriculum, but also apply what they know to solve authentic problems and produce results that matter. PBL students take advantage of digital tools to produce high quality, collaborative products. PBL refocuses education on the student, not the curriculum—a shift mandated by the global world, which rewards intangible assets such as drive, passion, creativity, empathy, and resiliency. These cannot be taught out of a textbook, but must be activated through experience. (Marham 2011)".

David Kolb introduced an Experiential Learning Method (ELM). In the ELM, he considered four mental activities taken by the learner and discusses the cycle of learning that develops from those stages. Kolb identified the four stages as: Concrete Experience (CE), Reflective Observation (RO), Abstract Conceptualization (AC) and Active Experimentation (AE) (Kolb 1984, Brown et al 2014).

Peer Evaluation was an approach to peer assessment of individual contributions to group work. Peer Evaluation should be understood as an educational

Course name	Software Testing Techniques	
Text book	Software Testing Methods and Techniques (second edition) Tsinghua University Press	
Class size	40	
Class hours	24 (lectures) +16 (labs)	
Credit	2.5	
Audience	Senior undergraduate students in SE	
When to offer	Each spring semester	
Grading	Attendance 5%	
	Lab performance 35%	
	Final closed exam 60%	

Table 1. Course information.

arrangement in which students comment on the quality of the work of their fellow students, for formative or summative purposes (Topping 2003, Poon 2011).

In our CDIO course design, the PBL was used and a project testing lab was proposed. Additionally according to ELM, the initial experiences (CE) of software testing through tutorial activities are provided. Students are then given the opportunity to reflect on that experience (RO) as they prepare to begin their regression testing. We provided a lecture as a framework for establishing the abstract concepts (AC) and benefits of software testing. Finally, they create new tests for their own project (AE), beginning the cycle anew. We also applied Peer Evaluation in the course. The group project mark partially awarded by the contribution of the individual student to the team as assessed by other team members.

3 SOFTWARE TESTING COURSE IN BIT

3.1 Course objectives

The course described in this paper is an undergraduate software testing course taught in the School of Software, Beijing Institute of Technology. The course is a 10-week, half-semester long course and meets for a two-hour session twice a week. The students enrolled in the course are software engineering majors in their senior year. It is a required course and taught after the student has completed programming, algorithm, database and software engineering theory classes. The basic course information is shown in Table 1.

The objectives of the course include:

This course will provide an introduction to software testing, bug detection, and maintenance techniques.

Students in this course will learn techniques and tools that could significantly improve their testing (and development) skills.

It teaches the students about engineering professionalism which is a very important subject that incudes soft skills, ethics, personal leadership, and team work.

3.2 Course design

The course was designed according to CDIO Initiative, Project Based Learning (PBL) and Experiential

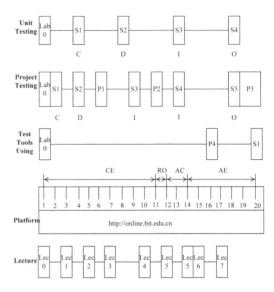

Figure 1. Structure of software testing course in BIT.

Learning Method (ELM). Figure 1 represents the design of the course structure discussed in this paper.

The class time was structured: some time was spent in traditional lectures, presenting topics such as Software Testing Foundation (Lec1 in Fig. 1), Software Testing Practice (Lec2 in Fig. 1), Software Testing Standards (Lec3 in Fig. 1), Test Design Techniques (Lec4 in Fig. 1), Test Application Techniques (Lec5 in Fig. 1) and Software Test Automation (Lec6 in Fig. 1). Time was also spent in hand-on lab activities, which were unit testing, project testing, and test tools using.

Three labs were designed in the course. One lab was unit testing and it was an individual work. Students must complete unit testing for a C/C++/Java program which was randomly allocated from a unit testing database. Another lab was project testing and it was a group work. Students should be divided into several project groups. General rule was to form heterogeneous groups with four students per group. The group was required to do a third party testing for a project which might be one of the postgraduate student research projects or undergraduate student graduation projects. Different projects would be guided by different teacher assistants who were postgraduate students or 4th year undergraduate students. The last lab was test tools using and it was also an individual work. Students were required to use a kind of test tool which was open or business test tool in unit testing or project testing at least.

Additionally in project testing the project groups were required to make three presentations during the project. In the first presentation (P1 in Fig. 1) the audience consisted of the instructor and teacher assistants in the role of a customer. Project groups presented their test plans, and then the instructor and teacher assistants would give instant feedback about the plans and their presentation skills. In the second presentation (P2 in

Fig. 1) approximately half way of the project, project groups presented the testing status of the project in the midterm interview. In the last presentation (P3 in Fig. 1) project groups presented a summary of the project. In test tools using, another presentation (P4 in Fig. 1) should be given. Voluntary students might share their experiences on using tools of unit testing or project testing. They were provided with instant feedback after their presentation.

The progress of the project groups was monitored in several project meetings. Project minutes must be submitted and added to a project portfolio. The project portfolio submitted for evaluation at the end of the project was inspected by an instructor and teacher assistants. The final documentation consisted of:

Test plan
Presentation documents
Project meeting minutes
Testing design
Testing implement
Bug report

Some associated important dates like document submission dates were presented to the students. After the kick-off meeting students prepared their preliminary test plans. Two weeks after the kick-off a tutoring meeting was arranged and the groups had an opportunity to ask questions about the test plan and any theoretical or practical issues on the project.

Unit testing was required to use a white-box test approach. The project third party testing focused on the functional testing and usability testing.

This kind of course design was based on CDIO, and Conceive-Develop-Implement-Operate were labeled in Figure 1. Project Based Learning (PBL) and Experiential Learning Method (ELM) were also instructive. Some of the topics were emphasized in order to support Project Based Learning. Some additional laboratory sessions were also included in the course to provide guidance for the project groups. Learning by doing and instant feedback help students get their testing experiences. This was a student-center and active leaning course. In this software testing course, the teacher's role had been changed. The teacher was not only a traditional lecturer-instructor, but also a tutor-facilitator. And the students' roles had also been changed from passive receiver to active member of a project group.

3.3 Course organization

Course organization is shown in Table 2.

We had allocated about 16 lab hours on this course, but the students must do much more than that. They must spend an extra 32 lab hours in their spare time at least. Some project groups even found themselves spending weekends in the project testing.

All of the course's resources, such as lecture notes and lab tutorials could be found on the platform (http://online.bit.edu.cn in Fig. 1) which was an online learning system in Beijing Institute of Technology. Documents submission and discussion

Table 2. Course schedule.

No.	Weeks	Hours	Contents
1	1st	1	Lecture 0 (Course Introduction and Overview)
		1	Lab 0 (Lab Schedule) Project Testing: Step 1: Project Introduction by Teacher Assistants
2	1st	2	Lecture 1 (Software Testing Foundation)
3	2nd	2	Unit Testing: Step 1: Unit Selection Project Testing: Step 2: Project Selection
4	2nd	2	Lecture 2 (Software Testing Practice)
5	3rd	2	Lecture 2 (Software Testing Practice)
6	3rd	1	Project Testing: Presentation 1: Test Plan
		1	Lecture 3 (Software Testing Standards)
7	4th	2	Lecture 3 (Software Testing Standards)
8	4th	2	Unit Testing: Step 2: Test Design Project Testing: Step 3: Project Test
9	5th	2	Lecture 4 (Test Design Techniques)
10	5th	2	Lecture 4 (Test Design Techniques)
11	6th	2	Lecture 4 (Test Design Techniques)
12	6th	1	Project Testing: Presentation 2: Midterm Interview
		1	Lecture 5 (Test Application Techniques)
13	7th	2	Lecture 5 (Test Application Techniques)
14	7th	2	Unit Testing: Step 3: Test Implement Project Testing: Step 4: Regression Test
15	8th	2	Lecture 5 (Test Application Techniques)
16	8th	2	Lecture 6 (Software Test Automation)
17	9th	1	Test Tools Using: Presentation 4: Share Experiences
		1	Lecture 7 (Course Review)
18	9th	2	Test Tools Using: Step 1: Tools Using Project Testing: Step 5: Project Summary
19	10th	2	Project Testing: Presentation 3: Final Presentation
20	10th	2	Project Testing: Presentation 3: Final Presentation

about the course also were put in the online learning system.

3.4 Course assessment

The traditional course assessment is a utility assessment, which results in the one-sidedness of the assessment items, ignores the overall developments of students, thus people pay excessive attention to the result, and neglects the daily and dynamic process assessment.

In this software testing course, the students' grade consisted of several parts-final closed examination 60%, attendance and participation 5% and laboratory sessions 35%, which included unit testing 10%, project testing 20% and test tools using 5%. Students must pass the final exam to pass the course.

61

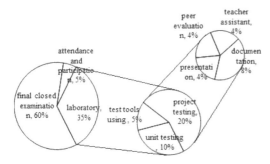

Figure 2. Grade distribution chart.

Project testing was a group work, so the grade of project testing should be a group score. It was presentation 4%, peer evaluation 4%, teacher assistant 4% and documentation 8%.

The following Figure 2 is a grade distribution chart of the software testing course.

The group project mark was awarded by the instructor, teacher assistants and the contribution of the individual student to the team as assessed by other team members. A kind of Peer Evaluation (Poon 2011) was used in the course. Peer Evaluation is considered an unbiased assessment approach because group members have been working together for a period of time and they themselves would seem in a natural position to provide reliable, valid evaluations of each other (Cederblom & Lounsbury 1980, Poon 2011).

4 RESULTS AND RESPONSES

The course was offered using CDIO approach in the spring 2012 and spring 2013 semesters. Enrollments in chronological order were 88 and 92 students. Currently, the course is being taught in the spring 2014 semester. The enrollment for the course is capped at 85 students.

Students responded well to the course organization. According to the result of an online survey from http://online.bit.edu.cn, students considered that they had learned the methods, techniques and tools of software testing from this course. And they came away with a better understanding of tester roles within three labs. Because the unit testing was a white-box testing and a code review was required to be done, some students even thought that their programming abilities were improved through the software testing course. And ethics, integrity, personal leadership, team work, and professionalism were built in the course learning process.

5 CONCLUSION

Software testing is a challenging course to teach. A course structure motivating students to improve their testing experiences was proposed. This paper described the course objectives, design, organization and assessment. Three labs which were unit testing, project testing and test tools using were introduced according to Conceive-Develop-Implement-Operate Initiative. Project Based Learning (PBL) and Experiential Learning Method (ELM) are the active learning methods. Students benefited from them. In the project evaluation, a group should document their work and their work would be evaluated not only by the instructor and teacher assistants, but also their peers in the group. The response of the course offered was active. From the course based on CDIO, students learned the importance of communication and coordination among project groups, and the tester roles that they might play in the future.

In 2013, the software engineering discipline of Beijing Institute of Technology became one of "A plan for Education of High Level Engineers" in China. The plan aims to train various high-level engineers, who had design abilities and creativity in engineering, match the social economic development needs. Some new requirements should be added in the software testing course according to this plan. As a next step, the researchers intend to provide the enterprise project for students' test. New teaching reform should be to design a school-enterprise cooperation course.

ACKNOWLEDGEMENTS

This work was supported by the teaching reform project of Beijing Institute of Technology.

We also would like to express our gratitude to all students, who had been taken part in studying software testing course in spring 2012, spring 2013 and spring 2014 semesters.

REFERENCES

[1] Brown, C., Pastel, R., Seigel, M., Wallace, C. & Ott, L. 2014. Adding unit test experience to a usability centered project course. In Proceedings of the 45th ACM technical symposi-um on Computer science education. ACM.

[2] Cederblom, D. & Lounsbury, J.W. 1980. An investigation of user acceptance of peer evaluations. Personnel Psychology, 33: 567–579.

[3] Fredrik, H. & Inger, E.K. 2014. SIGCSE '14 Proceedings of the 45th ACM technical symposium on Computer science education. ACM.

[4] Ikonen, A., Piironen, A., Saurén, K. & Lankinen, P. 2009. CDIO Concept in Challenge Based Learning. In Proceed-ings of the 2009 Workshop on Embedded Systems Educa-tion. ACM.

[5] Kolb, D.A. 1984. Experiential learning: Experience as the source of learning and development (Vol. 1). Englewood Cliffs, NJ: Prentice-Hall.

[6] Koo, T.J., Xiong, Z. & Qu, J. 2008. Curriculum reform for accommodating multi-core technology in Shantou University. In Proceedings of the 1st ACM Summit on Computing Education in China. ACM.

[7] Markham, T. 2011. Project Based Learning. Teacher Librari-an, 39(2): 38–42.

[8] Poon, J.K.L. 2011. Students' perceptions of peer evaluation in project work. In Proceedings of the Thirteenth Australasian Computing Education Conference-Volume 114. Australian Computer Society, Inc.

[9] Topping, K.J. 2003. Self and peer assessment in school and university: reliability, validity and utility. In M.S.R. Segers, F.J.R.C. Dochy, & E.C. Cascallar (Eds.), Optimizing new modes of assessment: In search of qualities and standards. 55–87. Dordrecht: Kluwer Academic.

Future Communication, Information and Computer Science – Zheng (Ed.)
© *2015 Taylor & Francis Group, London, 978-1-138-02653-7*

Optimization of parameters of the recognition method and quality analysis of images fine structures

S.V. Sai & N.Yu. Sorokin
Pacific National University, Khabarovsk, Russia

ABSTRACT: Image quality analyses have a lot of applications from HDTV broadcasting systems to the industrial vision systems. Mostly, known methods for image quality estimation are divided into two major groups – with and without reference images. The recent group has great advantages for the stand-alone applications and systems. In this work we propose optimization and application of no-reference method of image quality analyses based on the estimation of fine details. Parameters of the method – thresholds were selected through the heavy computations upon the sets of hundreds of photorealistic images. A criterion for the quality estimation of the photorealistic image is proposed.

Keywords: no-reference image analyses; image quality; fine details

1 INTRODUCTION

Advances in computer (and, particularly, in mobile) technology provides up to the current time unknown rates of the information delivery. This information is presented widely with digital images and video, and thus the quality control of such digital content is an actual task. Usually the distortion definitions of the static and dynamic test signals are the indirect values of the quality estimation of the broadband digital system. There exist different metrics (Lin & Jay, 2011), such as the traditional PSNR, MSE, or rather complex metrics like the Structural Similarity (SSIM) method and also Multi-Scale SSIM. SSIM method compares a pair of images in order to find the changes in luminance, contrast and structure of those images (Bovik & Mittal, 2012). This is done using pixel-wise comparison and image window analyses approach in order tom estimate an image quality value, expressed as a DMOS (Differential Mean Opinion Score). This method is a lot more popular than the PSNR and MSE estimations because it gets more information that is near to the visual perception of the human observer (Wang & Li, 2007). The drawback of all these methods is that the reference image should be presented.

Modern analyzers use the quality analyses methods that are based on a comparison of the differences between the test and distorted (corrupted) images. This comparison is carried out for some selected visual model. So, the absence of test images does not allow the application of such analyzers.

Finding the new methods of the no-reference (or blind) quality analyses is an actual problem. Such methods must provide the objective estimation of image definition quality and sharpness degradation without the test patterns. While the algorithms of the quality analyses using the reference images are well investigated the methods for the no-reference analyses are still an interesting task. One of such methods (Wang & Sheikh, 2002) deals with the estimation of the JPEG blurring and blocking effects in comparison with the MOS score. However, the estimation of the quality assessment model parameters is influenced by the type of the image and JPEG compression quality. The same idea, namely analyzing the noticeable horizontal and vertical distortions along with an estimation of blurring and blocking effects is proposed by several other authors, e.g. (Ming-Jun Chen & Bovik, 2011). Also, one of the possible methods of quality estimation is to identify the scene statistics (or the loss of naturalness) and therefore compute the score using BRISQUE technique. The review of the no-reference methods points out that almost all techniques are based on the a priori known distortion types in the compression systems (JPEG, JPEG2000, MPEG etc.) along with the known spectral and noise characteristics. However, there do not exist results for the no-reference analyses of the camera image definition.

This article deals with the strong optimization of the author's no-reference method for the analyses of the image quality of fine details, proposed recently (Sai & Sorokin, 2013). The major difference of this method from the well-known methods is the application of the equal color space and analyses of thin objects in the photorealistic images that were obtained from camera and/or after the pre-processing (filtering, scaling).

2 ALGORITHM FOR RECOGNITION AND ESTIMATION QUALITY

In order to estimate the image definition quality search and recognition of the fine structures must be carried out. Such structures include fine details like dot objects or thin line fragments with a size from one to several pixels. Author's work (Sai & Sorokin, 2013) presents a search algorithm which consists of several stages.

In the first stage the transformation of the primary color signals (RGB) to the equal color space ($W^*U^*V^*$) is produced for each pixel using the following equation:

$$W^* = 25Y^{1/3}\text{-}17; \; U^* = 13 \; W^*(u - u_0); \; V^* = 13 \; W^*(v - v_0),$$

where Y = luminance, changed from 1 to 100; W^* = brightness index; U^* and V^* = chromaticity indexes; u and v = chromaticity coordinates in Mac–Adam diagram. Equal color spaces ($W^*U^*V^*$, $L^*a^*v^*$, etc.) are used traditionally for the estimation of color transfer quality for the big details. Here the estimation of the color differences Δ is computed as:

$$\Delta = 3\sqrt{(\Delta W^*)^2 + (\Delta U^*)^2 + (\Delta V^*)^2} \; ,$$

where ΔW^* = brightness difference and $\Delta U^*, \Delta V^*$ = color difference between big details of the test image and of the distorted image. The color transfer quality is determined by the number of the minimum perceptible color difference (MPCD). If $\Delta < 1$, then the color differences are invisible to the human eye.

In our works (Sai, 2007), (Sai & Sorokin, 2009) we proposed the method of the color differences estimation between the fine detail ($W_o^*U_o^*V_o^*$) and the color coordinates of the background pixels ($W_b^*U_b^*V_b^*$) using the normalized equal color space. The following criterion was used:

$$\Delta K = \sqrt{\left(\frac{\Delta W^*}{\Delta W_{th}^*}\right)^2 + \left(\frac{\Delta U^*}{\Delta U_{th}^*}\right)^2 + \left(\frac{\Delta V^*}{\Delta V_{th}^*}\right)^2} \; , \tag{1}$$

where ΔK = color contrast of the fine detail relatively to background; ΔW_{th}^*, ΔU_{th}^*, ΔV_{th}^* = thresholds according to brightness and chromaticity indices for fine detail. Threshold values on brightness and chromaticity indices depend on the size of fine detail, background color coordinates, time period of object presentation and noise level. For fine details with sizes not exceeding one pixel the threshold values are obtained experimentally. In particular [15], for fine details of the test pattern located on a grey background threshold values are approximately $\Delta W_{th}^* \approx 6$ MPCD, and $\Delta U_{th}^* \approx \Delta V_{th}^* \approx 72$ MPCD.

In the second stage the criterion (1) is used for the fine details recognition and quality estimation. For this, the processed image is scanned with the 3×3 pixels window. Recognition algorithm of the fine structure is executed for each iteration step.

During the analyses we should recognize an object: is it a dot or a thin line. For this we use the binary images of the fine structures. Using images we can obtain the spatial coordinates of the object and background.

The analyses are started with the first structure – a dot. On the first step the mean values of the color coordinates of the background and object are computed.

After this, the background pixels are examined using the following condition:

$$\Delta_b = \frac{1}{N} \sum_{n=1}^{N} \Delta K_n < \eta \; , \tag{2}$$

where η = threshold value, when the color differences between the background pixels are invisible; ΔK_n = contrast of the n-th background pixel relative to the mean value. The color differences between the pixels of the object are computed in the same way:

$$\Delta_o = \frac{1}{M} \sum_{m=1}^{M} \Delta K_m < \eta \; , \tag{3}$$

where ΔK_n = color contrast of the m-th object pixel relative to the mean value.

The execution of conditions (2) and (3) means that the color differences between the neighbor background pixels and neighbor object pixels are invisible to the eye. Therefore, the image in the window corresponds to the first fine structure – a dot.

If the conditions (2) and (3) are fulfilled on the next step the contrast of the object relative to the background ($\Delta K_{o/b}$) is computed and the following condition is checked:

$$\Delta K_{o/b} \geq \xi \; , \tag{4}$$

where ξ = threshold value of the object-background contrast when the fine objects are distinguishable by eye.

If (4) is satisfied than the decision is made that the dot object is recognized and we save the special coordinates of the center of this object (i, j). After this we move the window further by three pixels and analyze another block of the processed image. If conditions (2–4) are not fulfilled the next structure is processed and so on. Also, if conditions (2–4) for each structure are not fulfilled we decide that the current window has no recognized objects and the window is shifted further by one element.

Hereby, the proposed method allows for recognition and selection of the fine structures – dots and thin lines fragments, noticeable to the eye.

In order to estimate the definition quality of the processed image we have to count the total number of recognized objects (N_R) relative (in %) to the total amount of pixels and then compare to the predefined threshold (Q_{th}):

$$Q(\%) = \frac{9 \cdot N_R}{N_x \cdot N_y} \geq Q_{th} \; , \tag{5}$$

where N_x and N_y = number of pixels in horizontal and vertical directions of image. Obtained parameter Q will be the quality estimation characteristic.

If the condition (4) is fulfilled – the processed image contains fine details noticeable to the eye. Hence, the image definition satisfies the named format or resolution that is used to present (store) this image. If the condition (5) is not fulfilled, i.e. the number of recognized fine details and structures is below the predefined threshold, than the definition of image does not satisfy the declared resolution. Also this fact means that the processed image does not have fine details and structures.

This result can be explained using the following reasons: a) image is of high visual quality (e.g., "Clouds"), but due to its nature the image has no fine structures; b) image is obtained from a camera with lower resolution than the currently represented format; c) image is blurred due to the wrong camera focus; d) image is blurred due to the low-pass filtering.

In the next sections we will show, that the condition (5) does not allow estimating the degradation of definition in the highly compressed images using JPEG and MPEG standards.

3 OPTIMIZATION OF THE METHOD

Apparently, the estimation of image quality definition using criterion (5) is dependent upon: 1) the threshold values η, ξ and Q_{th}, and 2) the internal structure of the image itself. Decrease of the threshold η in conditions (2) and (3) leads to the precise recognition of the fine structure, but there will be a reduction of the number of recognized fine structures. Increase of the threshold η leads to the change of the structure. Such changes can bring the method to the faulty recognition as an output. Threshold ξ in condition (4) is used to analyze the contrast of fine structures (including noise) that cannot be distinguished by the eye. Therefore, a decrease of this threshold leads to a faulty recognition. Increase of the threshold ξ allows analyzing fine details with the bigger contrast, at the same time low contrast structures will not be recognized. Selection of the threshold Q_{th} depends on the fine details and structures statistics for the high-quality images, i.e. without compression. Hence, in order to optimize parameters of the method we need to determine dependencies of quality estimation characteristic for some set of high definition images: $Q = F(\eta, \xi)$.

Figure 1 presents mean values of the quality estimation Q for a set of 500 different high-quality images with different resolution (from 4 to 10 megapixels). For the analyses we used three sets (each with 500 photorealistic images). The histogram in Figure 2 shows how the number of images K (in the set of 500 images) is distributed depending on their quality Q. The quality values were taken for the threshold parameter set $\eta = 0.5, \xi = 2$.

The following major conclusions can be drawn from the obtained results: 1. A number of fine details in

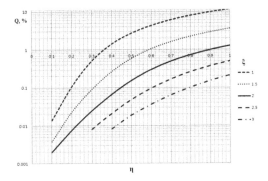

Figure 1. Mean values of the quality estimation Q for the set of 500 different high-quality images.

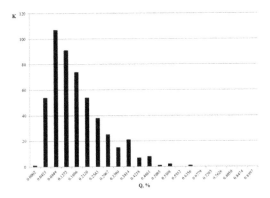

Figure 2. Distribution of the images number for the set of 500 images for the set of parameters: $\eta = 0.5, \xi = 2$.

high-definition images is bounded (in general) and is not greater than 1–10% from the total number of pixels; 2. A set of thresholds with values ($\eta = 0.5$, $\xi = 2$) may be considered optimal for the definition analyses of photorealistic images applying the proposed method; 3. The histogram in Figure 2 can be used for the estimation of optimal threshold of the parameter Q: if the condition $Q > 0.02\%$ is fulfilled, than we conclude that the number of recognized fine structures corresponds to the high definition image (in average for the whole set of images).

Experimentally obtained dependence of quality Q from different distortion types allows us to conclude that:

1. The proposed method allows estimating objectively the decrease of the image definition which is determined by: a) blur of the image due to the low-pass filtering and wrong camera focusing; b) increase of the initial image resolution.

2. JPEG and MPEG compressing artifacts lead to the appearance of false high-frequency components that are counted later as fine structures. In this case the proposed method outputs the good quality estimation for such an image. Therefore, it does not make sense to apply the proposed method to the quality estimation of highly compressed images.

a) Q=0.1931   b) Q=0.0011

c) Q=0.016   d) Q=0.1298

Figure 3. Examples of the processed images.

Figure 3 shows examples of the processed test portrait image with the initial resolution 256×340 pixels: original image (a), and processed images using the low-pass Gaussian filter with radius $= 0.7$ (b), scaling by 1.5 times (c), JPEG compression with low quality (d).

4 CONCLUSIONS

Now we emphasize main properties of the proposed no-reference method for recognition and quality estimation of fine structures of photorealistic images compared to the known methods. Analysis is carried out based on the algorithm for search and recognition of fine details of image. This algorithm has the following feature: recognition of fine details and structures is produced based on the visual perception of the color differences of the fine details. Here the author's method for the estimation of color differences in the normalized color space is applied. As an output of the method we receive the percent quality (Q) of the

recognized (and therefore noticeable by the eye) fine details and structures. The value Q is used in the proposed criterion (5) for the estimation of the real definition quality of the processed photorealistic image. Hence, compared to the known methods our method does not require the test images or tables with the fine structures (like dashed lines and so on). The feature of our method is that we cannot compare the parameter Q to the known metrics like SSIM, DMOS etc. because the value Q defines a number of identified fine details and structures. This number is not connected to the particular type of the distortion.

Results of the experiments allowed us to optimize the parameters of the proposed method – thresholds η and ξ. We selected those threshold values in such a way that the output of the method will contain only the fine details and structures authentically identified by eye (by form and by contrast). Analysis of the 500+ photorealistic images allowed to justify the selection of the quality value threshold $Q = 0.02\%$.

The proposed method can be used in various applications: a) Objective estimation of the image definition degradation in camera due to the low-pass filtering or wrong camera focusing; b) Objective estimation of the image definition degradation due to the increase of the initial image resolution; c) Expert analyses for the selection of test images with high definition; d) Robot's technical vision systems etc.

REFERENCES

[1] Bovik, A. & Mittal, A. 2012. No-Reference Image Quality Assessment in the Spatial Domain. *IEEE Transactions on Image Processing.* 21(12): 4695–4708.

[2] Hyunsoo, C. & Chulhee, L. 2011. No-reference image quality metric based on image classification. *EURASIP Journal on Advances in Signal Processing.* 1.

[3] Lin, W. & Jay Kuo, C-C. 2011. Perceptual Visual Quality Metrics. *A Survey. J. of Visual Communication and Image Representation.* 22(4): 297–312.

[4] Ming-Jun, Chen & Bovik, A.C. 2011. No-reference image blur assessment using multiscale gradient. *EURASIP Journal on Image and Video Processing.* 3.

[5] Sai, S.V. & Sorokin, N.Yu. 2013. Quality Analyses of Fine Structures Transfer in the Photorealistic Images. *Open Journal of Applied Sciences.* 3: 6–9.

[6] Sai, S.V. & Sorokin, N.Yu. 2009. Search Algorithm and the Distortion Analysis of Fine Details of Real Images. *Pattern Recognition and Image Analysis.* 19(2): 257–261.

[7] Sai, S.V. 2007. Methods of the Definition Analysis of Fine Details of Images. *In-Tech, In: Obinata, G., Dutta, A. (eds.) Vision Systems: Applications*: 279–296.

Future Communication, Information and Computer Science – Zheng (Ed.)
© *2015 Taylor & Francis Group, London, 978-1-138-02653-7*

Minimization of pump energy losses in dynamic automatic control of pressure in the main oil pipeline

D.P. Starikov, E.I. Rybakov, E.I. Gromakov & O.M. Zamyatina
Tomsk Polytechnic University, Tomsk, Russian Federation

ABSTRACT: Effective way of using electric power due to main pipeline pressure control is described in the article. Solutions of minimization of electric losses are explored and an algorithm and structure that allow reducing the non-effective electric power usage are proposed. It is based on an Advanced Process Control algorithm.

Keywords: energy efficient control; variable-speed pumps; valve position control; advanced process control, parallel; selective control

1 INTRODUCTION

Oil transportation presents one of the main operations in the oil and gas industry [1]. As a rule transportation from the place of extraction to the oil portal or consumer plant is provided by a land pumping with help of pump stations (CPS).

The main problem of oil transportation is the maintenance of assigned pressure value in the main oil pipeline. Different pieces of an oil pipeline system of automatic pressure control are implemented in different ways.

In this article we review the site which includes booster and main oil pump stations with oil supply control. At least one of the pumps is controlled by a frequency inverter (VFD is set on an electric drive synchronous type). Also this site includes valves with a position control on main pump inlet and outlet. According to the historical approach pressure or/and flow control in oil pipeline is provided by changing the motor rotating speed or by valve control.

Each of these variants has several disadvantages:

1) It is difficult to control the rotating speed in a powerful pump unit because of large inertia of movable parts. And continuous valve resetting leads to its rapid run-out.

2) Changing the motor rotating speed in forced modes leads to an increase in energy consumption. Pressure adjustment by a valve may be caused by a decrease of pumping efficiency. Energy, which is expended on choking, is lost and it decreases the general efficiency of pumps station. In some ways the expenses may reach up to 50%.

3) Overshoots and oscillations are possible with low quality of control transients in either case (pressure waves are possible).

2 RESEARCH

The main factors that change operating modes of a CPS-oil pipeline system are the following:

- Pipeline live load;
- Technological changes of pumps operating parameters and turning on and turning off additional pumps;
- Changing of rheological oil parameters caused by temperature seasonal changes;
- Emergency and repair situations appeared in an oil pipeline.

The objective of this research is a structure scheme test of main oil pipeline pressure automatic control system (PACS), which provides minimization of energy consuming by frequency-controlled pump during transients.

2.1 The reason of losses

A decrease of pumps power consuming that works to maintain a set value of pressure in the main oil pipeline is described in several works [2–6]. However, besides static loads the pump overcomes a dynamic resistance of motor internal masses in automatic mode.

$$\Delta W = 2J\Delta\omega^2$$

Loss of electric power during the transient process that is caused by dynamic acceleration and deceleration $\Delta\omega$ of pump and motor internal masses J in opposite mode may be expressed as a total of kinetic energy change:

$$\Delta W = 2J\Delta\omega^2$$

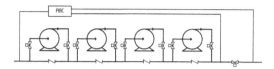

Figure 1. The scheme.

This expression implies that it is possible to decrease losses by reducing the number of pump switching at minor disturbance moments in the pipeline and by minimization of the pump acceleration during minor external influence regulating. This aim can be reached if an energy-efficient choice between regulation by inlet and outlet valves and pump speed will be implemented.

2.2 Solution

Whatever the technological mode of oil transportation, it does not impose higher requirements to speed of PACS. Continuous readjustment of pump speed leads to significant energy consumption even during controlling minor disturbances caused by dynamic resistance. Since the power of CPS motors can reach about 10 MW. In this way energy consumption can be significant in transient conditions. The typical scheme of a VFD-controlled pipeline system is shown in Figure 1.

This scheme of pressure of the automatic control system is an integration of several loops. The control loops of inlet and outlet valves allow the monitoring of oil pipeline pressure and pressure drops before and after the pump. And the pump regulation loop allows the efficient supply of oil to pipeline.

The following algorithms of Advanced Process Control (APC-algorithms) [9–11] are used to regulate mentioned multi-loop automatic control systems: FF/FB, Cascade, Override, Selective, Split range, Internal Model, Parallel, Model Predictive [13], Adaptive, Inferential control and also intellectual neuro- and fuzzy- control algorithms.

2.3 APC application

The rational choice of algorithm from the list for PACS is an independent problem. But it is clear that Selective, Split range and Override algorithms may provide structural solutions of PACS based on automatic regulation with reconfiguring structure. It allows separated control either by valves or directly by the pump.

Parallel-control can provide parallel-serial operation of several actuators. In such division operation is performed with the use of a fast and less inertial actuator, then in case of necessity it is continued with a more powerful but in the most cases more slower actuator. The schematic diagram of an algorithm is presented in Figure 2.

FF/FB and Cascade Control [8] can provide decrease of the effects of disturbance in inside and outside places of their uprise. Internal Model, Model

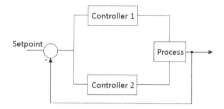

Figure 2. Parallel-control.

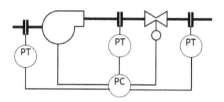

Figure 3. Parallel and Selective algorithms.

Predictive [7], Adaptive, Inferntial control and also intellectual neuro- and fuzzy- control provides smart-control that allows to compute and correct controlling with the consideration of disturbances and with using direct and indirect measurement and estimating.

2.4 Algorithm implementation

This article involves using an algorithm that is connected to Parallel and Selective algorithms. The structural diagram is shown in Figure 3. Switching between the actuators depending on energy calculations (changing the operating point) is performed as shown below:

- Let the position of the valve output provides 0% to 100% open pipeline. The special workspace introduced for position control valve, limited movement of the valve area near 0% and 100% of its opening.

We assume that in the normal state the positional valve position is close to 100% open pipeline. This corresponds to the flow and the pressure in the pipeline, setting the requirements of regulations. Such opening valve does not create serious obstacles for the pump unit, and provides a margin for moving, as downward opening and towards its increase.

Algorithm collaboration pump and valves in the system of automatic regulation provides that any pressure change arising in the main pipe installed regulatory requirements will be primarily offset by a movement of the valve (because of the relatively low inertia of its working body).

When approaching the limit values of the position of the working zone, the algorithm makes selective switching control valves to control the oil supply pump, which is accomplished by changing the rotational speed and the frequency of the drive.

Simultaneously, transfers to the initial state valve position. This algorithm is known as the compensation effect of the integral saturation valve (Anti-Windup). Changing the pump speed changes the position of the operating point of its characteristics "pressure-feed" that provides a standard set of pressure and flow of oil in the pipeline. Minimization is performed during the acceleration speed of the pump. When approaching the 100% open position valve lapping is made smoothly by using change in the speed of the pump.

Model researches structure of the automatic control system based on the use of Parallel and Selective algorithms, which, confirm that tuning algorithms provide transition process without overshoot is not difficult.

2.5 Results

Implementation of transients without overshooting in work as actuator of the pump provides a decrease of energy consumption, called by electrical braking of the actuator for the pump to slow down. In this case the the bandwidth of valve will retain at the expense of the integral shift of the valve in time of working out the disturbances at the expense of the compensation effect.

3 CONCLUSION

Using APC-algorithms in an automatic pressure-regulating pipeline is appropriate and promising according to results of modeling. Selected APC-algorithms are easily combined and perform many tasks for smart-pressure control in oil pipeline:

1) Algorithms provide a reduction of energy consumption and the drive of pumps in transition mode by operative positional restructuring the valve opposing external disturbances of pressure in the oil pipeline;

2) Continuous tuning of the pump speed that compensates integrated motion effects the valve during working off perturbations with low loss of electricity in the dynamic braking accelerations pump.

REFERENCES

[1] Roffel B. and Betlem B.H.L. "Advanced Practical Process Control" Springer, 2004.
[2] Carlos A. Smith. Principles and Practice of Automatic Process Control. 2nd edition John Wiley & Sons, Inc. 2006, 563 p.
[3] Charles L. Phillips and John M. Parr. Feedback Control Systems. 5th edition Prentice Hall PTR, 2011, 774 p.
[4] Harnefors L. H-P- Nee "Model-Based Current Control of AC Machines Using the Internal Control Model Method" IEEE Transactions on Industry Applications, Vol 34, No 1, January/February 1998, pp. 133–141.
[5] Jean Pierre Corriou. "Process Control: Theory and applications" Springer, 2004.
[6] Ma Z. and Wang S. Energy efficient control of variable speed pumps in complex building central air-conditioning systems", *Energy and Buildings*, Vol. 41, 2009, pp. 197–205.
[7] Pedersen G.K. and Yang Z. Efficiency Optimization of a Multi-pump Booster system. Proc. of Genetic and Evolutionary Computation Conference (GECCO-2008), Atlanta, Georgia, USA, Jul 12–16 2008, pp. 1611–1618.
[8] Perez M.A., Cortes P. and Rodriguez J. Predictive control algorithm technique for multilevel asymmetric cascaded h-bridge inverters, IEEE Transactions on Industrial Electronics, vol. 55, no. 12, pp. 4354–4361, Dec. 2008.
[9] Reeves D. Study on improving the energy efficiency of pumps, European Commission, 2001.
[10] Shiels S. "Optimizing centrifugal pump operation", *World Pumps*, Jan. 2001, pp. 35–39.
[11] Thomas O. Miesner and William L. Leffler Oil & Gas Pipelines in Nontechnical Language PennWell Corp, 2006, 357 p.
[12] Yang Z. and H. Borsting. Energy Efficient Control of a Boosting System with Multiple Variable-Speed Pumps in Parallel, 49th IEEE Conference on Decision and Control Atlanta, Georgia USA, December 15–17, 2010, pp. 2198–2203.
[13] Zhongwen Wang, Ruizhen Duan and Xiaoqiu Xu, Model Identification of Hydrostatic Center Frame Control System based on MATLAB, Journal of Networks, Vol. No. 6, pp. 1322–1328, June 2013.

Future Communication, Information and Computer Science – Zheng (Ed.)
© *2015 Taylor & Francis Group, London, 978-1-138-02653-7*

A study on flower image recognition service of wildflowers in Gwacheon nature-learning places

H.-H. Lee & K.-S. Hong

College of Information & Communication Engineering, Sungkyunkwan University Suwon, South Korea

ABSTRACT: We present a mobile flower image recognition service system to recognize various species of wildflowers in the natural environment. The wildflower were recorded in a mobile environment and in this paper, the objects were detected from the images of wildflowers in Gwacheon nature-learning places. A recognition technology was proposed by calculating the values of the morphological shape information, texture information, and color information. In order to verify the proposed method, a recognition experiment was conducted of images with a detection success of about 80 varieties of wildflowers and cultivars collected between September and October; the result of the experiment shows that, by using Multi-Class SVM, the flower image recognition rate was 93.67%, by using Random Forest the first priority recognition of the flower image was 96.45%, and the proposed method is planned for the future demonstration service.

Keywords: wildflower; flower recognition; mobile

1 INTRODUCTION

Convergence technology has recently become important in the cultural tourism and natural resources utilization services, corresponding to the popular trend of smartphones that can be used anytime and anywhere. Most existing mobile cultural tourism services mainly focus on the level of information share related to images and text. Therefore, in the field of image processing, it has now become important to classify the objects that correspond to the pertinent cognitive goals from the original image, including the area of multiple objects in the object recognition technology, based on the image, and to extract their features. In some cases, the objects included in the image are composed of similar colors, while in many other cases they are composed of various colors, shapes, and textures. In addition, some images have a number of objects, or cannot be particularly expressed with only the above features; thus, the stage of detecting the pertinent object area and extracting the pertinent features is very difficult to understand [1].

The difficulty involved in detecting a flower object is distinguishing it from the complex backgrounds such as other surrounding flowers, leaves and so on; thus, a method is used that can detect and separate the flower object by placing it on a black or white background [2] [3].

Therefore, in this paper, we attempt to realize the plant locations and a cultivar flower image recognition system. Additionally, in this paper, a system that recognizes the wildflower and cultivar flower images in mobile devices, targeting for tourist use, is suggested

by structuring the server-client environment to realize a plant image collection system to collect plant resources for the mobile environment.

2 FLOWER IMAGE RECOGNITION

2.1 Plant image collection system

Figure 1 shows a flow chart of the plant image collection system in the mobile environment. The plant image collection system is operated when the GPS is on; when the user inputs the collector's name and the scientific name, folders and sensor information log files are automatically generated. The plant images are saved in the memory, along with information including two sequential shots and the level of illumination. When the plant image has been taken or other species of plants have been collected, the process is operated from the scientific name input step. The flower information search service displays the results requested from the search in the flower image database previously collected by Google maps.

2.2 Flower image recognition system

As shown in Figure 2, in the mobile environment, the system of recognition of wildflowers and cultivars is classified into three areas: object detection to recognize the wildflowers, feature extraction, and recognition. After the users check the results, if incorrect detections have been made, the users correct the incorrectly detected objects by using the inclusion

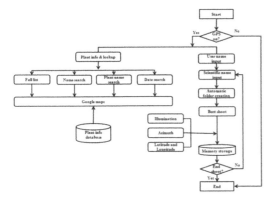

Figure 1. Plant image collection system.

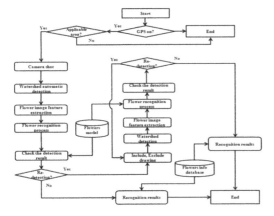

Figure 2. Flower image recognition system.

Figure 3. Images of wildflower and cultivated varieties collected.

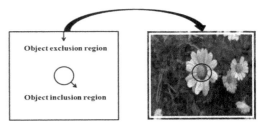

Figure 4. Object including and excluding set-up for automatic detection.

and exclusion drawings. After recognition through the redetection and the feature extraction steps, the results of the corrected objects are displayed to the users.

2.3 *Flower image object detection and feature extraction method*

By using the plant image collection system realized in this study, more than 50 flower images from each of the 80 species of wildflowers in the natural background environment of the Gwacheon nature-learning places and the cultivars were collected and used for an object detection, redetection, feature extraction, and recognition experiment.

Figure 3 shows an example of the flower image database used for the flower image recognition research in the mobile environment of this study.

The watershed algorithm is a morphological approach in which a topology is applied to the image and the image is segmented by applying a segmentation technique using the geomorphological characteristics of the ground surface for the same portion of the image. The region is then gradually expanded, similar to the way water fills up a vessel, from the pixel with the lowest value of inclination to the adjacent pixel

with the next highest value of inclination, after obtaining the value of inclination from the image by using the difference of brightness levels with each adjacent pixel of the image data [4].

These processes occur for the entire image, and the regions expand from the lowest value of inclination until the expansion stops, forming the watershed line where the regions meet. However, since the watershed algorithm is a method of expanding the regions, it may cause a problem by producing an over-segmented image. In this paper, regions that need to be excluded and regions where objects need to be included are assigned to minimize over-segmentation in the image.

In Figure 4, a region including the objects and a region of exclusion are assigned to improve the result of over-segmentation in the image of wild flowers and cultivation species of flowers. Since many shapes of wild flowers and cultivation species of flowers are similar to the original shapes, the region including the objects is set as having the original shapes and the region that is excluded is set based on the outline of the filmed image.

The Watershed algorithm was used in this paper to conduct flower image object detection, and the results are shown in Figure 5.

The object redetection method using the drawing suggested in this study involves drawing the flower

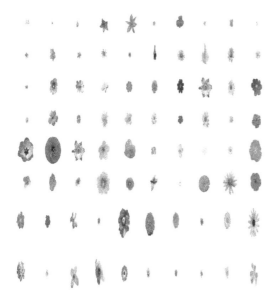

Figure 5. Wildflower and cultivated varieties image detection results.

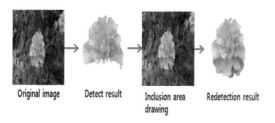

Original image → Detect result → Inclusion area drawing → Redetection result

Figure 6. Drawing detection results of inclusion area.

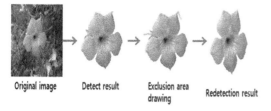

Original image → Detect result → Exclusion area drawing → Redetection result

Figure 7. Drawing detection results of exclusion area.

image inclusion and exclusion areas, correcting these areas, and then conducting the redetection step; Figure 6 and Figure 7 show the results. In the redetection step, after automatic detection of the flower image objects, the users check the detection results and can redetect them through the inclusion or exclusion area drawing.

A flower is morphologically composed of the stamen, pistil, petals, etc.; because the color information of each component value differs, the method of feature extraction used in this study involves dividing the image into three equal parts from the center of gravity of the detected object. Figure 8 shows an example

Figure 8. User drawing detection results of exclusion area.

of dividing the detected flower area into three partitioned areas from the center of the flower image [5]. For the values of YCbCr and HS used as the color feature information, the average color information of the entire area of the flower image object is used, as well as each area that was divided into three equal parts as the feature value.

In a shape feature, the number of petals of the object detected is used as the feature value by using the Zero Crossing Rate as an important feature that can determine the number of petals of the flower object. For the texture feature information, the Grey Level Co-occurrence Matrix(GLCM) suggested by Haralic K et al. is used. In this paper, four texture features were used, including contrast, correlation, entropy, and homogeneity [6].

2.4 Flower image recognition method

After detecting the flower object in order to recognize the flower image, a total of 25 items of feature information were detected by using color information, texture information, and shape information. For the recognition methods of flower images, Multi-Class SVM and Random Forest were used [7] [8].

3 EXPERIMENTS AND RESULTS

Android based JAVA was used as an experimental environment in this study; VC++ and OpenCV were used as a server environment and the TCP/IP communication environment was realized.

Figure 8 shows the initial screen of the plant image collection system, the camera guidelines used to take the plant image, and the plant information search. The plant image collection system makes it possible to search the plant distribution by supporting the various search methods.

Figure 9 shows the first screen of the flower recognition system and the drawing screen of the inclusive and exclusive areas for the recognition results of the flower images taken and the redetection after revising.

The database consists of about 80 images of species of flowers taken at the Gwacheon nature-learning places, Galhyeon-dong, Gwacheon-si, Gyeonggi-do, Republic of Korea. The recognition experiment was conducted using a total of 4,132 flower images of the 80 species. The automatic detection and redetection was completed by collecting more than 50 flower images from each species; the recognition experiment

Figure 9. Plant image collection and search system.

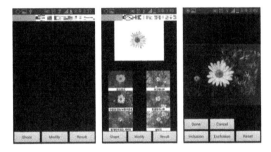

Figure 10. Initial screen (left), Recognition results (middle), and Drawing screen for the redetection (right).

Table 1. Flower image recognition results using Multi-Class SVM.

80 species	Multi-Class SVM
Recognition ratio	93.67%

Table 2. Flower image recognition results using Random Forest.

Rank-1~Rank-5	Recognition ratio
Rank-1 recognition ratio	96.45%
Rank-1~Rank-2 recognition ratio	99.52%
Rank-1~Rank-3 recognition ratio	99.85%
Rank-1~Rank-4 recognition ratio	99.95%
Rank-1~Rank-5 recognition ratio	100%

of the learned model and test images was conducted at a 70:30 ratio.

Table 1 shows the flower image recognition experimental results using the Multi-Class SVM, and Table 2 shows the flower image recognition experimental results using the Random Forest.

4 CONCLUSIONS

In this paper, the recognition results were drawn from more than 96% of the 1st recognition rate of the 4,132 total flower images of 80 species; the effectiveness and validity of this suggested method for future demonstration services were verified in this study. Under the goal of future demonstration services, various species of flowers will be collected, and the study of the object detection, feature extraction, and recognition method will continue to improve the recognition rate. A wildflower and cultivated varieties image recognition demo is available in [9].

ACKNOWLEDGMENTS

This research is supported by Ministry of Culture, Sports and Tourism (MCST) and Korea Culture & Tourism Institute(KCTI) Research & Development Program 2013.

REFERENCES

[1] Pornpanomchai, C. Sakunreraratsame, P. Wongsasirinart, R. and Youngtavichavhart, N. 2010. Herb flower recognition system. Electronics and Information Engineering, 2010 International Conference On(1): 123–127.

[2] Das, M. Manmatha, R. and Riseman, EM. 1999. Indexing flower patent images using domain knowledge. IEEE Intelligent Systems(4): 24–33.

[3] Hong, A. Chen, G. Li, J. Chi, and Z. Zhang, D. 2004. A flower image retrieval method based on ROI feature, Journal of Zhejiang University-Science(5): 764–772.

[4] Watershed(image peocessing) http://en.wikip edia.org/ wiki/Watershed_(image_processing)

[5] Lee, Hyo-Haeng, Li, Xianghua, Ching, Kwang-Woo and Hong, Kwang-Seok, 2012. Flower Image Recognition Using Multi-Class SVM. Innovation for Applied Science and Technology (284–287): 3106–3110.

[6] Haralick, R.M. Shanmugam, K. and Dinstein, I. 1973. Textural Features for Image Classification, IEEE Trans, on System, Man and Cybernetice(3): 610–621.

[7] Multi-Class Support Vector Machine, Http://svmlight. joachims.org/svm+multiclass.html

[8] OpenCV, Http://docs.opencv.org/modules/ml/doc/ random_trees.html

[9] Wildflower and cultivated varieties image recognition demo Http://hci.Skku.ac.kr/bbs/zboard.php?id=dwTest

Future Communication, Information and Computer Science – Zheng (Ed.)
© *2015 Taylor & Francis Group, London, 978-1-138-02653-7*

Data processing method of fault samples in analog circuits

L. Huang & C. Yao
School of Electronic and Information Engineering, Beijing Jiaotong University, Beijing, China

ABSTRACT: The quality and the quantity of fault samples will directly affect the results of fault diagnosis in analog circuits. This paper studied the mathematical methods of reducing the dimension of fault sample and normalizing fault samples, deduced the normalization algorithm of removing the dimension of the fault samples in analog circuits. By analyzing the circuit and fault samples, the input requirement of the diagnostic tools can be satisfied, the data amount of fault samples is reduced, and the accuracy and the efficiency of fault diagnosis are increased.

Keywords: analog circuit; fault sample; data processing; normalization processing

1   INTRODUCTION

The quality and the quantity of fault samples (Huang et al. 2013) will directly affect the results of fault diagnosis (Lang et al. 2013) of the analog circuits (Tan et al. 2013), because all methods of fault diagnosis that can execute the fault type identification are based on fault samples. Fault diagnosis of analog circuits is a very important project in electronics research. In recent years, scientists have gotten a great deal of research progress in the fault diagnosis of analog circuits.

Fault samples should completely describe the fault feature of analog circuits. The higher the quality of samples, the higher diagnostic accuracy we can get, in spite of small samples. Some diagnostic methods need a lot of fault samples as input data for simulation before testing or training (Yuan et al. 2010), so plenty of fault samples are also very important.

Fault samples of analog circuits can be collected from the scene of practical circuits, or obtained through software simulation. We can also get fault samples from theoretical derivation or expert experience. Original fault samples are made up of measured parameters of analog circuit test points. Before fault diagnosis of the fault samples collected from test points, we need to process and analyze the data, in order to satisfy the input requirement of diagnostic tools.

This paper studied the mathematical methods of reducing the dimension of fault sample in analog circuits, and the normalizing method of data processing. We can reduce the data amount of fault samples by reducing the dimension of fault samples in analog circuits and analyzing the analog circuits and fault samples, in order to improve the diagnostic accuracy and the efficiency in analog circuits. Then we deduced the normalization algorithm, which removes the dimension of fault samples in analog circuits. The qualified fault samples are prepared for fault diagnosis of analog circuits.

2   THE PRINCIPLES OF DATA PROCESSING OF FAULT SAMPLES

The fault samples which are collected from test points of analog circuits, without being processed and being disposed before, are called original samples. In order to get better diagnosis results and efficiency, we need to deal with the original samples and analyze them while conducting the fault detection in analog circuits.

We must do the following work while dealing with the fault samples.

2.1   *Choosing the best test points*

The optimum test points are selected in analog circuits. Some measuring points can not describe the fault features of analog circuits, which are useless for fault diagnosis results, so we should get rid of those test points.

2.2   *Improving the speed of fault diagnosis*

On the premise of retaining fault feature of analog circuits, we should try our best to reduce the amount and the dimension of fault samples in analog circuits. Thus they will be fewer but better. The other advantage is that the calculated amount can be decreased for computing equipment and the efficiency can be increased.

2.3   *The treatment of data*

If the diagnostic tools have some special requests for input data in analog circuits, the dimension and the

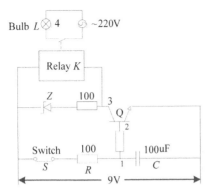

Figure 1. Circuit of controlling relay.

Table 1. Primary fault samples.

No.	U_1	U_2	I_2	U_3	L
1	8.99 V	0.715 V	82.9 μA	3.126 V	1
2	8.99 V	0.594 V	21.4 μA	7.357 V	0
3	8.99 V	0.689 V	67.6 μA	4.587 V	1
4	8.99 V	0 V	0 μA	9 V	0
5	8.99 V	0.732 V	225 μA	0.303 V	1

Table 2. Fault samples after reduced dimension.

No.	U_2	U_3	L
1	0.715 V	3.126 V	1
2	0.594 V	7.357 V	0
3	0.689 V	4.587 V	1
4	0 V	9 V	0
5	0.732 V	0.303 V	1

format of fault samples need to be disposed, such as normalization processing and removing dimension treatment. But the dimension and the arrangement of fault samples should remain the same if the dimension doesn't influence the classification results of diagnostic tools.

3 INSTANCES OF REDUCING DIMENSION TREATMENT OF FAULT SAMPLES

The analog circuit, as shown in Figure 1, is a relay control circuit. The type of the relay K we used is a DC normal opened electromagnetic relay. The switch S is a normally open tact button switch. The analog circuits in Figure 1 can achieve that weak current controls strong current by the relay. Every time you press the switch S, the bulb L will light for a while.

In Figure 1, when you press the switch S, the triode Q turns into the saturation conduction, the relay K will be closed and the bulb L will light. After you disconnect the switch S, the voltage of the capacitor C can hold on for a moment to make the triode still work in the conduction state. In consequence, the relay K and the bulb L can work on. In the meantime, the capacitor C discharges by the resistance R and the triode Q. So the voltage of both ends of the capacitor declines little by little. At last, the triode Q turns into the cut-off state, the relay turns into the disconnected state and the bulb turns into the dark state.

We choose four test points in the analog circuit, as shown in Figure 1. Then we collect five original data of it, such as the voltage U_1 of test point 1, the voltage U_2 of test point 2, the electric current I_2 of test point 2, the voltage U_3 of test point 3, the state of bulb L of test point 4. When the light is on, we write it as 1, while the light is closed, we write it as 0.

When the switch S is closed, we can simulate five kinds of states while the analog circuit is stable. Because this chapter deals only with reducing the dimension of fault samples and the normalizing processing, we do not need to diagnose the fault of analog circuit. Then five kinds of circuit states are chosen at

random. It can be normal, and can also be broken-down for resistance, the triode or the relay, and so on. Table 1 lists the five fault samples without treatment.

The dimension of every sample is five in Table 1. You will find that the voltage of test point 1 remains unchanged. When the switch S is closed, test point 1 is connected with the positive pole of power supply, and the voltage holds on for 9 V. We can delete the data U_1 on account of it being meaningless for fault diagnosis of the analog circuit.

We can conclude from these observations that the voltage of test point 1 is equal to the base voltage of triode Q. If the base voltage U_2 continues to increase, it means the positive bias of the triode increases. Then the base current I_2 will increase accordingly. The third column is related to the second column, so we can only keep one of them.

In general, the current data collection is much more complicated than the voltage data (Liao. 2009) collection, because you need to disconnect the wire and add an ampere meter into the circuit while measuring the current data. In fault diagnosis of analog circuits, you should always try to use the voltage signal as diagnostic data. Therefore the data of U_2 is retained and the data of third column I_2 is removed from Table 1.

The fourth column of Table 1 is U_3, the voltage of test point 3. U_3 stands for the collector voltage of the triode. We should reserve the data of U_3 because it directly affects the closed or opened state of the relay K. The fifth column data of Table 1 stand for the bulb state of test point 4. Because the analog circuit is a control circuit for the bulb, it is important and visualized whether the bulb will be alight or extinct, we need to retain the data L.

Through the above analysis, we can get five fault samples after reducing dimension treatment, as shown in Table 2.

There are only three columns data in Table 2 after circuit analysis and reducing the dimension treatment. So the dimension of fault samples declines from five to three. The amount of fault data is decreased. The advantage of reducing dimension treatment is obvious, such as decreasing the scale of fault samples, reducing the calculated amount of categorization tools and increasing the efficiency of fault diagnosis in analog circuit.

Table 3. Fault samples after normalization processing.

No.	x_1	x_2	x_3
1	0.977	0.325	1
2	0.811	0.811	0
3	0.941	0.493	1
4	0	1	0
5	1	0	1

4 NORMALIZATION PROCESSING OF FAULT SAMPLES

The dimension of data in Table 2 is not unified so those data do not conform to the requirement of the diagnostic tools. In order to make the data that has a different dimension comparable and meet the requirement of all diagnostic tools, we should accomplish the normalization processing and reduce the dimension treatment, unify the dimension and the magnitude of the data in analog circuit. It will be a great basis for the analog circuit fault diagnosis in the future.

Assuming that x is a vector that contains n variables.

$x = \{x_1, x_2, \ldots, x_n\}$.

The maximum variable in vector x:

$P = \max(x)$

The minimum variable in vector x:

$Q = \min(x)$

The biggest difference value between the minimum variable and the minimum variable is as follows.

$$\Delta = P - Q = \max(x) - \min(x)$$

The normalization formula is as shown in Equation 1.

$$x_i' = \frac{x_i - Q}{\Delta} = \frac{x_i - Q}{P - Q} = \frac{x_i - \min(x)}{\max(x) - \min(x)} \quad (1)$$

Finishing normalization processing to the data in Table 2 according to Equation 1, we can get Table 3, as follows.

The numerator in Equation 1 is the difference value between the variable x_i and the minimum variable of the vector x. The denominator is the biggest difference value between the minimum variable and the minimum variable. Data x_i, after normalization processing, is all between 0 and 1.

5 SUMMARY

Example analysis shows that we can reduce the data amount of fault samples, by reducing the dimension of fault samples of analog circuits by analyzing the circuit and fault samples, in order to improve the readability, diagnostic accuracy and efficiency of fault samples. The normalization algorithm we used, which removes the dimension of fault samples in analog circuits, prepares the qualified fault samples for fault diagnosis of analog circuits.

ACKNOWLEDGEMENTS

This work was financially supported by the National Natural Science Foundation of China (61201363) and the Fundamental Research Funds for the Central Universities (2013JBM016).

REFERENCES

[1] Huang, L. et al. 2013. A fuzzy quantifying algorithm of fault-degree in analog circuits. *Control Theory and Applications* 30(8): 1053–1058.
[2] Lang, R.L. et al. 2013. Data-driven fault diagnosis method for analog circuits based on robust competitive agglomeration. *Journal of Systems Engineering and Electronics* 24(4): 706–712.
[3] Liao. Y. 2009. Fault location observability analysis and optimal meter placement based on voltage measurements. *Electric Power Systems Research* 79(7): 1062–1068.
[4] Tan, Y. et al. 2013. Analog fault diagnosis using S-transform preprocessor and a QNN classifier. *Measurement* 46(7): 2174–2183.
[5] Yuan, l.f. Et al. 2010. a new neural-network-based fault diagnosis approach for analog circuits by using kurtosis and entropy as a preprocessor. *Ieee transaction on instrumentation and measurement* 59(3): 586–595.

Future Communication, Information and Computer Science – Zheng (Ed.)
© 2015 Taylor & Francis Group, London, 978-1-138-02653-7

A scheduling scheme based on cognition for UWB wireless personal area network

Y. Guo & L. Huang
Beijing Jiaotong University, Beijing, China

ABSTRACT: It is possible to increase the throughput of UWB WPAN based on IEEE 802.15.3 MAC by proper scheduling. IEEE 802.15.3 Medium Access Control (MAC) protocol is a TDMA-like one, in which no concurrent transmission is permitted. This paper proposes a scheme to schedule UWB nodes to transmit concurrently to increase the throughput of UWB WPAN. In our scheme, some pairs of UWB nodes can transmit concurrently to increase the throughput of the network if the mutual interferences satisfy certain conditions. For fairness, the conditions require that the throughputs of all pairs of nodes increase through concurrent transmission, that is, no pairs of UWB nodes sacrifice for the increase of throughput of the network. The conditions are derived and the scheduling algorithm is presented. The upper bound and lower bound of the throughput of concurrent transmission are also given. It is shown that the throughput of UWB WPAN can be improved by using our scheduling scheme.

Keywords: Cognition; scheduling; ultra-wideband; wireless personal area network

1 INTRODUCTION

UWB technology can support high data rate in short range with low power consumption (Roy et al. 2004). Ultra-Wideband (UWB) can find many applications in wireless personal area network (WPAN) or broadband home network (Kim et al. 2006). However, the existent WPAN medium access control (MAC) protocol was designed for narrow band communication networks (Liu et al. 2008). In order to make use of the potential of UWB in WPAN, MAC protocol should be specially designed for UWB WPAN while considering the specific features of UWB. Table 1. Margin settings for A4 size paper and letter size paper.

Devices in WPAN can autonomously form a piconet in which one device is selected as the piconet coordinator (PNC). The PNC is like the base station of cellular networks in respect of the network access coordination and radio resources allocation. The PNC of WPAN is in charge of synchronization, data request registration and channel time assignment. The PNC is not involved in the communication between source node and destination node. Source nodes and destination nodes in WPAN can make peer-to-peer communication without the PNC participation, which means that concurrent transmission is possible. This feature of WPAN provides possibilities for the improvement of the performance of WPAN. In (Liu et al. 2008) a scheduling scheme is proposed based on an exclusive-region to realize concurrent transmission. The exclusive region is a circle centered at the receiver and its radius is independent of link distance. Each UWB link is protected by the exclusive region in which no concurrent transmissions are allowed. If senders are outside the exclusive region of other flows, they can make a concurrent transmission. However, it is difficult to decide the optimal size of the exclusive-region, and the scheme is dependent on a path loss propagation model.

This paper proposes a scheduling scheme that can permit concurrent transmission based on the cognition of mutual interference. In our scheme, the increase of UWB WPAN throughput is realized by increasing the throughput of every pair of source-destination UWB nodes. The increase of throughput of WPAN is not at the cost of throughput of any pair UWB nodes. The conditions for concurrent transmission and the scheduling algorithm are presented. The upper bound and lower bound of the throughput of concurrent transmission are also given.

The remainder of this paper is organized as follows. In section 2, we derive our scheduling scheme based on the IEEE 802.15.3 standard MAC protocol. In section 3, a simulation and analysis are given. Section 4 is the conclusion.

2 SCHEDULING SCHEME

Paper (Liu et al. 2008) describes the suitability of IEEE 802.15.3 MAC standard for UWB WPAN, on which the scheduling scheme of (Liu et al. 2008) is based. Our scheme is also based on IEEE 802.15.3 MAC protocol. We will make a brief and necessary review of the 802.15.3 standard in this paper.

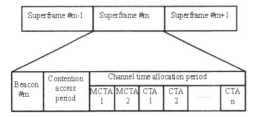

MCTA: Management Channel Time Allocation
CTA:　Channel Time Allocation

Figure 1.　Superframe structure defined in IEEE 802.15.3 MAC protocol.

The 802.15.3 standard adopts a hybrid contention and contention-free MAC protocol. The IEEE 802.15.3 standard defines a superframe structure as shown in Figure 1. Each superframe starts with a Beacon Period (BP) for network synchronization and control message broadcast. Then, the following are contention access period and channel time allocation period. In contention access period (CAP), devices send their resource request to the PNC using carrier sensing multiple access/collision avoidance (CSMA/CA) in conjunction with a backoff procedure. Channel time allocation period (CTAP) is a contention-free period. In CTAP, the PNC allocates channel time for data traffic using time division multiple access (TDMA) discipline. The reason is the ultra-low power spectral density of UWB. As we know, UWB devices are operating under the FCC regulation that the transmission power must not exceed −41 dBm/MHz, and the peak/mean power ratio must be less than 20 dB. The ultra-low power spectral density makes the widely-used CSMA/CA MAC unsuitable in UWB WPAN. However, the efficiency of TDMA UWB network is low because the acquisition time required of UWB.

(Radunovic & Le Boudec 2004) and (Cai et al. 2008) have suggested that power control is beneficial to reduce power consumption, but its gain in terms of total throughput in UWB network is minor, compared to that benefit from transmission scheduling. So the UWB devices simply use the maximum power level permitted for transmission when it is scheduled to transmit, which makes the cognition of mutual interference feasible.

Our scheduling scheme is based on cognition of mutual interference to improve the efficiency of UWB WPAN. Assume there are N pairs of nodes in communication. These pairs are numbered from 1 to N. The source node of first pair is called s1, and destination node is called d1, and so on. We first allocate each pair of these N pairs into one time slot out of N time slots. In the nth time slot, sn transmits, dn receives, other destination nodes listen. After first superframe, a gain matrix can be acquired as:

$$\begin{bmatrix} g_{11} & \cdots & g_{1N} \\ \vdots & \ddots & \vdots \\ g_{N1} & \cdots & g_{NN} \end{bmatrix}$$

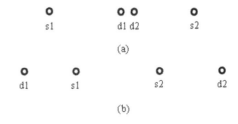

Figure 2.　Illustration of a): hidden terminal and b): exposed terminal.

In time slot 1, pair 1 are primary users, other nodes are secondary users. If g_{12} is rather small (it means s1 does not interfere d2 much), and at the same time, g_{21} is rather small too (it means s2 does not interfere d1 much), then pair 1 and pair 2 can make a concurrent transmission in time slot 1 and time slot 2.

Because this scheme is based on the cognition of destination nodes, it can avoid the problems of hidden terminals and exposed terminals. If we adopt CSMA/CA which is based on the sensing of senders, we can see from Figure 2, there probably will be a collision in case a. While in case b, two pairs can make a concurrent transmission, which will not be possible in CSMA/CA. In CSMA/CA, the senders sense the environment in the vicinity of themselves. However, it is the conditions near the receivers that reflect the situation of interference. In addition, carrier sensing is not suitable for UWB. As we know, the multipath effect is very severe in indoor environment. The channel is frequency selective. There may be severe fading in the UWB bandwidth range.

In one time slot, there can be multiple pairs making concurrent transmission as long as the throughput increases for any pair participating in the concurrent transmission. In an extremely ideal situation, all N pairs can concurrently transmit.

According to Shannon theory, channel capacity $R = W \log_2 (1 + SINR)$ bps, where W is the bandwidth, $SINR = \frac{P_r}{N_0 + I}$, P_r is the received power, N_0 is the Gaussian noise power, and I is the interference power from other users. From (Proakis 2001), for the UWB system, W is very large, capacity can be approximated by

$$R = \frac{P_r W}{N_0 + I} \log_2 e \quad \text{(bps)} \tag{1}$$

When there is no concurrent transmission, for pair i, in ith time slot, $R(i) = \frac{P_r(i)W}{N_0} \log_2 e$. Assume pair i makes concurrent transmissions in n time slots, then the throughput for pair i is

$$R_i^{Cn} = \frac{P_r(i)W}{N_0 + \sum_{j \neq i}^{n-1} I_{j,i}} \log_2 e \tag{2}$$

The throughput for pair i will be

$$G_i^{Cn} = \frac{nP_r(i)W}{N_0 + \sum_{j \neq i}^{n-1} I_{j,i}} \log_2 e \tag{3}$$

The condition for concurrent transmission should be: for every pair, the throughput should increase with the number of time slots for concurrent transmission. Assume pair i transmits in n time slots, the present data rate is

$$R_i^{Cn} = \frac{P_r(i)W}{N_0 + \sum_{j \neq i}^{n-1} I_{j,i}} \log_2 e$$

The data rate for $n + 1$ time slots case is

$$R_i^{C(n+1)} = \frac{P_r(i)W}{N_0 + \sum_{j \neq i}^{n-1} I_{j,i} + I_{n,i}} \log_2 e \tag{4}$$

where $I_{n,i}$ is the interference from the pair whose time slot is the $(n+1)$th time slot of pair i's concurrent transmission.

The condition for concurrent transmission in mathematical form is

$$nR_i^{Cn} < (n+1)R_i^{C(n+1)} \tag{5}$$

That is

$$\frac{nP_r(i)W}{N_0 + \sum_{j \neq i}^{n-1} I_{j,i}} \log_2 e < \frac{(n+1)P_r(i)W}{N_0 + \sum_{j \neq i}^{n-1} I_{j,i} + I_{n,i}} \log_2 e \tag{6}$$

This equation means that throughput of every pair increases through concurrent transmission. From equation (6) we can easily find that if the following condition is satisfied, the network throughput will increase.

$$nI_{n,i} - \sum_{j \neq i}^{n-1} I_{j,i} < N_0 \tag{7}$$

This is a recursive equation. In order to decide with which nodes source node i can make concurrent transmission, the computation needs to start from $n = 1$. For $n = 1$ (that means pair i makes a concurrent transmission in another one time slot beside his own time slot), we can have from equation (7)

$$I_{1,i} < N_0 \tag{8}$$

In order to increase throughput by concurrent transmission, the interference has to be less than the noise at least from equation (8).

There may be more than one link that satisfies this condition. The link having the least interference over pair i should be chosen first because this can make an increase in throughput of pair i most. Then, more links can be added until equation (7) is not satisfied.

According to equation (7) and the gain matrix, a scheduling matrix S is derived. The rows of the matrix stand for time slots. The columns of the matrix stand for pairs that communicate. In this scheduling matrix, 1 is for transmission and 0 is for no transmission. The diagonal elements are all 1s for diagonal elements are the primary time slots of every pair. The scheduling matrix is computed for every time slot row by row. Those elements satisfying the condition are set into 1, other elements into 0.

Below is an example of 6×6 matrix. The scheduling matrix is as follows:

$$\begin{matrix} 1 & 1 & 0 & 1 & 0 & 0 \\ 1 & 1 & 0 & 0 & 1 & 0 \\ 1 & 0 & 1 & 0 & 0 & 1 \\ 0 & 0 & 0 & 1 & 0 & 0 \\ 0 & 1 & 0 & 0 & 1 & 0 \\ 0 & 0 & 1 & 0 & 0 & 1 \end{matrix}$$

In order to support concurrent transmission, it must be ensured that the mutual interferences satisfy the condition in the time slots for concurrent transmissions. So, the non-diagonal elements of matrix S are given to the value according to the following conditions:

$$S_{i,j} = \begin{cases} 1 & if\ S_{i,j} = S_{j,i} = 1 \\ 0 & else \end{cases}$$

The S matrix becomes as follows:

$$\begin{matrix} 1 & 1 & 0 & 0 & 0 & 0 \\ 1 & 1 & 0 & 0 & 1 & 0 \\ 0 & 0 & 1 & 0 & 0 & 1 \\ 0 & 0 & 0 & 1 & 0 & 0 \\ 0 & 1 & 0 & 0 & 1 & 0 \\ 0 & 0 & 1 & 0 & 0 & 1 \end{matrix}$$

From this matrix, pair 1 can make a concurrent transmission with pair 2 in time slot 1 and time slot 2. Pair 2 can make a concurrent transmission with pair 1 and pair 5. Pair 3 and pair 6 can make concurrent transmissions as well.

Concurrent transmissions require that mutual interferences all satisfy the condition for concurrent transmission, so the matrix is symmetric. In some cases, we can relax the condition that $S_{i,j} = S_{j,i} = 1$ so that non-mutual concurrent transmission is possible. Here, we take the example of the above matrix, $S_{31} = 1$ that means pair 3's interference to pair 1 satisfies the condition. However, $S_{13} = 0$ does satisfy the condition. If we permit pair 1 to communicate in time slot 3, the throughput of pair 3 will drop compared with TDMA scheduling. But if the g_{31} is small enough, such as $-10\,\text{dB}$ compared with N, which is normal in UWB communication, the throughput of pair 3 drops very little when pair 1 transmits in time slot 3. The overall throughput will increase in the expense of the loss of throughput of pair 3. We can set a criterion for the loss of throughput to decide whether the non-mutual concurrent transmission is allowed.

Our scheme can guarantee the fairness by increasing the throughput of every pair to increase that of the network.

3 SIMULATION AND ANALYSIS

The increase of throughput by concurrent transmission is computed in different situations, from the simplest case ($n = 2$) to an infinite case. From equation (8),

	Algorithm
1.	For pair i, $i = 1$ to N
2.	From the smallest $I_{j,i}$, to the largest $I_{j,i}$
3.	If the condition of equation (7) is satisfied then
4.	$S_{j,i} = 1$
5.	else
	$\quad\quad i = i + 1$
6.	end
7.	for i=1 to N
8.	for j=1 to N
9.	$S_{i,j} = S_{j,i} = S_{i,j} \& S_{j,i}$
10.	end
11.	end

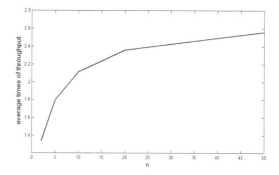

Figure 4. The average increases of throughput.

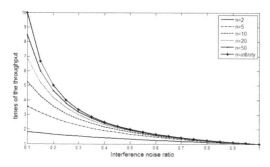

Figure 3. The increases of throughput by concurrent transmission.

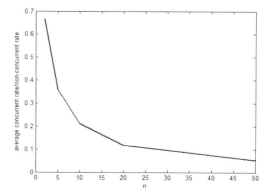

Figure 5. The average power efficiency.

interference from a single node has to be less than the noise. The interference ranges from $0.1 N_0$ to N_0. From equation (3), the minimal aggregate interference is $0.1(n-1) N_0$, and the maximal aggregate interference is $(n-1)N_0$. The times of throughput to that of non-current transmission is computed. The upper bound is 10 and the lower bound is 1. The increase of throughput is not notable when interference noise ratio approaches 1. It can be improved by adding more rigorous conditions to equation (7), such as interference less than $0.5N_0$.

The average increases of throughput is computed for a different number of time slots for concurrent transmission over the distribution of interference noise ratio. The interference distribution is assumed uniformly distributing from $0.1 N_0$ to N_0. The result is depicted in Figure 4. When n approaches infinity, the averaged value asymptotically goes to 2.735. It is apparent that the throughput increases with n. However, the power utilization efficiency does not increase with n monotonically because the slope of the curve gets smaller and smaller with the increase of n. The marginal benefit will decrease with the increase of n. That is the concurrent data rate decreases with the number of time slots for concurrent transmission as is shown in Figure 5.

4 CONCLUSION

This paper proposes a scheduling scheme for UWB WPAN based on IEEE 802.15.3 MAC protocol through cognition. The network throughput can increase by concurrent transmission scheduling. The condition for concurrent transmission is derived and the scheduling algorithm is presented. The algorithm is very simple and suitable for UWB WPAN PNC node. The upper bound and lower bound of the throughput of the concurrent transmission are given.

REFERENCES

[1] Cai, J., Liu, K., Shen, X., Mark, J. W. & Todd, T. D. 2008. Power allocation and scheduling for ultra-wideband wireless network. *IEEE Trans. Veh. Technol.*, vol. 57, no. 2, pp. 1103–1112.
[2] Kim, J., Lee, S., Jeon, Y. & Choi, S. 2006. Residential HDTV Distribution System using UWB and IEEE 1394. *IEEE Trans. Consum. Electron.*, vol. 52, no. 1, pp. 116–122.
[3] Liu, K., Cai, L. & Shen, X. 2008. Exclusive-region based scheduling algorithm for UWB WPAN. *IEEE Trans. Wireless Commun.*, vol. 7, no. 3, pp. 933–942.
[4] Proakis, J. G. 2001. *Digital Communicaion* (4th). McGraw-Hill, New York, USA.
[5] Radunovic, B. & Le Boudec, J. Y. 2004. Optimal power control, scheduling, and routing in UWB networks. *IEEE J. Select. Areas Commun.*, vol. 22, no. 7, pp. 1252–1270.
[6] Roy, S., Foerster, J. R., Somayazulu, V. S. & Leeper, D. G. 2004. Ultrawideband radio design: the promise of high-speed, short-range wireless connectivity. *Proc. IEEE*, vol. 92, pp. 295–311.

Future Communication, Information and Computer Science – Zheng (Ed.)
© 2015 Taylor & Francis Group, London, 978-1-138-02653-7

A data replica consistency maintenance scheme for cloud storage under healthcare IoT environment

B. Zhang, X. Wang & M. Huang
College of Information Science and Engineering, Northeastern University, Shenyang, Liaoning, China

ABSTRACT: In this paper, taking the inconsistency problem leaded by users' reading and writing operations of different data replicas into account, a multiple data replica consistency maintenance scheme for cloud storage under health-care IoT environment is proposed. It distributes the update task from "primary replica" to every replica node, thus the user's access requests can be responded to by each data replica and avoid data replica update collision problems by timestamp scheme. We have implemented the proposed scheme based on CloudSim and have done a performance evaluation. Simulation results have shown that it can not only improve user access time and promote load balance but also enhance system scalability and reliability.

Keywords: cloud storage; multiple data replica; consistency maintenance; timestamp

1 INTRODUCTION

With the development of IoT (Internet of Things) and digital medical equipment and their combination, the health-care IoT has emerged and has been widely applied in medicine and health-care area. Massive medical and health care data have been generated, asking for research and development of a cloud storage system. In order to improve its scalability and reliability and at the same time shorten user access time, the multiple data replica scheme should be adopted.

Cloud storage is derived from the basis of cloud computing and combined massive heterogeneous storage devices in the network by application software mainly through clustering technology, network technology, virtualization technology, distributed operating systems and distributed file system, etc. (Liu et al. 2011). Then the formed uniform storage system cooperatively works to provide data storage and data access service to users. Therefore, the users only need to acquire online storage space through a storage service provider without setting up their own data storage center, thus avoiding repeated construction of storage systems as well as saving the investment of hardware and software infrastructure.

Cloud storage is generally composed of massive storage devices distributed over different data centers. These devices are connected through the Internet and have different reliability and performance with temporary even permanent faults happening occasionally. The network status also has influence on the promptness and reliability of data access. In the face of a huge amount of access requests from massive cloud users, deploying multiple replicas of one single (probably very important or popular) data object, in a cloud storage system is generally needed in order to get prompt, reliable and efficient data access services, to improve cloud storage system response time to user access and enhance user QoE (Quality of Experience) (Lin et al. 2012). A multiple data replica scheme can deploy a right quantity of replicas for suitable data objects in due time and place. It needs an efficient management scheme to select proper locations for data replicas and assign user requests to them optimally so that the access and maintenance costs can be minimized and the performance of the cloud storage system improved. It mainly studies multiple data replica consistency maintenance scheme for cloud storage in this paper.

A whole cloud storage system connected by data center nodes are located in different positions through network, where different replicas of the same data distribute different nodes in the whole system. Replicas in different locations may respond read and write requests from cloud users and it needs data replica consistency maintenance in order to guarantee the reliability and credibility of data access of medical users.

The step of replica consistency maintenance (Mistarihi & Chan 2009) is as follows: a cloud user request to update data object according to application and submit to system. System transfer update in a certain way according to specified consistency maintenance strategy among all replicas and come to consistency state ultimately.

The related representative literatures are: Almeida et al. (2013) proposed a Geo-distributed key-value data store, named ChainReaction that offers causal+

consistency, with high performance, fault-tolerance, and scalability. ChainReaction enforced causal+ consistency which was stronger than eventual consistency by leveraging on a new variant of chain replication. MacCormick et al. (2008) designed and deployed a new replication protocol called Niobe which is in the primary-backup family of protocols, and shared many similarities with other protocols in distributed systems. Lv (2011) took vector timestamp based on log update, and an eventually consistency strategy based on gossip system structure was proposed. It improved "the home" replica management technology and update speed, shortened convergence time at the same time. In Ren et al. (2008), a replica clustering coefficient was defined for representing social capacity, by which replica nodes are classified into multi-levels to form one replica-tree, and constitute a peer-to-peer network with first-class replica nodes of different VOs. Dullmann et al. (2001) proposed a new Grid service, called Grid Consistency Service (GCS) that sat on top of existing Data Grid services and allowed for replica update synchronization and consistency maintenance. It gave models for different levels of consistency provided to the Grid user and discussed how they can be included into a replica consistency service for a data grid. Wang et al. (2010) presented a replica consistency maintenance strategy called replica information broadcast tree. The strategy stored the replica information in the binary tree, which made use of subscribe and unsubscribe algorithm to maintain replica consistency. Chen et al. (2005) presented SCOPE, a structured P2P system supporting consistency among a large number of replicas by building a replica-partition-tree for each key. SCOPE kept track of the locations of replicas and then propagates updated notifications. In Wu et al. (2009) aiming at the problem that it is hard to keep replica data consistency due to dynamic character in data gird, a keeping replica data consistency strategy in grid environments was presented. According to the dynamic character of grids, models for grid systems and replica data consistency respectively were established. In Yang et al. (2010) a node heterogeneous degree-based of consistency maintenance algorithm was proposed. It denoted the replica node capability by node heterogeneous degree and managed replica nodes using Chord protocol and collected those nodes node heterogeneous degree along with a replica heterogeneous degree collection tree built through the finger table-based ring partition method.

These works have studied and given valuable solutions that were enlightening to this paper about the problem of consistency maintenance from different aspects and points. Replicas residing in different place are possible to respond to read and write requests from cloud users. Taking the inconsistent issue of users read and write operations to different data replicas into account, we proposed a data replica consistency maintenance scheme in this paper to ensure the reliability and dependability of medical user accessing data. Every data replica could respond to access requests and avoid the collision problem of data

replica updates by timestamp mechanism. Therefore the system throughput and reliability of data access have improved.

2 PROBLEM DESCRIPTIONS

Data are dynamic changes in a cloud storage environment so the user can update data object in every replica node. Therefore it is necessary to consistently maintain the same data object and provide data accessor with a unified view.

Data consistency models are divided into strong consistency and weak consistency. Strong consistency means to modify data of some replica nodes and apply to all the other replica nodes at the same time, i.e. all replica nodes reach data consistency state at any time (Fekete & Ramamritham 2010). Weak consistency which is also known as eventual consistency refers to it being unnecessary to synchronize every data replicas after modifying some replica node. It can transfer to some replica node then to others asynchronously and realize every replicas data consistency ultimately (Vogels 2009).

Strong consistency could ensure avoiding the collision problem when concurrence update operation, but restrict the reliability, feasibility and tolerance of network partition (Ramakrishnan 2012). It is almost impossible to realize massive replica nodes synchronously by this method, because the request of hardware is relatively high. In addition the request of cloud storage stability and network connectivity is also relatively high. Therefore strong consistency maintenance in cloud storage system is impossible in conclusion.

Weak consistency has a low request to data consistency while improving the reliability of the system and the tolerance of network partition, but weak consistency exits the time window of data inconsistency. Weak consistency is the lowest request of copy algorithm, if ultimate consistency is unsatisfied, the content of the replica would keep the "old" state that leads to failure of data access. Then weak consistency always tries its best to transfer updates quickly among replicas to realize replica ultimate consistency. Weak consistency not only obtains the data which satisfy the practical request but also improve reliability and throughput of the system's aim at some scenarios.

Therefore, we proposed a weak data replica consistency maintenance method which is decentralized, i.e., disperses the "primary replica" update task to every replica node that is in charge of data access and update transmission, then realizes the load balance and improves the reliability of the cloud storage system. The consistency maintenance of data object in this paper involved two kinds of nodes respectively, storage server node and metadata storage node. The metadata storage node store metadata set M which is related to data replica consistency maintenance, there are mainly: the logical name of data object O,

Logical name of data object O	Physical address of metadata node	Maximum timestamp RTS_{max}
Replica numbering: R_i	Physical address of replica node R_i	Timestamp of replica node R_i: RTS_i
Replica numbering: R_2	Physical address of replica node R_2	Timestamp of replica node R_2: RTS_2
......	......	......
Replica numbering: R_n	Physical address of replica node R_n	Timestamp of replica node R_n: RTS_n

Figure 1. The metadata of consistency maintenance.

Replica numbering: R_i	Physical address of replica node R_i	Timestamp rts_i of replica node R_i
Timestamp: rts_1		Write operation: W_1
Timestamp: rts_2		Write operation: W_2
......		......
Timestamp: rts_k		Write operation: W_k

Figure 2. The replica data structure of consistency maintenance.

maximum timestamp RTS_{max} and relevant information of replica node including replica node numbering, physical address and current timestamp rts_i.

In addition, the metadata storage node maintains access queue AQ for every data object O as well, add the read and write requests from client C_i to queue AQ according to the sequence of the metadata storage node, then processes every read and write requests of queue AQ on the basis of FCFS (First Come First Service). Every replica node is equivalent to each other and responds to read and write requests from clients to data object independently. Data structure needs maintaining as follows: replica node numbering, physical address and current timestamp, write operation log when the replica object executea.

3 ALGORITHM DESIGN

The consistency maintenance algorithm will be different when the user request operations are different. The process of consistency maintenance will be explained from read and write operations respectively as follows.

3.1 Write operation

Step 1: Acquire metadata M of data object O by accessing metadata storage node, set the response message counter $k = 0$, select the replica node R_i which respond this write request by executing replica selection algorithm.

Step 2: If $RTS_i < RTS_{max}$, means this replica R_i lack part of write operations, then modify the parameter of replica node selection algorithm and return to step 1.

Step 3: If $RTS_i > RTS_{max}$, means some write operations cannot accomplish all replica nodes, then the operation will be hung up Δ_t time of which $\Delta_t = 4 * (transfertime\_$ max $+dealtime)$, return to step 1.

Step 4: If $RTS_i = RTS_{max}$, then send a message which timestamp is RTS_i to R_i, execute step 4.1; else, set $RTS_i = rts_i$ in metadata node and modify parameter of replica node selection algorithm, return to step 1.

Step 4.1: Find out set of replica node $R = \{R_j | RTS_j = RTS_i\}$ from M, send write request message to R which includes this write operation timestamp RTS_i. When replica node from R receives the message, if $rts_i = RTS_i$, then execute write operation and set $rts_i = rts_i + 1$, send response message to metadata storage node and current timestamp rts_i. Else, send failure message and current timestamp rts_i to metadata storage node.

Step 4.2: Select write operations that timestamp in the range of $[rts_{j+1}, rts_i]$ from write operation log of R_i for every replica node R_j that $RTS_j < RTS_i$. Set $rts_j = rts_i$ and send response message and current timestamp rts_j to metadata storage node after R_j executing corresponding write operations.

Step 4.3: The response message metadata storage node received does following treatment: set corresponding replica R_j timestamp $RTS_j = rts_j$, if $RTS_j = RTS_{max} + 1$, set $k = k + 1$. If $k = N$, update timestamp RTS_{max} of metadata and send write operation success message to data access offering point then notify replica node to delete write operation log and execute next operation; else, execute step 4.2.

3.2 Read operation

The inline equations (equations within a sentence) in the text will automatically be converted to the AMS notation standard.

Step 1: Acquire metadata M of data object O by accessing metadata storage node, execute replica selection algorithm to select replica node R_i which responds to this read request.

Step 2: if $RTS_i < RTS_{max}$, modify the parameter of replica node selection algorithm and return to step 1.

Step 3: Start counter $T = 0$, metadata node send read operation request message which includes timestamp RTS_i to replica node R_i. Compare timestamp after R_i received the read request message. If $rts_i = RTS_i$, return data object O which the client requested and send response message and timestamp rts_i; else, send read operation failure message and current timestamp rts_i to metadata node.

Step 4: If receive response message on $T < \Delta_t$ ($\Delta_t = 2 * (transfertime\_$ max $+dealtime)$), response message successful, end read operation and execute next access request; if type of response message is aborted, set $RTS_i = rts_i$ and modify the parameter of replica node selection algorithm the return to step 1. If it has not received the response message when $T = \Delta_t$, then modify the parameter of replica node selection algorithm and return to step 1.

There are some characteristics of the replica consistency maintenance scheme proposed in this paper: First, it allocated the update task of primary replica to all replica nodes and effectively solves the system bottleneck problem brought by single primary node. Secondly, it can always acquire the replica nodes that store latest data by metadata node accessing when client access data, the feasibility and reliability of data access has improved meanwhile. Thirdly, it unnecessary transmits the latest data to every replica node when data updates while it only needs to transfer corresponding write operation log simply. It can reduce the access delay and network burden effectively.

4 SIMULATION AND PERFORMANCE EVALUATION

The simulation experiment and performance evaluation on the proposed scheme in this paper have been done on the CloudSim (Calheiros et al. 2011). In order to measure the advantages and disadvantages of replication consistency maintenance strategy proposed in this paper, the work done in this article is compared with the Almeida et al. (2013) and Weil et al. (2007), and will now proceed to evaluate the performance of the following three aspects.

4.1 Cost of the consistency maintenance

Assuming the current node in the system comprises N copies. In terms of write operation, there are four message delays, i.e. timestamp request and response delay between the metadata storage nodes to the candidate nodes, write request and the response delay between the metadata storage nodes to the candidate nodes. In addition, two memory accesses delay as well as read and write operations on the replica node. So the resource costs including $2N + 2$ messages and $N + 1$ memory accesses. In terms of read operation, there are two message delays, i.e. read request and the response delay between the metadata storage nodes to the candidate nodes. In addition, one memory access delay of read operation on the replica node. So the resource costs including two messages and one memory access.

In terms of write operation in the primary method, delay includes four message delays, i.e. write request and response delay between client and the master node, write request and response delay between the master node and other replica nodes. In addition, there are two memory access latencies of write operation between the master node and replica node. So the resource costs including $2N$ messages and N memory accesses.

In terms of read operation, the delay includes two message delays, i.e. read request and response delay between client and the master node. In addition, one memory access delay of read operation on the master node. So the resource costs including two messages and one memory access.

In terms of write operation in the primary method, the delay includes four message delays, i.e. write request and response delay between client and the master node, write request and response delay between the master node and other replica nodes. In addition, there are two memory access latencies of write operation between the master node and replica node. So the resource costs including $2N$ messages and N memory accesses.

In terms of read operation, the delay includes two message delays, i.e. read request and response delay between client and the master node. In addition, one memory access delay of read operation on the master node. So the resource costs including two messages and one memory access.

The chain method in Almeida et al. (2013). In terms of write operation, the delay includes $N + 1$ message delays: request delay between client and the first node, transmission delay of write operation and response delay of the end node. In addition, N memory access delays of write operation on the replica node. So the resource costs including $2N$ messages and N memory accesses. In terms of read operation, the delay includes two message delays that read request and response delay between client and the end node. In addition, one memory access delay of read operation on the end node. So the resource costs including two messages and one memory access.

The splay method in Weil et al. (2007). In terms of write operation, the delay includes four message delays: request delay between client and the first node, diffusion delay of write operation and response delay of the end node. In addition, there are two memory access delays of write operation on the replica node. So the resource costs including $2N - 1$ messages and N memory accesses. In terms of read operation, the delay includes two message delays that read request and response delay between client and the end node. In addition, one memory access delay of read operation on the end node. So the resource costs including two messages and one memory access.

4.2 Time delay of access request completion

Time delay of access request completion is defined as the elapsed time that an access request issued until it receives the response of the replica node. Assuming each client submits the next access request only after the client receives the response message.

Under the consistency maintenance strategy proposed in this paper, the average time delay to complete a corresponding number of access requests is less than primary, chain and splay strategy. The average completion time corresponding to our scheme, splay, primary and chain are 113.3 ms, 150 ms, 186.7 ms and 326 ms respectively every time. This is because the consistency maintenance strategy designed in the paper puts the function of the master node into each replica node, so that each replica node can independently respond to the requests of read and write operation. This is very helpful to select the optimal replica node to respond to the access request under the consideration of the load conditions of replica node and the current status of replica node in the network, which reduces the average time delay of data access.

4.3 Access request throughput

The changed situation of throughput with the increase in the proportion of write request to total request under a different number of replica. The simulation result shows that with the increase of the proportion of write requests, various throughput consistency maintenance strategies are reduced, but throughput is higher using the strategy designed in this paper than using other strategies. Because the strategy designed in this paper allows that each replica node can autonomously respond to read and write requests, you can choose the best replica node for each access request, which would effectively reduce the average transmission delay of access request to replica node and improve the throughput of the access request.

By comparing the throughput of a different number of replica nodes with the number of replica node increase, the superiority of consistency maintenance strategy designed in this paper becomes more significant. When the replica number is 9, the average total throughput under this paper scheme, splay, primary and chain are 246, 150, 128 and 103. This is because the more the number of replicas, the more nodes can be selected to respond to the request, the average transmission delay and access delay are reduced, so the time delay of read operation is reduced. As for write operation, only members of the multicast increased, which has little effect on time delay as well as throughput of the access request.

For primary, splay algorithm, since the number of replica nodes responding to the read and write requests is fixed, there is no impact to add replica node. However, multicast members will increase for the write operation. It can only increase the completion time of write request, so the throughput of the access request will also reduce.

As for the chain algorithm, the increase of replica node has no effect on read request, but it will dramatically increase the transmission time between the various replica nodes of write operation, it also will lead to the throughput's dramatically decrease of the access request.

5 CONCLUSION

With the rapid development of network technology, a variety of application scenarios based on the Internet has emerged in an endless stream; health care is one of the hot areas. Healthcare data networking environments have the characteristics of being massive, heterogeneous, difficult to manage, etc, and the cloud storage system can effectively support data storage business with high-reliability, high-scalability and high-capacity. With a simple "pay for need" mode, it provides efficient, reliable and secure data access services for customers. Data replica management is an important mean for improving the efficiency of the cloud storage system. The optimized data replica management mechanism is critical for building an efficient and rational cloud storage system, further improving its scalability, reliability and shorting the access times of users especially for the cloud storage system under health-care IoT.

In the cloud storage, for a large number of users of data reading and writing requests, this paper proposes and implements a simulation of non-centralized and weak-consistency maintenance methods of data replica, so that each replica node can independently respond to the reading and writing requests, using the timestamp mechanisms to resolve data access conflicts, to achieve load balancing and effectively improve the throughput of data access. Based on CloudSim, it takes a performance evaluation and simulation implementation for the mechanism proposed in this paper, it suggests the consistency maintenance strategy by comparing it to other literature. By analyzing the time delay of completing access request and the throughput of access request, the result shows that this mechanism is feasible and effective. The focus of our research in the future is to achieve prototype implementation based on the cloud-based platform of Northeastern University to validate and improve the practicality of work proposed in this paper.

ACKNOWLEDGMENT

This work is supported by the National Science Foundation for Distinguished Young Scholars of China under Grant No. 61225012 and No. 71325002; the Specialized Research Fund of the Doctoral Program of Higher Education for the Priority Development Areas under Grant No. 20120042130003; the Fundamental Research Funds for the Central Universities under Grant No. N110204003 and No. N120104001.

REFERENCES

[1] Almeida S, Leitão J, Rodrigues L. 2013. ChainReaction: a causal+ consistent datastore based on chain replication. *Proceedings of the 8th ACM European Conference on Computer Systems.* Prague, Czech.: 85–98.

[2] AL-Mistarihi H H E, Yong C H. 2009. On fairness, optimizing replica selection in data grids. *IEEE Transactions on Parallel and Distributed Systems* 20(8): 1102–1111.

[3] Calheiros R N, Ranjan R, Beloglazov A, et al. 2011. CloudSim: a toolkit for modeling and simulation of cloud computing environments and evaluation of resource provisioning algorithms. *Software: Practice and Experience* 41(1): 23–50.

[4] Chen Xin, Ren Shansi, Wang Haining, Zhang Xiaodong. 2005. SCOPE: scalable consistency maintenance instructured P2P systems. *24th Annual Joint Conference of the IEEE Computer and Communications Societies* 3: 1502–1513.

[5] Dirk Dullmann, Wolfgang Hoschek, Javier Jaen-Martinez, Ben Segal. 2001. Models for replica synchronisation and consistency in a data grid. *High Performance Distributed Computing* 67–75.

[6]   Fekete A D, Ramamritham K (École Polytechnique). 2010. *Consistency models for replicated data*. Berlin: Springer Berlin Heidelberg: 1–17.

[7]   Lin Chuang, Hu Jie, Kong Xiangzhen. 2012. Survey on Models and Evaluation of Quality of Experience. *Chinese Journal of Computers* 35(1): 1–15.

[8]   Liu Tiantian, Li Chao, Hu Qingcheng.et al. 2011. Multiple-Replicas Management in the Cloud Environment. *Journal of Computer Research and Development* 48(Suppl.): 254–260.

[9]   Lv Yingnan. 2011. Stuy on eventually consistency strategy of distributed data. *Huazhong University of Science and Technology*.

[10] MacCormick J, Thekkath C A, Jager M, et al. 2008. Niobe: A practical replication protocol. *ACM Transactions on Storage* 3(4): 1–43.

[11] Ramakrishnan R. 2012. CAP and cloud data management. *IEEE Computer Society* 45(2): 43–49.

[12] Ren Xunyi, Wang Ruchuan, Kong Qiang. 2008. Efficient model for replica consistency maintenance in data grids. *Computer Science and its Applications* 159–162.

[13] Vogels W. 2009. Eventually consistent. *Communications of the ACM* 52(1): 40–44.

[14] Wang Yu, Zhao Yuelong, Hou Fang. 2010. Replica consistency maintenance strategy for P2P storage system. *Computer Engineering* 36(1): 27–29.

[15] Weil S A, Leung A W, Brandt S A, et al. 2007. Rados: a scalable, reliable storage service for petabyte-scale storage clusters. *Proceedings of the 2nd international worshop on Petascale data storage*. Nevada, USA: 35–44.

[16] Wu Changze, Tian Dong, Wu Zhongfu. 2009. Strategy of keeping replica consistency with dyllalnic voting mechanism in data grid. *Computer Science* 36(4): 172–189.

[17] Yang Lei, Li Renfa, Hu Yiming, Li Kenli. 2010. Node heterogeneous degree-based consistency maintenance method for unstructured overlay networks. *Journal on Communications* 31(10): 180–189.

Future Communication, Information and Computer Science – Zheng (Ed.)
© *2015 Taylor & Francis Group, London, 978-1-138-02653-7*

An effective game-theory based single-user handoff decision scheme in next generation internet

X. Wang
College of Information Science and Engineering, Northeastern University, Shenyang, China

H. Cheng
School of Computing & Mathematical Sciences, Liverpool John Moores University, Liverpool, UK

J. Zhang & M. Huang
College of Information Science and Engineering, Northeastern University, Shenyang, China

ABSTRACT: In Next Generation Internet (NGI), access networks become heterogeneous and diversified. Quality of Service (QoS) handoff with always best connected (ABC) support is essential. Meanwhile, the handoff should consider the profits of the user and the network service provider at the same time. In this paper, an effective game-theory based single-user handoff decision scheme is proposed with ABC support. It takes comprehensive factors into account, i.e., the status of access networks, the application QoS requirement, the preference of the user over the coding system of the access network, the preference of the user over the network service provider, the velocity and the residual electric quantity of the terminal. Based on the game theory, the scheme aims to find the Pareto-optimal solution under Nash equilibrium between the user utility and the network service provider utility. The simulation experiments demonstrate that the proposed scheme is effective.

Keywords: handoff; QoS; always best connected; next generation internet

1 INTRODUCTION

NGI will most probably develop into an integrated network environment which consists of multiple heterogeneous sub-networks. Therefore, it is possible that a user will face multiple heterogeneous and diversified access networks which are all available at the same time. The user needs to keep the ABC (Gustafsson & Jonsson, 2003) connectivity with the NGI in both the session initialization and on-going period, thereby achieving the QoS guaranteed global seamless roaming (Theodore et al. 2002).

The handoff schemes can be classified for vertical handoff and horizontal handoff (McNair & Zhu 2004). Under the commercialized operational environment of networks, ABC has to be achieved by considering the benefits at both the user side and the network service provider side (Bob et al. 2003). Another problem is that the application QoS requirements are very fuzzy (Shi et al. 2007) and cannot be accurately described. Moreover, supporting ABC should be able to avoid the ping-pong effect (Kim et al. 2007).

So far quite a few works have been done. In Navarro et al. (2008), a vertical handoff decision algorithm was proposed based on MarKov Process. In Chen et al. (2004), a flexible handoff decision scheme was proposed which consists of ranking phase and normal phase. In Hasswa et al. (2006), the vertical handoff decision was evaluated by the handoff cost function and the handoff valve function. In Song & Abbas (2008), a vertical handoff decision method based on QoS negotiation was proposed for the heterogeneous wireless network environment. In Shi & Chen (2008), a vertical handoff decision method was proposed based on the evaluation of the QoS satisfaction degree. In Liu et al. (2008), a speed sensitive vertical handoff decision algorithm was proposed which can adapt to different node speeds. In Liu et al. (2007), a pre-decision assisted vertical handoff decision algorithm was proposed.

In this paper, we propose a handoff decision scheme with the ABC supported. By considering multiple decision factors, the scheme tries to find an optimal handoff solution of assigning a mobile terminal to the most suitable access network based on game theory (Osborne 2004).

2 MODELS

2.1 *Application types, QoS requirements and their fuzzy degrees*

Based on the DiffServ idea (Blake 1998), assuming that there are K different application types in NGI, the set of application types is represented as $ATS = \{AT_1, \ldots, AT_K\}$. In this paper, six QoS parameters are considered, namely, bandwidth BW, delay DL, delay jitter JT, bit error rate BE, packet loss rate PL and security level SL. The mapping between the application types and the values of the QoS parameters is shown in (1). For the application type AT_i, the fuzzy degrees of the six QoS parameters are respectively defined in (2–7).

$$AT_i \rightarrow <[BW_i^l, BW_i^h], [DL_i^l, DL_i^h], [JT_i^l, JT_i^h], \\ [BE_i^l, BE_i^h], [PL_i^l, PL_i^h], [SL_i^l, SL_i^h]> \quad (1)$$

$$FB_i = \left(BW_i^h - BW_i^l\right) / \left(BW_i^h + BW_i^l\right) \quad (2)$$

$$FD_i = \left(DL_i^h - DL_i^l\right) / \left(DL_i^h + DL_i^l\right) \quad (3)$$

$$FJ_i = \left(JT_i^h - JT_i^l\right) / \left(JT_i^h + JT_i^l\right) \quad (4)$$

$$FE_i = \left(BE_i^h - BE_i^l\right) / \left(BE_i^h + BE_i^l\right) \quad (5)$$

$$FL_i = \left(PL_i^h - PL_i^l\right) / \left(PL_i^h + PL_i^l\right) \quad (6)$$

$$FS_i = \left(SL_i^h - SL_i^l\right) / \left(SL_i^h + SL_i^l\right) \quad (7)$$

2.2 Access network model

The following network parameters are described.

j, the access network No., $(1 \leq j \leq M)$, where M is the total number of the access networks.

PI_j, the ID of the provider of access network j, $PI_j \in PIS$, where $PIS = \{PI_1, PI_2, \ldots, PI_{|PIS|}\}$.

TI_j, the type ID of the access network j, $TI_j \in TIS$, where $TIS = \{TI_1, TI_2, \ldots, TI_{|TIS|}\}$.

Each type of access network TI_j has its coverage area CA_j and the highest mobility speed MV_j.

CS_j, the set of coding systems supported by the access network j, $CS_j \subseteq CIS$.

NAS_j, the set of application types supported by the access network j, $NAS_j \subseteq ATS$.

FR_j, the frequency range supported by j.

TB_j, the total bandwidth of the access network j.

AB_j, the current available bandwidth of j; $AB_j^{\min}$ the lower bound of the residual bandwidth of j.

TP_j, the lowest signal strength emitted by j.

ct_j, the cost of using one unit bandwidth per unit time of the access network j.

pr_{ij}, the sale price of one unit bandwidth per unit time of j for the AT_i type application. It includes the basic price pb_j and adjustive price pf_{ij}, shown in (8).

$$pr_{ij} = pb_j + pf_{ij} \quad (8)$$

pb_j is determined by the access network provider and divided into the low-rate pricing interval, the flat-rate pricing interval and the high-rate pricing interval (Andrew 1999). The load rate η_j of j is calculated as follows.

$$\eta_j = 1 - AB_j / TB_j \quad (9)$$

$$pb_j = \begin{cases} pb_j^{\min} & \eta_j < \eta_j^{\min} \\ A / \left(1 + \alpha \cdot \eta_j^{-\beta}\right) & \eta_j^{\min} \leq \eta_j \leq \eta_j^0 \end{cases} \quad (10)$$

$$pb_j = \begin{cases} pb_j^{\max} & \eta_j^{\max} < \eta_j < \eta_j^1 \\ B \cdot \left(2 - e^{-\delta \cdot (\eta_j - \eta_j^0)^2}\right) & \eta_j^0 \leq \eta_j \leq \eta_j^{\max} \end{cases} \quad (11)$$

When $\eta_j \geq \eta_j^1$, pb_j is determined using the auction (Patrick & Bruno 2006). In (10) and (11), η_j^0 and η_j^1 are predetermined empirical values, $0 < \eta_j^0 < \eta_j^1 < 1$. $pb_j^{\min}$, pb_j^0 and $pb_j^{\max}$ are the starting price of the low-rate pricing interval, the starting price and the highest price of the flat-rate pricing interval respectively. We have $pb_j^{\min} \leq pb_j^0 \leq pb_j^{\max}$, where $\eta_j^{\min}$, η_j^0, $\eta_j^{\max}$ are the corresponding load rates. When $\eta_j^{\min} \leq \eta_j \leq \eta_j^0$, pb_j is similar as semi-Cauchy distribution; when $\eta_j^0 \leq \eta_j \leq \eta_j^{\max}$, pb_j is similar as semi-Normal distribution (Mordeson & Nair 2001). Therefore, we can calculate A, α, B and δ. pf_{ij} is determined by AT_i to reflect the effect on the sale price of the QoS requirement parameters except the bandwidth.

$$AT_i \rightarrow pf_{ij} \quad (12)$$

2.3 Terminal model

The following terminal parameters are described.

t, the terminal No..

TAS_t, the set of application types supported by the terminal t, $TAS_t \subseteq ATS$.

MCS_t, the set of the coding systems supported by the terminal t, $MCS_t \subseteq CIS$.

WF_t, the working frequency of the terminal t.

RS_t, the lowest signal strength received by t.

CV_t, the current velocity of the terminal t.

RC_t, the current residual electric quantity of t.

C_{tc}, the current residual electric quantity valve of t.

HP_{ti}, the unit time unit bandwidth price that a terminal user is willing to pay for AT_i application.

$PC_t = \{PC_{t1}, \ldots, PC_{tq}\}$, the preference sequence of the terminal user over the access network coding systems, $PC_{tp} \in CIS$, $1 \leq q \leq |CIS|$, $1 \leq p \leq q$.

$PP_t = \{PP_{t1}, \ldots, PP_{tm}\}$, the preference sequence of the terminal user over the access network providers, $PP_{tn} \in PIS$, $1 \leq m \leq |PIS|$, $1 \leq n \leq m$.

2.4 The user QoS and price satisfaction degrees

If the QoS provided by j for the AT_i type application on terminal t is $< bw_{t,j}, dl_{t,j}, jt_{t,j}, be_{t,j}, pl_{t,j}, sl_{t,j} >$, the user's evaluations on the actual QoS are separately calculated as follows.

$$EB_{t,i} = \begin{cases} 0 & bw_{t,j} < BW_i^l \\ \varepsilon & bw_{t,j} = BW_i^l \\ \left(\dfrac{bw_{t,j} - BW_i^l}{BW_i^h - BW_i^l}\right)^\lambda & BW_i^l < bw_{t,j} < BW_i^h \\ 1 & bw_{t,j} \geq BW_i^h \end{cases} \quad (13)$$

$$ED_{t,j} = \begin{cases} 0 & dl_{t,j} > DL_i^h \\ \varepsilon & dl_{t,j} = DL_i^h \\ 1-e^{-\left(\frac{DL_i^h - dl_{t,j}}{DL_i^h - DL_i^l}\right)^k} & DL_i^l < dl_{t,j} < DL_i^h \\ 1 & dl_{t,j} \le DL_i^l \end{cases} \tag{14}$$

$$EJ_{t,j} = \begin{cases} 0 & jt_{t,j} > JT_i^h \\ \varepsilon & jt_{t,j} = JT_i^h \\ 1-e^{-\left(\frac{JT_i^h - jt_{t,j}}{JT_i^h - JT_i^l}\right)^k} & JT_i^l < jt_{t,j} < JT_i^h \\ 1 & jt_{t,j} \le JT_i^l \end{cases} \tag{15}$$

$$EE_{t,j} = \begin{cases} 0 & be_{t,j} > BE_i^h \\ \varepsilon & be_{t,j} = BE_i^h \\ 1-e^{-\left(\frac{BE_i^h - be_{t,j}}{BE_i^h - BE_i^l}\right)^k} & BE_i^l < be_{t,j} < BE_i^h \\ 1 & be_{t,j} \le BE_i^l \end{cases} \tag{16}$$

$$EL_{t,j} = \begin{cases} 0 & pl_{t,j} > PL_i^h \\ \varepsilon & pl_{t,j} = PL_i^h \\ 1-e^{-\left(\frac{PL_i^h - pl_{t,j}}{PL_i^h - PL_i^l}\right)^k} & PL_i^l < pl_{t,j} < PL_i^h \\ 1 & pl_{t,j} \le PL_i^l \end{cases} \tag{17}$$

$$ES_{t,j} = \begin{cases} 0 & sl_{t,j} < SL_i^l \\ \varepsilon & sl_{t,j} = SL_i^l \\ \left(\frac{sl_{t,j} - SL_i^l}{SL_i^h - SL_i^l}\right)^k & SL_i^l < sl_{t,j} < SL_i^h \\ 1 & sl_{t,j} \ge SL_i^h \end{cases} \tag{18}$$

The user's satisfaction degrees over the corresponding QoS parameters and the overall QoS satisfaction degree are calculated as follows.

$$SB_{t,j} = \left(\alpha_1 \cdot EB_{t,j}\right)/\left(\beta_1 \cdot FB_i + 1\right) \tag{19}$$

$$SD_{t,j} = \left(\alpha_2 \cdot ED_{t,j}\right)/\left(\beta_2 \cdot FD_i + 1\right) \tag{20}$$

$$SJ_{t,j} = \left(\alpha_3 \cdot EJ_{t,j}\right)/\left(\beta_3 \cdot FJ_i + 1\right) \tag{21}$$

$$SE_{t,j} = \left(\alpha_4 \cdot EE_{t,j}\right)/\left(\beta_4 \cdot FE_i + 1\right) \tag{22}$$

$$SL_{t,j} = \left(\alpha_5 \cdot EL_{t,j}\right)/\left(\beta_5 \cdot FL_i + 1\right) \tag{23}$$

$$SS_{t,j} = \left(\alpha_6 \cdot ES_{t,j}\right)/\left(\beta_6 \cdot FS_i + 1\right) \tag{24}$$

$$\begin{aligned} SQ_{t,j} = {} & \omega_1 \cdot SB_{t,j} + \omega_2 \cdot SD_{t,j} + \omega_3 \cdot SJ_{t,j} + \\ & \omega_4 \cdot SE_{t,j} + \omega_5 \cdot SL_{t,j} + \omega_6 \cdot SS_{t,j} \end{aligned} \tag{25}$$

Here, k is a constant. ε is a pure decimal far less than 1. $\alpha_1, \ldots, \alpha_6, \beta_1, \ldots, \beta_6$ are dimensional adjusting coefficients. $\omega_1, \ldots, \omega_6$ are weighting systems for reflecting the relative importance of the corresponding QoS parameters. More, $\sum_{n=1}^{6} \omega_n = 1$.

If the sale price of one unit bandwidth per unit time of the access network j for the AT_i type application is pr_{ij}, the price satisfaction degree of the user over the access network is defined as follows.

$$SP_{t,j} = \begin{cases} 1 & pr_{ij} \le HP_{ti} \\ 0 & \text{otherwise} \end{cases} \tag{26}$$

2.5 The user preference satisfaction degrees

The user preference satisfaction degree over the coding system of access network and the access network provider are separately defined as follows.

$$ST_{tj} = \begin{cases} 1/x^2 & CI_{tj} \in PC_t \\ 0 & \text{otherwise} \end{cases} \quad SR_{tj} = \begin{cases} 1/y^2 & PI_{tj} \in PP_t \\ 0 & \text{otherwise} \end{cases} \tag{27}$$

where $CI_{tj} \in CIS$ is the coding system ID used by the terminal t user after the handoff into the access network j, x is the sequence number of CI_{tj} in PC_t, $1 \le x \le |PC_t|$, $PI_{tj} \in PIS$ is the ID of the access network j's provider which the terminal t user makes a handoff into, and y is the sequence number of PI_{tj} in PP_t, $1 \le y \le |PP_t|$.

2.6 The suitability degrees of the velocity and the battery electric quantity

The suitability degrees of the access network j to the terminal t user's current speed and the residual battery electric quantity of the terminal t are separately defined as follows.

$$SM_{tj} = \begin{cases} 1/z^2 & CV_t \le MV_j \\ 0 & \text{otherwise} \end{cases} \quad SW_{tj} = \begin{cases} 1/d^2 & RC_t \le C_{fc} \\ 1 & \text{otherwise} \end{cases} \tag{28}$$

where z is the sequence number of TI_j when all the access network types are sorted by the size of the coverage area (from big to small), d has a same meaning (but from small to big), $1 \le z, d \le |TIS|$.

3 SINGLE-USER HANDOFF DECISION SCHEME

3.1 Gaming analysis

A two-person game is used and the two players are the access network j and the terminal t user. j has two strategies a_1 and a_2. a_1 represents that the access network agrees to admit t while a_2 represents the opposite. t has two strategies b_1 and b_2. b_1 represents that the terminal is willing to be admitted by j while b_2 represents the opposite. The payoff matrices of j and t are denoted as NP and TP, defined in (29–30).

$$NP = \begin{bmatrix} pr_{ij} - ct_j & pr_{ij} - ct_j \\ -\mu \cdot (pr_{ij} - ct_j) & -(pr_{ij} - ct_j) \end{bmatrix} \tag{29}$$

$$TP = \begin{bmatrix} HP_{ti} - pr_{ij} & -\mu \cdot (HP_{ti} - pr_{ij}) \\ HP_{ti} - pr_{ij} & -(HP_{ti} - pr_{ij}) \end{bmatrix} \tag{30}$$

where the top and bottom row vectors represent the two strategies a_1 and a_2 of j, respectively, and the left and right column vectors represent the strategies b_1 and b_2 of t, respectively. μ is a penalty factor (Cao et al 2002) denoting that if the access network refuses the terminal who is willing to join it, the negative effect will be brought to the user when it makes the decision about the future handoff. μ is greater than 1.

If $np_{c^*d^*} \geq np_{c^*d}, tp_{c^*d^*} \geq tp_{cd^*}$, then $\{a_{c^*}, b_{d^*}\}$ $(c^*, d^*, c, d = 1, 2)$ is the strategy pair to achieve the Nash equilibrium among the gains of the respective parties.

3.2 Utility

A weight coefficient array $\Lambda = [\lambda_1 \ \lambda_2 \ \lambda_3 \ \lambda_4 \ \lambda_5 \ \lambda_6]$ is defined to reflect weights of the factors in selecting the access network for handoff. $\sum_{n=1}^{6} \lambda_n = 1$. These factors are the application QoS requirement, the sale price of the unit bandwidth per unit time and the cost that the user is willing to pay, the preference over the coding system of access network, the preference over the access network service provider, the velocity of the terminal, and the residual electric quantity of the terminal. They can be empirically determined or calculated by AHP method. The adjustive coefficient Ω is used to reflect the effect of Nash equilibrium on the utility. The evaluation matrix $G_{t,j} = [SQ_{tj} \ SP_{tj} \ ST_{tj} \ SR_{tj} \ SM_{tj} \ SW_{tj}]^T$ is denoted.

$$\Omega = \begin{cases} 1 & \text{Nash equilibirum} \\ <1 & \text{non-Nash equilibirum} \end{cases} \quad (31)$$

The user utility and the network service provider utility are defined below.

$$uu_{t,j} = \begin{cases} \Omega \cdot \Lambda \cdot G_{t,j} \cdot (HP_{ti} - pr_{ij})/HP_{ti} & \text{the terminal } t \text{ enters} \\ & \text{the access network } j \text{ due to handoff} \\ 0 & \text{otherwise} \end{cases} \quad (32)$$

$$nu_{t,j} = \begin{cases} \Omega \cdot \Lambda \cdot G_{t,j} \cdot (pr_{ij} - ct_j)/pr_{ij} & \text{the terminal } t \text{ enters} \\ & \text{the access network } j \text{ due to handoff} \\ 0 & \text{otherwise} \end{cases} \quad (33)$$

where uu_{tj} and nu_{tj} represent the user utility and the network service provider utility when the access network j is selected and the application type of mobile terminal t is AT_i.

3.3 Problem description

Terminal t is covered by M access networks simultaneously. The handoff problem is to find a suitable access network for t and determine the appropriate service level which matches the application requirements and the user cost expectation. $< AN_t, sl >$ denotes the handoff solution for t, that is, AN_t is selected as the access network and sl is the service level provided by AN_t. The objectives of our scheme are formulated as below.

$$\text{Maximize } uu_{t,j}, nu_{t,j}, \left(uu_{t,j} + nu_{t,j}\right) \quad (34)$$

s.t. satisfying the QoS requirements of the user.

3.4 The handoff procedure

Step 1: Initialize $AN_t = 0$ and $sl = 0$.

Step 2: Initialize the final handoff solution as $AN_{final} = 0$, $sl_{final} = 0$ and $U(AN_t, sl) = MAXIMUM$.

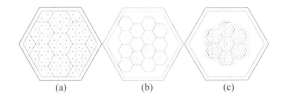

Figure 1. Topology 1 of 82 nodes (a), Topology 2 of 66 nodes (b) and Topology 3 of 107 nodes (c).

Step 3: Check the candidate solution $< AN_t, sl >$ by the following criteria:

Step 3.1: Can the price of this solution meet the user's requirement? If not, turn to Step 6;

Step 3.2: Can the access network AN_t support the user's speed? If not, turn to Step 6;

Step 3.3: Can the access network AN_t support the user's application type? If not, turn to Step 6;

Step 3.4: Do the mobile terminal t and the access network AN_t use the same working frequency? If not, turn to Step 6;

Step 3.5: Can the mobile terminal support the coding system of AN_t? If not, turn to Step 6;

Step 3.6: If AN_t accepts the mobile terminal t, will the residual network bandwidth be less than the lower bound? If yes, turn to Step 6.

Step 4: Let the mobile terminal and the access network perform the two-person game. The game analysis is made and the game factor Ω is updated.

Step 5: Calculate $U(AN_t, sl) = (1/uu_{t,AN_t}) + (1/nu_{t,AN_t})$. If $U(AN_t, sl) < U(AN_{final}, sl_{final}), AN_{final} = AN_t$ and $sl_{final} = sl$, $U(AN_{final}, sl_{final}) = U(AN_t, sl)$; otherwise, keep AN_{final}, sl_{final}, and $U(AN_{final}, sl_{final})$ unchanged.

Step 6: $sl = sl + 1$.

Step 7: If $sl < |SL|$, turn to Step 3.

Step 8: $AN_t = AN_t + 1$, $sl = 1$.

Step 9: If $AN_t < M$, turn to Step 3; otherwise, the algorithm ends and outputs $< AN_{final}, sl_{final} >$.

4 PERFORMANCE EVALUATION

The proposed handoff decision scheme is implemented through simulation in network simulator 2 (NS2). In the simulation experiments, we have used three different hexagon cellular topologies as shown in Figure 1.

Topology 1 has deployed four layers of networks. In most areas of topology 2, three layers of networks are deployed whilst four layers of networks are deployed in the simulated hot spots. Topology 3 has deployed five layers of networks.

For comparison purposes, we have chosen two benchmark algorithms. One is the handoff decision scheme based on the use of multiple attribute decision making (MADM) method: VIKOR for vertical handoff decision (Medina et al. 2009). Another is the handoff decision scheme based on utility and game-theory (Chang et al. 2001). We name our proposed

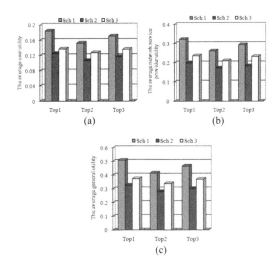

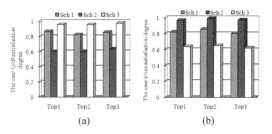

(a)　　　　　　　(b)

Figure 3. Comparisons of the user QoS satisfaction degree (a) and the user price satisfaction degree (b).

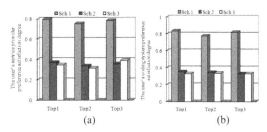

(a)　　　　　　　(b)

Figure 4. Comparisons of the user's service provider preference satisfaction degree (a) and the user's coding system preference satisfaction degree (b).

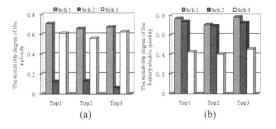

(a)　　　　　　　(b)

Figure 5. Comparisons of the suitability degree of the velocity (a) and the battery electric quantity (b).

Figure 2. Comparisons of the average user utility (a), the average network service provider utility (b) and the average general utility (c).

scheme as Scheme 1, the VIKOR based as Scheme 2, and the utility and game-theory based as Scheme 3. Under topologies 1-3, these three schemes are all executed 500 times. Under each topology, we compare these three schemes in terms of the average value for each performance metric.

4.1 Utility

In Figure 2, since Scheme 2 only considers QoS parameters, price, network load, and network coverage areas, its utility is low. And Scheme 3 has considered the user velocity additionally and it achieves a slightly better utility. However, it has not considered the user preferences and the residual electric quantity of the terminal. Thus, its utility is lower than Scheme 1. Due to that topology 2 has deployed less layers of networks, the overall three average utility values in topology 2 are the lowest in these three topologies for all the schemes.

4.2 The user QoS satisfaction degree and the user price satisfaction degree

Figure 3(a) shows the performance of the schemes is basically the same over the three different topologies. Scheme 3 can achieve the best user QoS satisfaction degree because it does not consider the price. As for Scheme 2, it considers the price and tends to select the cheapest service level. Therefore, it has the lowest user QoS satisfaction degree. In Scheme 1, multiple decision factors have been considered comprehensively, so it can make a good trade off. Figure 3(b) shows these schemes have similar performance over the three topologies. Scheme 2 achieves the best user price satisfaction degree aiming at the cheapest price. Due to that Scheme 3 has not considered the price factor, it selects the handoff strategies with better QoS satisfaction degrees but lower price satisfaction degrees.

4.3 The user preference satisfaction degrees

Figure 4 shows Scheme 1 achieves much better performance than the other two schemes, attributing to considering these two user preferences. All the three schemes perform worst on topology 2 among the three topologies.

4.4 The suitability degrees of the velocity and the battery electric quantity

Figure 5 shows Scheme 1 can satisfy the requirements of the users with low battery electric quantities. Scheme 2 has the worst suitability degree of velocity because it has not considered the velocity at all. However, it selects the access network with small coverage area for the terminal with low battery electric quantity. Scheme 3 achieves good results in terms of the suitability degree of velocity, but it has low suitability degree on this metric due to ignoring the effect of the battery electric quantity.

5 CONCLUSIONS

In this paper, an effective ABC supported single-user handoff decision scheme is proposed based on the game theory. It is compared and evaluated through simulation experiments in NS2. We have compared it with two other schemes which have used different decision methods and the results show that the proposed one is effective. Currently, we are extending and evaluating the ABC supported handoff decision schemes for multiple users based on some novel swarm intelligence techniques, e.g., free search algorithm and quantum-inspired immune clonal algorithm. These efforts will help to further evaluate the performance of our proposed models. The future work will also focus on the practical applications of the proposed models and schemes.

ACKNOWLEDGMENT

This work is supported by the National Science Foundation for Distinguished Young Scholars of China under Grant No. 61225012 and No. 71325002; the Specialized Research Fund of the Doctoral Program of Higher Education for the Priority Development Areas under Grant No. 20120042130003; the Fundamental Research Funds for the Central Universities under Grant No. N110204003 and No. N120104001.

REFERENCES

[1] A. Hasswa, N. Nasser, and H. Hassanein. 2006. Tramcar: a context-aware cross-layer architecture for next generation heterogeneous wireless networks. *Proc. ICC'06*, Istanbul, Turkey.

[2] B. Bob, D. Vasilios, and H. Oliver. 2003. A market managed multi-service internet. *Computer Communications*, 2003, 26(4): 404–414.

[3] B. Theodore, G. Konstantinos, P. Christos, et al. 2002. Global roaming in next-generation networks. *IEEE Communications Magazine*, 2002, 40(2): 145–151.

[4] C. Chang, T. Tsai, and Y. Chen. 2009. Utility and game-theory based network selection scheme in heterogeneous wireless networks. *Proc. IEEE Wireless Communications and Networking Conference*, Budapest.

[5] E. Gustafsson and A. Jonsson. 2003. Always best connected. *IEEE Wireless Communications*, 2003, 10(1): 49–55.

[6] E. Stevens Navarro, Y. Lin, and V. Wong. 2008. An MDP-based vertical handoff decision algorithm for heterogeneous wireless networks. *IEEE Transactions on Vehicular Technology*: 1243–1254.

[7] J. McNair and F. Zhu. 2004. Vertical handoffs in fourth-generation multinetwork environments. *IEEE Wireless Communications*, 2004, 11(3): 8–15.

[8] J. N. Mordeson and P. S. Nair. 2001. *Fuzzy Mathematics: An Introduction for Engineers and Scientists*, 2nd edition. Physica-Verlag HD.

[9] J. Gallardo Medina, U. Pineda Rico, and E. Stevens Navarro. 2009. VIKOR method for vertical handoff decision in beyond 3G wireless networks. *Proc. 6th International Conference on Electrical Engineering, Computing Science and Automatic Control*, Toluca.

[10] L. J. Chen, T. Sun, and B. Chen, et al. 2004. A smart decision model for vertical handoff. *Proc. ANWIRE'04*, Athens, Greece.

[11] M. J. Osborne. 2004. *An Introduction to Game Theory*. Oxford University Press.

[12] M. Patrick and T. Bruno. 2006. Pricing the Internet with multibid auctions. *IEEE/ACM Transactions on Networking*, 2006, 14(5): 992–1004.

[13] M. Liu, Z. C. Li, and X. B. Guo. 2008. A speed sensitive vertical handoff algorithm. *Acta Electronica Sinica*, 2008, 30(6): 1198–1201.

[14] O. Andrew. 1999. Paris metro pricing: The minimalist differentiated services solution. *Proc. IEEE/IFIP IWQoS'99*, Basking Ridge, NJ, USA.

[15] Q. Y. Song and J. Abbas. 2008. A quality of service negotiation-based vertical handoff decision scheme in heterogeneous wireless systems. *European Journal of Operational Research*, 2008, 191: 1059–1074.

[16] S. Blake. 1998. An architecture for differentiated services. *IETF2475*, December, 1998.

[17] X. Cao, H. Shen, R. Milito R, et al. 2012. Internet pricing with a game theoretical approach: concepts and examples. *IEEE/ACM Transactions on Networking*, 2002, 10(2): 208–216.

[18] X. Liu, L. G. Jiang, and C. He. 2007. A novel vertical handoff algorithm based on fuzzy logic in aid of pre-decision method. *Acta Electronica Sinica*, 2007, 30(10): 1989–1993.

[19] Y. Shi and S. Z. Chen. 2008. A QoS satisfaction degree evaluation based vertical handoff decision method for mSCTP. *Microel Ectronics & Computer*, 2008, 25(2): 9–13.

[20] Y. Shi, S. Z. Chen, Y. H. Li, et al. 2007. An application-oriented cooperative vertical handoff decision method for multi-interface mobile terminals. *Proc. ITST'07*, Sophia Antipolis, France.

Future Communication, Information and Computer Science – Zheng (Ed.)
© *2015 Taylor & Francis Group, London, 978-1-138-02653-7*

Spatial and activity data-collection and visualization in WWW environment

J. Masner, M. Stočes & J. Vaněk
Department of Information Technologies Czech University of Life Sciences Prague (CULS), Prague, Czech Republic

ABSTRACT: The paper addresses the issue of spatial and activity data in the www environment (in a web application). The acquired knowledge and principles were used to evaluate the spatial activity of forest game in the selected regions of the Czech Republic (Europe). The data base for the analysis and processing consists of large data sets obtained on the basis of mutual co-operation among the Faculties of the Czech University of Life Sciences (CULS) and other external entities. The analytical and software solution was developed by the Department of Information Technologies of the Faculty of Economics and Management (CULS) that administers and operates the system. The solution addresses the collection, validation, and various types of visualization and interpretation of activity data in a web browser.

Keywords: web application; spatial data; Google maps; activity data; GPS, GSM; Czech Republic

1 INTRODUCTION

Owing to the development of modern information technologies it is technologically relatively very easy and inexpensive to obtain spatial data and transmit them online for processing. Spatial data can be used in a number of industries, from security technologies and logistic systems to agriculture and environmental protection. Spatial data are also used in the research.

In most cases, they are collected via the GNSS (Global Navigation Satellite System) radio positioning system. At the moment, there are two functional systems available to the public: GPS (*Global Positioning System*) that has been established and controlled by the US Department of Defence and GLONASS (*Globalnaya navigatsionnaya sputnikovaya sistema*) controlled by the Russian air force. There are other systems under preparation – Chinese BeiDou and European Galileo. Data can be acquired by using many mobile devices equipped with chips for satellite data processing (animal collars, smartphones, wearables). Measured data can be stored in the device memory and subsequently transmitted in batches for processing. The collected data can be obtained from the device via wireless communication technologies and protocols (GSM, UHF, VHF or satellite communication).

2 MATERIALS AND METHODS

There are several ways that can be used for the spatial data collection, transfer and processing. In our opinion the following seems to be optimal: online data collection, batch data transfer in line with the chosen technology and time period with the subsequent validation of the archived data having an option of their further processing and display, e.g. in map outputs. By analysing the needs we obtained the principles and knowledge for the online collection, validation, archiving and displaying of spatial data (Stillwell & Clarke 2004).

These principles and knowledge were used e.g. for the evaluation of the spatial activity of game in the Doupovské Mountains and the Šumava National Park in the Czech Republic. Databases for the analysis and processing were created using a large data set obtained on the basis of mutual co-operation among the Faculties of the Czech University of Life Sciences (CULS) and other external entities. The main objective of the presented solution is the visualisation of continuously acquired spatial and activity data – game monitoring (Jarolímek et al., 2012). The whole solution is based on our web application (Fig. 1), which processes validated spatial data kept in the database (data archive) and makes the outputs available to users. This web application has been developed in the PHP 5 environment using libraries of the Nette framework (Nette, 2013). The application obtains data from the database server via the Dibi database layer (Dibi, 2013). Google Maps mapping services developed by Google Inc. are used to display spatial and activity data of game animals. Communication with Google Maps is ensured by Google Maps JavaScript API V3 (Google Maps API, 2014). The information on the monitored game animals is displayed by means of the client-side programming language (*JavaScript*) using the jQuery framework. The application is optimised for most widely used web browsers that run on various operating systems and devices, e.g. desktops, laptops as well as on mobile devices such as tablets or smartphones. The application

Figure 1.   Web application Game Online (zver.agris.cz/en).

Figure 2.   Collar from Vectronic Aerospace.

runs on the Microsoft IIS (*Internet Information Services*) Internet server. Data from the application are further provided to other applications, namely for mobile devices (Masner et al., 2014).

2.1   *Game telemetry*

At present (April 2014), we have already monitored four game species – red deer (*Cervus elaphus*), sika deer (*Cervus nippon*), wild boar (*Sus scrofa*) and most recently also the European bison (*Bison bonasus*). The designated animals are monitored via GPS, by means of a collar (Fig. 2), which records the animal's position with the accuracy of several meters; the location; records of the animal's GPC receiver; the date and time in programmed intervals (i.e. usually 1 hour). In addition, the sensor records the temperature and accuracy of measurements. Newer collars are equipped with GSM modules, which contain telephone SIM (*subscriber identity module*) cards allowing for the transfer of data to the user's computer. Generally, it is useful to validate data and to store them in a database system for the subsequent processing and display (Fan, 2004).

2.2   *Activity data*

The collar also contains a sensor recording animal activity. Movement values are recorded onto two axes by means of the accelerometer. Data are measured

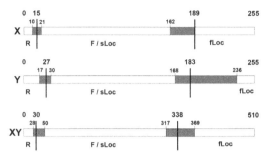

Figure 3.   Intervals in red deer behaviour models.

Table 1.   Monitored species.

Specimen	Number of monitored animals	Number of positions	Monitoring start
Roe deer	21	>71200	28. 1. 2013
Sika deer	10	>53800	31. 8. 2009
Wild boar	12	>39000	16. 1. 2013
European bison	1	>160	25. 3. 2014

approximately 6 to 8 times per minute, averaged in five-minute intervals and recorded. Compared to spatial data, their processing is much less demanding, namely as far as electric energy is concerned. Collar batteries are not thus depleted so much. Data reading is done by using UHF/VHF technologies.

There is a methodology for determining particular behaviour models for the red deer. They can be classified into the following categories based on the measured values: resting, feeding or locomotion and fast locomotion (Lötker, et al., 2009).

Figure 3 includes intervals for determining particular behaviour models (R – resting, F/sLoc – feeding/slow locomotion, fLoc – fast locomotion).

3   RESULTS AND DISCUSSION

Information on the game movement is obtained via GPS in the collar equipped with a GSM module. Data are received through the ground station, thereafter validated and stored on a database server. This process is implemented using software, Ground Station Harvester (*GSH 1.0*) that was developed specifically for this purpose (GSH, 2012).

The first game monitoring and data collection were carried out in 2009 on the sika deer and they have been gradually expanded to other species. At the moment, the database contains more than 160,000 spatial records and approximately one million pieces of activity data. Data from 42 animals are available (not all animals have a collar with a GSM module or a collar that is still in operation). Activity data are available only from selected animals (Table 1). The number of monitored animals has been constantly growing.

Spatial data on game are recorded several times a day and they are transmitted in batches via GSM network services or recorded by means of UHF/VHF. In addition to spatial information, other data acquired via additional sensors (in particular the temperature sensor) are recorded as well.

3.1 Ground Station Harvester (GSH 1.0)

GSH was developed by the Department of Information Technologies of the Faculty of Economics and Management, Czech University of Life Science in Prague, Czech Republic. The application provides automated loading and validation of spatial data using GPS Plus X (GPS PLUS X, 2013) from devices such as a ground station (e.g. a device equipped with a communication GSM module for receiving spatial data from mobile devices – animal collars). Collected spatial data are processed – adjusted and cleaned (e.g. validated based on the accuracy of the determined position) and then saved onto a database server; after that they can be used by various subsequent applications (standard, web and mobile apps). This is a general SW component usable for the collection and validation of spatial data in various industries. The SW solution has been developed in the standard PHP 5 environment using dBase and Dibi database layer (Dibi, 2013) libraries. The application supports namely work in the MS SQL database server environment but due to the use of the Dibi database layer it can be also used in other commonly available database systems.

3.2 Data visualisation

Data can be displayed online using a web application with several functions.

3.2.1 Visual outputs
– Simple displaying of points – projection of game positions in the given period (period of time, and the time of day);
– Route of the minimum distance travelled;
– Projection of the home range – as of the selected date;
– Heat map – animal occurrence density with respect to their home range;
– Activity data – allocation of activities to points in the given time period and to the location (these data are available only internally due to high demands on performance when selecting data; the public can view only an example).

3.2.2 Statistical outputs
– Calculation of the occurrence areas (home range size); and
– Calculation of the movement route length within the given time period (at least two distances travelled);
– Displaying of sufficient information on the position (e.g. temperature or altitude).

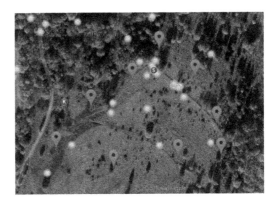

Figure 4. Determination of the area for displaying activity data.

3.3 Activity data evaluation

The existing methodology for the evaluation of red deer behaviour models must be verified for the sika deer.

In order to verify the respective algorithm (Section 2.2) it is necessary to specify the areas of typical resting and grazing habitats (Fig. 4). Thereafter we have examined the percentage of expected activities in the given area. This experiment is based on the assumption that if the error rate is similar to the algorithm of the red deer it could be also used on the sika deer.

It was necessary to establish SW support for this solution. It is possible to mark out the area (polygon) in the web application where spatial data are selected. A simple algorithm is used to select positions (*spatial data*) from the database. Each point is tested, i.e. a straight line is fitted away from it in any direction so that:

– If it passes through an odd number of the selected polygon sides – the point is inside the polygon;
– If it passes through an even number of its sides or through no side – the point is outside the polygon and it is discarded.

Due to the different measurement intervals of activity and spatial data, it is necessary to assign several activity data to each selected position in an interval of several minutes.

Measured values of the accelerometer in individual co-ordinates (X, Y, X + Y) are compared in order to assign an activity (resting, slow locomotion – grazing, fast locomotion). If all of them are within the specified interval an activity is assigned (Fig. 5).

3.4 Requirements for the web application

Basic requirements for the application include: simple and easy to use, open to the general public and ability to view the stored data. Due to safety reasons (namely for the protection of animals observed) only selected history data (last occurrences with a one-month delay) can be viewed by the general public. Some individual animals or species cannot be viewed by the public

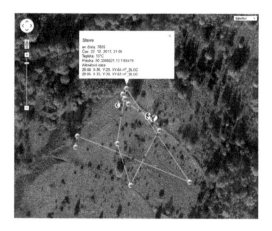

Figure 5. Assignment of activity data to particular positions.

at all. All history and current data are available after logging into the internal section of the web application. At the same time, we have defined user roles that can see only some individual animals or species.

4 CONCLUSION

The solution can be used for visualizing the spatial activities of various game species in the selected environment but also generally for the monitoring and visualization of any moving objects. It has been used for the evaluation of spatial activities (online application) of the sika deer, red deer and wild boar in selected localities of the Czech Republic.

The resulting application allows for the use of validated data in the database for further scientific research in various fields, for educational purposes and for the popularisation of scientific research.

In the future, it should be used for comparing individual JavaScript map APIs (e.g. Google Maps API, OpenLayers, Leaflet), taking into consideration the application performance when displaying large data collections. Emphasis should be placed on various types of devices (PCs, mobile phones and mobile devices in general).

This solution has been further improved (a mobile application for Android OS has been developed) and it should be used in other areas in the future.

ACKNOWLEDGMENTS

The results and knowledge included herein have been obtained owing to support from the following institutional grants.

Internal grant agency of the Faculty of Economics and Management, Czech University of Life Sciences in Prague, grant no. 20141048, "Zpracování velkých kolekcí dat v JavaScriptových mapových API pro www prostředí".

University internal grant agency of the Czech University of Life Sciences in Prague, grant no. 20134312, "Analysis and visualization of GPS telemetry outputs of cloven-hoofed animals in Doupov Mountains and the Bohemian Switzerland National Park".

REFERENCES

[1] Dibi – Database Abstraction Library for PHP 5 (2013) [on-line], available: http://dibiphp.com/en/.

[2] Fan, F. and Biagioni, E. S. (2004) 'An approach to data visualization and interpretation for sensor networks', Paper presented at the Proceedings of the Hawaii International Conference on System Sciences, 37 999-1008.

[3] Google Maps API (2014), [on-line], available at: https://developers.google.com/maps/

[4] GPS PLUS X (2013), [on-line], available at: http://www.vectronic-aerospace.com/wildlife.php?p=wildlife_downloads.

[5] GSH – Ground Station Harvester 1.0. (2012), [on-line], available at: http://www.agris.cz/kit/gsh/ in czech.

[6] Jarolímek, J., Masner, J., Ulman, M, Ulman a M., and Dvořák, S. (2012) 'Cloven-hoofed animals spatial activity evaluation methods in Doupov Mountains in the Czech Republic', AGRIS on-line Papers in Economics and Informatics. 4(3), s. 41–48. ISSN: 1804-1930.

[7] Lechner, W. and Baumann, S. (2000) 'Global navigation satellite systems'. Computers and Electronics in Agriculture, 25(1-2), s. 67–85.

[8] LÖTTKER, Petra, Anna RUMMEL, Miriam TRAUBE, Anja STACHE, Pavel ŠUSTR, Jörg MÜLLER a HEURICH. New Possibilities of Observing Animal Behaviour from a Distance Using Activity Sensors in Gps-Collars: An Attempt to Calibrate Remotely Collected Activity Data with Direct Behavioural Observations in Red Deer Cervus elaphus. Wildlife Biology. 2009, vol. 15, issue 4, s. 425–434. DOI: 10.2981/08-014.

[9] Nette Framework (2013) [on-line], available at: http://nette.org/en/.

[10] Owen-Smith, N., Goodall, V., and Fatti, P. (2012) 'Applying mixture models to derive activity states of large herbivores from movement rates obtained using GPS telemetry', Wildlife Research, 39(5), pp. 452–462.

[11] Stillwell, J. and Clarke, G. (2004) 'Applied GIS and Spatial Analysis', Wiley, Chichester, 420 pages, ISBN: 978-0-470-84409-0.

[12] MASNER, Jan, Jiří VANĚK and Michal STOČES. Spatial Data Monitoring and Mobile Applications – Comparison of Methods for Parsing JSON in Android Operating System. AGRIS on-line Papers in Economics and Informatics. 2014, vol. 6, č. 1.

Future Communication, Information and Computer Science – Zheng (Ed.)
© *2015 Taylor & Francis Group, London, 978-1-138-02653-7*

Use of ICT in education for mitigation of social exclusion

J. Jarolímek, M. Ulman, J. Vaněk, J. Masner & M. Stočes
Department of Information Technology, Faculty of Economics and Management, Czech University of Life Sciences in Prague, Czech Republic

ABSTRACT: The topic of the paper is based on the long-term research activity of the Department of Information Technologies at the Faculty of Economics and Management CULS in Prague. Main goal of the paper is to analyse the current state of the art of evaluation of impacts of the education of given groups of socially disadvantaged people. The selected topic covers many aspects the most significant among which are: life long learning, social impacts of digital divide, influence of technologies and their usability in the educational process.

Keywords: Lifelong learning; digital divide; social exclusion; web site usability; online study materials

1 INTRODUCTION

At CULS, we teach regular university students and we also provide lifelong learning to various groups of people. Among our most successful lifelong learning educational activities belong teaching of women on maternity leave, the virtual university of the third age, courses for young new entrepreneurs, help to handicapped people and education of farmers and people living in rural areas. We have employed information and communication technologies (ICT) in all mentioned cases to a certain extent starting from an online study material distribution to a full-scale online learning. According to our experience, there are differences in the acceptation of diverse teaching methods among different groups of people. The selected topic covers many aspects the most significant being life long learning, social impacts of digital divide, influence of technologies and their usability in the educational process.

Lifelong learning is of growing importance in contemporary society. Rapid technological change calls for adjustments in the competence needed in working life as well as in retirement and leisure time. Learning is nowadays not limited to the educational institution. Compared to some decades ago where it was possible to have one education that qualified an employee for his or her entire working life, we can see now a different situation when professions rapidly change, which is often directly or indirectly spurred by the technological development. Life long learning is required partly due to technological changes and changes in organizations and in society.

Lifelong learning according to the European Commission was stated as "all learning activities undertaken throughout life with the aim of improving knowledge, skills and competences within a personal, civic, social and / or employment related perspective." (EC, 2001). Lifelong learning paradigm is focused on student-centred, active, autonomous learning (Bryderup et al, 2009) that is demand driven and flexible (Kendall et al, 2004).

Digital divide, as defined by the OECD (the Organization for Economic Cooperation and Development), is the term that refers to "the gap between individuals, households, businesses and geographic areas at different socio-economical levels with regard both to their opportunities to access information and communication technologies and to their use of Internet for a wide variety of activities" (OECD, 2001).

According to Cilan, Cigdem & Aricigil (2009), there is a significant level of digital divide in the EU. Shortcomings perceived in current digital divide research (van Dijk, 2006) are such as lack of theory (merely descriptive level emphasizing demographics of income, education, age, sex, and ethnicity), lack of interdisciplinary research (digital divide cannot be understood without addressing issues such as attitudes toward technology (eg. technophobia and computer anxiety), lack of qualitative research, lack of dynamic approach, and lack of conceptual elaboration and definition.

2 DIGITAL AGENDA FOR EUROPE 2020

Overcoming digital divide and mitigation of the social exclusion of threatened groups of citizens belongs to the priority areas of Digital Agenda for Europe 2020 (DAE 2020). The Pillar VI and the Pillar VII of DAE 2020 cover some issues related to our project context and identifies groups of people that are threatened with social exclusion due to digital illiteracy and lack of skills.

Actions to be taken to enhance digital literacy, skills and inclusion are projected under the Pillar VI of DAE 2020. Issues related to our research are such as:

– Lack of skills excludes people from modern society (Pillar VI, Action 57): Some 30% of Europeans have never used the Internet. These people – mostly the elderly, unemployed or low incomes – lack the skills, confidence and means to use digital media and are thus unable to participate in today's society. Digital skills and media literacy play a huge role in employability and equal societal participation (EC, 2010);

– Digital literacy and new skills for jobs (Pillar VI, Action 58): With the European labour market radically changing it is clear that new skills will be needed for the jobs of tomorrow. Shortages of adequate skills in some sectors or occupations already co-exist with unemployment across the EU (EC, 2010);

– Increase participation of women in the ICT workforce (Pillar VI, Action 60): The number of young people studying and choosing careers in ICT is decreasing and is not keeping up with growing demand. Women are under-represented at all levels in the ICT sector, especially in decision-making positions (EC, 2010);

– EU-wide indicators of digital competences (Pillar VI, Action 62): The inability to access or use ICT (Information and Communication technology) has become a barrier to social integration and personal development. The digitally illiterate are missing out on social and economic opportunities and on easy access to online public services that can save time and money. Those without sufficient ICT skills are disadvantaged in the labour market and have less access to information to empower themselves as consumers or as citizens (EC, 2010);

– Member States to implement digital literacy policies (Pillar VI, Action 66): 150 million Europeans have never used the Internet. This group is largely made up of older people or people on low incomes, the unemployed, immigrants, and the less educated and those at risk of social exclusion in general. In many cases the take-up gap is due to a lack of user skills, such as digital and media literacy and competences (EC, 2010);

– ICT for Ageing Well (Pillar VII, Action 78): Demographic ageing, putting strong pressure on sustainability of health and social care, on labour participation, on quality of life for older people and persons with disabilities, and on national budgets in general.

3 MOTIVATION VERSUS CONTENT AND FORM OF EDUCATION

The elderly as the group of learners, are motivated to continue learning by different ambitions than job promotion or improved qualifications. Their interest lies in learning to know more and to continue improving as individuals; individuals who should be included in today's changing society (Escuder-Mollon, 2012).

Figure 1.   Illustration of multimedia lectures and other study materials (VU3A).

As example of elderly education we can give virtual courses of the University of the Third Age (VU3A) that represent a new alternative to the classical attendance education of U3A. They are based on use of new communication technologies, multimedia and the Internet; they have elements of distance education and eLearning, and they are adapted didactically to characteristic specifics of a target group – the seniors with an interest in education at university level. The distant education is determined above all for education of seniors in regions who cannot participate in UVA lectures in the attendance form in seats of universities for various reasons (distance, health and time reasons, financial costs for transportation and so on). Availability anywhere and any time is the biggest advantage of the virtual courses (Jarolímek, Vaněk & Šimek, 2010). Positive impacts on learning outcomes and sustainability of success rate were even noticed with application of online study materials for distant learning of bachelor first year students in ICT courses at CULS in Prague (Vasilenko et al, 2012).

Companies tend to requalify their older employees (50+) by using distance learning methods. Employees over the age of 50 demonstrate a high degree of motivation and interest in further vocational training by means of distance learning methods (Hoenig & Stummer, 2013), a willingness to contribute and to learn ICT skills to retain active for both social and employment reasons (Lam and Chung, 2010). Factors in distance learning are required in varying degrees (Sieber, 2008).

The growing frequency of use of online study materials puts higher demands on user interface usability at various types of end devices. Research of the usability of web sites for education of groups with different needs, various computer and information literacy and various level of technical equipment is up to date.

The web site usability is defined in literature as "the user optimization" in terms of appearance optimization, items layout and page navigation such as change of size and font style, size of headings and buttons, etc. (Krug, 2006). Usable web sites are supposed to be clear, comprehensible and easy to use. The users should be able to orientate themselves in a couple of seconds and retrieve the information without long searching. The useable web sites are those that do not force the user to think and decide, but operate intuitively and logically (Krug, 2006; Foraker Labs,

2011). The situation is getting more complex because of specific requirements of particular users and a large variety of end user devices. The testing of a web site must be focused on the user, not the product. Then it is possible to obtain information about what the users like, what makes them confused, or what bothers them, and most of all how to improve the product so that its use becomes more convenient for them (Rubin, 2008; Krug, 2006; Barnum, 2010).

There are various ways of usability testing. Their use differentiates according to the user group that the web application is designed for. Many web site usability tests were not focused on particular user group or subject matter (Agarwal & Venkatesh, 2002). Nathan & Yeow (2011) proposed the method for web application usability assessment that enables to identify key factors of the web site usability for a given area and brings recommendations for web sites from users' perspective (students).

An extensive usability study of e-learning portals for education proved lacks in design and usability of learning portals and confirmed an intensive need to propose and improve approaches in the assessment of e-learning platforms (Granica & Cukusic, 2011).

The evaluation of effects of mobile platforms on learning process (Hyman, Moser & Segala, 2014) is an up to date topic. As perspective research topics appear the evaluation of user (learner) experience with mobile platform in context of preferred type of study (Hyman, Moser & Segala, 2014), identification of main interests and requirements of learners and teachers, and factors affecting successful acceptation of mobile learning platforms (Sarab, Al-Shihi & Hussain, 2013).

4 CONCLUSION

Modern didactical means of education develop very fast. In close connection with that development, new areas of the so-called media didactics and media pedagogy are born. The media didactics deal with the integration of component media into the education process in order to reach its optimization. The media pedagogy creates the media as such and deals with their usage as the object of the analysis. Both of the before-mentioned relate to each other and blend together. Virtual education uses both the theories and practical experience that stems from university pedagogy.

Following current global trends, media tools are expected to be more and more used in all types of education. Interactive technologies and creation of systems that enable all users to be actively involved have been analyzed. For instance, Yves Bertrand, the world-famous pedagogue, indicates the creation of open models (thus true virtual courses) as a basic general principle of media background arrangements.

The paper presents issues that need to be taken into account in implementation of life long learning systems based on information and communication technologies (Virtual education), and that are designed for various social groups. The most important issues are:

- The development of computer and information literacy is expected to bring higher acceptance of new technologies which requires reflection of new types of education;
- Fast advancements in information and communication technologies, namely types and parameters of end user devices, educational software and forms of electronic documents;
- Arrangements of parameters, forms and the usability of online study materials in life long learning courses are only expected but have not been empirically verified yet;
- Mitigation of the social exclusion through life long learning is one of the priorities of the European Union within Digital Agenda for Europe 2020;
- Current economic and demographic situation in the Czech Republic and most of European countries underlines the need to look for solutions of the above mentioned issues;
- Learning needs of various groups in life long learning are different;
- There is an information gap in scientific literature and in application area in the field of education of socially disadvantaged groups of people.

REFERENCES

[1] Agarwal, R. & Venkatesh, V. 2002. Assessing a firm's web presence: a heuristic evaluation procedure for the measurement of usability. *Information Systems Research*, 13: 168–186.

[2] Barnum, C.M. 2010. *Usability Testing Essentials: Ready, Set ...Test!* San Francisco, CA: Morgan Kaufmann.

[3] Bryderup, I.M., Larson, A. & Trentel, M.Q. 2009. ICT-use, educational policy and changes in pedagogical paradigms in compulsory education in Denmark: From a lifelong learning paradigm to a traditional paradigm? *Education and Information Technologies*, 14(4): 365–79.

[4] Çilan, Ç.A., Bolat, B.A. & Coskun, E. 2009. Analyzing digital divide within and between member and candidate countries of European Union. *Government Information Quarterly*, 26(1): 98–105.

[5] Escuder-Mollon, P. 2012. Modelling the Impact of Lifelong Learning on Senior Citizens' Quality of Life. *Procedia – Social and Behavioral Sciences*, 46(0): 2339–2346.

[6] European Commission. 2001. *Making a European area of lifelong learning a reality*. Brussels: European Commission.

[7] European Commission. 2010. Communication from the Commission to the European parliament, the Council, the European economic and social committee and the Committee of the regions. *A Digital Agenda for Europe*.

[8] Foraker Labs. 2011. *Website Design*. [online]. 2011, [cit. 2014-04-06]. Retrieved at: http://www.usabilityfirst.com/about-usability/website-design/.

[9] Granića, A. & Ćukušić, M. 2011. Usability Testing and Expert Inspections Complemented by Educational

Evaluation: A Case Study of an e-Learning Platform. *Educational Technology & Society*, 14(2): 107–123.

[10] Hoenig, W. & Stummer, H. 2013. Training for older employees in Germany – opportunities for distance learning? An exploratory study. *Global Business and Economics Review*, 15(1): 59–75.

[11] Hyman, J.A., Moser, M.T. & Segala, L.N. 2014. Electronic reading and digital library technologies: understanding learner expectation and usage intent for mobile learning. *Educational Technology Research and Development*, Vol. 62(1): 35–52.

[12] Jarolímek, J., Vaněk, J. & Šimek, P. 2010. Virtual University of Third Age – Education of Seniors in Regions. *7th International Conference on Efficiency and Responsibility in Education (ERIE 2010)*, pp. 154–163.

[13] Jarolímek, J., Vaněk, J., Černá, E., Šimek, P. & Vogeltanzová, T. 2010. Conditions and Limitations of Multimedia Senior Education in Regions. *Journal on Efficiency and Responsibility in Education and Science*, 3(2): 66–78.

[14] Kendall, M., Samways, B., Weert, T.J.V. & Wibe, J. 2004. IFIP TC3 Lifelong learning position paper (pp.193-4). In Kendall, M. & Weert, T.J.V. (Eds.) *Lifelong learning in the digital age: sustainable for all in a changing world*. Boston, MA: Kluwer Academic Publishing.

[15] Krug, S. 2006. *Don't make me think: A Common Sense Approach to Web Usability*, Berkeley, California USA: New Riders Publishing.

[16] Lam, S. & Chung, W. 2010. The Changing Landscape of Ageing Workforce in Hong Kong – The Importance of ICT Training in Lifelong Corporate Learning. *International Journal of Advanced Corporate Learning (iJAC)*, 3(1): 11–16.

[17] Nathan, R.J. & Yeow, P.H.P. 2011. Crucial web usability factors of 36 industries for students: a large-scale empirical study. *Electronic Commerce Research*, 11(2): 151–180.

[18] Organization for Economic Cooperation and Development (OECD). 2001. *Understanding The Digital Divide*. Paris: OECD.

[19] Rubin, J. 2008. *Handbook of Usability Testing*. Indianapolis, IN: Wiley Publishing.

[20] Sarrab, M., Al-Shihi, H., & Hussain Rehman, O.M. 2013. Exploring major challenges and benefits of M-learning adoption. *British Journal of Applied Science & Technology*, 3(4): 826-n/a.

[21] Sieber, H. 2008. Methoden fuer die Bildungsarbeit, pp. 17–18, 3rd ed., *Leitfaden fuer aktivierendes Lernen*, Bertelsmann, Bielefeld (Perspektive Praxis).

[22] van Dijk, J.A.G.M. 2006. Digital divide research, achievements and shortcomings. *Poetics*, 34: 221–235.

[23] Vasilenko, A., Očenášek, V., Ulman, M. and Vaněk, J. 2012. Effects of the Use of Multimedia Study Materials on Students Examination Results in Undergraduate ICT Course. In *EFFICIENCY AND RESPONSIBILITY IN EDUCATION 2012*, Prague: Czech University of Life Sciences Prague, Dept. of Systems Engineering. Pp. 593–601.

Future Communication, Information and Computer Science – Zheng (Ed.)
© *2015 Taylor & Francis Group, London, 978-1-138-02653-7*

An online distributed VM migration scheme for congestion avoidance in data center networks

G. Deng, H. Wang & Z.H. Gong
College of Computer, NUDT, Changsha, China

ABSTRACT: Networks are usually the critical resource in data center. Imbalance of applications and traffic distributions in data center networks usually causes local hotspot and congestion. Recently, many network-aware solutions have been proposed for network performance optimization. Some of them try to find an optimal VM-to-host mapping offline, while the others rebalance network load in an online fashion by VM migration. However, due to the high complexity of VM placement, the first category is usually time-consuming. The second category, which usually adopts a centralized manner and highly relies on accurate traffic forecast, still suffers from some shortages.

In this paper, we propose DVMS, a novel distributed VM migration scheme. Instead of trying to search for an optimal VM placement, we just seek a practical solution which can effactually avoid congestion. DVMS performs VM migration periodically. In each migration period, DVMS tries to minimize the VM migration number and thus minimize the impact to the network. Furthermore, instead of using aggregate or average traffic as a measurement of the weight between VM pairs, we introduce a statistical method, which is a more accurate evaluation of the traffic impact between VM pairs. Simulation results show that our scheme can decrease congestion near 80% with only less than 0.2% extra traffic imposed on the network.

Keywords: data center networks; congestion avoidance; traffic balancing; VM migration

1 INTRODUCTION

Virtualization now becomes an indispensable technology in data centers. It offers tremendous conveniences for network performance optimization. Since Clark, et al. [1] first proposed and implemented a live VM migration in 2005, lots of works have been proposed for energy saving, server consolidation, traffic balancing and so on. Network is the stringent resource in data centers. The imbalance of its utilization may induce local congestion, and thus affect application performance and network efficiency. Practical measurements show that data center networks usually suffer from long-lived congestion [9]. Virtualization offers opportunities for the data center manager to balance the network load by VM migration, However, VM migration is an expensive operation [1][10]. It not only takes a long time for migration, but also brings pressure for the network due to the additional traffic generated by migration. Therefore a solution to this problem should balance the tradeoff between migration profit and cost.

Recently, many works have been proposed for traffic balancing by VM migration. Generally speaking, they can be classified into two categories: offline and online. Offline methods usually adopt a centralized method and require complete information about the traffic matrix between VM pairs. They usually search for a full placement for all VMs. It is time-consuming and may cause large amounts of extra traffic, which will further increase network unbalance and congestion. Offline solutions may be suitable for finding an optimal placement in the initial stage, but it's not applicable in a dynamic real-time situation. In contrast, online solutions gradually adjust VM placement according to online traffic measurement and try to minimize the impact on networks. However, current solutions also suffer from some efficiency or practical problems: First, most of them adopt a centralized fashion. To obtain the traffic matrix, VM or hypervisor has to transfer traffic information periodically with the centralized manager, which imposes additional traffic to the networks. Second, these algorithms usually use an aggregate traffic between VM pairs as the measurement of influence to network congestion. However, aggregate traffic may cover the variety of traffic in time. Take figure 1 as a simple example. All three flows have the same aggregate traffic in the period time 0 to 5. But, evidently they are significantly different in the variance. Assuming the link bandwidth is 10 Mbps and larger or equal to 8Mbps is congestion, then, in this period, flow A will not encounter congestion, flow B may encounter 20% congestion in time, and flow C may reach 40%.

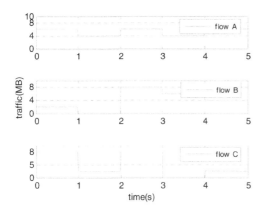

Figure 1. A example of flows with same aggregate traffic but different effects to congestion.

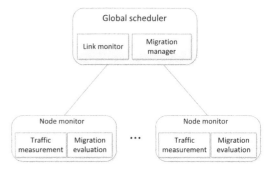

Figure 2. The DVMS framework.

In this paper, we present a novel distributed VM migration scheme DVMS. From a high-level view, DVMS includes two main components: centralized controller (CCTR) and distributed local strategic maker (DLSM). CCTR is responsible for the globe decision making and DLSM performs the local information collection and local decision making. By the cooperation of CCTR and DLSM, DVMS can perform VM migration to decrease congestion in data centers effectually. Moreover, DVMS usually just requires a small portion of VMs to be migrated.

Specifically, we make the following main contributions in this paper.

- We propose a novel VM migration framework. In this framework, the VM migration decision is decomposed into two stages: distributed self-decision making phase and global migration decision phase, which alleviates the extra workload imposed by traffic information collection and thus quickens the decision making process.
- Instead of using the aggregate traffic as the measurement of weight between VM pairs, we introduce a statistical method, which is more accurate to reflect the actual influence of the traffic between VM pairs.
- We evaluate DVMS via large scale simulation. We present the experiment results and show its efficiency for congestion avoidance.

The rest of this paper is organized as follows: Section 2 proposes the distributed VM migration framework and its details. section 3 validates our design via extensive simulations. We summarize related work in section 4 and conclude the paper in section 5.

2 DVMS' DESIGN

2.1 Design goals and challenges

Indeed, DVMS can be classed into the second category. Many former works have proved that it's a NP-hard and varieties of heuristic algorithms have been proposed

[4][5][11]. Most of them perform in a centralized fashion and try to find an optimal solution. However, as is mentioned above, finding an optimal solution is costly. Instead, DVMS adopts a distributed manner and just tries to find a practical solution, which is equal to striving for balance between the cost and efficiency. The main challenges to the design and implementation of DVMS include:

- **Making a decision based on incomplete information.** While the number of VMs and hosts in a modern data center is usually large, a centralized method may be time-consuming. To address this problem, we introduce a distributed framework. However, how to make a decision based on incomplete information becomes a challenge.
- **Evaluating the impact of a time-varying flow.** As is mentioned in *section 1*, the former works usually use aggregate traffic as a measurement of the weight between VM pairs. However, that may cover the variance of traffic and cannot reflect the different impact of those flows. How to accurately evaluate the influence of traffic between VM pairs is another challenge.

2.2 Congestion avoidance VM migration algorithm

The top view of DVMS is shown in figure 2. It includes two main components: a centralized global scheduler and many node monitors spreading over each VM. Accordingly, our congestion avoidance algorithm includes two main stages: first, each VM computes its best migration destination independently according to its traffic measurement, and then when a scheduling interval is reached or a scheduling condition is satisfied, the global scheduler selects a set of VMs to migrate to their destinations. We will discuss the detail functions of each component in the following.

A. Node monitor

Node monitor component runs in each VM. It continually monitors the communication traffic with other VMs and evaluates which host is its best mapping destination. So the node monitor component contains two modules, namely traffic measurement module (TMM) and migration evaluation module (MEM).

0.2	0.3	0.2	0.2	0.1	0	0	0	0	0.1
20	35	50	65	80	95	110	125	140	155	170

Figure 3. The segment partitions and their relative ratios.

The traffic measurement module is responsible for periodically measuring the traffic with other VMs. Usually, the former works adopted a centralized method and used aggregate traffic as a measurement of the connection weight between VM pair. However, as is mentioned in *section 1*, such a method suffers from some shortages: On one hand, aggregate traffic may cover the variance of traffic and thus cannot reveal the real impact on network congestion. On the other hand, it's impractical for a centralized monitor to measure the traffic in a short period. Then, in DVMS, we introduce a different method. Instead of using aggregate traffic, we adopt a statistical method. Since the traffic measurement modules perform in a distributed manner, each module is just responsible for one VM's traffic collection, it can run in a relative short period. Thus, we divide a VM migration period into n subperiods (e.g. ten subperiods). We collect the traffic in each subperiod and get a traffic set TS in each migration period. Then, in each migration period, the traffic is divided into ten segments from the minimum to the maximum according to their sizes. The segment length is computed by formula 1.

$$Length = \frac{max(TS) - min(TS)}{10} \quad (1)$$

The traffic whose size fall into the segment is classed into the same group. We compute the ratio of each group and sort them in descending order. Then, we use the average of the top 80% traffic as the measurement of traffic between VMs. For example, assuming the subperiod number is ten and the measurement results are (20, 68, 42, 55, 47, 83, 61, 170, 39, 26), then, the segment length will be 15 and each segment's ratio is shown in figure 3.

The average of the top 80% traffic will be:

$$avg = (20 + 26 + 39 + 42 + 47 + 55 + 61 + 68)/8$$
$$= 44.75 \quad (2)$$

then, we use 44.75 as the weight between the two VMs.

The migration evaluation module then is responsible for evaluating which host is the most desirable destination. Here, we assume MEM has the knowledge of the network topology, routing information and VM-to-host mapping information. In fact, such information can be obtained from global scheduler with little cost, for MEM just needs a small portion of VMs location information as we will discuss in the following.

We model a data center network as an undirected graph $G(V, E)$, where V is the vertex set and E is the edge set. There are two types of vertices in V: the physical hosts and the network switches. The edges represent the communication links. The physical host set is defined as $P = (p_1, p_2, \ldots, p_m)$ and the VM set is defined as $V = (v_1, v_2, \ldots, v_n)$. We further define a mapping function as:

$$Map(v_i, p_j) = \begin{cases} 1, if \ v_i \ is \ assigned \ to \ p_j \\ 0, otherwise \end{cases} \quad (3)$$

We define the cost between virtual machines V_i, V_j as $Cost\ (v_i, v_j) = D\ (p_l, p_m) \times W(v_i, v_j)$. Where p_l, p_m are the relative hosts of v_i, v_j. $D\ (p_l, p_m)$ is defined as the number of hops between physical machine p_l, p_m.

Then, to find an optimal destination host for v_i is equal to the following optimization framework:

$$min \sum_{j < |V|} Cost(v_i, v_j) \times map(v_i, p_k) \quad (4)$$

However, this is also hard. If we consider all VMs and all possible p_k, its hardness may approach a global optimal algorithm. Then, we further make some simplifications. First, it's obvious that it's no sense to consider a VM with no traffic with v_i. Then, for the other VMs, we first sort them by traffic with v_i in decreasing order. We just take the top VMs which contain 80% traffic into consideration. According to [9], this will significantly reduce the number of VMs further, for usually only a small portion of VMs contribute to most of the traffic. Besides, we constrain p_k in the scope which has VM corresponding with v_i (obviously, in typical data center network structures such as tree, Fat-tree, VL2, BCube and DCell, it is easy to find a better host than the one which does not host a VM corresponding with v_i). Here, we do not take the host resource constraints into consideration. We will address this in the global scheduling phase.

B. *Global scheduler*

The global scheduler component is responsible for link monitoring and VM selecting for migration. It contains two main modules: Link monitor and VM manager. the Link monitor is responsible for link state monitoring. Though some works used an agent implementation, we adopt a SNMP monitoring fashion, because in most of the commercial switches, it's impractical to wedge a custom agent into them. Link monitor periodically collects and evaluates the extent of congestion. We use a link utilization threshold 80% as a representation of congestion. Similarly, we divide a VM migration period into some sub-periods. We collect the link load in each sub-period and then calculate congestion ratio in each migration period. Then, the VMs which contribute to the top n congestion links are candidates for migration.

But there may be too many VMs whose traffic traverse across those congestion links. Then, which one should be migrated? That is the main function of Migration manager module. Certainly, it's best to select an optimal solution which makes the traffic balanced across the whole network. However, as is

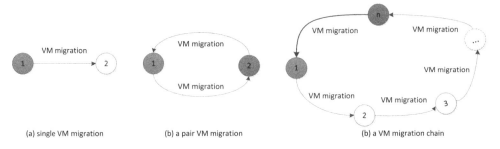

(a) single VM migration   (b) a pair VM migration   (b) a VM migration chain

Figure 4.   A VM migration chain.

mentioned in section 1, that is time-consuming and expensive, and may induce frequent VM migration. Instead, for each congested link, we just choose one VM which contributes most of the traffic for migration. All these VMs to be migrated are denoted as $M = (m_1, m_2, \ldots, m_k)$ for each element in M. We move it to its desirable destination selected in the migration module, assuming it is p_d. We don't take the host resource constraints into consideration here. But the actual situation may be that: there is not enough resource in the destination host. In this case, assuming there is a VM v_i that $map(v_i, p_d) = 1$ and $v_i \in M$, then, we only need to migrate v_i to its destination host further. This process continues until we encounter a host which all the VM's hosts don't include in the set M and it has not enough resource to host the VM to be migrated in. In this case, we select a VM which has minimal traffic with other VMs in this host and migrate it to the initial host. This process is like a chain shown in figure 4(c). The blue node 1 is the initial host with a VM to be migrated from, and the node 2 to $n-1$ are the intermediate nodes that both act as destination host and source host. The last node n is the destination of VM_{n-1}, but it has no VM in the migration VM set M and not enough resource for VM_{n-1} then, we select a VM in host n which has the minimal traffic with other VMs and migrate it to the initial host 1. The other two cases where destination p_d has enough vacant resource and destination p_d has not enough vacant resource and no VM to be migrated are shown in figure 4(a) and (b). They are similar to the case above-mentioned but more simple. A pseudo code of the whole migration process is shown in figure 5.

3   PERFORMANCE EVALUATION

In this section, we evaluate DVMS via extensive simulation. We investigate DVMS in two aspects: first, we evaluate the congestion ratio before and after using DVMS; then, we further investigate DVMS on the number of migration VMs and show the cost imposed by VM migration. In our simulation, the size of VMs are randomly chosen from 512 MB to 1 GB following uniform distribution, and the ratio of migration traffic to VM size is fixed to 1.2. We adopt a three-layer tree data center with 992 physical hosts for simulation. We assume each server can host at most 16 VMs.

Migration(CLS) /* CLS: Congested Link Set */

---

1.   M ← get VM set to be migrated from CLS;
2.   Foreach VM_i in M
3. $dh_i = Best\_destination(VM_i)$;
4.   Chain_migration(VM_i, dh_i, M);
    END

Chain_migration (vm, dh, M)/*vm: VM to be migrated;
   dh :destination host; M: VM set to be migrated*/
1. ch_i ← current host of vm;
2.   M ← M-{ vm };
3.   Migrate_to(vm, dh); /*perform VM migration */
4.   If(Vacant_resource(dh)>= Require_resource(vm))
5.   return;
6.   Else if(Map(M)<1)
7.   vm ← Minimal_traffic_VM(dh);
8.   Migrate_to(vm_j, ch_i);
    Else
9. vm_j ← select a VM whose host is dh from M;
10. $dh_j = Best\_destination(vm_j)$;
11.   Chain_migration (vm_j, dh_j, M)
12.   End If

Figure 5.   The VM migration algorithm.

All links are 1 Gbps except that the links between core switches and aggregate switches are 10 Gbps. Initially, each server hosts 10 VMs and each VM randomly corresponds with 16 to 64 other VMs. We model the arrival rate of flow between VM pair with Poisson distribution with parameters $\lambda = 60$, which gives median inter-arrival time of roughly one minute. The amount of bytes transmitted within a flow follow a uniform distribution $U(80,120)$ MB. All the simulations are performed on a windows sever with an Intel 2.7 GHz dual-core CPU and 8G DRAM. The language we use is C++.

We first investigate the link congestion ratio before and after using DVMS. We simulate the link utilization ratio in 2 hours with one minute interval. Then, we divide the link utilization ratio into ten segments with the interval of 0.1. For each segment, we compute the ratio of link number. We compare the ratio before and after using DVMS. The migration period in DVMS is ten minutes. The results are shown in figure 5. We can see that the ratio of links whose utilization ratio

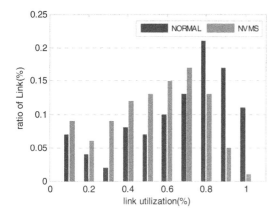

Figure 6. The ratio of links under different link utilization ratios.

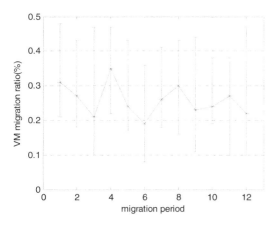

Figure 7. VM migration ratios.

is above 0.8 decreases obviously after using DVMS, which means that our VM migration can efficiently alleviate the hotspot load and decrease congestion.

Then, we repeat the above simulation ten times, we record the number of VMs migrated in each migration period and observe the number of migration VMs. Figure 6 plots the average, minimal and maximal migration ratio in each segment. We can see that the VM migration number is relatively low: for a data center with 9920 VMs, the migration ratio is no larger than 0.5% and only contributes less than 0.2% traffic, which may have little impact on the network.

4 RELATED WORK

Virtualization has become an indispensable technology in modern data centers. It decouples applications from physical resources and allows elastically distributing resources upon applications. Virtualization is also the basic technology of cloud computing. In recent years, many studies have been proposed on this subject.

Migration implementation: [1] first started a study on live migration. It presented a design, implementation and evaluation of VM migration on top of the Xen VMM and showed that the service downtimes were as low as 60 ms. Another work [10] also presented a VM migration system in the Intel x86-based operating system and showed the application downtime caused is less than a second. These study results show that live VM migration is available in practical systems.

VM placement: To find a VM-to-host mapping which not only satisfies each VM's host resource requirement but is also resilient to the time-variations network demand, the recent work [6] presented MCRVMP, which transform the traffic balancing problem into a Min Cut Ratio-aware VM Placement. MCRVMP is NP-hard, so [6] introduces two heuristics. But both of them are time-consuming. For a network with 3430 VMs, both algorithms take thousands of seconds to find a mapping. Other works [3][7][8][12] try to improve the network resource utilization by optimizing VM placement. However, improvement of the network resource does not necessarily decrease congestion. Meanwhile, all of them are of high time complexity.

Online VM migration: Comparing it with VM placement algorithms which focus mainly on the initial placement of VMs, VM migration algorithms mainly take the dynamic state change into consideration. Our method belongs to this category. The most similar work may be [5]. It presented an automatic VM migration system MWLAN. MWLAN can detect the network hotspots and dynamically remap VMs to avoid congestion. Though MWLAN also adopts a distributed load information collection manner, it adopts a centralized migration scheduling method. Our scheme differs from MWLAN in two aspects: First, instead of seeking an optimal solution, DVMS just tries to find an applicable one, because accurately forecasting the traffic is hard and seeking an optimal solution is of high complexity. Second, to reduce the complexity, we just select a portion of VMs whose communication traffic is larger than a certain threshold for consideration. Other works [2][4][11] also aim to avoid congestion by VM migration, however, different from DVMS, all of them adopt a fully centralized manner.

5 CONCLUSION

In this paper, we present DVMS, a distributed VM migration scheme to avoid long-lived congestion. DVMS contains two main components: a global scheduler and a distributed node monitor. The node monitor dwells in each VM and periodically collects the traffic with other VMs. Each VM computes its most desirable destination host and informs the global scheduler. The global scheduler periodically monitors the link utilization information. When it finds some links suffer from continued congestion, it chooses one VM which makes the most traffic contribution for each congested link to migrate it to its desirable host. Simulation results show that DVMS can effectively decrease congestion with only a few VM migration costs.

To simplify the description, in this paper, we assume the node monitor components dwell in each VM. However, a VM should be transparent to the migration operation. Fortunately, this can be obtained by a slight modification of DVMS. For a number of VMs a physical host is usually limited, we can simply combine the functions of these node monitors in a same physical host into their relative hypervisors.

ACKNOWLEDGEMENTS

This work is supported by Program for National Basic Research Program of China (973 Program) 'Reconfigurable Network Emulation Testbed for Basic Network Communication'. Program for Changjiang Scholars and Innovative Research Team in University (No.IRT1012), 'Network technology' Aid program for Science and Technology Innovative Research Team in Higher Educational Instituions of Hunan Province and Hunan Provincial Natural Science Foundation of China (11JJ7003). The National High Technology Research and Development Program of China (2011AA01A103). Research on Trustworthy Cross-Domian Information Exchanging Technology Based M-TCM (KJ-12-07).

REFERENCES

[1] Christopher Clark, Keir Fraser, Steven Hand, et al. Live migration of virtual machines // Proc of NSDI '05.

[2] Daniel S. Dias and Luís Henrique M. K. Costa. Online Traffic-aware Virtual Machine Placement in Data Center Networks // Proc of Global Information Infrastructure and Networking Symposium (GIIS), 2012.

[3] Vivek Shrivastava, Petros Zerfos, Kang-won Lee, et al. Application-aware Virtual Machine Migration in Data Centers // Proc of INFOCOM'11.

[4] Xitao Wen, Kai Chen, Yan Chen, et al. VirtualKnotter: Online Virtual Machine Shuffling for Congestion Resolving in Virtualized Datacenter // Proc of IEEE 32nd International Conference on Distributed Computing Systems (ICDCS), 2012.

[5] Jun Chen, Weidong Liu, Jiaxing Song. Network Performance-Aware Virtual Machine Migration in Data Centers // Proc of Cloud Computing, 2012.

[6] Biran O, Corradi A, Fanelli M, et al. A Stable Network-Aware VM Placement for Cloud Systems. Proc of 12th IEEE/ACM International Symposium on Cluster, Cloud and Grid Computing, 2012.

[7] Meng X, Pappas V, and Zhang L, Improving the scalability of data center networks with traffic-aware virtual machine placement // Proc of INFOCOM'10.

[8] Jiang J, Lan T, Ha S, et al. Joint VM Placement and Routing for Data Center Traffic Engineering // Proc of INFOCOM'12.

[9] Kandula S, Sengupta S, Greenberg A, et al. The nature of data center traffic: measurements and analysis // Proc of IMC'09.

[10] Michael Nelson, Beng-Hong Lim, and Greg Hutchins. Fast Transparent Migration for Virtual Machines // Proc of USENIX Annual Technical Conference, 2005.

[11] Vijay Mann, Akanksha Gupta, Partha Dutta. Remedy: Network-Aware Steady State VM Management for Data Centers // Proc of IFIP Networking'12.

[12] Joe Wenjie Jiang, Tian Lan, Sangtae Ha, et al. Joint VM Placement and Routing for Data Center Traffic Engineering // Proc of INFOCOM'12.

Future Communication, Information and Computer Science – Zheng (Ed.)
© 2015 Taylor & Francis Group, London, 978-1-138-02653-7

Mobile application development options for news and information portals

M. Stočes, J. Vaněk, J. Masner & J. Jarolímek
Department of Information Technologies, Czech University of Life Sciences Prague, Czech Republic

ABSTRACT: This paper is about the issue of mobile application development on established information web portals. A Solution is demonstrated on the choice of suitable technologies and methods to develop mobile applications displaying content of AGRIS web portal (accessible on www.agris.cz). AGRIS – agrarian web portal – was developed on the Microsoft platform and the data have been saved in relation database and XML files. AGRIS – agrarian web portal – was developed and has been run by the Department of Information Technology, Czech University of Life Sciences Prague.

Keywords: C# ; NET; JSON; XML; cross-platform development; mobile application; agrarian WWW portal; rural development; PhoneGab; hybrid application;web application; native application

1 INTRODUCTION

Inequality of economic and social relations between towns and rural regions is generally recognized; it is caused by quite a few historical, geographical, political and economic phenomena. In the period of information society development, the use of information and communication technologies is considered one of the crucial tools for rural development and use of its potential. (Vaněk a et al, 2010).

Smart mobile devices with Internet connection have become more and more popular. They enable rapid access to information for low costs. GSM mobile Internet covers most of the Czech Republic and some areas are covered by more rapid LTE Internet technology. It is supposed that 98% of the Czech Republic will be covered by LTE technology – rapid Internet connection – by 2017 (Peterka, 2013).

2 MATERIAL AND METHODS USED

2.1 AGRIS - agrarian web portal

The aim of the AGRIS agrarian web portal is to create a unified on-line information place on Internet for the agrarian sector (agriculture, food industry, forest industry, water management) and countryside. The target users are enterprise managers, officers of state administration and regional self-government institutions, students and all food consumers and countryside residents. The portal is accessible in Czech and English mutation. The portal belongs to the most visited non-government servers with an agrarian theme. The project arose in 1999 on the basis of cooperation between the Czech University of Life Sciences

Prague and the Ministry of Agriculture of the Czech Republic.

Basic functions of the portal: (Figure 1)

- Aggregation of agrarian news.
- Aggregation of prices of agrarian commodities from different stock markets.
- Aggregation of notices of chosen agrarian events (trade fairs, conferences …).
- Daily summary of important happenings.
- Weekly summary of important happenings in English.
- List of links to important sources.
  (Vaněk a et al, 2012)

The portal is accessible on www.agris.cz. Nowadays, the fifth version is running. The original version from 1999 was developed in PHP 4 programming language. The current version is running on Microsoft platform (Windows Server 2008, IIS 7.0, MSSQL database server) its web application is written in MVC framework, version 2. (Šimek a et al, 2011) The portal data has been partly saved in file system in XML documents and in MSSQL database server. (Masner a et al, 2014).

The content of the web portal is described by metadata (DublinCore standard and VOA3R application profile v.2 with use of AGROVOC term) which is further shared through OAI-PMH. Metadata is uploaded and accessible e.g. on VOA3R portal (Open Access Agriculture & Aquaculture Repository: Sharing Scientific and Scholarly Research related to Agriculture, Food, and Environment.) (Šimek, 2013).

The database contains unique assemblage of papers and prices related to the agrarian sector from 1999 up to the present. The portal database contains over 182 000 papers and over 620 000 prices.

Figure 1. Current version of WWW portal Agris (Source: own processing).

2.2 Portal visitors

The profile of the typical user was analyzed by Google analytics tools and by a component for the analysis of visits of particular pages. The typical portal user is very "conservative", he or she usually visits web pages between 8am and 4pm every working day. In the last year, more visits to the web portal were monitored from mobile devices and even outside of working time. (Alpar, 2002)

On the basis of this knowledge editorial council decided to create mobile application which will enable both on-line and off-line access to the AGRIS agrarian portal. On the basis of analysis of access it was decided to develop applications for iOS and Android systems.

2.3 Application requirements

Editorial council, developers and users set the following demands on mobile application:

- Off-line version (batch loads of articles for subsequent off-line work with articles).
- Implementation of all basic modules of web application
  - Articles – display of news and actualities
  - Prices – display of prices of agrarian commodities and formation of graphs
  - Events
- "Conservative look" coming out from web application (Jeong & Han 2012).

2.4 Mobile application types

There are three basic attitudes to the development of mobile applications. (Figure 2.)

- Native Application – Apps developer using native SDK (software development kit). Different application for every platform is necessary. Installation via App store is necessary. (4js.com, 2014)

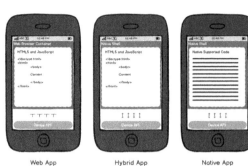

Figure 2. Native, Web, and Hybrid Applications (4js.com, 2014).

- Hybrid Application – Apps develop using HTML5, CSS, JavaScript and native shell, combination of native application and web application. Development common for more platforms. Installation via App store necessary. (Xanthopoulos & Xinogalos, 2013)
- Web Application – Apps developer using Web technologies. Application runs on Web browser. Connection to Internet is necessary permanently. Rapid actualization of application. (Kulkarni & Klemmer, 2011)

3 RESULT AND DISCUSSION

Chosen parameters evaluated during the analysis of the used methods of mobile application development.

- Native Application
  - development for every platform separately with use of different tools and technologies,
  - actualization of application for every platform separately,
- Hybrid Application
  - rapid development
  - rapid actualization
  - easy extension of application
- Web Application
  - mobile version of Agris portal already run,
  - application accessible online only.

  (Charland & LeRoux, 2011)

To create mobile hybrid application with use of methods of multiplatform development seems to be the best solution in the defined conditions. (Xanthopoulos & Xinogalos, 2013)

3.1 Environment for hybrid development

Accessible applications/frameworks were analysed for the purpose of creating hybrid applications and it was decided to use PhoneGab platforms (Figure 3). Platform enables a developer to create an app that runs

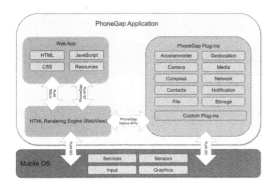

Figure 3. PhoneGab Architecture (Galimbikov, 2013).

on a variety of mobile devices. The developer accomplishes this largely by writing the user interface portion of their application with Web technologies such as HTML, CSS and Java Script PhoneGap's development tools then bundle the HTML, CSS and JavaScript files into platform-specific deployment packages. (Traeg, 2014)

PhoneGap supports a wide variety of platforms:

- iOS
- Android
- Windows 8
- Windows Phone 7 and 8
- BlackBerry 5.x+
- WebOS
- Symbian
- Tizen
  (Palmieri et al, 2012)

4   CONCLUSION

On the basis of analysis of the current agrarian web portal system and analysis of demands of the editorial council, technical group and end users of the portal it was decided to create hybrid application with use of multiplatform development.

Mobile application will be accessible for free on chosen systems – Android and iOS. In future, it will be possible to create applications for other platforms, e.g. Windows Mobile, WEB OS and others.

Open-source Framework PhoneGab was chosen as a platform for future development which will enable rapid development with use of HTML CSS, JavaScript technology. Communication with the server will be realized with the use of JSON and XML technologies which are supported by the server.

Application will contribute to information support of the agrarian sector and regions and will extend the user base of Agris agrarian portal.

Realization of the application is set to 9/2014.

It is about a universal solution and approach which can be used generally for mobile application developments concerning news servers.

ACKNOWLEDGEMENTS

The results and knowledge included herein have been obtained owing to support from the following institutional grant.

Internal grant agency of the Faculty of Economics and Management, Czech University of Life Sciences in Prague, grant no. 20141043, "Nativní a multiplatformní prostředí klient/server informačních zdrojů".

REFERENCES

[1] Alpar, P., Porembski, M. & Pickerodt, S. 2002, "Measuring the efficiency of Web site traffic generation", International Journal of Electronic Commerce, vol. 6, no. 1, pp. 53–74.

[2] Charland, A. & LeRoux, B. 2011, "Mobile application development: Web vs. native", Communications of the ACM, vol. 54, no. 5, pp. 49–53.

[3] Galimbikov, I. 2013, "Using PhoneGap Plugin to Expand BlackBerry Feature Support", [online], available at: http://cases.azoft.com/phonegap-blackberry-plugin/

[4] Jeong, W. & Han, H.J. 2012, "Usability study on newspaper mobile websites", OCLC Systems and Services, vol. 28, no. 4, pp. 180–198.

[5] Kulkarni, C. & Klemmer, S.R. 2011, "Automatically adapting web pages to heterogeneous devices", Conference on Human Factors in Computing Systems – Proceedings, pp. 1573.

[6] Masner, J., Vaněk, J. & Stočes, M. 2014, "Spatial data monitoring and mobile applications – Comparison of methods for parsing JSON in android operating system", Agris On-line Papers in Economics and Informatics, vol. 6, no. 1, pp. 37–46.

[7] Palmieri, M., Singh, I. & Cicchetti, A. 2012, "Comparison of cross-platform mobile development tools", 2012 16th International Conference on Intelligence in Next Generation Networks, ICIN 2012, pp. 179.

[8] Peterka, J. 2013, "Aukce mobilních kmitočtů skončila, na nového hráče zapomeňte!", [on-line], available at: http://www.lupa.cz/clanky/aukce-mobilnich-kmitoctu-skoncila-na-noveho-hrace-zapomente/ (in czech).

[9] Šimek, P., Vaněk, J., Jarolímek, J., Stočes, M. & Vogeltanzová, T. 2013, "Using metadata formats and AGROVOC thesaurus for data description in the agrarian sector", Plant, Soil and Environment, vol. 59, no. 8, pp. 378–384.

[10] Šimek, P., Vaněk, J., Jarolímek, J., Stočes, M. & Vogeltanzová, T. 2011, "New version of the AGRIS web portal-overcoming the digital divide: By providing rural areas with relevant information", Agris On-line Papers in Economics and Informatics, vol. 3, no. 4, pp. 71–78.

[11] Traeg, P. 2014, "Four Ways To Build A Mobile Application, Part 3: PhoneGap" [on-line], available at: http://mobile.smashingmagazine.com/2014/02/11/four-ways-to-build-a-mobile-app-part3-phonegap/

[12] Vaněk, J., Červenková, E., Jarolímek, J. & Šimek, P. 2010, "State and evaluation of information and communication technologies development in agricultural

enterprises in the Czech Republic", Plant, Soil and Environment, vol. 56, no. 3, pp. 144–147.

[13] Vaněk, J., Stočes, M., Šimek, P. & Hrbek, I. 2012, "Information support of regions and possibilities of its further development", Agris On-line Papers in Economics and Informatics, vol. 4, no. 3, pp. 71–78.

[14] Xanthopoulos, S. & Xinogalos, S. 2013, "A comparative analysis of cross-platform development approaches for mobile applications", ACM International Conference Proceeding Series, pp. 213.

[15] What is the GWC hybrid mode?, 2014 [on-line], available at: http://www.4js.com/online_documentation/fjs-gas-manual-html/c_gas_ghc_what_is.html

Future Communication, Information and Computer Science – Zheng (Ed.)
© 2015 Taylor & Francis Group, London, 978-1-138-02653-7

Chinese microblog sentiment classification considering users' reviews based on Naive Bayesian algorithm

B. Yan, B. Zhang, H.Y. Su & H. Zheng
Key Lab of Intelligent Information Technology, Beijing Institute of technology, Beijing, China

ABSTRACT: Nowadays, with the rapid development of social networks, community-oriented Web sentiment analysis technology has gradually become a hot topic in the field of data mining. Being concise and flexible, Chinese microblog poses new challenges for sentiment analysis. This paper proposes an approach to classify Chinese microblog sentiments into positive and negative by the plain Naive Bayesian algorithm. Based on data preprocessing, sentiment lexicon construction, combining element of users' reviews, this research posits this Sentiment Classification which is a novel method of attaching microblog users' reviews to the target microblog in order to improve the accuracy of sentiment classification.

Keywords: Chinese microblog; sentiment classification; Naive Bayesian algorithm; users' reviews

1 INTRODUCTION

Microblog is a very popular networking tool for obtaining and publishing information. Users can instantly share 140-Word text messages. With the development of microblog, there has been a lot of microblogs with a perspective, so that there is an urgent need to analyze and get information on the Semantic Orientation (SO) of these. It is of great importance to study the sentiment classification of microblog, for it is beneficial for the monitoring, discovery and guidance of public opinions.

In this paper, sentiments expressed by microblog are divided into positive and negative. Combining the machine learning method the sentiment classification of text is studied. Considering the characteristics, we divide it into main bodies of messages and subdivided opinions. We use the users' reviews as a part of the additional features related to the features of the main bodies. We get the subjective microblog of hot topics for feature extraction and calculation of feature weight. Then we build classifiers to compare and analyze the performance of different classifiers under different conditions.

2 RELATED WORKS

With regard to traditional text research, Pang *et al.*, 2002 applied different feature election methods such as NB, Maximum Entropy (ME) and SVM to classify the movie reviews. The experiment indicates that Boolean value as feature weights works yields more accurate results than term frequency. Mullen *et al.*,

2004 accomplished more by taking the evaluative factor, potency and activity values of adjectives, along with the SO value of the valuable phrases as features, based on Turney's five combination model of polarity words. Employing CHI and information gain, Ni 2007 implement sentiment classification based on NB, SVM and Rocchio's algorithms.

In research on the Chinese language, Tang Huifeng *et al.*, 2007 studied the text features and sentiment classification based on supervised learning under the training set of various quantities and sizes. Considering the impact of negative adjectives and adverbs on semantic tendency, Xu Jun 2007, analyzed and classified the news reviews corpora with the NB and ME model.

3 CLASSIFICATION WITH USERS' REVIEWS

3.1 *Basic framework of sentiment classification*

Chinese grammar and language convention values the context. The microblog contains more than one sentence in a specific context. Although it can accommodate up to 140 characters, the information that microblog contains still does not accurately describe the SO. We introduced the features related to the theme into the microblog, to explore whether it is conducive to classification. In order to fully cater to its special features, we have used the information of users' reviews extended target microblog to increase classification accuracy.

Figure 1 shows a basic framework of sentiment classification consisting of two phases: training and users' testing phase. These two stages include similar

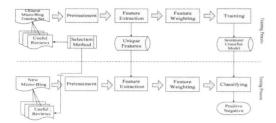

Figure 1. Classification framework considering the reviews.

processes, such as the selection process of users' reviews, microblog pretreatment, feature extraction, feature weight calculation. The machine learning method is then used to train and get text classification model. In the classification process, it is used to predict a new microblog text sentiment classification of polarity.

3.2 Selection process of reviews

3.2.1 Microblog reviews

A challenge for sentiment classification is to make machines understand these expressions. In order to take full advantage of the characteristics possessed, we find that the subjective reviews can be viewed as a basis for the classification of target microblog. An effective part of the review is actually a basic understanding of the emotions of the public for target microblog, which is synonymous with antisense transformation or conversion. So reviews can be treated as a specific sentence in the context. For example, a target microblog is as follows: "The hunger marketing way of XiaoMi mobile is too successful, too talented!" Valid review "I lost it again! We really cannot tolerate it anymore." Mere analysis of target microblog may produce inaccuracy. The valid review is negative related to this microblog. So sentiment of the target microblog is likely to be negative.

3.2.2 Selection process

It is found that users' reviews, which contain the main characteristics, are an important part of microblog. Two kinds of valuable information, one of which is holding the same point as target microblog, the other on the contrary is expected to be extracted from the reviews collection intermixed with extraneous ones. We sequence the users' reviews based on the relevancy, quality and time. Interception within a certain number of reviews is perceived as valuable on the target microblog.

The first factor that needs consideration is the relevance between the reviews and the target microblog, which refers to the degree of similarity between these two. It can effectively filter out a lot of review spam and advertising reviews. Relevance alone cannot guarantee the accuracy. Thus some peculiar characteristics of microblog production can be applied to delve in more significant issues, which often consist of the number of browsing users, number of "good" as well as

the relevance between the latter and time. To further explain, a review with more browsing users, which means more attention, obviously is of greater significance and worth for the research. A review which gains a lot of "good" is more likely to own higher quality, with more approval from the users. Moreover, the earlier the review is released, the more attention it tends to attract.

The calculation method based on the semantic similarities of HowNet is introduced in this paper. The formula of the similarities between two sentences is:

$$\text{Sim}(s_t, s_i) = \frac{(\sum\limits_{u=1}^{n} \max\limits_{1 \leq v \leq m} \text{Sim}(w_{1u}, w_{2v}))/n + \sum\limits_{v=1}^{m} \max\limits_{1 \leq u \leq n} \text{Sim}(w_{1u}, w_{2v}))/m)/2}{} \quad (1)$$

In target microblog sentence (S_t), there are n words: $w_{11}, w_{12} \ldots w_{1n}$. In the review sentence (S_i), there are m words: $w_{21}, w_{22} \ldots w_{2m}$. $Sim(w_{1u}, w_{2v})$ refers to the similarities between two sentences' words.

The quality of the reviews include: quality of reviews (Q) and the time dimension (T). It represents the number of user's reviews to be read within a certain time, and as well as the extent to which the public would find it helpful, it can objectively reflect influence of this microblog. Formulas follow as:

$$Q_i = \frac{useful_i}{\sum\limits_{j=1, t=t_i}^{n, t_{now}} view_j} + \frac{useful_i}{view_i} \quad (2)$$

$$T_i = 1 + \alpha \left(\frac{t_i - t_0}{t_{now} - t_0} \right) (\alpha \in 0,1) \quad (3)$$

In the equation Q_i, $useful_i$ represents that the users regard the review as useful and $view_i$ means the times that the review is browsed. As to T_i, t_i is the time at which the review is published, t_{now} refers to the time when the review is extracted and stands for a const from 0 to 1. Sort the evaluation score based on the relevance, quality and time. Formula is as follows:

$$rank_i = \alpha \times sim_i + (1 - \alpha) \times fun(T_i, Q_i, \beta) \quad (4)$$

For $i = 1 \ldots n$, where a, β are constants of 0 to 1. The $fun(T_i, Q_i, \beta)$ refers to the formula the quality of the $Review_i$, which requires certain alteration based on different conditions. A simple method adopted in this paper.

3.3 Pretreatment works

3.3.1 Chinese word segmentation

Chinese text needs word segmentation before further processing. In this paper, we use NLPIR, a Chinese word segmentation system. The next step is the removal of stop words in text and special symbols. These special symbols include punctuation marks, numbers and letters, which can have an impact on word segmentation. The privative will be removed from the stop word because it will change the semantic orientation of polarity words.

3.3.2 Features extraction and features weight

Privative and polarity words require special handling. The privative words can be viewed by setting the sliding window and then be added to the feature set. In this paper, we use the HowNet to calculate the SO of the polarity words. The results of the calculation served as the principal evidence of classification. In every microblog, including those of the training and testing set, the TF and TF-IDF score the highest before a word as an attribute value is recorded as:

$$\text{features}(t) = \{\omega_1, \omega_2, \cdots, \omega_N, \forall i \neq j, \omega_i \neq \omega_j\} \quad (5)$$

For test set, on per microblog and corresponding valid reviews sets from which the top n most frequently occurring words are selected as features' values. For train set, select features values in different classifications as the sum of the values as:

$$\text{features}(\text{category}Y) = \{features(t)|\forall t, s.t.category(t) = Y\} \quad (6)$$

The feature weight calculated in accordance with the size of TF-IDF and TF. TF refers to the number that the feature appears in the document, whereas TF-IDF modified TF calculation method.

3.4 Naive Bayesian algorithm

After the texts are represented as a document-term matrix, we used a NB method to train a sentiment classifier. Bayes theorem is applied to predict the of an unknown category sample. Then the category with the most is selected as a sample. Within the given category y, under the polynomial algorithm, it is observed that joint $W_1, W_2 \ldots W_n$ probability is as follows:

$$P(\omega_1, \omega_2, \cdots, \omega_N | Y) = \prod_{i=1}^{n} P(\omega_i | Y) \quad (7)$$

The microblog t belongs to a certain category of probability, locating the eigenvalues of t as $W_1, W_2 \ldots W_N$, m categories $C_1, C_2 \ldots C_M$. M = 2 in this case considering the duality classification of positive and negative. Based on Bayes formula, the probability of t, $P\{C_i|t\}$ belonging to the class of C_i is:

$$P\{C_i|t\} = \frac{P\{t|C_i\}P(C_i)}{P(t)} = \frac{\prod_{j=1}^{N} P\{\omega_j|C_i\}P(C_i)}{\sum_{j=1}^{M} \prod_{j+1}^{N} P\{t|C_i\}P(C_i)} \quad (8)$$

4 EXPERIMENT RESULTS ANALYSIS

4.1 Data collection

We collected the Chinese microblog data of hot topic, from weibo.com. A data set including 4,000 pieces of microblog is picked out and divided into a training set and a testing set. They mark consistent results an emotional tone. Training set contains 3,000 target microblog, with 1429 positive ones and 1571 negative ones; test set contains 1,000 records, with positive ones numbered 578, and negative 422.

4.2 Evaluation indicators

According to the evaluation scores, the method posited is analyzed. The performance evaluation indexes of the classifier mainly consist of recall, precision, F1-Meatrue. The *recall* is defined as the percentage of correctly classified texts among all texts belonging to that category, and the *precision* is defined as the percentage of correctly classified texts among all texts that were assigned to the category by the classifier. And *F1-measure* evaluates the performance of sentiment classifier. Evaluation Indicator as follows:

$$F1 = \frac{2 \times precision \times recall}{precision + recall} \times 100\% \quad (9)$$

4.3 Experiment process

Experiment 1: basic classification experiment. Original corpus is preprocessed. Pretreatment includes noise reduction, segmentation and POS tagging under different conditions in which are different classifiers, features and feature weight calculation. Better classifier and features are selected based on experimental results; the text is split into a number of clauses, and obtained sentence polarity in the test data. Finally, the whole microblog eventual polarity is obtained by adding up the polarity numbers of the sentence. Experiment 2: Users' reviews are attached. After processing, they are attached to the training set and testing set, with the same procedures as experiment 1.

4.4 Experiment results

4.4.1 Basic classification results

This experiment applied the combination of three methods to describe the features, which are Unigram (U-gram), Polarity words (Pol), Privative stand for Negative words (Neg). This experiment compared the effects of TF and TF-IDF, two methods of feature weighting on experimental results.

Figure 2 shows the performance of U-gram features classifier, which is based on NB algorithm, in different ways of weights calculation and different feature numbers. In the initial stage, the error often occurred due to the lack of features. As the number of features increased, the classification results raised at first then reduced. The reason of the reduction is the interference from certain irrelevant words as the number of features became larger. It indicates that features selected a number of 3,500 taking the classifier stability in various combinations.

4.4.2 Classification results of attached reviews

Regarding the reviews as a single sentence in a specific context, this experiment adds the reviews of the

117

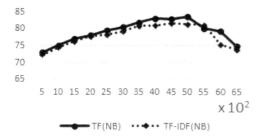

Figure 2. Comparison of different weights under NB.

Table 1. Results of Microblog-lever Classification (F1%).

| Weight | Basic Classify | | Attached Reviews | |
Feature	TF	TF-IDF	TF	TF-IDF
Unigram	77.3	75.6	80.1	80.3
U-gram + Polarity	80.5	80.1	82.4	81.2
U-gram + Pol + Neg	82.7	82.0	83.7	84.5

Table 2. Results of Microblog-lever Classification (%).

| Type | Recall | | Precision | |
	Sentence	M-Blog	Sentence	M-Blog
Positive	80.3	83.4	82.7	85.6
Negative	83.1	83.6	82.8	84.2

subjective microblog as the criteria. Table 1 shows the classification of comparison between the basic and attached-reviews of the combination among different features weighting with NB algorithm under different features. It is observed that adding polarity words and dealing with privative, can improve the accuracy. It also shows that NB classifier with TF feature weighing performs the best. It also performs better than the ones without users' reviews information attached. It can be concluded that the attachment of reviews can improve the accuracy.

As the table 2 shows, precision of the review reached 85.6%, almost 3% higher than the sentence-level precision. One error in the sentence prediction on the whole has little effect, which also verifies the function of the attached reviews to promote accuracy.

4.4.3 *Experimental summary*
Based on the NB classifier, as well as a variety of features, this paper implemented the sentiment classification of the Microblog data sets, and attached reviews. In terms of feature selection, unigram is the most important feature. And experimental results show the NB classifier with the F1-measure of 84.5%. With the unigram feature alone, NB classifier reached the F1-measure of 82.7%. After adding sentiment polarity

words and privative after two features, it rose by about 2%, to the degree of 84.5%. In terms of feature weight calculation, though simple, TF works much better. So attaching the processed reviews, this paper obtained a precision rate of 85.6% and the F1-measure of 84.5%, which is a satisfactory result.

5 CONCLUSION

Chinese microblog sentiment analysis originates from the sentiment analysis of traditional text. But its peculiarities are obvious, worth an in-depth study. At present, a developed filtering technique is needed to filter out the spam interfering with the sentiment classification. To optimize the classification results of classifiers, the microblog is divided into different categories. The best learning strategy would be adopted according to the characteristics of each category. In addition, sentiment classification of text should be combined with other text-mining technologies, which would dig out more valuable information, which can improve the accuracy.

REFERENCES

[1] Pang B & Lee L. 2002. Thumbs up? Sentiment classification using machine learning techniques. The Empirical methods in natural language processing-Volume 10. 79–86.
[2] Turney P D. 2002. Thumbs up or thumbs down? Semantic orientation applied to unsupervised classification of reviews. The 40th annual meeting on association for computational linguistics. 417–424.
[3] Mullen T & Collier N. 2004. Sentiment Analysis using Support Vector Machines with Diverse. Information Sources: 412.
[4] Ni X & Xue G R et al. 2007. Exploring in the weblog space by detecting informative and affective papers. The 16th international conference on WWW. ACM, 281–290.
[5] Tang H & Tan S 2007. Research on sentiment classification of Chinese reviews based on supervised machine learning techniques. Chinese information processing, 21(6): 88–94.
[6] Xu J, Ding Y X & Wang X L. 2007. Sentiment classification for Chinese news using machine learning methods Chinese Information Processing, 21(6): 95–100.
[7] Tian XIA. 2005. Study and Application on Similarity Computation in Chinese Information Processing. Computer Engineering: 33–35.
[8] Go A & Bhayani R. 2009. Twitter sentiment classification using distant supervision. Project Report, Stanford: 1–12.
[9] Zhai Z, Xu H & Jia P. 2010. An Empirical Study of Unsupervised Sentiment Classification of Chinese Reviews. Tsinghua Science & Technology 15(6): 702–708.
[10] Zhai Z & Xu H et al. 2011. Exploiting effective features for Chinese sentiment classification. Expert Systems with Applications, 38(8): 9139–9146.

Future Communication, Information and Computer Science – Zheng (Ed.)
© 2015 Taylor & Francis Group, London, 978-1-138-02653-7

A crowdedness measurement algorithm using an IR-UWB radar sensor

J.W. Choi & S.H. Cho
Department of Electronics and Computer Engineering, Hanyang University, Seoul, Korea

ABSTRACT: We propose a crowdedness measurement algorithm using an impulse radio ultra-wideband (IR-UWB) radar sensor without any private issue. The crowdedness measurement algorithm is composed of three stages. In the first stage, we remove the signal of clutters which mean obstacles in the environment and decrease the effect of noise. In the second stage, an in-bound/out-bound analysis of the human who is going in or going out from the sensor's region is done. Finally in the third stage, signal detection and final crowdedness measurement using the information about the number of humans follows. The performance of the proposed algorithm is proved by experiments about multi-human in the ordinary class room.

Keywords: IR-UWB; UWB; impulse radar; people counting; crowdedness measurement

1 INTRODUCTION

UWB system is defined as a radio technology which uses the bandwidth of more than 500 MHz or which fractional bandwidth is more than 25%. Various advantages such as high range resolution, multipath immunity, good penetration and simple hardware structure of the UWB system are caused mainly by the large bandwidth. In 2007, a standard for the UWB communication is developed by the standardization group IEEE .802.15.4a. Based on the UWB standard, diverse research is being studied in many areas.

An IR-UWB radar sensor is one of the applications of the UWB radio technology. The IR-UWB radar sensor uses a very short impulse signal whose time duration is ns to ps. By transmitting the impulse signal without any modulation, we can receive the signal reflected by the multi-human and clutters of the environment. Using the received signal, we can extract some useful information such as the number of humans and the time of arrival (ToA) of multi-human. The IR-UWB radar sensor has the many advantages of the UWB system. Thus, IR-UWB radar sensor has fine range resolution and good penetration property as we said.

The IR-UWB radar sensor can be used to measure the vital signs such as breathing rate and heart beat rate and also can be used for human detection in the place of a disaster. Besides, indoor global positioning system (GPS) and counting the number of humans are the applications of the IR-UWB radar sensor. In a paper of the references, the counting algorithm of multi-human for the IR-UWB radar system is proposed. That algorithm works on the principle of counting the effective peaks from the instantaneous power signal of the background subtracted signal. But using the multiple IR-UWB radar sensors for the crowdedness measurement up to 4 humans is not practical.

In this paper, we propose a new crowdedness measurement algorithm using an IR-UWB radar sensor. Experiment shows that the proposed algorithm can count up to the nine randomly waled humans.

2 SYSTEM AND CHANNEL MODEL

Received signal of the IR-UWB radar sensor can be modeled as

$$r_k(t) = \sum_{n=1}^{N_{path}} a_{kn}s(t - \tau_{kn}) + n(t) \qquad (1)$$

In equation (1), $s(t)$ is a transmitted impulse signal and the subscript k represents a slow time index which means the k-th received signal from the k-th transmitted signal. The a_{kn} and τ_{kn} are the amplitude and delay of the pulse of n-th path of the k-th received signal respectively. The $n(t)$ is an observation noise of the channel.

The received signal of the IR-UWB radar sensor is composed of the delayed and scaled impulse signals from the multiple paths. If there are more humans in the environment, there will be the more paths and the more impulse signals will be received. Using this basic property, we can measure the crowdedness using the IR-UWB radar sensors.

But if the monostatic IR-UWB radar sensor is used, there are two physical limitations. First, we cannot distinguish the multi-human who are in the same distance from the IR-UWB radar sensor, because the received impulse signals are summed and the signal seems to be the same as the reflected signal from a human. Furthermore, the human who are in the non-line of sight (NLOS) position cannot be detected. Though the IR-UWB radar system has good penetration property, the transmission power of the impulse signal is strictly regulated and the signal of humans who are in the NLOS position cannot be received. To measure the crowdedness measurement, we should overcome these physical limitations by some statistical techniques such as averaging.

3 A CROWDEDNESS MEASUREMENT ALGORITHM

3.1 Overall algorithm

The crowdedness measurement algorithm is composed of three stages. In the first stage, we remove the signal of clutters which mean the obstacles in the environment. And we attenuate the signal of noise to increase the signal to noise ratio (SNR).

In the second stage, we do in-bound/out-bound analysis as a supplementary algorithm of the crowdedness measurement algorithm. The in-bound/out-bound analysis means sensing the human who is going in or going out from the IR-UWB radar sensor's region and updating the crowdedness information according to the results. The effect of the in-bound/out-bound analysis is more dramatic if the number of humans is large. Because it is difficult to distinguish the difference of one human if there are many humans.

As the final stage, the processes of finding the effective peaks and deciding the crowdedness are followed. The detail descriptions are in the next section.

3.2 Clutter removal and noise suppression

As a first process of the first stage, we do background subtraction using the received signal of the IR-UWB radar sensor. The background subtraction means removing the signal of the clutters. There are several algorithms for background subtraction using the singular value decomposition (SVD), short time Fourier transform (STFT) and the running average.

The algorithm using the SVD need so much time for calculation that the method is not proper in the real time operating system (RTOS). Generally, the algorithm using the running average is used because the algorithm shows high performance in comparison with the small calculation time. The background subtraction algorithm using the running average can be written as

$$c_{k-1} = \alpha \cdot c_{k-1}(t) + (1 - \alpha) \cdot r_k(t)$$
$$y_k(t) = r_k(t) - c_k(t) \qquad (2)$$

The c_k represents a clutter signal called as clutter map which is made by using k received signals. The $y_k(t)$ is the background subtracted signal which contains only the signal of moving objects, mainly humans.

The α is a parameter which determines the ratio of applying the current received signal to the clutter signal. If the α is larger, the calculation time for the generation of the clutter map is also larger. But the

made clutter map is more stable. If the α is smaller, the opposite case is true.

Using the background subtraction algorithm, we can remove the signals of the clutters. But despite the background subtraction algorithm, some signals of the clutters have remained still in the received signal. If there is clutter which has a large radar cross section (RCS) or if a human is moving in front of clutter, more clutter's signal remain in the $y_k(t)$. Thus, to attenuate the noise signal, we use a filter such as an averaging or median filter.

3.3 In-Bound/Out-Bound analysis

In-bound/out-bound analysis is a supplementary algorithm of the overall crowdedness measurement algorithm. The in-bound/out-bound analysis means sensing the human who is going in or going out of the IR-UWB radar sensor's region.

In a paper of the references, the algorithm of the in-bound/out-bound analysis is proposed. The farthest fan-shaped boundary of the IR-UWB radar sensor's observation region is separated as the two regions A_1 and A_2. And we set the variable s_1 as the indicator which lets us know if a human exists or does not exist in region A_1. And we also set s_2 as an indicator which let us know if a human exists or does not exist in region A_2. Using these two indicators we can construct the state table about the various situations relating to the movement of the human. And to know in what time we should decide the situation of the region, A_1 and A_2, we set the $s_3 = s_1 + s_2$ as an indicator which means the crossing of the mid line of region A_1 and A_2. The state table about the several basic situations is in Table 1.

Using this algorithm, we can detect the human who is going in or going out from the IR-UWB radar sensor's region. If we separate the boundary area as more regions, $A_1, A_2, \ldots A_n$, the algorithm can handle more complicated situations. But the complexity of the algorithm increases rapidly with the number of regions.

3.4 Signal detection and crowdedness measurement

Using the background subtracted and noise attenuated signal, we do signal detection and final crowdedness measurement in the third stage. Generally there are

Table 1. State Table of Moving Human.

Time	t − 4	t − 3	t − 2	t − 1	t
State	s_1 s_2 s_3	s_1 s_2 s_3	s_1 s_2 s_3	s_1 s_2 s_3	s_1 s_2 s_3
→	0 0 0	1 0 1	1 1 2	0 1 1	0 0 0
←	0 0 0	0 1 1	1 1 2	1 0 1	0 0 0
U-Turn (⇒)	0 0 0	1 0 1	1 1 2	1 0 1	0 0 0
U-Turn (⇐)	0 0 0	0 1 1	1 1 2	0 1 1	0 0 0

120

many impulse signals from one human. The adjacent received impulse signals make clusters of humans. To measure the crowdedness, we should regard the cluster as a human.

In a paper of references, an algorithm which detects the clusters of multi-humans is proposed. The algorithm is composed of three main processes. First, we detect maximum value from the signal. And as a second process, we verify the validity of the maximum value. Verifying the validity of the maximum value is determining firstly what the maximum value is made by the human, or not by the human, mainly the clutter and determining secondly what the maximum value is caused by the same human who already is found in the previous iteration step or the maximum value is caused by a new human. If the two conditions are satisfied the found maximum value is regarded as a new human. The detail sequence of the signal detection algorithm is listed as

1. Set the threshold level T_{hd}, the left marginal time N_{left} and the right marginal time N_{right}.
2. Compute the instantaneous power $p(t)$ of the received signal $p(t) = r^2(t)$.
3. Initialize the dirty map $d_0(t)$ by the power signal. Set the TOA array $A_{toa}(1:end) = 0$, the human counter $k = 0$ and the iteration counter $n = 0$. ('$a{:}b$' means from a to b).
4. Find the index $\hat{\tau}_n = \text{argmax}_t(d_n(t))$ and the amplitude $\hat{a}'_n = d_n(\hat{\tau}_n)$.
5. If $\hat{a}'_n < T_{hd}$, go to step 9.
6. Clean the dirty map by padding zeros such that $d_n(\hat{\tau}_n - N_{left} : \hat{\tau}_n + N_{right})$ and update the dirty map $d_{n+1}(t) = d_n(t)$.
7. If $\hat{a}'_n < \max\{r(\hat{\tau}_n - N_{left} : \hat{\tau}_n + N_{right})\}$, update iteration counter, $n \leftarrow n + 1$ and go to step 4.
8. Store the TOA $\hat{\tau}_n$, $A_{toa}(k) = \hat{\tau}_n$ and update iteration, human counter, $n \leftarrow n + 1$, $k \leftarrow k + 1$ and go to step 4.
9. The TOAs of human are $A_{toa}(1), A_{toa}(2), \ldots A_{toa}(k)$.

After the signal detection algorithm, we measure the crowdedness using the information of the number of humans. If we use only the instantaneous information of the signal detection algorithm, the NLOS situation and same distance of the multi-humans problems cannot be handled. Thus we use an averaging filter to decide the final crowdedness for stable performance of the overall crowdedness measurement algorithm.

4 EXPERIMENT RESULTS

4.1 *Experiment environment*

We implement the overall crowdedness measurement algorithm and we experiment with actual IR-UWB radar devices. We use the board which is named as NVA-R661 made by NOVELDA. Vivaldi antennas of opening angle 40° are used for transmitting and receiving.

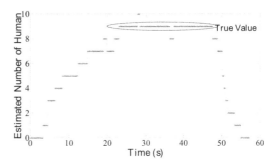

Figure 1. Experiment Result of Standing Multi-Human.

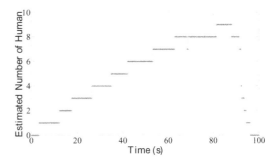

Figure 2. Experiment Result of Randomly Walking Multi-Human.

The experiments are done in the ordinary class room of the concrete wall. And we set the region of the IR-UWB radar sensor as 6 × 6 (m) rectangular region.

4.2 *Results of standing multi-humans*

To verify the algorithm's performance first, we experiment on the multi-humans who are in the LOS positions and positions of different distances from the IR-UWB radar sensor. The humans go inside the region one by one up to nine, and after that all the humans go out from the region simultaneously. The result of the experiment is in Figure 1. In Figure 1, the result shows that the performance of the crowdedness measurement is almost perfect.

4.3 *Result of randomly walking multi-humans*

We experiment on the randomly walking multi-humans in the sensor's region. All the nine humans go inside simultaneously and walk around randomly. The result of the experiment is in Figure 2.

There is some time delay of about 20 seconds until the IR-UWB radar sensor recognizes the nine humans because if nine humans are randomly walking in the 6 × 6 (m) region, NLOS and same distance situations occur so frequently. Though, some time delay and errors exist, we can see that the performance of the crowdedness measurement algorithm is fine up to nine humans.

5 CONCLUSION

We proposed the crowdedness measurement algorithm which counts the number of humans in the IR-UWB radar sensor's region. The algorithm is mainly composed of three stages and we described each stage specifically. The performance of the proposed algorithm is proved by some experiments.

As a future work, we have a study plan about the more effective noise suppression algorithm and data processing algorithms to handle the NLOS and same distance problems of the multi-human cases.

ACKNOWLEDGEMENT

This research was supported by a grant (13RTRP-B067918-01) from Railroad Technology Research Program funded by Ministry of Land, Infrastructure and Transport of Korean government.

REFERENCES

[1] Cassioli, D.; Win, M.Z.; Molisch, A.F., 2002, The ultra-wide bandwidth indoor channel: from statistical model to simulations, *Selected Areas in Communications, IEEE Journal on*, vol. 20, no. 6, pp. 1247,1257.

[2] Jeong Woo Choi; Sung Ho Cho, 2013, A new multi-human detection algorithm using an IR-UWB radar system, Innovative Computing Technology (INTECH), *2013 Third International Conference on*, vol., no., pp. 467,472.

[3] Jeong Woo Choi; Jin Ho Kim; Sung Ho Cho, 2012, A counting algorithm for multiple objects using an IR-UWB radar system, Network *Infrastructure and Digital Content (IC-NIDC), 2012 3rd IEEE International Conference on*, vol., no., pp. 591,595.

[4] Joon-Yong Lee; Scholtz, R.A., 2002, Ranging in a dense multipath environment using an UWB radio link, *Selected Areas in Communications, IEEE Journal on*, vol. 20, no. 9, pp. 1677,1683.

[5] Poon, M.W.Y.; Khan, R.H.; Le-Ngoc, S., 1993, A singular value decomposition (SVD) based method for suppressing ocean clutter in high frequency radar, *Signal Processing, IEEE Transactions on*, vol. 41, no. 3, pp. 1421,1425.

[6] Piccardi, M., 2004, Background subtraction techniques: a review, Systems, Man and Cybernetics, *2004 IEEE International Conference on*, vol. 4, no., pp. 3099, 3104.

[7] Xuanjun Qua; Jeong Woo Choi; Sung Ho Cho, 2014, In-bound/Out-bound Detection of People's Movements Using an IR-UWB Radar System, *International Conference on Electronics, Information and Communication (ICEIC)*.

[8] Zhao Li; Wenzhe Li; Hao Lv; Yang Zhang; Xijing Jing; Jianqi Wang, 2013, A Novel Method for Respiration-Like Clutter Cancellation in Life Detection by Dual-Frequency IR-UWB Radar, *Microwave Theory and Techniques, IEEE Transactions on*, vol. 61, no. 5, pp. 2086,2092.

Future Communication, Information and Computer Science – Zheng (Ed.)
© 2015 Taylor & Francis Group, London, 978-1-138-02653-7

Distributed collaborative filtering recommendation model based on two-phase similarity

C.Q. Wang, H.Y. Su, Y. Zhu, F.X. Li & B. Yan
Key Lab of Intelligent Information Technology, Beijing Institute of technology, Beijing, China

ABSTRACT: The recommendation system based on collaborative filtering is one of the most popular recommendation mechanisms. However, with the continuous expansion of the system, several problems that traditional Collaborative Filtering recommendation algorithm (CF) faced such as cold startup, accuracy, and scalability are worsened. In order to address these issues, a Distributed Collaborative Filtering recommendation model based on Two-Phase similarity (DCF-TP) is proposed. DCF-TP is based on Weighted Distance Similarity Measure (WDSM), a new measure created in this paper. According to WDSM, the similarity of the users is calculated and the similarity matrix of users is obtained, meanwhile, in line with the co-occurrence matrix method, the similarity of items is counted, getting the co-occurrence matrix of the items. With regard to the similarity matrix of users, their preferences are endowed with weights and the new preferences matrix of users is received. Besides, on the basis of the co-occurrence matrix of the items and the new preferences matrix of users, the nearest neighbor item is found and a more accurate recommendation to the target user is given. Furthermore, in terms of the parallel computing framework, the distributed implementation of DCF-TP is completed. All these experiments are done on MovieLens dataset. The results show that DCF-TP overcomes the problem of cold startup and has a qualitative leap both in the aspects of precision and recall ratio. With the increasing numbers of the computing nodes, the distributed algorithm has achieved higher linear speedup.

Keywords: data mining; double similarity; collaborative filtering; distributed applications

1 INTRODUCTION

Nowadays, people are becoming more and more accessible to various social networks and recommender systems are used to provide personalized services, such as the recommendation of the videos in YouTube [1]. The collaborative filtering [2] which was proposed by Goldberg has become one of the most widely used methods to recommend items for users. In most cases, a recommender system providing accurate recommendations can attract the interest of customers. However, some problems (cold startup, sparsity and scalability) the traditional collaborative filtering recommendation algorithm (CF) is always confronted with have placed constraints on creating a large maintenance program. To solve these problems, many optimization methods are proposed. Malucelli et al. [3] presented a two-phase method. Both the distance between items on a suitable graph and the estimation of the information reliability were considered. Experimental results showed that the approach outperformed other algorithms.

In this paper, a new model is introduced to address three key issues. The innovations of this paper are embodied in the following aspects: A distributed collaborative filtering recommendation model based on

two-phase similarity (DCF-TP) is proposed; A similarity measure named Weighted Distance Similarity Measure (WDSM) is presented. The problem of cold startup is overcome therefore; According to the MapReduce framework [4], the distributed implementation of DCF-TP is completed; On the MovieLens data set, the contrast tests of DCF-TP and CF are made, the experimental results show that the new approach is obviously valid.

2 RELATED WORKS AND BACKGROUNDS

In this section, we review the background of the CF and elaborate the WDSM in detail.

2.1 Collaborative filtering recommendation algorithm

As the main recommendation technology, the fundamental assumption of CF is that if user X and user Y rate n items similarly, they tend to act on other items similarly [5]. In general, there are three types of CF, which are CF based on item, CF based on user and CF based on model. In the following, CF based on user is taken as an example in detail:

2.1.1 Fetching data

In a typical CF scenario, there is a list of M users $\{u_1, u_2, \ldots, u_m\}$ and a list of N items $\{i_1, i_2, \ldots, i_n\}$, and each user, u_i, has a list of items, Iu_i, which the user has rated. The list of the items which users like or dislike can be converted to an $m \times n$ user-item ratings matrix B. In matrix B, as in (1), p_{ij} represents the preference of user i to item j as below.

$$B_{m \times n} = \begin{bmatrix} p_{11} & \cdots & p_{1n} \\ p_{21} & \cdots & p_{2n} \\ \vdots & & \vdots \\ p_{m1} & \cdots & p_{mn} \end{bmatrix} \tag{1}$$

2.1.2 Searching nearest neighbor

For CF based on user, we mainly calculate the similarity between users who have both rated the same items, and then we give the nearest neighbor list of the target user by sorting the results. The most popular measure to compute similarity between objects is Tanimoto Coefficient Similarity. It's the number of items that two users express some preference for, divided by the number of items that either user expresses some preference for. Stated plainly, Tanimoto Coefficient Similarity can be shown in formula (2):

$$sim_{(u,v)} = \frac{|I_u \cap I_v|}{|I_u \cup I_v|} \tag{2}$$

I_u represents the set of the items that user u has provided the preference, $sim_{(u,v)}$ means the value of the similarity between user u and v. After finishing all these related operations, sort the sim and then the nearest neighbor of the target user can be received.

2.1.3 Making recommendations

In formula (3), $R_{(v,j)}$ is used to estimate the rating that user v may give to item j. It is estimated based on nearest neighborhood's preference value.

$$R_{(v,j)} = \frac{1}{|I_u|} \sum_{j' \in I_u} sim_{(j,j')} P_{(u,j')} \tag{3}$$

Specifically, I_u represents the set of the items which user u has given the preference. Then, the best recommendation (the most similar to item j') can be chosen according to the value of the $R_{(v,j)}$.

2.2 Weighted distance similarity measure

Although the similarity measure most-widely used for CF has been proved to be successfully, they are limited to be used in cold startup situations. In order to address the problem, a weighted distance similarity measure (WDSM) is presented. The measure is composed of two factors, similarity and weight. With the WDSM, the process of the similarity calculating

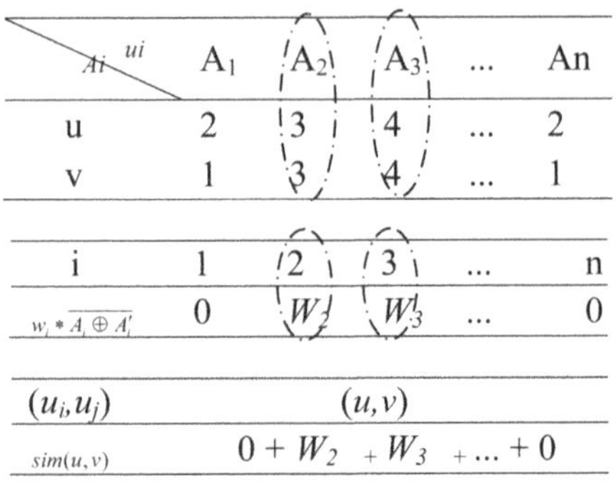

A_i ＼ ui	A_1	A_2	A_3	...	A_n
u	2	3	4	...	2
v	1	3	4	...	1
i	1	2	3	...	n
$w_i * A_i \oplus A_i'$	0	W_2	W_3	...	0
(u_i, u_j)		(u,v)			
$sim(u,v)$		$0 + W_2 + W_3 + \ldots + 0$			

Figure 1. Weighted distance similarity measure.

between two users u and v is presented as illustrated in the formula (4) and figure 1:

$$sim_{(u,v)} = \sum_{i=1}^{n} w_i * (\overline{a_i \oplus a'_i}) \tag{4}$$

Suppose the attributes of user u can be expressed as the vector $A_u = (a_1, a_2, \ldots, a_n)$ and allocate each attribute with a weight, then the corresponding vector of weight is $W_u = (w_1, w_2, \ldots, w_n)$, $\sum_{i=1}^{n} w_i = 1$, the value of W_i is adjusted based on the related parameter adjusting method and the achievements, getting the most optimal allocation of value. Similar to user u, the attribute vector of user v is $A_v = (a'_1, a'_2, \ldots, a'_n)$. Then the formula (4) can be used to obtain similarity between user u and user v. In the formula, $\overline{a_i \oplus a'_i}$ is the complementation process of XNOR operator. In other words, the value of a_i xor a'_i is 0 when a_i and a'_i are not equal. On the contrary, the value of a_i xor a'_i is 1 when they are the same. Thus, if the two users' value of the ith attribute is equal and the weight of which is W_i, the final similarity value between user u and user v is $\sum_{i=1}^{n} w_i$. In conclusion, the more the same value of the user attributes, the greater the Weighted Distance Similarity values.

3 DISTRIBUTED COLLABORATIVE FILTERING RECOMMENDATION MODEL BASED ON TWO-PHASE SIMILARITY

In this section, DCF-TP is displayed for detail.

3.1 Compute two-phase similarity

1) Get User Similarity Matrix: Compute the similarity between users to get the similarity matrix of users by WDSM and users' attributes profile. Then, save the matrix and users' preference fileon HDFS (Hadoop Distributed File System [6]).

2) Get User Preference Matrix: do MapReduce operation on preference file to get the preference matrix as "userId$_i$: itemId$_1$, p; ...; itemId$_n$, p".

3) Get Item Similarity Matrix: do MapReduce operation based on the co-occurrence matrix method to get the items' similarity matrix.

Algorithm	The user's new preference mtrix

Input: Output from step*1)* and output from step*2)*,the number of user **M** and the number of the item **N** ;

Output: The user's new preference matrix as "userId_i : itemId_1, new p; …; itemId_n, new p" ;

Require: Do MapReduce parallel matrix multiplication

1: **class Mapper**
method Map (key doc*ID* , value Text)
  if(it is user similarity file)
    for each line as (userID1, userID2, similarity) do
    int i=0
    while(i<=N)do
    EMIT((userID1, i), ("similarity", userID2, similarity))
  if(it is user preference file)
    for each line as (user*ID*, item*ID*, p) do
    int i=0
    while(i<M)do
    EMIT((userID, i), ("preference", itemID, preference))
2: **class Reducer**
method Reduce (key (i, j), value(Iterator<fileName, k, m>)
  double[] similarity = new double[N] and initialized every element as 0
  double[] preference = new double[N] and initialized every element as 0
for each element e in value
  if(e.name equals to "simialrity")   similarity[k] = m;
  if(e.name equals to "preference")   preference [k] = m;
  int x=0
  double finalPreference=0
  while(x<=N)do
  finalPreference += similarity[x]* preference[x]
  EMIT(i, (j, finalPreference))

Figure 2. The algorithm of getting the new preference matrix.

4) Get the User's New Preference Matrix: do parallel matrix multiplication operation on the user preference matrix and the user similarity matrix to obtain the user's new preference matrix. The detail is presented in figure 2.

3.2 *Get nearest neighborhood*

Do parallel matrix multiplication operation of Map-Reduce on the new user preference matrix and item similarity matrix to obtain the user's expected preference matrix. Then, place the file on the HDFS server steadily.

3.3 *Make recommendations*

Select the top-N items to recommend to the target user and the recommendation is finished. Citing a simple example, as in figure 3, the correspondent value of itemId_2 and itemId_3 in the user's new preference matrix is 0, and then value of item$Id3$ is larger than that of item$Id2$ by the calculation. If N = 1, the best recommendation (the itemId_3) will be chosen to recommend to the userId_i.

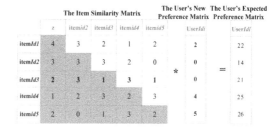

Figure 3. Making recommendations.

Table 1. Precision and recall ratio of CF and DCF-TP.

	Precision		Recall	
Algorithm	100 k	1 M	100 k	1 M
CF	0.001 169	0.001 094	0.003 770	0.003 890
DCF-TP	0.062 147	0.063 875	0.437 550	0.323 601

4 EXPERIMENTAL AND RESULTS

In this section, we evaluate DCF-TP in comparison with CF.

4.1 *Datasets*

The three data sets (ML-100K, ML-1M, and ML-10 M) of MovieLens (http://www.movielens.umn.edu) are used. With regard to the user profile, it includes the users' age, sex, and profession. Besides, each data set is divided into two parts: 20% of the users are selected to be testing model, and the remaining as training model.

4.2 *The accuracy evaluation of DCF-TP*

This section introduces the precision and recall ratio to evaluate the recommendation quality. The detailed results are shown in table 1:
  In table 1, compared with CF, the precision and recall ratio of DCF-TP have qualitative leaps and can obtain a better performance in the whole range.

4.3 *The experiment about cold startup*

To the problem about cold startup, we choose some users in ML-100K randomly, and the total number is 10, 20, 30, 40 and 50 respectively, removing the users' rating. Then DCF-TP is used to predict the rating. Compared with the original rating, the precision is shown in table 2.
  As is shown in table 2, the problem of cold startup is resolved. This is due to the new user's similarity matrix which can join in the similarity calculation.

Table 2. Recommendation precision of the new user.

	Precision				
Algorithm	10	20	30	40	50
DCF-TP	0.44142	0.42661	0.39467	0.38232	0.39101

Table 3. Run time of Tasks on single machine.

	Run Time (sec)		
Configuration	100 K	1 M	10 M
Single machine	1196.8	1338092.3	11593854.7
HOD(one node)	1 014	1 031	4 604
HOD(two nodes)	886	890	2 912
HOD(three nodes)	832	835	2 191

Table 4. Speedup of Tasks on HOD cluster.

	Speedup		
Node Number	100 K	1 M	10 M
1	1	1	1
2	1.144 470	1.158 427	1.581 044
3	1.218 750	1.234 731	2.101 324

4.4 The efficiency evaluation of DCF-TP

To verify the efficiency of DCF-TP, the test is done on a single machine and a small HOD (Hadoop on Demand) cluster respectively. The machine is configured as follows: Dell Power Edge SC1430, Intel Xeon 5110 (1.6 GHz), Dual-core Processor, 4 GB of Physical Memory. The cluster is composed of 3 machines which are served as pbs_server, namenod and datanode respectively. All nodes are installed in the executable program of Hadoop distributions. The detailed results are shown in table 3~4:

See from table 3~4, as to the same data set, the run time of a single node has been reduced greatly compared with a single machine, which reflects the MapReduce's advantage of data processing. Moreover, concerning the HOD cluster, with the increasing numbers of the computing nodes, the run time of all recommendation tasks are shortened and a higher linear speedup is achieved. In short, DCF-TP has achieved an effective speed performance.

5 CONCLUSIONS

The paper firstly analyzed the disadvantages of the existed similarity measures, and then in order to overcome these shortages, a distributed collaborative filtering recommendation model based on two-phase similarity is proposed. Then the experimental results reveal the efficiency of DCF-TP. What's more, DCF-TP can overcome the problem of cold sartup, which is faced by CF. In the future, we will investigate different techniques to improve the performance of our system, for example, using the data classification algorithm to alleviate the data sparse problem and using GPU to make our system more available.

ACKNOWLEDGMENT

Funding for this research was provided by the National Science Foundation of China under Grant No. 91024030.

REFERENCES

[1] Shumeet, B., Seth, R., Sivakumar, D., Jing, Y., Yagnik, J., Kumar, S., Ravichandran, D., Aly, M. 2008. Video suggestion and discovery for YouTube: taking random walks through the view graph. International Conference on World Wide Web: 895–904.

[2] Goldberg, D., Nichols, D., Oki, B., et al. 1992. Using collaborative filtering to weave an information tapestry. Communications of the ACM. Volume 35(12):61–71.

[3] Malucelli, F., Cremonesi, P., Rostami, B. 2012. An application of bicriterion shortest paths to collaborative filtering. Computer science and information systems (FedCSIS): 423–429.

[4] Dean, J., Ghemawat, S. 2008. MapReduce: simplified data processing on large clusters. Communications of the ACM. Volume 51(1): 107–113.

[5] Goldberg, K., Roeder, T., Gupta, D., Perkins, C. 2001. Eigentaste: a constant time collaborative filtering algorithm. Information Retrieval. Volume 4: 133–151, 2001.

[6] Borthakur, D. HDFS architecture guide. HADOOP APACHE PROJECT http://hadoop.apache.org/common/docs/current/hdfsdesign.pdf, 2008.

[7] Koren, Y., Bell, R. 2011. Advances in collaborative filtering //Recommender Systems Handbook. Springer US. 145–186.

[8] Zhang, Y., Liu, H. 2013. A Distributed Collaborative Filtering Recommender System based on Cloud Computing for Mobile Commerce. DCFRS. Appl. Math, 1: 1–10.

Future Communication, Information and Computer Science – Zheng (Ed.)
© *2015 Taylor & Francis Group, London, 978-1-138-02653-7*

Blind source separation of QAM signals by using a novel optimality of the constant modulus algorithm

H. Zheng, B. Yan, H.Y. Su & Y.J. Liu
Beijing Institute of Technology, Beijing, P.R.China

Z. Abderrahmen
Beijing Institute of Technology, Algiers, Algeria

ABSTRACT: Blind Source Separation (BSS) is used to estimate N unknown transmitted signals from the only knowledge of a set of P observation signals. This paper presents a new approach called CMA-norm based on the polynomial norm functions. It combines the CMA term with a Constellation Matching Error term to force the estimated symbols to converge to the original constellation symbols. Through several simulations performed and comparing to CMA, the proposed algorithm not only compensates the phase, but also presents good separation performance in terms of SINR and speed of convergence.

Keywords: blind source separation; BSS; constant modulus

1 INTRODUCTION

Blind source separation is a fundamental problem in signal processing. Blind source separation can be modeled in general regardless of the application domain as follows: signals (sources) emitted by a finished N independent transmitters, through a mixture of dimension (P×N) number, is received by a finite number of sensors P (observations). At the reception, we only have access to the received signals, the emitted signals and mixed signals are all unknown.

Most of linear BSS problems can be modeled in the following matrix form:

$$y = A\,\underline{S} + \underline{v} \qquad (1)$$

where $\underline{S} \in C^{N}$ represents the vector of sources, $y \in C^{P}$ is the vector of the observations received, $\underline{v} \in V^{P}$ is an unknown matrix representing errors or noise, and A is the mixture matrix defined from C^{N} to C^{P}.

In summary, we can write:

$$\begin{pmatrix} y_1 \\ y_2 \\ \vdots \\ y_P \end{pmatrix} = \begin{pmatrix} a_{11} & a_{12} & \cdots & a_{1,N} \\ a_{12} & a_{22} & \cdots & a_{2,N} \\ \vdots & \vdots & \ddots & \vdots \\ a_{P,1} & a_{P,2} & \cdots & a_{P,N} \end{pmatrix} \times \begin{pmatrix} S_1 \\ S_2 \\ \vdots \\ S_N \end{pmatrix} + \begin{pmatrix} v_1 \\ v_2 \\ \vdots \\ v_P \end{pmatrix} \qquad (2)$$

where a_{ij} defines the processing performed by the propagation channel, respectively, the source j propagating to the sensor i.

To recover the original signals using only the signals received, consider the case where we have as many sensors as sources, which means N = P. Additive noise is assumed to be negligible (v = 0). The mixing matrix A is therefore a square matrix and assuming it is invertible, there exists a solution such as:

$$\underline{S} = A^{-1}\,\underline{y} \qquad (3)$$

The challenge of the separation is to estimate a separation matrix B such that:

$$\underline{z} = B\,\underline{y} \qquad (4)$$

where z ∈ CN is the vector of estimated source signals. B is the separation matrix (inverse of A).

2 RELATED WORK

There are different methods of blind signal separation. These methods can typically be classified into three families (Puigt & Mouchtaris 2013), which are Independent Component Analysis, Factorization Non-negative Matrices and Sparse Component Analysis.

Comon & Jutten (2010) generalized the concept of Independent Component Analysis (ICA) to obtain statistically independent output signals equal to the source signals, but to an indeterminacy of scales and permutations close, this is in the case of linear instantaneous mixtures. For this type of mixture, this result is very useful and is a major method of the separation of sources.

A second family of methods of BSS assumes that the sources and the coefficients of the mixing matrix are positive. These methods are called Non-negative Factorization Matrix methods (NFM). They took attention of the scientific community and are particularly suitable for the separation of spectroscopy signals (Heittola et al. 2011).

Another family of methods, which is the subject of great interest over the last ten years, is known as Sparse Component Analysis (SCA) (Bofill & Zibulevsky 2001).

Some other methods are based on one or more of the following properties: independence, temporal coherence, parsimony and / or non-stationary. Tong et al. (1990) formulate mathematically the problem of source separation taking into account the presence of an additive noise n(t). They propose a simple algorithm (AMUSE) for the identification based on the covariance matrix of the sources.

In the algorithm SOBI (Second Order Blind Identification), Belouchrani et al. (1997) use second-order statistics. This method is based on multiple covariance matrices associated with non-zero delays. Mukai et al. (2006) have studied the problem of separation and 'reverberation' of convoluted mixtures of sources. They take into account the acoustic properties of the signal processing problem in frequency (set of frequency ICA).

3 CONSTANT MODULUS ALGORITHM (CMA)

Generally, blind techniques require knowledge of the statistical properties of the transmitted symbols to perform the task of separation. The latter often means minimizing a contrast function (cost function). In this context, several criteria have been proposed. The most famous is the Godard, or constant modulus, criterion, the algorithms are known as CMA (Van Der Veen, & Paulraj 1996).

The idea of CMA is to reduce the spread by attaching a gain, for example, that the issued and reconstructed constellations correspond to signals of the same variance. In the case of phase modulation, the constant modulus forces the output of the equalizer necessarily cancels the inter-symbol interference.

When the modulation is not a simple phase modulation, being as close as possible to the module output $z(t)$ has a constant modulus under certain assumptions the same effect.

Most of Godard algorithms are based on the following cost function:

$$J(W) = \frac{1}{4}\sum_{i=1}^{N} E\{(|z_{ik}|^2 - \gamma_i)^2\} \qquad (5)$$

where N is number of sources, γ_i is the dispersion coefficient for the i source and z_{ik} the estimated i source at the moment k defined by

$$z_{ik} = e_i^T W y_k \qquad (6)$$

where e_i a column vector of N-size ($e_i = 1$, $e_j = 0$ for $i \neq j$), W(N, P) the separation matrix, P the number

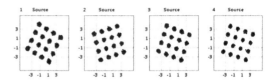

Figure 1. 16-QAM signals Constellations recovered by CMA (N = 4, P = 4, SNR = 30 dB and $\mu = 10^{-5}$).

of captors and y_k the mixture vector of size P at the instant k. The formula for updating the separation matrix W is:

$$W(k+1) = W(k) - \mu\left(\sum_{i}^{N}(|z_{ik}|^2 - \gamma_i)(z_{ik})(y_k e_i^T)\right) \qquad (7)$$

Before giving simulation results for the CMA algorithm we will present an example of the transmitted QAM signals before and after entering the transmission channel.

For all simulations presented in this section, we have considered the mixing matrix defined by Belouchrani (1996) as follows: p-th element of the q-th column of this matrix is given by $a_{pq} = \exp\{2i\pi pa_q\}$ with $a_q = 0.2q$.

Figure 1 shows the 16-QAM constellations of the recovered signals by CMA.

We can see clearly that the CMA cannot recover the phase of the signals. This is because the QAM signals do not have a constant modulus (which is the case for PSK signals), for that reason a PLL (Phase-Locked Loop) is often integrated in the CMA.

4 PROPOSED ALGORITHM

The algorithm presented above is based on cost functions to constrain estimated symbols to converge to either a circle (PSK modulation) or to an average mean square (QAM). This can lead to relatively high residual errors because the modulus is not constant and even for the PSK signals it becomes too difficult to separate the symbols especially for higher order modulations. To overcome this problem, we introduced a new hybrid algorithm called CMA-norm. The idea of CMA-norm is to combine the term CMA with a penalty term called CME for Constellation Matching Error. The CME's role is to force the estimated symbols to converge to the points of the constellation symbols.

The name of this criterion came from the fact that we use the product of L1-norms (polynomial norm functions) of the deviation between the equalized symbols and the points of the constellation. The cost function is written as:

$$J(B) = CMA + CME \qquad (8)$$

$$CMA = \frac{1}{4}\sum_{i=1}^{N} E\{(|z_{ik}|^2 - \gamma_i)^2\}. \qquad (9)$$

The proposed term CME is:

$$CME = \beta \sum_{i=1}^{N} (g_r(z_{ik}) - g_{im}(z_j k)) \qquad (10)$$

where β is a real constant, $g_r(z_{jk}) = |R^2(z_{jk}) - 1||R^2(z_{jk}) - 9|$ and $g_{im}(z_{jk}) = |I^2(z_{jk}) - 1||I^2(z_{jk}) - 9|$ (for 16-QAM).

Note that $R(z_{jk})$ and $I(z_{jk})$ are the Real and imaginary parts of the vector z_{jk}. So for the 16-QAM constellation, the cost function to minimize will be:

$$J(W) = \frac{1}{4} \sum_{i=1}^{N} E\{(|z_{ik}|^2 - \gamma_i)^2\} +$$
$$\beta \sum_{i=1}^{N} (|R^2(z_{ik}) - 1||R^2(z_{ik}) - 9| - |I^2(z_{ik}) - 1||I^2(z_{ik}) - 9|) \qquad (11)$$

The formula for updating coefficients of the separation matrix is:

$$W^{(n)} = W^{(n-1)} - \mu \nabla_W J(W) \qquad (12)$$

where μ is the adaptation step size of the gradient algorithm. $\nabla_W J(W)$ is the gradient of the cost function $J(W)$ relative to the separation matrix W. It can be written as:

$$\nabla_W J(W) = \frac{\partial J(W)}{\partial W} = \frac{\partial CMA}{\partial W} + \frac{\partial CME}{\partial W} \qquad (13)$$

Then we have the following formula for updating the separation matrix for 16-QAM:

$$W(k+1) = W(k) - \mu(\sum_{i=1}^{N} (|z_{ik}|^2 - \gamma_i)(z_{ik})(y_k e_i^T) +$$
$$2\beta(\sum_{i=1}^{N} \{(R(z_{ik})\{sign(R^2(z_{ik}) - 1)|R^2(z_{ik}) - 9| +$$
$$sign(R^2(z_{ik}) - 9)|R^2(z_{ik}) - 1|\} + \qquad (14)$$
$$(j)I(z_{ik})\{sign(I^2(z_{ik}) - 1)|I^2(z_{ik}) - 9| +$$
$$sign(I^2(z_{ik}) - 9)|I^2(z_{ik}) - 1|\}\}(yke_i^T)))$$

To simulate the behavior of the CMA-norm algorithm, we considered the steps below:

(1) Generating a sequence of symbols of QAM modulation.
(2) Select a transmission channel.
(3) Applying the impulse response of the channel to the generated sequence symbols.
(4) Add white Gaussian noise (AWGN) to the transmitted symbols.
(5) Initialize the equalizer coefficients.
(6) Use stochastic gradient to adjust the equalizer coefficients.
(7) Estimate symbols.
(8) Decision on symbols.

The constellations of the recovered signals by CMA-norm are shown in the following Figure 2.

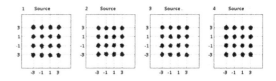

Figure 2. 16-QAM signals Constellations recovered by CMA-norm (N = 4, P = 4, SNR = 30 dB, $\mu = 3 \times 10$-5 and $\beta = 1$).

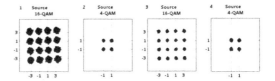

Figure 3. 4 and 16-QAM signals Constellations recovered by CMA-norm (N = 4, P = 4, SNR = 30 dB).

In order to investigate the efficiency of the algorithm in the case of different types of signals, we performed some tests using mixtures of different modulations order. We used two sources of 4-QAM and two sources of 16-QAM. Our algorithm separates the sources efficiently as is shown in Figure 3.

From the pictures above, we can see clearly that the CMA-norm algorithm recover the all the sources, and moreover it compensates the signals phase because the cost function takes in account both modulus and phase. This represents a great gain compared to the CMA algorithm. By using the CMA-norm algorithm we don't need to add a module of phase compensator at the receiver side.

5 PERFORMANCE COMPARING

Maybe the best criterion of comparison is known under the acronym SINR for Signal to Interference Noise Ratio. It measures the contribution of other sources of noise in one of the outputs of the separator. It is defined by:

$$SINR_k = \frac{|g_{kk}|^2}{\sum_{l, l \neq k} |g_{lk}|^2 + W_k^T R_n W_k^*} \qquad (15)$$

where $SINR_k$ is signal-noise and interference ratio associated with the k-th output of the separator. $g_{ij} = A_i^T W_j$ where A_i^T and W_j the i-th and the j-th column of the matrices A and W respectively, $R_n = E[nn^H] = \sigma_N^2 I_N$ the noise covariance matrix.

In the simulations and for congestion avoiding, we rather consider the average SINR on all outputs. It is given by:

$$SINR = \frac{1}{M} \sum_{k=1}^{M} SINR_k \qquad (16)$$

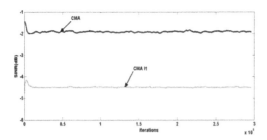

Figure 4. Comparison between CMA and CMA-11algorithms using SINR (N = 2, P = 2, SNR = 30 dB).

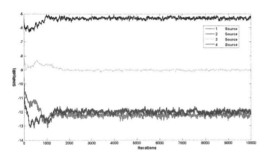

Figure 5. Comparison between SINR of sources separated by CMA-norm (N = 4, P = 4, SNR = 30 dB).

Figure 4 shows that the algorithms CMA and CMA-norm have different SINR levels. In this case the algorithm CMA-norm is the best with a SINR = −4.5 dB. Furthermore, we can see that in terms of speed of convergence algorithm CMA-norm is better (500 iterations).

The fact that the algorithms mentioned above extract the sources one by one, we thought it would be interesting to study the performance of separation between the sources in terms of SINR. Figure 5 shows the result of the separation algorithm CMA-norm in the case N = 4, P = 4.

We note that the first source is estimated with a SINR = −12.5 dB, while the fourth is estimated with SINR = −6 dB. This difference is explained by the accumulation of the estimation error caused by successive update lines of separation matrix W. This can be seen as a drawback in the case where the number of sources is high.

6 CONCLUSION

In this paper we have introduced a new blind source separation algorithm called CMA-norm. It is based on the calculation of the deviation between the equalized symbols and the points of the transmitted constellation. Through the numerical experiments, we can see the benefit of using hybrid algorithms that add the penalty term to the CMA cost function. With this novel technique, not only the phase is compensated, but also it presents good performances in terms of SINR and speed of convergence. In future work, we intend to study the automatic setting of the algorithm parameters.

REFERENCES

[1] Belouchrani, A. & Abed-Meraim, K. 1996. Constant Modulus Blind Source Separation Technique: A New Approach. *Signal Processing and Its Applications, ISSPA 96. Fourth International Symposium on. Australia, August 1996.*

[2] Belouchrani, A., Abed-Meraim, K., Moulines, E. & Cardoso, J. 1997. A Blind Source Separation Technique Using Second-Order Statistics. *Signal Processing, IEEE transactions on* 45(2): 434–444.

[3] Bofill, P. & Zibulevsky, M. 2001. *Signal processing*, 81(11): 2353–2362.

[4] Comon, P. & Jutten, C. 2010. *Handbook of Blind Source Separation: Independent Component Analysis and Applications*, Burlinton: Academic Press.

[5] Heittola, T., Mesaros, A., Virtanen, T. & Eronen, A. 2011. *Sound Event Detection in Multisource Environments Using Source Separation, Workshop on Machine Listening in Multisource Environments (CHiME 2011), Florence, Italy, 2011*, 36–40.

[6] Mukai, R., Sawada, H., Araki, S. & Makino, S. 2006. Frequency-Domain Blind Source Separation of Many Speech Signals Using Near-Field and Far-Field Models. *EURASIP Journal on Applied Signal Processing, 2006*, 200–200.

[7] Puigt, M. & Mouchtaris, A. 2013. Real-time Multiple Sound Source Localization and Counting Using a Circular Microphone Array, *Audio, Speech, and Language Processing, IEEE Transactions on* 21(10): 2193–2206.

[8] Tong, L., Soon, V.C., Huang, Y. & Liu, R. 1990. AMUSE: A New Blind Identification Algorithm. *Circuits and Systems, IEEE International Symposium on* 3: 1784–1787.

[9] Van Der Veen, A.-J. & Paulraj, A. 1996. An Analytical Constant Modulus Algorithm. *Signal Processing, IEEE Transactions on* 44(5): 1136–1155.

Future Communication, Information and Computer Science – Zheng (Ed.)
© 2015 Taylor & Francis Group, London, 978-1-138-02653-7

An equidistance multi-human detection algorithm based on noise level using mono-static IR-UWB radar system

D.H. Yim & S.H. Cho
Department of Electronics Engineering, Hanyang University, Seoul, Korea

ABSTRACT: On the use of Impulse Radio Ultra-Wide Band (IR-UWB) system, it is impossible to detect the number of the targets, which are equidistant, from the IR-UWB Radar because we can measure only the distance of each target. From these facts, IR-UWB Radar has limits and needs more antennae for application such as the people counting algorithm in specific area. This paper proposes an algorithm of detecting the number of people, who are equidistant, using noise-level of received signal.

Keywords: IR-UWB; Multi-human; Ranging; Equidistance

1 INTRODUCTION

Mono-static IR-UWB radar can be used for a ranging system using the time of arrival (ToA) method. It has good penetrability, high accuracy, low power consumption and high resolution. Because of these advantages IR-UWB radar can be used in many fields such as vital signal detection, internal material inspections, baby condition monitoring, intruder surveillances, through wall surveillance and distance measurements of moving objects.

When the mono-static IR-UWB detects multi-human in the detecting area, we need to observe only the signal of people without the signal of a fixed object. It is the reason of using the background subtraction loopback (BSL) filter. The radar can detect the number of people, who are not equidistant, as a signal passed the BSL filter because the measured data type is not direction but only distance. Also, the signal of multi-human is no different from the signal of one human even with amplitude contrary to the ideal. In an ideal environment, however, it is possible to detect the number of people as the amplitude of signal. It is shown in Figure 1.

This paper proposes a novel algorithm of detecting the equidistance multi-human based on noise level. The noise is made by many factors such as diffraction, interference, and multi-path signal, and the noise has a low level if nothing is detected. The specific relationship between the noise level and the number of people is shown, we can decide the number of people as the noise level of the received signal.

2 SIGNAL MODEL & BACKGROUND KNOWLEDGE

2.1 *Signal model*

Received signal of IR-UWB radar is presented as

$$r(t) = \sum_{n=1}^{N} a_n s(t - \tau_n) + n(t) \qquad (1)$$

When there are N paths including a direct path, we define the meaning of the each variable as follows: $s(t)$ is template signal of IR-UWB radar, $n(t)$ is white gaussian noise (WGN), a_n is amplitude of n-th path signal, and τ_n is delay of received n-th path.

As the ToA ranging method, the distance is calculated as $d = \frac{\tau_n}{2} \cdot c$. c is light speed. Then the received impulse array of i-th frame is defined as

$$a[k] = \sum_{n=1}^{N} a_n \delta \left[k - \frac{\tau_n \cdot c}{2 d_s} \right], \qquad d_s \cdot k = \frac{\tau_n \cdot c}{2} \quad (2)$$

d_s is the distance of each sample and k is distance parameter. Sampled template signal is defined as $s[t]$. And then i-th frame is shown as

$$S_{r,i}[k] = s[n] * a[k] \qquad (3)$$

2.2 *Background knowledge*

If two objects are equidistant, these signals have i-th path and j-th path each other, however, τ_i and τ_j is same

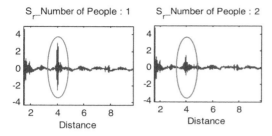

Figure 1. Compare the n-th frame signal of one and two people who are equidistant.

as τ_0. Then this signal is presented as

$$a_i s(t - \tau_i) + a_j s(t - \tau_j) = (a_i + a_j) \cdot s(t - \tau_0)$$
$$= M \cdot s(t - \tau_0) \quad (4)$$

It can be seen as the signal of a path because M is often smaller than a_{other} even if M is the sum of a_i and a_j. Thus, the amplitude cannot be used to decide the number of people.

As an electromagnetic wave the IR-UWB signal can be affected by interference, diffraction, multi-path and so on. All of these factors acted as noise, particularly the multi-path is the most influential factor. Admittedly, the noise level has a certain relationship with the number of people. So the noise level can be used to detect the number of people.

3 BACKGROUND SUBTRACTION LOOPBACK FILTER & CALCULATED AREA

3.1 BSL filtering

The received signal including the signal of the target, signal of the background clutter and WGN is shown as

$$S_{recive,i}[k] = S_{Target,i}[k] + S_{clutter,i}[k] + n(k) \quad (5)$$

Thus, it is necessary to remove the signal of the background clutter for observing the target. The clutter of signal can be represented by the received signal and can be shown as

$$S'_{clutter,i}[k] = \alpha \cdot S_{recive,i}[k]$$
$$+ (1 - \alpha) \cdot S'_{clutter,i-1}[k] \quad (6)$$

And the received signal without the clutter of signal is shown as

$$S'_{Target,i}[k] = S_{recive,i}[k] - S'_{clutter,i}[k] \quad (7)$$

Also the received power signal (RPS) of the target is shown as

$$P_{T,i}[k] = \{S'_{Target,i}[k]\}^2 \quad (8)$$

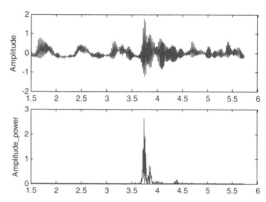

Figure 2. The Received Signal (top), Received Target Power Signal (bottom).

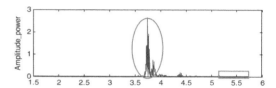

Figure 3. The peak of Received Target Power Signal (circle), Calculated Area (bottom).

3.2 Calculated Area (CA)

The variation of the target signal's peak value is greater than noise, thus it is necessary to limit the observation area. The circle which is shown in Figure 3 is the sum of the direct path and the strong multi-path. The rectangle in Figure 3 is the area which is limited to access and defined as CA. We assume that there is no direct and strong multi-path in the calculated area.

4 CALCULATING NOISE LEVEL

The noise level is an indicator of deciding the number of people. With the increase of the number of people, the multi-path becomes denser and it can be seen as the noise. Thus we can use this point to decide the number of people.

In CA, the variance of each sample can be calculated with the data in last M frames. If the current frame is i_0-th frame and the length of CA is N, the noise level can be shown as

$$N[i_0] = \frac{1}{M} \sum_{k=L-M}^{L} \left[\frac{1}{N} \sum_{i=i_0-N}^{i_0} \left\{ P_{T,i}[k] - \frac{1}{N} \sum_{i=i_0-N}^{i_0} P_{T,i}[k] \right\}^2 \right]$$
$$= \frac{1}{M} \sum_{k=L-M}^{L} \left[\frac{1}{N} \sum_{i=i_0-N}^{i_0} \{P_{T,i}[k]\}^2 - \left\{ \frac{1}{N} \sum_{i=i_0-N}^{i_0} P_{T,i}[k] \right\}^2 \right] \quad (9)$$

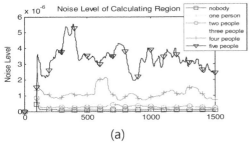

(a)

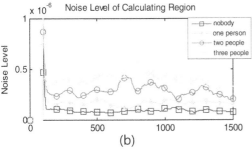

(b)

Figure 4.   The Noise Level of each case (0~5 peoples, top), The Noise Level of each case (0~3 peoples, bottom).

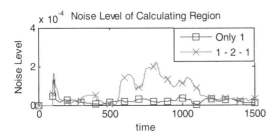

Figure 5.   The blue line is the noise level of one people and the red line is the noise level when the number of people is changed (0~500 & 1000~1500: one person, 500~1000: two people).

5   EXPERIMENTS RESULTS

The NVA-R661 chip of IR-UWB Radar from Novelda is used for generating and receiving impulse signals with a commercial mono-static antenna which has property of opening angle $20°(V) \times 50°(H)$.

As shown in Equation 9, the noise level of each case is shown.

All of the cases are experimented in the same position. The noise level of the data is proportionate to the number of people although in another experimental environment.

The real time detection experiment is also shown in Figure 5. The equipment used in this experiment is the same as the previous experiment. The changing of the noise level shows the changing of the number of people.

6   CONCLUSION

An equidistance multi-human detection algorithm is proposed in this paper. This algorithm is different from the previous detection algorithm which is based on peak value. The algorithm based on peak value cannot detect the humans who stay in the same distance from the mono-static radar, because the peak values of the received signal from humans are similar. The noise level of the target power signal in a specific area is clearly depending on the number of people who are detected by IR-UWB radar. As a future work, we should improve the proposed algorithm to use in practice. Therefore the calculated area becomes smaller, and the number of frames for calculating variance should also be reduced.

ACKNOWLEDGEMENT

This research was supported by a grant (13RTRP-B067918-01) from Railroad Technology Research Program funded by Ministry of Land, Infrastructure and Transport of Korean government.

REFERENCES

[1]   Jeong Woo Choi, Sung Ho Cho, 2013, A new multi-human detection algorithm using an IR-UWB radar system, *Innovative Computing Technology (INTECH), 2013 Third International Conference on*, vol., no., pp. 467–472.

[2]   NING, Xiaoyan, SHA, Xuejun, WU, Xuanli, 2010, A TOA Estimation Method for IR-UWB Ranging Systems, *Pervasive Computing Signal Processing and Applications (PCSPA), 2010 First International Conference on. IEEE*, pp. 684–687.

[3]   KIM, JinHo, et al, 2012, A ranging algorithm for multiple targets using an IR-UWB radar system, *Proc. Int. Tech. Conf. on Circ./Syst., Comput. and Commun.(ITC-CSCC)*.

[4]   Jeong Woo Choi, Jin Ho Kim, Sung Ho Cho, 2012, A counting algorithm for multiple objects using an IR-UWB radar system. *Network Infrastructure and Digital Content (IC-NIDC), 2012 3rd IEEE International Conference on. IEEE*, pp. 591–595.

Future Communication, Information and Computer Science – Zheng (Ed.)
© *2015 Taylor & Francis Group, London, 978-1-138-02653-7*

The application research of campus network remote access solution based on redundant architecture

S. Jing
Shandong Provincial Key Laboratory of Network Based Intelligent Computing, University of Jinan, Jinan, China

Z. Zheng
School of Information Science and Engineering, University of Jinan, Jinan, China

R. Sun
Shandong Provincial Key Laboratory of Network Based Intelligent Computing, University of Jinan, Jinan, China

ABSTRACT: With the rise of informationization level of campuses, the demand of remote network access is increasing. This paper, set in the demand of remote campus network access of University of Jinan, analyzed correlative technique of Easy VPN and DMVPN. Using the old routers based on a modified roll polling algorithm to build a cost-effective and efficient campus network remote access solution based on redundant architecture.

Keywords: redundancy; campus network; VPN; roll polling algorithm; cost-effective

1   INTRODUCTION

With the rise of informationization level of campus network, the easy-to-use of campus network internal resources is becoming more important. A problem that needed to be solved for the campus network administrator is providing a secure, effective remote access solution for sub-campus users and external users [1]. Many campuses built a solution to solve this problem by purchasing new VPN Firewalls, but there are still some problems such as single point of failure, high cost and so on. The old router that is replaced in the process of campus network upgrading can be used to build the remote access VPN by upgrading the router's operation system, but there are still some features like poor stability, poor access capability and so on. There are some replaced Cisco routers in the campus network of the University of Jinan. How to take advantage of these routers to build a secure, efficient, and cost effective campus network remote access solution is the focus of this research [2].

2   VPN TECHNOLOGIES OVERVIEW

VPN based on the public network achieved the security of the network, simplified the network design and management and reduced the costs by using the tunnel encryption and decryption algorithm and identity authentication technologies [3–4].

2.1   Easy VPN

Easy VPN, also known as EzVPN is a remote access VPN solution for remote users and remote branch offices. It provides central VPN management, dynamic policy distribution and it also reduces the complexity of deployment of remote access VPN, increases the expansibility and flexibility.

Easy VPN has the following characteristics: (1) Easy VPN is a private technology of Cisco, it can only be used with Cisco equipment. (2) Easy VPN is appropriate for the environment of fixed central site and the client gets the address dynamically, in addition, the behind networks of the client should be as simple as possible. (3) The policy of Easy VPN is configured in the central site, they will be pushed to the client when the client connects to the central site. So, it can reduce the management stress of the branch site [5].

2.2   DMVPN

DMVPN (Dynamic Multipoint VPN) is a kind of technology that is used to build VPN tunnel dynamically, its purpose is resolving the address of the opposite end that needs to build the VPN tunnel by using Next Hop Resolution Protocol (NHRP) in the Hub-And-Spoke network environment and it uses Multipoint GRE tunnel port to build multipoint GRE over IPsec VPN tunnel. On account of IPsec VPN is the most widely used VPN technology at present and Hub-and-Spoke

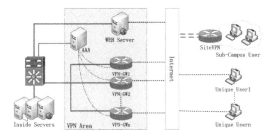

Figure 1. VPN System Topology.

is the most used network topology, So DMVPN technology is valuable in application and reference.

3 CAMPUS NETWORK VPN SOLUTIONS

3.1 *System design*

Campus network remote access users can be divided into two categories, unique users and sub-campus users. Unique users' use of remote access service is temporary, they establish connections before using and then disconnect immediately after using. In this circumstance, the administrator needs to grant permanent or temporary permissions, it's fit for the teachers and students who are out of the campus network to access the campus' network inside resources. For those sub-campus users, the establishment of permanent and stable VPN tunnels is needed to let users who are in different campuses easily access other campus inside resources at any time. VPN provides a secure encryption link between multi-campuses [6].

This project uses three replaced routers from campus networks to build VPN service area to achieve relatively stable and low cost VPN solutions by adopting data redundancy, link redundancy and certification redundancy. The project takes a full consideration of the stability of old routers, adopts modified roll polling algorithm to allocate VPN resources reasonably to improve the system efficiency. The solution diagram is shown in figure 1.

3.2 *Unique user solution*

Unique users need to login to the Web server to register before using VPN service. An administrator will audit the user's registration information. The approved user data will be written to an AAA authentication server and all of the VPN routers at the same time. When the management server assigns the VPN address information for the user, it usually assigns a main VPN and a backup VPN router address to reduce the effects of a single point of failure.

The distribution method of VPN routers: the system selects one router by a roll polling mechanism and then gets the VPN router's CPU usage by SNMP protocol. If the current selected router's usage is the highest, the system will select the next router by roll polling mechanism and make it become the main VPN router

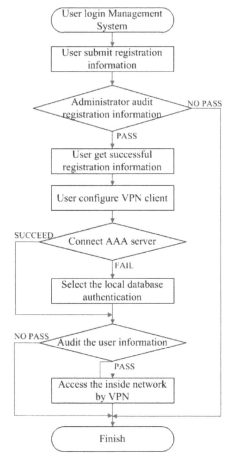

Figure 2. Unique User Authentication Flow Chart.

and then select one other router to become the backup VPN router randomly.

Users will receive the successful registration mail after the administrator's auditing and then get the VPN connection's addresses. Users use VPN client to connect to VPN routers. The login information will be sent to an AAA server to be authenticated. If the connection is failed, the login information will be authenticated locally. Double authentication policy improves the reliability of the service.

The configuration information on the routers:
aaa new-model
aaa authentication login vpn-login group radius local
Users who are authenticated successfully can access the campus' network inside resources. The authentication flow chart is shown figure 2.

3.3 *Sub-Campus user solution*

For those sub-campus users, remote access adopts DMVPN. The three campus network VPN routers configure same security policy and every router builds a virtual tunnel with the branch site router. In the branch site routers, they have built one virtual tunnel with the three campus network VPN routers at the same time and notice the three virtual tunnels by dynamic routing protocol. So, the three virtual tunnels can realize load balancing in the routing entry toward central site to provide central VPN gateway's availability of single point of failure. To reduce the stress of the central VPN gateway, the branch site routers also need to enable tunnel segmentation.

3.4 *IP addresses and routing planning*

When Easy VPN and DMVPN is established, external network users will be in the same campus network logically with the original campus network users. If the IP addressing is unreasonable, there will be some problems like routing loops, inaccessible routing and so on. So IP addresses and routing should be planned reasonably in the beginning of the design. Within the campus network VPN project, the design uses the address block of 172.20.0.0/15 which is not in use by the current campus network as the VPN addresses. The design of routing should consider two aspects. The first aspect is the routing from campus core to the VPN subnets. For this aspect, campus core can be configured as static routes to direct the addresses that assigned Easy VPN and DMVPN users to the VPN gateways. The second aspect is the routing from remote subnets to the campus network. For this aspect, Easy VPN users can use ACLs to match interesting traffic, while DMVPN users can use dynamic routing protocol to define interesting traffic.

3.5 *Management system design*

The management system is based on B/S architecture. To integrate the management system database and AAA authentication server database, the Cisco ACS system that can invoke external databases should be installed in the AAA authentication server so that the management system can control the VPN's usage by managing the ACS database. The management system can provide the administrators functions such as querying the VPN gateway addresses assigned to users, the current users and login history, the server's operational status and adding, auditing or managing users and user groups.

4 CONCLUSIONS

With the rising of demands of remote external users to access the campus network, this paper which is on the basis of the related technologies research has taken advantage of old equipment to build a remote access solution based on redundant architecture. This solution has features of extensibility, redundancy and effectiveness. After verifying, this solution is feasible, it can provide some reference value for the remote access solution design of other campus networks to a certain extent.

ACKNOWLEDGEMENTS

This work was supported by the Program for sub-project of the National Sci-Tech Support Plan (2012BAF12B07-3), and the next generation of Internet technology research and development, industrialization and large scale commercial projects (CNGI-12-03-017).

REFERENCES

[1] Ye Guojian, Pan Keyu, An Hong, "VPN Technology Application in Campus Network," Computer Engineering, Vol. 26, Sup, pp. 210–214, October 2000.
[2] Ni Jianhong, Lv Guanghong, "Research on VPN based Different Implementation," Application Research of Computers, No. 7, pp. 257–260, July 2005.
[3] Li Fang, "VPN-based Research and Application of Campus Networks," Ocean University of China, 2011.
[4] D. Farinacci, T. Li, S. Hanks, D. Meyer, and P. Traina, "Generic Routing Encapsulation," RFC 2784, March 2000.
[5] Qin Ke, "Cisco IPsec VPN Practical Guide," POST & TELECOM PRESS, 2012.
[6] WEI Nianzhong, "Study on the Network Safety of Multi-district Campus Based on VPN Technology," Vol. 24, No. 10, pp. 108–110, October 2007.

Future Communication, Information and Computer Science – Zheng (Ed.)
© 2015 Taylor & Francis Group, London, 978-1-138-02653-7

Information and knowledge engineering in agricultural enterprises in the Czech Republic

M. Ulman, J. Tyrychtr, V. Vostrovský & J. Jarolímek
Department of Information Technology & Department of Software Engineering, Faculty of Economics and Management, Czech University of Life Sciences in Prague, Czech Republic

ABSTRACT: The paper presents the current state and overview of information and knowledge management in small agricultural enterprises in the Czech Republic. The focus is on the level of hardware and software equipment at farms, and the uptake of e-services provided by national authorities. Results of a survey among Czech farmers (n = 135) are presented and analyzed with basic descriptive statistics. The paper identifies information gaps in knowledge management among farmers and formulates further research topics as conclusions.

Keywords: Knowledge management; farm; Czech Republic; information support; information needs; agricultural e-government

1 INTRODUCTION

Since the accession of the Czech Republic in the European Union in 2004, the competitiveness of the Czech agriculture has become an even more important issue. Many opportunities for agricultural commodities and product exports have arisen, while the imports of cheap food from abroad have increasingly started to threaten the Czech farmers' production. The Czech agricultural trade is characterized by the orientation on the EU market, and by a long-term negative agricultural trade balance that is still increasing (Vološin et al., 2011; Bielik, Smutka, Svatoš & Hupková, 2013). The overall profitability of the agricultural production is endangered by a number of threats and risks such as diseases, infection, pests and climate changes (Xing & Lu, 2010; Ju, Lin, Wheeler, Challinor & Jiang, 2013). These threats pose challenges that need to be faced through the increase of immunity of the breeds, growing more resistant crops, or improving the gene pool of species, which inevitably requires considerable financial and labor costs, and adequate knowledge as well.

It is evident that the worsening of production periods in past years and even more resistant varieties will put higher demands on timely and effective implementation of the relevant agronomic and veterinary measures. The agricultural enterprise needs to have an appropriate set of available knowledge for the optimal choice and use of measures (Nuthall, 2011). In agriculture, especially at small farms, there is a real danger that this kind of knowledge is usually closely tied with its owner and is likely to be lost after his or her retirement (Šubrt, Zuzák et al., 2010; Nuthall, 2011). It is a priority to find ways how to prevent losses, and how

to store, distribute and use necessary knowledge in agricultural enterprises.

1.1 *Information and knowledge management*

Businesses in agriculture have also integrated ICT into knowledge management and decision-making. The decision-making processes in farming are highly tacit knowledge based (Eastwood & Kenny, 2009; Nuthall, 2011). Tacit knowledge is a form of experiential knowledge that is 'context-specific and often used intuitively and unconsciously' (Hoffmann, Probst & Christnick, 2007). Explicit knowledge is a more objective form based on theory and rationality (Nonaka, 1994) which is more readily expressed and exchanged with others, for example ration formulation for concentrate feeding of cattle. Knowledge is constantly exchanged between the tacit and explicit forms (Nonaka, 1994).

The priority should focus especially on dissemination and further use of knowledge while the main requirement is to ensure that knowledge is not only passively accumulated, but also transmitted and effectively used. The basic characteristics of knowledge workers are such: (1) they are costly, (2) their number is scarce because it takes many years to form his or her profile, (3) they cannot be fully substituted, so their sudden departure poses a loss for the enterprise as the knowledge is gone with the person. Regarding these facts, there is a need to start immediately with the corresponding externalization of knowledge of such workers so that the knowledge does not leave, is not forgotten or dies (Šubrt, Zuzák, et al., 2010). Beckman (1999) states the knowledgeexternalization is the process of formalizing tacit knowledge to explicit knowledge which allows their easier storage and subsequent dissemination in the particular organization. The success of any business entity is then predetermined among others by whether it is able to select correctly the appropriate knowledge strategy and apply this strategy in practice (Šubrt, Zuzák, et al., 2010; Payne & Birchall, 2003). Prerequisite for the success of such strategies must be a meaningful effort to include and subsequently use the knowledge-based solutions in existing information support of these businesses.

1.2 Information support by public authorities

The deployment of information and communication technologies into operation and management of the agricultural sector has been denoted as eGovernment, or agricultural eGovernment (Ntaliani et al., 2010). E-government in agriculture could be understood as a type of e-government that is based on the use of information technology by the state administration to facilitate reciprocal information exchange between the involved agricultural public authorities such as the Ministry of Agriculture and agricultural enterprises to improve efficiency of its internal use and to provide fast, accessible and quality information services (Ulman, Vostrovský & Tyrychtr, 2013). Agriculture in Europe still remains under strong government control (Havlíček & Tichá, 2003). However, the detailed information about the uptake and the quality of electronic public services among agriculturists, forestry, fishing and aquaculture is still lacking (Rysová et al., 2013).

Goal of the paper is to present an overview of the current state of information needs, support and government authority provisions for small agricultural enterprises in the Czech Republic.

2 RESULTS AND DISCUSSION

The descriptive analysis of information needs, support and the level of the use of e-government services was based on a questionnaire survey conducted by the Department of Information Technologies and Department of Software Engineering at the Faculty of Economics and Management at Czech University of Life Sciences in Prague in 2013.

There were 135 enterprises that participated in the survey. The respondents were asked basic characteristic questions as to the location of the farm, the number of hectares, the number of employed people and whether or not they are recipients of subsidies.

A vast portion of the sample were representatives of an agricultural business with less than nine employed persons (96%), and with plant (90%) and animal (52%) production. Despite the small human resources of farms, half of them were operating on land of over 100 hectares (100–499 hectares, 50%). The Czech Republic has one of the highest numbers of farms operating on land of over 100 hectares in Europe (Martins

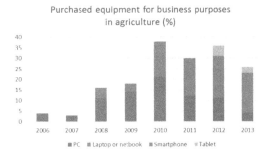

Figure 1. Purchased hardware for business purposes in Czech agricultural enterprises (own survey, 2013).

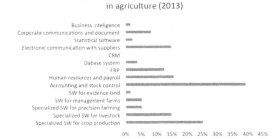

Figure 2. Types of software used for operation of Czech agricultural enterprises (own survey, 2013).

& Tosstorff, 2011), which was inherited from former state-owned cooperatives. The strong ties to national and EU subsidies were also revealed among farmers (98% were recipients of direct payments per land).

2.1 Information support of the Czech agricultural enterprises

In general we can conclude that the information support of enterprises in the Czech Republic and information and communication technologies that are used by them are on a relatively high level (see Figure 1).

However, the current state does not correspond to demands of knowledge economy and knowledge management. Practically, it is needed to provide not only relevant data on demand, but also knowledge to apply data effectively. The trend line of hardware equipment is depicted at Figure 1. As to the software, the farmers put importance to specialized software for animal and plant production, and also to accounting software (see Figure 2).

It is obvious that applications for knowledge management were not represented in the responses of Czech agriculturists. The computerized storage of knowledge is fatally missing in the current information support of agricultural enterprises. They have sets of data and information available for managerial decisions, but there is no knowledge support for their decision-making. The utilization of knowledge management needs to become an integral part of the current ICT infrastructure of enterprises. The efficient

Table 1. The rate of use of electronic services among agricultural enterprises in the Czech Republic.

E-services	Rate(n)	Rate(%)
E-mail	115	86%
E-signature	25	19%
eAGRI website	72	53%
Land registry*	99	73%
Animal registry*	61	45%
Registry of plants* protection products and fertilizers	57	42%
Portal for subsidies*	71	53%
Environ. rel. submissions	13	10%
Czech Statistical Office	23	17%
Secured e-mail**	29	21%
Portal of Public Admin.	17	13%
Tax portal	29	21%
E-procurement	4	3%
E-customs	7	5%
New company registration	2	1%

*The e-service is a part of the eAGRI portal.
**Secured and authorized electronic communication system provided by the Czech government.

Table 2. The rate and type of information received by farmers from Czech national authorities and institutions.

Information	Rate(n)	Rate (%)
News in agriculture	85	63%
Reminders	27	20%
Invitation to events and meetings	103	76%
Public tender offers	17	13%
Agricultural production	64	47%
Subsidies and EU funds	48	36%
Other	8	6%

Table 3. The institutions and national authorities that send information to farmers by e-mail.

Institution	Rate(n)	Rate (%)
Ministry of Agriculture	82	61%
Other ministries*	26	19%
State Phytosanitary Admin.	57	42%
State Veterinary Admin.	24	18%
Associations of breeders	31	23%
Associations of planters	30	22%
Czech Trade Inspection Authority	3	2%
Customs Administration	17	13%
Czech Office for Surveying, Mapping and Cadastre	15	11%
Czech Agriculture and Food Inspection Authority	9	7%
Other	9	7%

*Ministries of environment, industry and trade, interior affairs, finance, regional development and labor and social affairs.

implementation of ICT in enterprise processes might help in better access of farmers to agricultural product markets (Wang, Yu, Mo & Fu, 2013).

2.2 E-government in Czech agriculture

There were 13 different electronic services offered in the questionnaire plus electronic mail and electronic signature. Five of them were e-services specific for agriculture. They are located at eAGRI Portal which is operated by the Czech Ministry of Agriculture and provides specific information and e-services for agriculturists at www.eagri.cz. Good awareness and frequent use of information portals among Czech agriculturists was also confirmed by Šimek, Vaněk, Jarolímek, Stočes & Vogeltanzová (2011). The rest were general online services for businesses. All thirteen services were then evaluated with focus on their frequency of use, importance, usefulness and quality (see Table 1).

The most prevailing service among farmers was e-mail as a tool for communication with public authorities (86%). All major agriculture specific e-services at eAGRI portal turned out to be profound for more than 40% of all participants varying only by the type of production (plant or animal). The use of electronic submission to environmental institutions was rather low (10%). General public e-services such as e-submission to the Czech Statistical Office, Tax portal and the Portal of Public Administration fluctuated only between 10–21%. Their users among farmers do not perceive high benefits in these services. The obvious disproportion between traditional e-mail and secured e-mail is caused by a voluntary basis of use of the latter among business owners. There is also insufficient promotion

of the use of secured e-mail in the Czech Republic and limited applications for the agricultural sector.

2.3 Information needs of farmers

The electronic communication and computer operation were done mostly by the farm owner or maintainer (86%). In 26% of the cases the accountant or another employee were responsible for the operations. Most of the farmers communicate electronically by themselves and on a regular basis, which puts high demands on the level of their computer and information literacy.

As to the frequency of e-communication, 59% stated the contact with public authorities at least once a week or more, 25% once a month, 11% once a year, and only 4% did not use it at all. 82% admitted that they receive e-mails from various authorities and professional institutions. Only 13% expressed which other information they lack and need from public authorities. Types and rates of the information is in Table 2.

The structure of institutions that farmers receive the information from is depicted in Table 3. The rate of information inflow from the governmental bodies is relatively high. Besides the dominating role of the Ministry of Agriculture, connections to the State Phytosanitary Administration are strong, as well as to professional associations. The lower rate of use of the e-mail information from the land authority is due to online land registry application at eAGRI portal which is also highly used by farmers (see Table 1).

3 CONCLUSIONS

The review and the survey results presented in the paper identified the following conclusions that also point out further possible research topics. There is an information gap in the measurement of information and knowledge needs, and in the evaluation of potential agricultural production. The agricultural production is based on information and knowledge processes that are complex. The need for high skilled information workers who are able to do decision-making in a precise, timely and qualified manner is still growing. Then the danger of an irreversible loss of knowledge at a farm or an agricultural enterprise is real. Plenty of key processes are already connected and served by state-operated electronic applications that simplify the administration of certain activities in the agricultural production.

However, the uptake and quality of e-services provided by the state administration to farmers need to be monitored and analyzed to enhance the utilization of their potential. Regarding the growing computer and information literacy among farmers and people working with information and communication technologies at agricultural enterprises, we propose to conduct continuous analysis and review of current e-services by their providers (ie. state authorities).

ACKNOWLEDGEMENT

The results and knowledge included herein have been obtained owing to support from the Internal grant agency of the Faculty of Economics and Management, Czech University of Life Sciences in Prague, grant no. 20141036, "Analýza a přístupy k řešení informačních a znalostních potřeb v resortu zemědělství v kontextu zemědělského eGovernmentu".

REFERENCES

[1] Beckman, T.J. 1999. The current state of knowledge management. *Knowledge management handbook*, 1(5). CRC Press.

[2] Bielik, P., Smutka, L., Svatoš, M. & Hupková, D. 2013. Czech and Slovak agricultural foreign trade – two decades after the dissolution. *Agriculture Economics – Czech*, 59 (10): 441–453.

[3] Eastwood, C. & Kenny, S. 2009. Art or science? Heuristic versus data driven grazing management on dairy farms. *Extension Farming Systems Journal*, 5(1): 95–102.

[4] Havlíček, J. & Tichá, I. 2003. One-Stop Government in agriculture. *Agriculture Economics – Czech*, 49: 201–207. ISSN 1805-9295.

[5] Hoffmann, V., Probst, K. & Christnick, A. 2007. Farmers and researchers: How can collaborative advantages be created in participatory research and technology development? *Agriculture and Human Values,* 24(3): 355–368.

[6] Ju, H., Lin, E., Wheeler, T., Challinor, A. & Jiang, S., 2013. Climate Change Modelling and Its Roles to Chinese Crops Yield. *Journal of Integrative Agriculture*, 12(5): 892–902.

[7] Martins, C. & Tosstorff, G. 2011. Agriculture and fisheries. *Statistics in Focus*, 18. Eurostat. ISSN 1977-0316.

[8] Ntaliani, M., Costopoulou, C., Karetsos, S., Tambouris, E. & Tarabanis, K. 2010. Agricultural e-government services: An implementation framework and case study. *Computers and Electronics in Agriculture*, 70(2): 337–347.

[9] Nonaka, I. 1994. A Dynamic Theory of Organizational Knowledge Creation. *Organization Science*, 5(1): 14–37.

[10] Nuthall, P.L. 2011. *Farm business management: Analysis of farming systems*. Oxfordshire: CABI.

[11] Payne, J. & Birchall, D. 2003. Knowledge management – the key to business success. *Proceedings of the Institution of Civil Engineers-Civil Engineering*, 156(4): 148–148.

[12] Rysová, H., Kubata, K., Tyrychtr, J., Ulman, M. & Vostrovský V. 2013. Evaluation of electronic public services in agriculture in the Czech Republic. *Acta Universitatis Agriculturae et Silviculturae Mendelianae Brunensis*, LXI(2): 473–479. ISSN 1211-8516.

[13] Šimek, P., Vaněk, J., Jarolímek, J., Stočes, M. & Vogeltanzová, T. 2011. New Version of the AGRIS Web Portal – Overcoming the Digital Divide by Providing Rural Areas with Relevant Information. *AGRIS on-line Papers in Economics and Informatics*, III(4): 71–78.

[14] Šubrt, T, Zuzák, R. et al. 2010. *Mastering knowledge*. Prague: Alfa publishing.

[15] Ulman, M., Vostrovský, V. & Tyrychtr, J. 2013. Agricultural e-government: design of quality evaluation method based on ISO SQuaRE quality model. *Agris on-line Papers in Economics and Informatics*, V(4): 211–222.

[16] Vološin J. et al. 2011. Analysis of external and internal influences on CR agrarian foreign trade. *Agricultural Economics – Czech*, 57: 422–435.

[17] Wang, S., Yu, W., Mo, Z. & Fu, G. 2013. Study on the Current Situation and Countermeasures for Agricultural Information Construction in China. In *International Conference on Advanced Information Engineering and Education Science (ICAIEES 2013)*, Atlantis Press: 123–127.

[18] Xing, L. & Lu, K. 2010. The Importance of Public-Private Partnerships in Agricultural Insurance in China: based on Analysis for Beijing. *Agriculture and Agricultural Science Procedia*, 1(0): 241–250.

Future Communication, Information and Computer Science – Zheng (Ed.)
© 2015 Taylor & Francis Group, London, 978-1-138-02653-7

Measurement system for wide range frequency at nonlinear devices

N. Suresh Kumar
GIT, GITAM University, Visakhapatnam, India

D.V. Rama Kotireddy
College of Engineering, Andhra University, Visakhapatnam, India

ABSTRACT: Most of the Digital Measurement Systems face a clock skew problem due to improper synchronization between initial state and the port-stand. The clock skew leads to an increase of relative error in Digital Measurement System (DMS). In Real time the digital signals from the measurement are fed to the processor for processing and recording purpose. But this digital output may not reflect the actual value of the measurement. This happens due to the propagation delay between initial state and port-stand of the processor, secondly due to Not Ready Sequence (NRS) of Digital Systems. In the present work an improved DMA method is proposed to minimize the relative error and to attain a good stability of measurement.

Keywords: clock; measurement system; latches; propagation delay; counter, pipeline; synchronization

1 INTRODUCTION

The present work describes a digital system which is able to measure the pulses from a tachometer with minimum relative error. One more important factor one here needs to consider is constant relative error for long frequency range to attain good stability in the study of a System Design.

The speed measurement can be achieved using different methods [Suresh].

1. Time measurement-determines time interval between pulses [dicenzo] [McCarthy] [Habibullah] [Huber]
2. Pulse counting-counts input pulses within sampling time [Lin] [Maloney]
3. Combined method [Khanneche] [ohmae] [Lin. A]
4. Constant Elapsed Time method (CET) [Bonert] [Milon]
5. DMA Transfer method. [Milon] [Milon]
6. Synchronous Frequency Measurement [Hong]

Merits and limitations of the existing methods are discussed in the paper [Suresh].

1.1 *Significance of the research*

A digital speed system is superior in that there is no nonlinearity in the speed transducer. In the conventional method, the counting times become excessively long for a reasonable accuracy, while in the latter case the transient variations may be lost due to the averaging technique used. In such cases it becomes necessary to measure the frequency instantaneously, i.e., on a cycle-to-cycle. But this is not possible in all cases of digital systems.

2 ENHANCED METHOD

In the present DMA method the measured frequency is limited because at least one rising edge of the input pulses should be detected during the sampling time [Suresh]. So in a not ready sequence the DMA may not be able to receive the input pulses with synchronizing clock pulse. The new DMA based frequency measurement system [Suresh] solved this NRS problem using pipeline technique. In the new system a new clock scheme is implemented with the pipeline to achieve a good hit ratio [Suresh][Suresh]. In the present work accurate measurements are achieved with small modifications in the circuit design of a New DMA based method [Suresh]. In the present paper detailed measurements are given with neat graph analysis. In the new method the timer is set to one minute to count the pulses from the parity checker. In the new method the timer values are compared with free running timer. The counting ratio of free running timer [Suresh] and internal timer are proportional to each other. The number of pulses counted by Timer 0 directly represents the number of pulses generated at digital Tachometer.

3 SYSTEM HARDWARE

3.1 *Tachometer*

Pulses are fed to the tachometer at a frequency to be measured. In this model, the range of speed control is approximately 1 rpm to 3000 rpm. The digital approach eliminates measurement nonlinearity and makes the speed setting exactly reproducible. The system operates on one second sampling interval and is

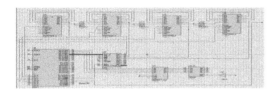

Figure 1. System hardware Clock Scheme and pipeline Interface.

scaled so that one binary bit corresponds to 1 pulse. This gives the system a steady state speed error of 1rpm over the entire control range. In the prototype, the disk is constructed such that each revolution of the shaft creates 8 pulses at the output of the parity checker [Suresh].

Parity checking is used for a greater number of pulses which can be generated for a given maximum photocell switching speed. This improves the resolution of the measurement [Mohammad].

3.2 Clock scheme for digital circuit

Based on a worst case analysis, clocking schemes for high performance systems are analysed. These are 1- and 2-phase systems using simple clocked latches.

When traditional registers are used, single-phase clocking is neither safe nor fast [Stephen] and a multiphase clock is required. This is due to the lower bound constraint of the minimum short path delay of the combinational circuits that could result in a data race through problem. In the present work a microcontroller is used to generate a clock pulse. The clock period can be varied depending on the signal arrived from Interrupt logic. And the clock pulse $T_{\mu cf}$ comes after the rising edge of the clock at the microcontroller. This clock pulse is then distributed to different stages of the pipeline as shown in figure 1.

After the arrival of the first valid data at the interrupt controller the interrupt controller interrupts the microcontroller. In response to this interruption the microcontroller sends a clock signal to the next stage register. Similarly after receiving a valid signal from the second register the interrupt controller again interrupts the microcontroller. In the same way the microcontroller activates the next stage in a different path.

The main goal in most of the digital systems is to design a clocking scheme to make the period as small as possible; this is to maximize the speed of the system. While designing it, it is highly essential to consider a flip-flop based or latch based clock scheme to alleviate clock skew [Hyein]. It is obvious that, minimizing D_{max} is the basic factor needed to consider minimizing the clock period. But, as pointed out above, it is also important to keep the smallest path delay D_{min} as large as possible. But it is not that easy to make the logic path delay uniform in value. For this reason, a system design is proposed here to control the clock skew in multiphase with the help of a programmable controller. And it is very difficult to manage the clock

system between the pipeline stages and delay elements by satisfying all the 2-phase constraints [Stephen].

Here the width of the pulse generated from the port pins ($T_{\mu pcw.i}$) of the microcontroller must be greater than the propagation delay of data from one register to an other register. The propagation delay ($D_{prop\_delay} = D_{min} - D_{max}$) between any two registers depends upon holding and setting the time of the internal register and small fractional delay element (Y_{fd}).

i.,e., $$D_{min} - D_{max} \geq T_h + T_s + 2\Delta_{clk} + y_{fd}$$

$$T_{\mu pcw.i} \geq D_{prop\_delay}$$

The amount of Y_{fd} indirectly depends on the addition of IRQ response (R_{IRQ}) and INTR response (T_{INTR}) at microcontroller $R_{IRQ} + T_{INTR}$. Because this sum creates an addition delay in the clock path generated by ports. This additionally added delay is balanced by attaching small fractional delay elements between two stages. So here the delay produced by $R_{IRQ} + T_{INTR}$ is balanced by Y_{fd}. The delay element produces a delay between events on its input and events on its output. When the enabled input at the delay buffer is inactive, events are passed without delay.

4 RESULTS

The register Operations and events are counted and recorded using CAD tools and Proteus. Some simulated results are shown in the following figures. The positive pulses represented by the light source is detected through disk slots. Where the negative pulses represented by a light source are not detected as the disk passes from one slot to an other slot. At the nth clock pulse the pipeline output is fed to DMA controller in 'n-bit' D type flip flop based circuit. Even in the NRS state of the DMA controller the pipeline fetches the pulses and buffers in the pipeline. And some readings are observed at the digital output of the system. Very small relative errors which are almost zero are observed at low frequency ranges as shown in Table 1. Some results are studied at different frequency ranges. In the simulation results the yellow and blue colours represents the clock pulses for the first stage and second stage of the pipeline respectively. The green colour and pink colour represent the input data and output data respectively. At higher data rates at 1 M to 100 Mhz the conventional digital methods face severer problems in reading higher data rate pulses. Data loss occurs due to a mismatch of speed between input data and measurement systems set time.

At high speed data rates the clock frequency must be very high and must be greater than the 1/10th of the data frequency. And hence the data loss will be small and hit ratio will be more. Hence small clock width can be achieved, power consumption will be small. Hence efficiency of the measurement circuit is increased.

At low frequencies data losses are minimized by using pipeline and compared with traditional non

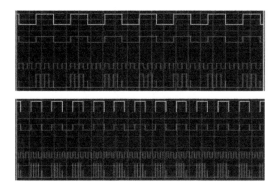

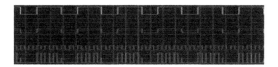

Figure 2. Data with 50 Hz speed in simple methods without pipeline and conventional pipeline respectively.

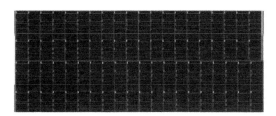

Figure 3. Data with 50 Hz speed in new pipeline methods.

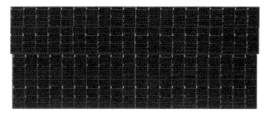

Figure 5. The data with 500 Hz speed in 4 bit 4 stagepipeline (last 2 stages).

Figure 4. The data with 500 Hz speed in 4 bit 4 stage pipeline (first 2 stages).

pipeline based methods as shown in figure 2. Some data with low frequency data rates are shown in figures 2. The data losses further minimized with new pipeline based digital method which is shown in figure 3. Data with higher frequency rates propagated with minimum losses is shown in figures 4 and 5. The data losses are further reduced by increasing pipeline stages. The data propagation is tested through various stages starting from single stage to 4 stage pipeline. And in each stage, four and eight bits are stored at a single clock pulse.

In the first stage the data input and output are in the same phase. That means in the second phase the input and output appear in different phases of the clock. Because the data arrived at in the second stage input is after crossing logic circuit between the two stages. So it took one clock pulse to come to the input pin of the second stage. And the clock pulse, clock2 arrives after one clock pulse. So the output appears after one clock pulse after the input appears at the second stage. The output will appears at the second stage after two clock pulses after the first input appears at the first stage. At the same time, while the second stage processes the first data wave the first stage receives the second data wave. That means while processing the one data wave the circuit is fetching the second data wave [Suresh].

In the table 2 the first four values are measured on a series pfDC Shunt motors. The remaining values are measured on a DC shunt motor. In the measured values of other methods an average 0.5% to 0.02% of relative error is observed [Suresh]. This relative error is minimized and almost minimized to zero in the enhanced method. Some variations (error) around 0.25 to 0.375 are observed in the new method readings. And 0.0125% of relative error is observed between the ranges of 1000 rpm to 3000 rpm. A constant relative error is observed in the new method as significantly different from other methods. In graph 1 armatures current (I_a) is plotted on x-axis and Torque is plotted on y-axis.

In the present system a stepper motor is used to measure Low speed rpm. And 0.01% of relative error is observed at the range of 1 rpm to 1000 rpm.

Relative Error = (Actual Value – Measured Value) / Actual Value.

Table 1. Comparision of Different Measured values of motor rpm with theoretical value.

Si No	V (Volts)	I_a (Amps)	I_f (Amps)	T_1 (Kg)	T_2 (Kg)	P_{out}	N (rpm) Theoretical	N (rpm) Other Method	N (rpm) CET Method	N (rpm) DMA Method	N (rpm) New Method
1	215	0.3	0.45	0	0	0	2985	2975	–	2984.375	2984.625
2	215	11.3	0.45	9	4	1305.14	2230	2220	–	2229.75	2229.75
3	215	0.6	0.45	0	0	0	2000	1990	1999	1999.5	1999.75
4	215	3	0.45	3.0	1.2	408.75	1940	1931	1938	1939.5	1939.75
5	215	6	0.45	4.0	1.8	484.06	1880	1870	1878	1879.375	1879.625
6	215	10	0.45	7.0	3.0	842.78	1800	1791	1799	1799.5	1799.625

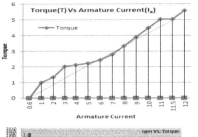

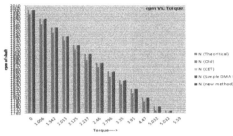

Graph 1.   Graph between armature current and torque and Plot between rpm (N) in different methods and Torque.

5   CONCLUSION

The relative error of the frequency measurement is almost zero when compared with other measurement methods. Constant relative error is achieved at a wide range frequency measurement. The system can also measure the small readings which are near to zero value. The clock distribution becomes simpler by controlling clock signals using internal ports of microcontroller. Small clock widths are achieved and so the power consumption is small when compared to other methods. While DMA processes the pulse from the last stage of pipeline, the first stage can read the next data pulse from the measure end. When measure end frequency is higher than the DMA counter, the pipeline will take part to synchronize the speed between DMA and measure end and vice versa. Unlike other microprocessor interfacing methods it is able to fetch input data in Not Ready Sequence also.

REFERENCES

[1] dicenzo, D.C., et al., "Digital measurement of angular velocity for instrumentation and control," IEEE Trans. Industr. Electr. Con. Instr., vol. IECI-23, pp. 83–86, 1976.

[2] Lin, S.C., et al., "A microprocessor-based incremental servo system with variable structure," IEEE Trans. Industr. Electr., vol. IE- 31, pp. 313–316, 1984.

[3] Bonert, R., "Design of a high performance digital tachometer with a microcontroller," IEEE Trans. Instcm. Meas., vol. 38, pp. 1104–1108, Dec. 1989.

[4] Suresh Kumar, N., et al., S, "A New Method to Enhance Performance of Digital Frequency Measurement and Minimize the Clock Skew", IEEE Sens J, vol 11. No 10. Oct. 2011.

[5] Hong Q unhuan et al.," Design of Digital Frequency Meter Based on Synchronous Frequency Measurement Method", 2010 2nd International Conference on Computer Technology and Development (ICCTD 2010), IEEExplore.

[6] Suresh Kumar, N., et al., "Effect of Interrupt Logic on Delay Balancing Circuit", International Journal of Computer Application (US), vol 27 no 4. September, 2011.

[7] Mohammad Maymandi, "A digital programmable delay element: Design and analysis", IEEE transaction VLSI systems, vol. 11, no. 5, October 2003.

[8] Stephen h. unger, "Clocking Schemes for High- Speed Digital Systems", IEEE Transactions on computers, vol. c-35, no. 10, october 1986, pp. 880–895.

[9] McCarthy, E.P., "A digital instantaneous frequency meter," IEEE Trans. Instrum. Meas., vol. IM-28, no. 3, pp. 224–226, Sept. 1979.

[10] Maloney, T.J. and Alvarado, F.L. "A digital method for dc motor speed control", IEEE Trans. Znd. Electr. Con. Znstr., vol. IECI-23, no. 1, pp. 44–46, Feb. 1976.

[11] Milan Prokin, "DMA transfer method for wide range and frequency measurement", IEEE Transactions on instrumentation and measurement, vol. 42, no. 4, August 1993.

[12] Milan prokin, "Speed measurement using the improved DMA transfer method", IEEE Transactions on industrial electronics, vol. 38, no. 6, Dec 1991.

[13] Habibullah, B., et al., "A new digital speed transducer". IEEE Transactions on Industrial Electronics Contr. Instrum. 25, 339–342, 1978.

[14] Huber, et D. Wal., "A digital device to measure angular speed and torque angle". IEEE Transactions on Industrial Electronics Contr. Instrum., 22, 186–188, 1975.

[15] Khanniche, M.S., et al., "A microcontroller-based real-time speed measurement for motor drive systems", Journal of Microcomputer Applications, (Elsevier) (1995) 18, 39–53.

[16] Ohmae, T., et al., "A microprocess-controlled high accuracy speed regulator for motor drives". IEEE Transactions on Industrial Electronics, 29, 207–211, 1982.

[17] Lin, A.K., et al., "A microprocessor speed control system". IEEE Trans. Ind. Electron. Contr. Instrum. 24(3)241–247, 1977.

[18] Hyein lee, et al., " Pulse width allocation and clock skew Scheduling: Optimizing Sequential Circuits Based on Pulsed Latches", Iee Transaction on Computer aided design of integrated circuits and systems, vol 29, no 3, March 2010.

[19] Suresh Kumar, N., et al., "Clock synchronization in digital circuits", IJETT, vol 2, no 2, 2011, pp. 42–44.

Future Communication, Information and Computer Science – Zheng (Ed.)
© *2015 Taylor & Francis Group, London, 978-1-138-02653-7*

Adaptive bidirectional relay selection and optimal power allocation of cognitive networks

J.L. Jiang & R.H. Qiu
College of Information Sciences and Technology, Donghua University, Shanghai, China
Engineering Research Center of Digitized Textile & Fashion Technology, Ministry of Education, Donghua University,
Shanghai, China

ABSTRACT: In order to improve the average total rate of secondary users, this paper proposes a new Bidirectional Relay Selection scheme. Under the consideration of the joint relay node and the bidirectional relay link transmission characteristics, this paper is also considering the direct link transmission characteristics. When the direct link transmission characteristics are superior to relay channel, it does not use the relay transmission. According to the channel Condition, it will choose the right way to form an Adaptive Bidirectional Relay Selection scheme (ABRS). On the basis of the primary user restrictions, this paper proposes a strategy of the Optimal Power Allocation (OPA). Under the basis of maximizing the total rate of one-way links, this paper adds Lagrange factor model to optimize. The simulation results show that the proposed method can significantly improve transmission rate of the secondary users.

Keywords: average total rate; bidirectional relay selection; power allocation

1 INTRODUCTION

The spectrum utilization for wireless communication is more concerned in recent years. Improving spectrum utilization has been the direction of technological development, which has been prompted by the cognitive radio [1]. Recently, a cognitive two-way relay system has aroused the attention of academics [2–4]: how to improve the communication performance of secondary users in cognitive networks has been a research focus. Its key technology is the collaborative communications technology, because of its half-duplex mode of operation, One-way relay system will result in a low spectral efficiency. However, Shannon has proposed a two-way relay system. Research shows that on the basis of the one-way relay system two-way relaying system spectrum efficiency can further improve the performance of the system [5–8]. It proposed a multiple relay nodes relay selection strategy [9], and strict BER analysis proved that the strategy can get the full diversity. It proposed a two-way AF relay selection system (BRS) and secondary users power allocation strategy [10].

2 SYSTEM MODEL

2.1 *Cognitive radio network model*

Considering the existence of primary and secondary users in cognitive radio networks, in the main network,

a primary user (PT / PD) sends data to another primary user (PT / PD), in the second user network, this paper considered a direct link in certain conditions, the two secondary users SD and ST communicate with each other through M relay, which is shown in Figure 1. The dashed line in the Figure 1 is the mutual interference between the primary and secondary users. In a general cognitive radio, we learned that the primary users and secondary users may be able to exchange data in the same time slot or the same frequency. The premise is that secondary users are not compromising the primary user of normal communication. To ensure the QOS of the primary user in a certain range, the limited interference is allowed.

2.2 *Two-way relaying transmission system model*

This paper considers a bidirectional secondary user cooperative communication system which has a choice of M cognitive relay. The transmitter and the receiver can both send or receive information. They can perform a two-way communication by means of cognitive relay, they can also be a direct link transmission under certain conditions. Secondary user's communications are carried out using a single antenna, the second user can share the spectrum with a specific primary user and needs two time slots during the communication.

Figure 2 shows that this model consists of ST and SD and the M-relay cognitive node. h_{i1}, h_{i2}, h, which

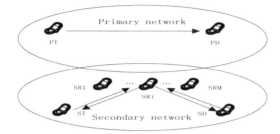

Figure 1. Cooperative cognitive radio communication system model.

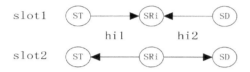

Figure 2. Two-way collaborative communication slots.

is respectively the channel coefficients of SRi to ST and SRi to SD and ST to SD, and they are subject to complex Gaussian distribution.

The direct link is adopted only if a certain condition $\psi = \{h^2 > \max\{h_{i1}^2, h_{i2}^2\}\}$ can be carried out, but the method mentioned in this paper is not involved in the merging of the signal receiving end and other issues. We consider only a single direct communication of collaboration and two relay cases.

2.3 Systems analysis

In the cognitive network, the second user's communication may be restricted by and interfered from the primary user, and the cognitive relay is not only receiving the message from the second user, but also receiving an interference information from the primary user; these problems we need to take into account.

1. When the event ψ does not occur, we take a two-way relay link. Assuming that the signals X_1 is transmitted by ST, X_2 is transmitted by SD the relay terminal SRi is expressed by y_{SRi}

$$y_{SRi} = \sqrt{P_{ST}}\, h_{i1} X_1 + \sqrt{P_{SD}}\, h_{i2} X_2 + I_i + N_i \quad (1)$$

Where P_{ST} and P_{SD} are the transmit powers of ST and SD respectively; h_{i1} denotes the channel gain from ST to the i-th relay; h_{i2} denotes the channel gain from SD to the i-th relay; I_i is the interference from the PU transmitter, and N_i is an independent and identically distributed (i.i.d.) complex Gaussian random variable, the selected relay transmits the signal it received in the first phase after scaling it, and the transceiver nodes receive the signals from the relay nodes. Assuming the transceiver-to-relay and the reverse channel gains are reciprocal, the signal received at the transceiver ST can be expressed as y_{ST};

$$y_{ST} = \sqrt{p_{ST}}\, \beta\, h_{i1}^2 X_1 + \sqrt{P_{SD}}\, \beta\, h_{i1} h_{i2} X_2 + W \quad (2)$$

$$W = h_{i1}\beta I_i + h_{i1}\beta N_2 + \overline{I}_i + \overline{N}_i \quad (3)$$

where β is the amplification factor at the i-th relay, $\beta = \sqrt{\frac{P_{SRi}}{P_{ST}h_{i1}^2 + P_{SD}h_{i2}^2 + N_i}}$, $\overline{I}$ represents the interference resulting at ST from the PU transmission, and $\overline{N}_i$ represents the additive white Gaussian noise (AWGN) at transceiver ST. Similarly, the signal received at the transceiver SD can be expressed as y_{SD}

$$y_{SD} = \sqrt{P_{ST}}\,\beta h_{i1} h_{i2} X_1 + \sqrt{P_{SD}}\,\beta h_{i2} h_{i2} X_2 + V \quad (4)$$

$$V = h_{i2}\beta I_i + h_{i2}\beta N_i + \overline{\overline{I}}_i + \overline{\overline{N}}_i \quad (5)$$

where $\overline{\overline{N}}_i$ represents the AWGN at transceiver SD and $\overline{\overline{I}}$ represents the interference resulting at SD from the PU transmission. The SU transceivers can remove the self-interference from the received signals using the knowledge of the CSI, in the end, we use $\overline{y}_{ST}$ and $\overline{y}_{SD}$ to express.

$$\overline{y}_{ST} = \sqrt{P_{SD}}\,\beta h_{i1} h_{i2} X_2 + h_{i1}\beta I_i + h_{i1}\beta N_i + \overline{I}_i + \overline{N}_i \quad (6)$$

$$\overline{y}_{SD} = \sqrt{P_{ST}}\,\beta h_{i1} h_{i2} X_1 + h_{i2}\beta I_i + h_{i2}\beta N_i + \overline{\overline{I}}_i + \overline{\overline{N}}_i \quad (7)$$

When the event ψ occurs, we take the direct link transmission, the following analysis:

$$\overline{y}_{ST} = \sqrt{p_{SD}}\, h\, X_2 + \overline{I} + n_0 \quad (8)$$

$$\overline{y}_{SD} = \sqrt{p_{ST}}\, h\, X_1 + \overline{\overline{I}} + n_0 \quad (9)$$

where h is Fading coefficient in the direct link, and n_0 is additive white Gaussian noise in the direct link. So we will come to the SNR.

3 OPTIMAL RELAY SELECTION AND POWER ALLOCATION

3.1 An improved two-way relay selection strategy

1. In the first step, setting a threshold to filter the larger relay link SNR candidate relay;

2. In the second step, selecting the biggest gain relay from the selected candidate relay;

3. Under the case of choosing the best relay links, comparing the channel gain at this time of the channel gain maximum of two-way relay link and direct link channel gain, that is to say, if an event $\psi = \{h^2 > \max\{h_{i1}^2, h_{i2}^2\}\}$ occurs, we know the relay link channel condition is not as good as a direct link, in which we take a separate direct link to transport; if the event does not occur, the two-way relay link is selected for collaborative communication, and always informed of perfect channel state information to adaptively change the transmission mode, this method can effectively increase the total rate of information and improve the communication performance.

3.2 Two-way relay system to optimize power allocation strategy

3.2.1 An optimized power allocation strategy

Under the limited power of user and interference from the main precondition for the user, through effective power distribution to improve the overall performance of the two-way relay system. In the limit of the total system power, this paper presents an optimized power allocation strategy. Under the limit of primary user, the system can achieve the max power of P, I_{ST} I_{SD} I_{SRi} respectively represents User ST. SD. SRi interference to the primary user.

$$P_{ST} + P_{SRi} + P_{SD} \le P \qquad (10)$$

$$p_{ST} \le \frac{I_{ST}}{h_3^2}, \quad p_{SRi} \le \frac{I_{SRi}}{h_4^2}, \quad p_{SD} \le \frac{I_{SD}}{h_5^2} \qquad (11)$$

The optimal power allocation is obtained in the formula to improve the overall average speed of the system by the effective power distribution of each node. In the constraints of the main user, we define the average total rate, we define average total rate $E[R_{sum}]$ represented:

$$E\lfloor R_{sum} \rfloor = E\lfloor R_{ST} \rfloor + E\lfloor R_{SD} \rfloor \qquad (12)$$

1. When an event ψ does not occur, the above equation can be described as:

$$P_{ST} + P_{SRi} + P_{SD} \le P \qquad (13)$$

For convenience, we assume that this two-way link system is equal to two one-way links, that is ST to SD and SD to ST, which must maintain cognitive relay selected the same in two unidirectional links. We put this formula into two sub-optimization model:

Sub-optimization model 1:

$$\text{Max} \quad E\lfloor R_{ST} \rfloor \quad \text{St} \quad P_{ST} + P_{SRi} \le P_1 \qquad (14)$$

Sub-optimization model 2:

$$\text{Max} \quad E\lfloor R_{SD} \rfloor \quad \text{St} \quad P_{SD} + P_{SRi} \le P_2 \qquad (15)$$

In the Sub-optimization model, where P_1, P_2 are expressed in a single link ST to SD and SD to ST transmission link total power.

$$L\left(P_{ST}, P_{SRi}, \lambda\right) = \frac{1}{2}\log\left(1+U\right) + \lambda\left(P_1 - P_{ST} - P_{SRi}\right) \qquad (16)$$

$$U = \frac{P_{SD} P_{SRi} G_1 G_2}{G_1 P_{SRi} + G_1 P_{ST} + G_2 P_{SD} + 1} \qquad (17)$$

It is similar to optimization model 2, adding Lagrangian to the optimization

$$L\left(P_{SD}, P_{SRi}, \lambda\right) = \frac{1}{2}\log_2\left(1+Z\right) + \lambda\left(P_2 - P_{SD} - P_{SRi}\right) \qquad (18)$$

$$Z = \frac{P_{SD} P_{SRi} G_1 G_2}{G_2 P_{SRi} + G_1 P_{ST} + G_2 P_{SD} + 1} \qquad (19)$$

When the event ψ occurs, indicating the relay link condition is very poor, then using the direct link can increase the average total rate at this time;

$$E\lfloor R_{sum} \rfloor = E\lfloor R_{ST \to SD} \rfloor + E\lfloor R_{SD \to ST} \rfloor \qquad (20)$$

$$\text{max} \quad E\lfloor R_{sum} \rfloor \quad \text{St} \quad P_{SD} + P_{ST} \le P \qquad (21)$$

The above analysis shows that depending on the channel gain values to select different transmission modes, the system can improve the rate; while the poor link channel gain can get much more power.

4 SIMULATION RESULTS

In the simulation, the channel is following the Rayleigh flat fading, and assuming the parameter value of all the corresponding noise variance and signal interference noise from the main use was 1, the number of cognitive M-relay is 8, the thresholds for the SNR is 2.5.

As can be seen in Figure 3, under the limit of the total power, two-way relaying system selected on the basis of the adaptive two-way relay selection can increase the overall rate of the secondary user; because you take into account the variability of the wireless channel and may cause the condition that a two-way channel relay channel gain is less than the direct link channel gain. It is a good way to choose the direct link. When the channel gain is greater than the the direct link channel gain, the two-way channel relay channel was selected. By this method, we can further improve the transmission rate user. Under the limit of QOS of the primary user, when the maximum total power of the secondary user

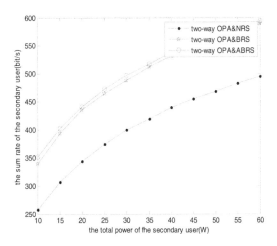

Figure 3. OPA-based different relay selection corresponding to User average total rate.

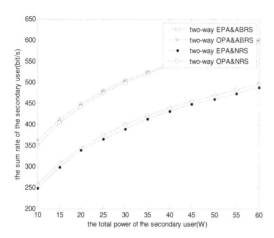

Figure 4. NRS and ABRS based on different power allocation strategy corresponding to rate.

is 15 w, This paper presents an adaptive relay selection strategy, which can improve the secondary users total transmission rate of 1/5. In figure 4, we are comparing the user rate in a two-way relay system based on the OPA and EPA, this method can improve the total rate of secondary users, thus proving the power allocation strategy can improve communication rate of secondary users, for example, when the value of the total power is 15 w. This method this paper proposes can improve the overall rate of 1/10.

5 CONCLUSIONS

Goal of this paper is to improve the average total transmission rate of secondary users. This paper presents an adaptive bi-cognitive relay selection method and also joins the Lagrange factor for power distribution to optimize power allocation strategy, by comparing the two-way link with a straight trunk channel gain links. This method can have adaptive changes to adapt to select the channel of transmission, in order to further improve the transmission rate of the second users; by introducing Lagrange factor, we solved the power

optimization to further improve the transmission rate of secondary users; the simulation results show that the proposed method can improve the overall average transmission rate and improve the cognitive radio secondary users.

REFERENCES

[1] A. Goldsmith, S. Jafar, I. Maric, and S. Srinivasa, "Breaking spectrum gridlock with cognitive radios: an information theoretic perspective," Proc. IEEE, vol. 97, pp. 894–914, May 2009.

[2] S.H. Safavi, M. Ardebilipour and S. Salari, "Relay beam forming in cognitive two-way networks with imperfect channel state information," IEEE Wireless Commun. Lett., vol. 1, no. 4, pp. 344–347, August 2012.

[3] R. Wang, M. Tao and Y. Liu, "Optimal linear transceiver designs for cognitive two-way relay networks," IEEE Trans. on Signal Process., vol. 61, no. 4, pp. 992–1005, Feb. 15, 2013.

[4] Li Qiang, Ting S H, Pandharipande A. Cognitive spectrum sharing with twoway relaying systems. IEEE Transactions on Vehicular Technology, 2011, 60: 1233–1240.

[5] S. Talwar, J. Yindi, and S. Shahbazpanahi, "Joint relay selection and power allocation for two-way relay networks," Signal Processing Letters, IEEE, vol. 18, no. 2, pp. 91–94, 2011.

[6] S.-I. Chu, "Performance of amplify-and-forward cooperative communications with the best-relay selection scheme" IEEE Commun. Lett, vol. 15, no. 2, pp. 172–174, Feb. 2011.

[7] Shannon C.E. Two-way communication channels. Proc. 4th Berkeley Symp. Math. Statist. Stat. Prob. California, America, 1961: 611–644.

[8] V. Havary-Nassab, S. Shahbazpanahi, and A. Grami, "Optimal distributed beam forming fort wo-way relay networks," IEEE Trans. Signal Process, vol. 58, no. 3, pp. 1238–1250, Mar. 2010.

[9] Jing Yin-di. A relay selection scheme for two-way amplify-and-forward relay networks. IEEE wireless communication & signal processing. Nanjing, China, 2009:1–5.

[10] Donlin, Liu Tong, Chen Qian-bin, etc. bidirectional relay systems and power distribution cooperative node selection strategy. Electronics and Information Technology, 2010, 32(9): 2077–2082.

Future Communication, Information and Computer Science – Zheng (Ed.)
© *2015 Taylor & Francis Group, London, 978-1-138-02653-7*

Research on practice education method of foundation courses in information specialty

Yang Li, Y.J. Tang & J. Wang
School of Information Engineering, Beijing Institute of Graphic Communication, Beijing, China

ABSTRACT: In traditional foundation practice courses, the contents and difficulty of practice are not adapted with the current requirements from students and society. To improve the effectiveness of practice education, and increase practice interests from students, the practice education system should be innovated. In this paper, some innovated practice education methods, which include lab course integrating, lab course and course project fusing, practice and academic competition bonding, are introduced. The effectiveness of these innovations is proven.

Keywords: practice education; innovation; education method

1 INTRODUCTION

The foundation courses in information specialty include the courses of circuits, analog electronics, logic electronics, microcomputer principle, signal and system. Similar to this major, these courses are practical and abstract. How to raise the interests of students, and improve their ability of practice is the question that educators face for a long term. However, belonging to the progress of electronics and information engineering, are the electronic system trend to be modular currently and the basic circuit design seems to be less important and less attractive. In this way, the education method of relative foundation courses should also consider how to draw students' attention on basic circuit design. Considering the practical characters of these courses, the practice education method of these courses should be innovated.

2 TRADITIONAL PRACTICE EDUCATION METHOD

Based on the actual experience of practice education in college, the education plan, contents and difficulty of traditional practice education method is discussed in the following.

2.1 *Education plan*

In the education plans of these foundation courses, the classroom teaching and lab teaching are individual from each other. The practices are applied to enhance the important and difficult knowledge points that take place in classroom teaching. However, these knowledge points are discrete, and the practices are also disconnected from each other. Both contents and time of practices in one course are dispersed in one semester. It's difficult for students to connect these contents logically, and the teaching effect is not ideal. There is another problem that some practices in different courses are similar, such as the practice of the A/D converter which appeared in both principles of microcomputer and single-chip microcomputer as an individual practice. It's undoubted that the A/D converter is an important device in electronics design, but the repeating of one practice leads to time and resource waste. This problem has come from the course programming of majors and the education plan of relative courses. In one word, the traditional course programming and education plans for foundation courses do not pay enough attention to practice education, which loads the teaching effect of practice and cannot reach the educators want.

2.2 *Contents of practice*

The period of practice normally takes about 20 percent of the whole course period of these foundation courses. The practices in foundation courses are mostly validation practices, which are applied to confirm the basic circuit theories such as Kirchhoff's Current Law (KCL), or the basic circuit analysis such as BJT Amplifier. There is no doubt that the knowledge referred to in these practices is necessary and important. However, these practices have been taught and practiced for a long time without changing and improving. The practice contents are similar to those that took place 30 years ago, and cannot keep in touch with the current progress of science.

2.3 Difficulty of practice

On the other hand, along with the expanding student quantity, the student quality is incomparable to earlier. To help current students master the basic and necessary knowledge of their major, the difficulty of practice is also reduced. In the lab, most students don't need to assemble circuits by discrete devices. In the practice kit, all circuits have been manufactured using PCB, and all the test points have also been reserved. What students can do is only to use voltmeter and ampere meter to measure and calculate relative electric quantities. In this case, students don't even need to know the architecture and method of circuits. Along with simplifying the practice procedure, the teaching effect is unfortunately discounted.

3 INNOVATION FOR PRACTICES OF FOUNDATION COURSE

Based on the shortages of the traditional practice education method, the education plan, contents and difficulty of lab teaching should be improved to raise the teaching effect. Considering the actual lab teaching procedure in the School of Information Engineering in Beijing Institute of Graphic Communication (BIGC), the following innovations are introduced.

3.1 Integrating lab course

From the above discussion, in traditional practice teaching methods, lab teaching plays the subordinate role compared with classroom teaching. This causes students to ignore the importance of practice, and pay less attention. To change this condition, the practices of foundation courses might be integrated. For example, the practices for the courses of circuit, analog electronics, and digital electronics could be integrated into one practice course, which is named as electronic circuit design and practice. In this course, both the unit circuit and the electronic circuit system are introduced and practiced.

In the past, every practice in a certain course mostly focused on one limited topic, such as the BJT amplifier circuit or low pass filter. Students could study and know how to build and analyze one unit circuit related to the topic by experience, however, they might not know how to use this circuit in the electronics system. The reason for this problem is that every practice topic only concentrates on the knowledge in a certain course, but not in the electronics system. The integrated lab course might change this situation. Considering the practice of the low pass filter, in the past, the topic covered the structure of the filter, the stop frequency of the filter, the transfer character of the filter. In current integrated lab course, the practice of the low pass filter is combined with the practice of the multiplier. Thus, students also know the function of the low pass filter is to select the wanted signal and stop the unwanted part in the system, and particularly realize the function of frequency spectrum shifting in the wireless communication system.

In the real operation, the practice courses of circuit, analog electronics, and digital electronics are integrated. The practice courses of microcomputer principle, and microcontroller are also integrated. In the simplified and optimized new education plan, the practice courses help students master the relative knowledge from both unit circuit and whole system view, and the teaching effect is improved.

3.2 Optimizing lab course system

In the traditional education plan, the practice education included lab course and course project. Normally, the lab course is processed with a classroom course in the semester, and the course project is arranged at the end of the semester after the classroom course is finished. This arrangement is aimed to provide a longer block of practice education for students in this major. However, considering the contents and difficulty of the current practice, the effective of this lab course system isn't ideal. To optimize this lab course system, the contents and configuration of lab course should be rearranged.

3.2.1 Fusing lab course and course project
Considering some outdated contents of lab courses, the lab course and the course project can be fused together to introduce modern engineering technique and knowledge. For example, in the lab course of electronics, some analog and digital circuit modules are introduced separately. To know the thinking of a circuit system, students should learn the course project of electronics circuit design at the end of the semester. In fact, some contents of electronics lab course are similar to or included in the course project. To improve the efficiency of practice education in electronics, the lab course can be simplified to avoid repeating. For example, the practice of BJT amplifier is seldom used in current actual engineering application, and this practice should take a longer time than other practices. Then the BJT amplifier can be practiced with simulation software (such as Multisim, Pspice...). In this case, students can not only learn the circuit theory of BJT amplifier, but also the application of circuit simulation software, and learning time is also saved. For another example, the operation amplifier is one basic practice target, and appeared in both the lab course and course project. In actual education procedure, the lab course of operation amplifier can absolutely be removed, because this device should be used in the course project many times and in different circuit architecture. In BIGC, the lab courses of analog electronics and digital electronics are integrated. Some practices that are not easy to complete in time are changed to use simulation software to realize, some practices that might be repeated are moved to the course project of electronics system design.

3.2.2 Practices bonding with competitions
To improve the practice quality, the practice system should also be optimized. The reasonable practice system can increase student's interest in engineering

experience, and provide the road to explore specialized knowledge. In BIGC, the practice education is bonded with the academic competition. For the lower grade undergraduate students, the electronic process practice is bonded with the electronic process competition. This competition encourages students to master the basic skill of circuit soldering and circuit assembling. For the upper grade undergraduate students, the electronic design competition is provided to encourage students to combine the analog and digital electronics knowledge to design and manufacture electronic systems. Based on these two academic competitions in campus, students are also encouraged to attend academic competitions in Beijing and the nation. For example, in recent years, it has been regular that students in BIGC attended the Beijing undergraduate electronic design contest, the national undergraduate electronic design contest, and the national undergraduate students "Freescale cup" intelligent car competition. Through the competitions, students' interest in practice is increased obviously, and the experience ability is also improved.

4 CONCLUSION

Based on the innovation on practice education system, the lab course system is optimized, and the contents of lab courses are also improved. Through these changes, the students' interest and ability of experience are both increased. In recent years, students in BIGC have achieved plenty of rewards from academic competitions, which proves the good effective of practice education innovation.

ACKNOWLEDGEMENTS

Supported by Key Education Innovation Program of BIGC (Grant No. 22150114014), and General Education Innovation Program of BIGC (Grant No. 22150113017)

REFERENCES

[1] Burkhalter, B.B. 2002. *How can institutions of higher education achieve quality within the new economy?* In Total Quality Management, 2002: 367–377.
[2] Evans, James R. 1996. *What should higher education be teaching about quality?* In Quality Progress, 1996(8): 83.
[3] Gopal K. Kanji, Abdul Malek Bin a Tambi. 1999. *Total quality management in UK higher education institutions.* In Total Quality Management, 1999(1): 129.
[4] Ho, Samuel K, Wearne, Katrina. A. 1996. *Quality Assurance in Education.* Bradford, 1996(2): 35.
[5] John J. Lawrence, Michael A. McCullough. 2001. *A Conceptual Framework for guaranteeing higher education.* In Quality Assurance in Education, 2001(3): 139.
[6] Peggy Brewer, Terri Friel, William Davig, JudithSpain. 2002. *Quality in the classroom.* In Quality Progress, 2002(1): 67.
[7] Sharples, Kathleen A., Slushier, Michael, Swain, Mike. 1996. *How TQM can work in education.* In Quality Progress, 1996(5): 75.

Future Communication, Information and Computer Science – Zheng (Ed.)
© *2015 Taylor & Francis Group, London, 978-1-138-02653-7*

Estimation of nonlinear dynamics for ECG with Premature Ventricular Contraction (PVC) arrhythmia by polar coordinate mapping

J.-H. Kim, S.-E. Park & K.-S. Kim
School of Biomedical Engineering, College of Biomedical & Health Sciences, Konkuk University, Korea

ABSTRACT: This research proposes a new method to characterize the discriminative feature of Premature Ventricular Contraction (PVC) arrhythmia heartbeat from the normal one. With this aim, the ECG data segment consisting of three consecutive R-peaks in the MIT-BIH PVC and Normal Sinus Rhythm (NSR) database is resolved to convert its time-evolved trend into the portrait trajectory by utilizing polar coordinated mapping. Our simulated results show that the polar coordinated trajectory can offer the discriminatory portrait to differentiate PVC beat from normal one without referring to time delayed statistics.

Keywords: ECG attractor; R-Peak; QRS Complex; Arrhythmia; Premature Ventricular Contraction; Polar Coordinates; MIT-BIH database

1 REPRESENTATION OF ECG ATTRACTOR BY EMBEDDING A LAG TIME

1.1 *Fiducial features of Electrocardiogram*

The Electrocardiogram (ECG) signal that represents the varying electric potential of a human heart can reveal out of the clinical test parameters in diagnosing most heart diseases. The primary features includes the characteristics of R peak, QRS complex, S and T fiducial points in terms of time-location and morphological features. The additional-clinical parameters can be sought by the combinations of the fiducial features such as RR, PR, PT, PP, TT, ST and QRS, PQ or ST segment (Fig. 1) [Blinowska, 2012].

1.2 *ECG attractor for estimating the time trend of ECG data*

One of the methods is to describe time-behavior embedded in the ECG data, $x(t)$ is to reconstruct two or three dimensional state trajectories [Hundewale, 2012] by constructing the data vectors $E_2(t)$ or $E_3(t)$ which are mapped by a time delay element d:

$$E_2(t) = \left(x(t), x(t-d) \right) \tag{1}$$

$$E_3(t) = \left(x(t), x(t-d), x(t-2d) \right) \tag{2}$$

To investigate the time-evolving trend of the ECG data, we utilize the MIT-BIH database [Goldberger, 2000] which recorded the long-term ECG readings from the subjects with the sampling rate of 360 Hz and 180 Hz in text file format. Especially the record number of

ECG-119-1 dataset is considered because it includes the normal sinus rhythm and Premature Ventricular Contraction (PVC) arrhythmia. In this study, the interested ECG data segment is defined by detecting three consecutive R-peaks as follows:

- The baseline wandering noise is initially eliminated by designing low-pass Finite Impulse Response (FIR) filter specifications (Table 1) and converting it into the high-pass filter parameters by selecting the center frequency of the two-sided filter shape [Kim, 2014][Vegte, 2002].
- The first R-peak position, R_0 is sought in terms of time location by finding the maximum amplitude within the pre-defined size of a local window region which is decided by the sampling frequency. The location of the second R-peak, R_1 is resolved by moving the window region from R_0 position to $R_0 + 0.3 \times$ sampling frequency. The third R-peak is detected by adjusting the size of the window in the amount of time interval from R_0 to R_1 and by shifting the window by $R_1 + 0.3 \times$ sampling frequency.
- Construct the data segment which is a part of the ECG data including three R-peaks and obtain a data vector in time delay coordinates as in equation (1) or (2).

Figure 2 depicts a part of the ECG-119-1 dataset marking three R-peaks positions with a symbol of 'O' and the reconstructed attractor by embedding lag of $d = 3$.

Compared to normal beats, PVC rhythm [Rajendra, 2007] in which ventricular beats occur earlier than anticipated characterizes an anchor-like shape which has the tendency to approach closely the basal

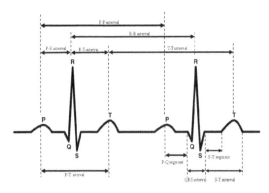

Figure 1. Illustration of the primiary and secondary fiducial features of the ECG signal.

Table 1. Low-pass FIR design specification for eliminating baseline drift in the ECG data.

Parameters	Value
Window function	Hanning
Sampling frequency	180 Hz
Filter order	1,759
Pass band cutoff frequency	89.16 Hz
Stop band cutoff frequency	89.5 Hz
Center frequency	90 Hz

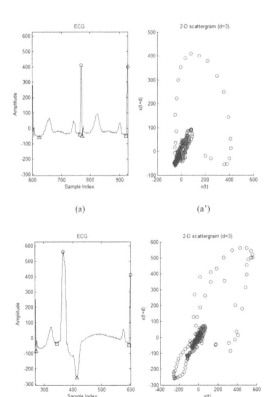

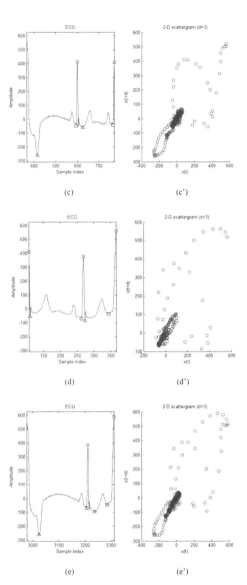

Figure 2. The ECG segment (the part of ECG-119-1) consisting of three consecutive R-peaks and the reconstructed phase portrait: (a) three normal heart beats with its attractor-a' which is illustrated by two dimensional scatter plot (b) normal-PVC-normal beats (c) PVC-normal-normal beats (d) normal-normal-PVC beat (e) PVC-normal-PVC beats.

point. Due to the dependency of the shape of scatter plot on the value of delay time, the appropriate value of d is usually required to address features that ascertain PVC beat embedded in the interested ECG segment. The discriminative features might include the standard deviation of projection of the Poincare plot on the line of identity with positive or negative slope [Piskorski, 2005] and the degree of tendency towards the base. To illustrate the dependency on the lag time, the phase portraits are redrawn for the ECG segments considered in Figure 2 by choosing $d = 5$ (Figure 3).

156

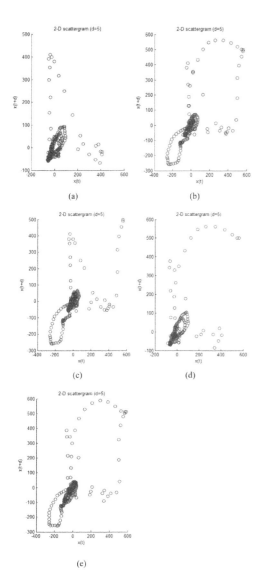

(a)   (b)

(c)   (d)

(e)

Figure 3. The reconstructed phase portrait ($d = 5$) for: (a) three normal heart beats with its attractor-a' which is illustrated by two dimensional scatter plot (b) normal-PVC-normal beats (c) PVC-normal-normal beats (d) normal-normal-PVC beat (e) PVC-normal-PVC beats.

2   REPRESENTATION OF ECG ATTRACTOR BY POLAR COORDINATED MAPPING

2.1   Interpretation of a time-series ECG segment in polar coordinates

To nullify the effects of selecting a lag time in estimating the phase portrait, polar coordinated mapping scheme is proposed to convert the time series data of an ECG segment into radius r and angle θ (Figure 4).

The time interval of an ECG segment is arranged into the angle values ranging from 0 to 2π and a voltage level is converted to the radius value as follows:

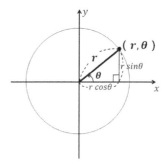

Figure 4.   Illustration of Polar Coordinates (r, θ).

1. Define the data vector consisting of N samples in an ECG segment including three R peaks:

$$ECG(n) = \left(x_1, x_2, x_3 \bullet \bullet \bullet x_{N-1}, x_N \right)$$

2. Transform the time-series data $ECG(n)$ into polar coordinated series $ECG_p(n)$ by:

$$ECG_p(n) = \begin{pmatrix} r_1 e^{j0}, r_2 e^{j(2\pi/N)}, r_3 e^{j2\bullet(2\pi/N)} \\ \cdots r_k e^{j(k-1)\bullet(2\pi/N)} \cdots r_N e^{j(N-1)\bullet(2\pi/N)} \end{pmatrix}$$

3. Normalize the radius value of $ECG_p(n)$ by the maximum amplitude M in an ECG segment:

$$\begin{pmatrix} (r_1/M)e^{j0}, (r_2/M)e^{j(2\pi/N)}, \\ \cdots (r_k/M)e^{j(k-1)\bullet(2\pi/N)} \cdots (r_N/M)e^{j(N-1)\bullet(2\pi/N)} \end{pmatrix}$$

2.2   Representations of ECG attractor by polar coordinates without embedding a delay time

Figure 5 shows the trajectories which represent the normalized value of $ECG_p(n)$.

Figure 5 demonstrates the sharpness of the anchor-shaped portrait and is dulled when the ECG segment contains at least one PVC beat. Thus, a kurtosis value which measures the concentration of a distribution around its mean and standard deviation of projection on the line of identity with positive or

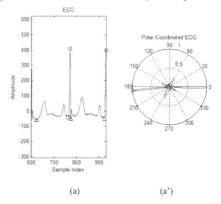

(a)   (a')

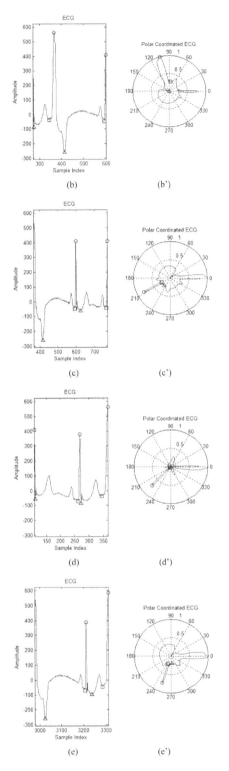

negative slope can be used to discriminate the characteristics of a PVC beat from a normal one. Also, the covered area A in phase plot can be computed quadrant-wise such as

$$A_i = \left(\frac{1}{M}\right) \cdot \sum_{j=1}^{l} r_j, \quad i \cdot \frac{\pi}{2} - \frac{\pi}{2} \le \theta \le i \cdot \frac{\pi}{2}, \quad i = 1, \ 2, \ 3, 4 \ (3)$$

where l is a number of samples in i^{th} quadrant.

3 CONCLUSIONS

To investigate the chaotic behavior of ECG time series, the state-space portrait is usually sought by forming the data vector with embedding a lag time d. However, the appropriate selection of d value is required to estimate the possible trend. In this work, a new method to evaluate the dynamics of ECG data with PVC beats is suggested and the experiments on MIT-BIH ECG database show that the characteristics of the PVC beats can be revealed without the dependency on the selection of a lag time d value.

ACKNOWLEDGEMENT

This research was supported by Basic Science Research Program through the National Research Foundation of Korea (NRF) funded by the Ministry of Education, Science and Technology (NRF-2013R1A1A2012393).

REFERENCES

[1] Blinowska, K.J. & Zygierewicz, J. 2012. *Practical Biomedical Analysis Using MATLAB*. Boca Raton: CRC Press.
[2] Goldberger, A.L. 2000. PhysioBank, PhysioToolkit, and Physionet: components of a new research resource for complex physiologic signals. *Circulation* 23(101): 215–220.
[3] Hundewale, N. 2012. The application of nonlinear dynamics for ECG in Normal Sinus Rhythm. *International Journal of Computer Science Issues* 9(1): 458–467.
[4] Kim, J.H., Park, S.E., Lee, J.W. & Kim, K.S. 2014. Design and implementation of digital filters for mobile healthcare applications. *International Journal of Electronics and Electrical Engineering* 2(1): 75–79.
[5] Piskorski, J. & Guzik, P. 2005. Filtering Poincare plots. *Computational Methods in Science and Technology* 11(1): 39–48.
[6] Rajendra, A., Suri, J.S., Spaan, J.A.E. & Krishnan, S.M. 2007. *Advances in Cardiac Signal Processing*. Berlin Heidelberg: Springer.
[7] Vegte, J.V. 2002. Fundamentals of Digital Signal Processing. Upper Saddle River, New Jersey: Prentice Hall.

Figure 5. The ECG segment (the part of ECG-119-1) and the normalized trajectory in polar coordinates: (a) three normal heart beats with its trajectory-a' (b) normal-PVC-normal beats-b' (c) PVC-normal-normal beats-c' (d) normal-normal-PVC beat-d' (e) PVC-normal-PVC beats-e'.

Future Communication, Information and Computer Science – Zheng (Ed.)

Taint propagation path analysis based on program dependency

Y. Wang, C. Wang, C.Z. Hu & C. Shan
School of Software, Beijing Institute of Technology, Beijing, China

ABSTRACT: Taint propagation path analysis is an important pre-processing part to software vulnerability detection. This paper presents a static taint propagation path analysis method based on program dependency. After transforming the source code to the SSA (Static Single Assignment) form, we make a program dependency analysis, which consists of data dependency analysis within basic blocks, control dependency analysis between basic blocks, and Phi function dependency analysis. In particular, the branch control variables calculation from control dependency analysis and Phi function dependency analysis can transform the dependency between basic blocks into dependency between variables. As a result, we obtain the global program variable dependency graph, based on which, we can further find taint propagation paths through marking tainted data. We have implemented the method on the LLVM (Low Level Virtual Machine) compiler, and the experiment results show its validity.

Keywords: taint propagation path analysis; SSA; data dependency; control dependency; Phi function; branch control variables; value dependency graph

1 INTRODUCTION

Taint propagation analysis which originated from Perl's security feature called taint mode refers to the process of marking the external input data as tainted data and tracking the propagation of those data. From the obtained taint propagation paths, we can clearly understand tainted data propagating process and whether tainted data flow into sensitive functions. There have been a great number of studies in the literature to apply taint analysis to software vulnerability detection (Wang, T.L. & Wei, T. 2010), automatic vulnerability signature generation (ZhuGe, J.W. et al. 2013), and software testing. These studies can broadly be categorized under static (Lattner, C. & Adve, V. 2004, Wang, T.L. & Wei, T. 2009, Yu, H. & Xue, J. 2010) and dynamic taint analysis approaches (Enck, W. & Gilbert, P. 2010, Lai Z. 2010, Wang, T.L. et al. 2011). Static approaches analyze the source code or its immediate representation and program information flows without executing programs. Static analysis is capable of analyzing all the possible paths from the tainted source to potentially vulnerable statements. Dynamic analysis is performed at each step of execution of a single run, and needs to get the useful information about taint through code instrumentation (Wang, T.L. 2011), i.e. inserting its own analysis code on an instrument platform to monitor the program's execution. Obviously, inserting extra code leads to lower performance. Furthermore, since only one path is analyzed per execution, the path coverage is much lower than that of static analysis.

Static taint analysis is based on information flow analysis (Huang, Q. & Zeng, Q. 2011), which includes data flow analysis and control flow analysis. In essence, taint propagation is a particular type of variable propagation, because taints propagate through variable dependency. Therefore, this paper proposes a taint propagation path discovery approach, which combines program dependency analysis and static taint analysis.

The rest of this paper is organized as follows. Sec. 2 surveys related work, Sec. 3 presents the program dependency based taint propagation analysis method, and Sec. 4 describes the experiments. In the last section we draw our conclusions.

2 RELATED WORKS

In this section, we mainly overview the research works on static taint analysis. ARCHER (Wang, T.L. & Wei, T. 2009) and IPSSA (Yu, H. & Xue, J. 2010) both use path sensitive analysis methods, and can detect memory access errors and report vulnerabilities positions, but no specific description of error paths is given; however, IPSSA extends SSA by applying define-using relationship. Huang, Q. & Zeng, Q. (2011) proposed a taint propagation analysis and dynamic verification based on information flow policy, to transform source code into a specific SSA representation, and obtain the information flow path of variables and other entities. However, too much irrelevant information flows needed to be maintained which had a bad effect on the performance. In order to reduce false taint alarm rate and negative taint rate, Chen et al. (2013) proposed an implicit taint propagation system, which defined implicit taint propagation by the form

```
1    int test_dep(int m, int n, int k)
2    {
3        int   i = 0;
4        int   x = 1;
5        while( i < 10 ){
6            if( k == 5 ){
7                x = x * m;
8            }else{
9                x = x / n;
10           }
11           i++;
12       }
13       return x;
14   }
```

Figure 1.   Example source code.

of "if … else …". However, its control dependency analysis stops at the basic block level, and it does not deal with the dependency of Phi function at the basic block. Scholz, B. et al. (2008) proposed user input dependency analysis via user graph reachability.

In order to solve the above problems, this paper proposes a taint propagation path analysis method based on program dependency. Particularly, the introduction of computing branch control variables and analyzing both data dependency and control dependency of Phi function can transform the dependencies between basic blocks into dependencies between variables.

3 TAINT PROPAGATION PATH ANALYSIS METHOD BASED ON PROGRAM DEPENDENCY

Our method involves three main steps: (1) transform source code into SSA based intermediate representation; (2) analyze its program dependency, which includes data dependency analysis within basic blocks, control dependency analysis between basic blocks, both data dependency and control dependency analysis of Phi function; (3) find out taint propagation paths.

3.1 Source code transformation

In order to improve efficiency and accuracy, we first transform the source code into SSA form, in which, all variables have a single assignment. Specially, we use LLVM complier framework (Lattner, C. & Adve, V. 2004) to do this transformation. Besides the SSA representation, LLVM also generates (not explicitly) a control flow graph for each function, which can help further analysis.

The example source code used in this paper is shown in Figure 1 and the SSA immediate representation is given in Figure 2. Figure 3 shows the corresponding Control Flow Graph (CFG) generated by LLVM.

3.2 Program dependency analysis

The program dependency analysis consists of data dependency within basic blocks, control dependency between basic blocks, and dependency of Phi function

```
define i32 @test_dep
(i32 %m, i32 %n, i32 %k) nounwind {
entry:
    %m.addr = alloca i32, align 4
    %n.addr = alloca i32, align 4
    %k.addr = alloca i32, align 4
    %i = alloca i32, align 4
    %x = alloca i32, align 4
    store i32 %m, i32* %m.addr, align 4
    store i32 %n, i32* %n.addr, align 4
    store i32 %k, i32* %k.addr, align 4
    store i32 0, i32* %i, align 4
    store i32 1, i32* %x, align 4
    br label %while.cond
while.cond:       ; preds = %if.end, %entry
    %0 = load i32* %i, align 4
    %cmp = icmp slt i32 %0, 10
br i1 %cmp,label %while.body,label %while.end
while.body:       ; preds = %while.cond
    %1 = load i32* %k.addr, align 4
    %cmp1 = icmp eq i32 %1, 5
br i1 %cmp1, label %if.then,label %if.else
if.then:        ; preds = %while.body
    %2 = load i32* %x, align 4
    %3 = load i32* %m.addr, align 4
    %mul = mul nsw i32 %2, %3
    store i32 %mul, i32* %x, align 4
    br label %if.end
if.else:         ; preds = %while.body
    %4 = load i32* %x, align 4
    %5 = load i32* %n.addr, align 4
    %div = sdiv i32 %4, %5
    store i32 %div, i32* %x, align 4
    br label %if.end
if.end:          ;preds = %if.else, %if.then
    %6 = load i32* %i, align 4
    %inc = add nsw i32 %6, 1
    store i32 %inc, i32* %i, align 4
    br label %while.cond
while.end:        ; preds = %while.cond
    %7 = load i32* %x, align 4
    ret i32 %7
}
```

Figure 2.   The SSA IR form of example program.

at basic block aggregation points. After the dependency analysis, a global program variable dependency graph can be generated.

3.2.1 Data dependency within basic blocks
In this part, we do a data dependency analysis for statements within basic blocks. The taint attributes of variables propagate directly through assignment and

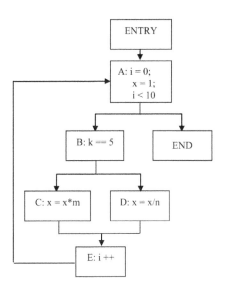

Figure 3. Control flow graph of example program.

arithmetic operations, so we mainly focus on binary instructions, such as (algebraic operations, shift operations, logic operations), type conversion instructions, and compare instructions. Their left-hand values have data dependency on their right-hand side.

Taking the dependency of basic block C as an example (see Fig. 3), variable m is a parameter of function test_dep which is called by the main function, so mark m as tainted, then variable x becomes tainted after statement 7 (see Fig. 1) is executed, i.e. the taint attribute passes from variable m to variable x. For the IR form in Figure 2, variable $\%x$ data depends on variable $\%mul$, while $\%mul$ depends on $\%m$, therefore the taint attribute of variable m delivered to variable x through a series of arithmetic and assignment statements.

3.2.2 Control dependency between basic blocks

The control dependency analysis is essentially computing the post dominance frontier of each basic block in a CFG. Given a CFG G, node B is said to post dominate node A if each path from A to END contains B, we denote such relationship as: B PDom A. The post dominance frontier can be described as follows:

PDF(Y) = {X | Z is a successor of node X, i.e. $Z \in$ succeed (X) and Y PDom Z and Y + (PDom) X}. Control dependency is analyzed between basic blocks, that is the taint attributes of variable x indirectly pass to variable y through the conditional statements such as "if ... else" and "switch ... case".

We say basic block Y depends on basic block X by control dependency, if they satisfy the following conditions:

(1) There exists a path P from X to Y : $X \rightarrow Z_1 \rightarrow Z_2 \rightarrow \cdots \rightarrow Z_n \rightarrow Y$, Y post dominate all Z_i, $i \in N$;

(2) Y does not post dominate X, i.e. there is another path from Y to X.

According to the definitions of control dependency and post dominance frontier, searching control dependency between basic blocks can be converted to computing post dominance frontiers. Since taint propagation is essentially tainted variable propagation, and control dependency is between basic blocks, therefore control dependency needs to be transformed to control dependency between variables in conditional statements and branch statements. So we introduce branch control variable (BCV) as follows:

Assuming that basic block B contains definition statement of variable v or use variable v, the post dominance frontier of B is a set PDF(B). For any $B_i \in$ PDF(B), if the last statement of B_i is a conditional statement P, the condition is C_i, then the branch control variables of v is constituted by condition C_i, denoted by BCV(v) = {$C_i | i \in N$}.

Now we can get control dependency between basic blocks by computing post dominance frontiers and can further find the branch control variables (BCV) of variables defined or used in the basic blocks. In such a way, control dependency between basic blocks can be transformed into dependency between variables.

For the example CFG in Figure 3, the post dominated nodes for each node are as follows:

A PDom {A, B, C, D, E}
B PDom {B}
C PDom {C}
D PDom {D}
E PDom {E, B, C, D}

The PDFs for basic blocks are shown below:

PDF(A) = { }
PDF(B) = {A}
PDF(C) = {B}
PDF(D) = {B}
PDF(E) = { }

Variable k is used in basic block B, the branch control variables of k is BCV(k) = {i}, and variable x is defined and used in basic block C and D, so we know BCV(x) = {k}.

3.2.3 Dependency of Phi functions at basic block aggregation points

When a basic block encounters a branch, and the branch has operations on a variable, we cannot determine which version of the variable to use. Therefore, we introduce the function Phi to connect all branches and generate a new definition of the variable. Figure 4 gives an example CFG with Phi function. The definition of $x4$ comes from function Phi, which takes previous definitions $x2$ (from the left branch) and $x3$ (from the right branch) as operands, i.e. $x4 = \Phi(x2, x3) = \Phi(x2) \cup \Phi(x3)$. So the dependency of $x4$ consists of three parts: control dependency of basic block where $x4$ is located, data dependency of $x2$ and control dependency of basic block where $x2$ is located, data dependency of $x3$ and control dependency of basic block where $x3$ is located. Thus, the dependency of Phi function can be calculated using Equation 1.

Dependency(ϕ)

$$= \left(\text{DataDep}(\phi) \cap \text{ControlDep}(\phi)\right) \cup \text{ControlDep}(B) \quad (1)$$
$$= \cup \left(\text{DataDep}(v_i) \cap \text{BCV}(B_i)\right) \cup \text{BCV}(B)$$

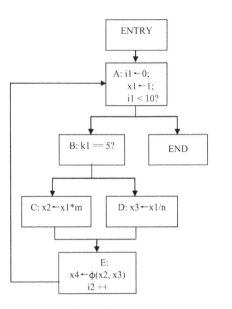

Figure 4. Example CFG with a branch.

In which, B is the basic block where function Phi is located, B_i is precursor basic block of Phi node, v_i is defined or used in basic block B_i.

As shown in Figure 2 and Figure 3, both basic block C and D use variable x, and the successor block of E has assignment of variable x, so we insert a Phi function with variable x in block A as $x = \%7 = \Phi(<\%mul, B>, <\%div, C>)$. The dependency of x can be expressed as Equation 2.

$$
\begin{aligned}
&\text{Dependency}(x) \\
&= \text{Dependency}(\%7) \\
&= \big(\text{DataDep}(\%mul) \cap \text{BCV}(C)\big) \cup \big(\text{DataDep}(\%div) \cap \text{BCV}(A)\big) \quad (2) \\
&= \{m, n, k, i\}
\end{aligned}
$$

We can see that variable x depends on variable m, n, k, and i, in which, i is not tainted, while m, n, k are the parameters of function test_dep and are marked as tainted. Therefore, variable x can be determined to be tainted through data dependency and control dependency analysis.

3.2.4 Generation of program variable dependency graph

Based on the above analysis, program variables can be divided into two categories: variables assigned by Phi functions; variables assigned by functions other than Phi and their dependency include data dependency and control dependency. The program variable dependency can be calculated using Equation 3.

$$
\text{Dependency}(v) =
$$
$$
\begin{cases}
\text{DataDep}(v) \cup \text{BCV}(B), v \notin \text{Phi function} \\
\cup\big(\text{DataDep}(v_i) \cap \text{BCV}(B_i)\big) \cup \text{BCV}(B), v \in \text{Phi function}
\end{cases} \quad (3)
$$

If variable v does not belong to function Phi, basic block B is where the assignment statement of v is

Table 1. Algorithm for generating program variable dependency graph.

Algorithm1: generating program variable dependency graph.

```
//Initialize dependency set LIST
for each basic block B in function F
    if the last statement of B is a branch instruction
        then add branch condition into set LIST
    end if
end for
//Compute dependency relationship until LIST no //longer
changes
changed = true;
while(changed)
    changed = false;
    for each variable v in LIST
        compute Dependence(v);
        add Dependence(v) into dependency set of v: Set(v);
        for each variable vDep in Set(v)
            if vDep does not belong to LIST
                then add vDep into LIST;
                    changed = true;
            end if
        end for
    end for
end while
```

located; if variable v belongs to function Phi, the basic block B is where Phi function is located, B_i is precursor basic block of Phi node, v_i is defined or used in basic block B_i.

After calculating program variable dependency, we can generate global program variable dependency graph using Algorithm1 shown in Table 1. The algorithm first collects branch control variables BCV and adds BCV into a dependency set LIST, then calculates Dependence(v) for each variable v in LIST, until LIST does not change, that is reaching the fixed point, i.e. the program variable dependency analysis is complete. Since the program variable dependency determines how tainted variables propagate, we can further mark every tainted variable to obtain taint propagation paths.

3.3 Discovery of taint propagation path

Based on the global program variable dependency graph in the above section, we use abstract interpretation (Cousot, P. & Cousot, R. 1977) type inference to discover taint propagation paths. We first adopt the definitions of Poset, complete lattice, and Galois connection from (Cousot, P. & Cousot, R. 1977). Then we define abstract domain $\mathcal{R} = <\text{L}, \sqsubseteq>$, where L = {Tainted, Untainted}, the partial order of the complete lattice is defined as Untainted $\sqsubseteq$ Tainted, the smallest element $\bot$ = Tainted, the largest element $\top$ = Untainted. Galois connection is defined as: $<\sigma, \sqsubseteq>$ is a specific domain, $<\text{L}, \sqsubseteq>$ is an abstract domain, abstract function α is a mapping function from specific domain to abstract domain, specific function γ is

Table 2. Algorithm for type inference of taint data.

Algorithm 2: type inference of taint data

```
//Mark tainted data in LIST
    for each variable v in LIST
        if variable v is user input data
            then α(v) ← Tainted
        end if
        else α(v) ← Untainted
        end else
    end for
//Update the taint attribute of variables in LIST
    changed = true;
    while(changed)
        changed = false;
        for each variable v in LIST
            if variable v depends on tainted data w and α(w) is
            Tainted
                if α(v) is Untainted
                    α(v) ← α(w) ⊔ α(v);
                    changed = true;
                end if
            end if
        end for
    end while
```

a mapping function from abstract domain to specific domain, as shown in Equation 4:

$$\langle \sigma, \sqsubseteq \rangle \underset{\gamma}{\overset{\alpha}{\rightleftarrows}} \langle L, \sqsubseteq \rangle \qquad (4)$$

In a value dependency graph, nodes represent variables, edges represent dependencies between nodes, if a node v depends on node c_i, $i \in N$, then abstract function can be defined as Equation 5:

$$a(v) = \underset{i \in N}{\text{ò}} a(c_i) \qquad (5)$$

According to the type theory, Equation 5 can also be described as:

$$c_1:\text{Untainted ò } c_2:\text{Untainted} \rightarrow \text{Untainted} \qquad (6)$$

$$c_1:\text{Untainted ò } c_2:\text{Tainted} \rightarrow \text{Tainted} \qquad (7)$$

$$c_1:\text{Tainted ò } c_2:\text{Tainted} \rightarrow \text{Tainted} \qquad (8)$$

According to Equations 4–8, we know that as long as variable node v depends on c whose taint attribute is Tainted, v is tainted data, i.e. its taint attribute is Tainted. The type inference algorithm for taint data is shown in Table 2. The algorithm first marks the input data (variables in LIST) as tainted, then does abstract interpretation based type inference for taint attribute iteratively, until the program reaches the fixed point. Finally, we can obtain a value dependency graph by marking all tainted nodes gray, and marking all untainted nodes white, thus the directions of arrows are reverse to the propogation directions of tainted data.

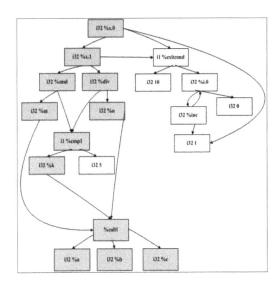

Figure 5. Taint propagation graph.

4 EXPERIMENTAL RESULT

We implement this method in Ubuntu Linux and use LLVM compiler framework as an experimental platform.

The example source code (see Fig. 1) is transformed to SSA-based intermediate representation (see Fig. 2), then we do a data dependency analysis, control dependency analysis and analysis of Phi funtion (see Fig. 3 and Fig. 4), finally we obtain a taint propagation graph by generating value dependency graph and solving taint propagation paths. In the taint propagation graph shown in Figure 5, each node represents a variable, each edge represents a dependency between variables, so we can find taint propagation paths in the graph.

From Figure 5, we can see that variable x depends on the "while" condition and "$k == 5$" judgment statement by control dependency, meanwhile, x depends on values of variable m and n by data dependency. The main function transfer untrusted input a, b, c by calling function test_dep, as the result of which, variable x is tainted. $x.0$ and $x.1$ are two SSA representation renames to variable x, the taint attribute of $x.1$ is passed to $x.0$, while $x.1$'s taint attribute is determined parameter m and n. Thus we can find out taint propagation paths in the graph clearly.

5 CONCLUSIONS

The discovery of taint propagation paths in the program is important to software vulnerability detection. In this paper, we propose a program dependency based static taint propagation analysis method. The method gurantees its validity by introducing both branch control variable calculation and data dependency and control dependency of Phi function. Moreover, the taint propagation graph generated by the method can make subsequent works of vulnerability detection easier and more intuitive.

ACKNOWLEDGMENTS

This work is supported by the Key Project of National Defense Basic Research Program of China under Grant No. B1120132031.

REFERENCES

[1] Chen, H. & Zhang, X. et al. 2013. An implicit taint propagation system and its implementation. China patent: CN103412750A.

[2] Cousot, P. & Cousot, R. 1977. Abstract interpretation: a unified lattice model for static analysis of programs by construction or approximation of fixpoints. In *Proceedings of the 4th ACM SIGACT-SIGPLAN symposium on Principles of programming languages; New York, NY, USA, 1977.* ACM.

[3] Enck, W. & Gilbert, P. 2010. TaintDroid: An Information-Flow Tracking System for Realtime Privacy Monitoring on Smartphones. *OSDI* 10: 1–6.

[4] Huang, Q. & Zeng, Q. 2011. Taint propagation analysis based on information flow policy and dynamic verification. *Journal of Software* 22(9): 2036–2048.

[5] Lattner, C. & Adve, V. 2004. LLVM: A compilation framework for lifelong program analysis & transformation. In *Code Generation and Optimization. Proc. Intern. symp., San Jose, March 2004.* IEEE.

[6] Lai Z. 2010. Dynamic taint analysis based on state protocol implementations of software fuzzing method. [D]. *Changsha: National University of Defense Technology.*

[7] Scholz, B., Zhang, C. & Cifuentes, C. 2008. User-input dependence analysis via graph reachability. Technical Report, Sun Microsystems, Inc. Mountain View, CA, USA.

[8] Wang, T.L. 2011. Research on Binary-Executable-Oriented Software Vulnerability Detection. [D]. *Beijing: Peking University.*

[9] Wang, T.L. & Wei, T. 2009. IntScope: Automatically Detecting Integer Overflow Vulnerability in X86 Binary Using Symbolic Execution. *In Proceedings of the 16th Annual Network and Distributed System Security Symposium, San Diego, CA, February 2009.*

[10] Wang, T.L. & Wei, T. 2010. TaintScope: A checksum-aware directed fuzzing tool for automatic software vulnerability detection. In *Security and Privacy (SP), 2010 IEEE Symposium on., Oakland, 16–19 May 2010.* IEEE.

[11] Wang, T.L., Wei, T. & Zou, W. 2011. RoBDD-Based Fine-Grained Dynamic Taint Analysis. *Acta Scientiarum Naturalium Universitatis Pekinensis* 47(6): 1003–1008.

[12] Yu, H. & Xue, J. 2010. Level by level: making flow-and context-sensitive pointer analysis scalable for millions of lines of code. In *Proceedings of the 8th annual IEEE/ACM international symposium on Code generation and optimization.*

[13] ZhuGe, J.W. & Chen, L.B. & Tian, F. 2013. Dynamic taint analysis based on type technology. *Acta Scientiarum Naturalium Universitatis Pekinensis* 52(10): 1320–1328.

Future Communication, Information and Computer Science – Zheng (Ed.)
© 2015 Taylor & Francis Group, London, 978-1-138-02653-7

Research on computer network courses teaching based on constructivist learning theory

L.P. Feng, T. Li, H.Q. Liu & K. Xie
Beijing Institute of Graphic Communication, Beijing, China

ABSTRACT: With the speeding up of the internet, the related course of computer network becomes a basic and important course of Computer Science and Technology Specialty in universities. But it is difficult for students to realize the abstract concepts and the intricacies regarding network communication. In order to probe into the effective theoretical and experiment teaching method of computer network courses, we reform computer network courses according to constructivist learning theory, and enhance the interactive teaching, cultivating the desire of autonomous learning and active learning. This paper discusses the methods of computer network courses teaching reform, and brings forward some of the measures of combined school teaching practice to promote computer network courses teaching applications. With the research and practice of the teaching reform of computer network courses, the students feel computer network is no longer boring but challenging and fun, and their interest in learning computer network courses is improved greatly.

Keywords: computer network courses; constructivist learning theory; teaching reform; teaching resources; experiment teaching

1 INTRODUCTION

With the rapid development of information, computer network courses are essential parts of the college computer science curriculum. These courses not only contain abstract and profound theoretical knowledge such as network architecture, the principles of network communication, TCP/IP protocols, etc., but also involve a large number of network technologies such as operation and configuration of the typical network interconnect equipment, from LAN, WAN to Internet. All these make it difficult for students to fully comprehend and appreciate the intricacies regarding network communication without implementing some portion of the network logic [1].

From 90 years of the 20th century on, constructivist learning theory and situated cognition theory have been a significant development in Europe and the United States academia, as an important branch of cognitive learning theory. Constructivist teaching is based on the belief that learning occurs when learners are actively involved in a process of meaning and knowledge construction rather than passively receiving information. Learners are the makers of meaning and knowledge. Constructivist teaching fosters critical thinking and creates motivated and independent learners [2].

We probe into the effective theoretical and experiment teaching method of computer network courses based on constructivist learning theory, and bring forward some of the measures combining schools'

teaching practice to promote computer network courses teaching applications.

This paper discusses the methods of computer network courses' teaching and describes the research and practice of our teaching reform.

2 THE EXPLORATION OF TEACHING

2.1 The teaching goals

The overall goal of computer network courses is to understand the principles of computer network communication, and realize the protocols of TCP/IP architecture for the students. And on this basis, the students can make use of various network devices and solve the problems in the network. Furthermore, they can do network programming and network development.

We set up three stage teaching goals based on an analysis of the social demands and discussion among teachers.

The first stage goal focuses on the basic theory of computer network, to make students grasp the basic technical ability of network construction and network maintenance.

The second stage goal is the ability to configure and manage network equipments. The students can plan and design networks according to the demand of the enterprise, detect and deal with various network intrusions.

The third stage goal emphasizes the mastery of network programming.

To achieve these goals, we divide computer network courses into three modules: computer network infrastructure, computer network technology and computer network programming [3], and cultivate the students from the two aspects of theory and practice. The aspects of theory comprises the lectures used to teach the principle and theoretical understanding of computer network. The other intends to develop the student's practical network design skills.

2.2 The teaching theory

In the traditional teaching mode, the teacher is in an active position of learning activities, students are passively taught.

With the rapid development of multimedia technology, the effect of a constructivist idea generates increasing attraction to teaching design.

Constructivist learning theory thought that students are the subject of learning, and knowledge is a cognitive process and construction experience. Constructivism emphasizes that every learner should form his own experiences through interaction with the world, and construct his own knowledge based on the experiences rather than wait for knowledge transfer. That means, the students should construct their own knowledge actively.

According to constructivist learning theory, we take a variety of teaching methods to enhance the interactive teaching and cultivate the desire of autonomous learning and active learning.

2.3 The teaching methods

We should avoid being boring, dull and cramming during theoretical teaching. In order to ensure the students understand and master the abstract concepts of computer network in the class, we apply multimedia teaching to make the lecture vivid, and promote interactive teaching mode. There are several teaching manners:

(1) Instantiation based teaching

In the computer network courses teaching, some concepts such as network protocol, network architecture, especially the exchange of network data behind the protocol, are so abstract that most students feel it is difficult to understand. We use instantiation to correlate abstract theory with practice.

We apply some animation about protocols to the lecture, let the students have a deeper realization on the computer network principles, and improve the learning effect greatly. From a cognitive point of view, instantiation based teaching helps to stimulate students to get pleasure in learning, have a sense of achievement, and increase the learning internal driving force.

(2) Multimedia teaching

The implementation of the multimedia teaching is the modern teaching method, and it is a challenge to the traditional teaching pattern. Constructivist learning theory thought, in order to obtain the best teaching effect, it should be learner centered. We need to use not only the modern teaching methods and tools, but also teaching design, reasonable use of teaching resources, teaching courseware and teaching scheme. We did some beneficial exploration in computer network courses innovation, and achieved some satisfactory results.

(3) Virtual reality teaching

Because the experimental laboratory is limited, it is difficult to be open for every one. The development of virtual network technology provides a new idea for the reform of experiment teaching. We construct a network virtual experiment platform with simulation software, and take advantage of a virtual network experiment platform, combined with the real experimental environment, to promote the reform of experiment teaching and cultivate students' comprehensive practical ability and innovation ability.

(4) The top-down approach

In the past, computer network courses teaching used the bottom-up approach for the network protocol, from the underlying protocols of physical layer, data link layer and network layer, to the protocol of transport layer and application layer. Since 2010, we have been dealing with the network protocols in the top-down approach, that means to explain network protocols from application layer downwards until physical layer. This is also a new concept in the computer network courses teaching in recent years.

The top-down approach has several important benefits. First of all, many revolutionary innovations in the field of computer network has occurred in the application layer. The students can pay close attention to these new technologies application if we emphasise the application layer early. Secondly, the students are eager to understand how network application of Email and Web work. That will stimulate the students to study the services provided and implemented by the relatively low layers. Finally, the top-down approach enables teachers to introduce the development of Web applications in the early stage of teaching, and the students can experience creating their own Web application and protocol, and feel it is not a very difficult thing.

3 THE EXPERIMENT TEACHING

3.1 The experiment platform

As the network experiments involved the implementation of the underlying system, it is difficult to implement the experiments in the real environment. The experiment teaching depends on the laboratory construction.

Because of the ageing equipments of our laboratory, it can only set up the basic network experiments and network technology experiments in the past. The condition of our network laboratory can not meet the requirement of computer network courses. In order to carry out the experiments better and meet the demand for fostering computer network talents, according to the new curriculum system of computer network, we upgrade the network laboratory, build "SimpleNPTS Network Protocol Teaching System" [4] and "Network Information Security Experiment Teaching System" [5].

SimpleNPTS Network Protocol Teaching System is a network experiment simulation platform developed by Beijing Simpleware Technology Co., Ltd. It combines the actual situation of university education, realizes the theoretical knowledge of networks through software. Students can understand the basic network theoretical knowledge well, comprehend the internal structures and protocols of network clearly in the process of practice. SimpleNPTS Network Protocol Teaching System helps students to study the internal network principle clearly and aids them to debug network program by editing the various protocol packets.

Network Information Security Experiment Teaching System is developed by CSS-JLU INFO-TECH CO., LTD. It covers the whole network security course practice teaching from the information security technology education to the technology application, provides encryption algorithm and the application, security of communication, audit, network attack, virus attack, firewall, information hiding etc, to help the students master the skills of network security.

3.2 *Experiment teaching method*

In general, the experiments content of computer network courses includes two parts: one is basic and confirmatory experiments to simulate and verify principals, and the other is integrity and designing experiments in which students are the main body to accomplish an overall design scheme in accordance with experiment condition and requirement [6].

The basic and confirmatory network experiments include TCP/IP protocol installation, IP address allocation, subnet partition, routing configuration, Internet application, etc. The students can implement basic configurations of network service, network programming, website construction and webpage design.

The integrity and designing network experiments use network equipments including mainly router, switch to help the students realize the virtual secret network technology, configuration of routing, network address translation on the hardware device. Comprehensive experiments also involve a variety of network attack technology, selection and configuration methods of different equipments and software, being an important process to enhance the students' ability of engineering practice.

4 CONCLUSION

In this paper we have discussed a reform of computer network courses, including theoretical teaching and experiment teaching. Our research brings forward some of the measures that combined schools teaching practice to promote computer network courses teaching applications.

The teaching reform shows that the introduction of constructivist teaching theory in the teaching of computer network, is conducive to the cultivation of the learners' ability for innovation and practice, to stimulate interest in learning, improve the learning effect. The students feel computer network is no longer boring but challenging and fun, and the interest of the students in learning computer network courses is improved greatly.

Network technology is developing rapidly, and the methods and manners are also progressing. We need to update teaching ideas to arouse the students' interest and creativity.

ACKNOWLEDGMENTS

This paper is supported by the course construction project of Beijing Institute of Graphic Communication (No. 22150113077).

REFERENCES

[1] X. H. Li, R. L. Zhang, L. Yang, et al. Optimization of Computer Network Teaching and Learning Behavior Using Virtual Experiment Technology. International Conference on Information, Business and Education Technology (ICIBET). Atlantis Press, 2013: 1058–1061.

[2] D. Alt. Constructivist Teaching Methods. Changes in Teachers' Moral Role, Springer, 2012.

[3] L. B. Zheng. Discussion on the construction of computer network curriculum. Proceedings of teaching reform and research of application oriented universities, Electronic Industry Press, 2009: 45–47.

[4] SimpleNPTS Network Protocol Teaching System Experiment Guidebook. Beijing Simpleware Technology Co., Ltd., 2010.

[5] Network Information Security Experiment Tutorial. CSS-JLU INFO-TECH CO., LTD., 2007.

[6] P. Qi, L. Y. Shi. Construction of Experiment Teaching on Computer Network Course. Conference on Education Technology and Management Science (ICETMS 2013). Atlantis Press, 2013.

Future Communication, Information and Computer Science – Zheng (Ed.)
© 2015 Taylor & Francis Group, London, 978-1-138-02653-7

Adaptive access mechanism for IEEE 802.11 WLANs in non-saturated conditions

C. Shi, Z. Deng, S. He, H. Lin & J. Wang
School of Information Science and Technology, Hainan Normal University, Haikou, P. R. China

ABSTRACT: How to adjust parameters to adapt to the dynamic network conditions is a main problem in wireless communications. There are no effective methods to deal with this problem, especially in non-saturation conditions. By analyzing the feedback signal of channel, we introduce a linear adjustment rule of CW about idle slot intervals of channel, which can be used in non-saturation conditions. And then, we propose a fully adaptive access mechanism, which adjusts parameters dynamically according to the traffic loads in the network. With the linear adjustment of CW and measurement of idle slot intervals, the mechanism separates the adjustment rule of parameters from the status of data transmissions, which is distinguished from that in the DCF method. By sensing the channel idle slot intervals, this mechanism does not need to know or judge whether it is in saturation conditions or not, which simplifies the process of access mechanism. The simulation results demonstrate the validity and good scalability of the proposed access mechanism.

Keywords: IEEE 802.11; medium access control; non-saturation condition; backoff algorithm.

1 INTRODUCTION

The applications of wireless communications have been used greatly. Existing works on the performance of the 802.11MAC have focused primarily on its throughput [2] and network delay [3] under saturated conditions. Bianchi proposed a foundation model to analyze the performance of the IEEE 802.11 Distributed Coordination Function (DCF), and gave a throughput performance evaluation algorithm in saturation conditions [2]. The performance of MAC delay of DCF in saturation conditions is analyzed in [3].

The IEEE 802.11 is likely used for small scale networks and the access mechanism can result in high collision probability during the data transmission in large scale networks. The main reason is that the DCF method cannot deal with the dynamic network conditions with its adjustment rules and constrained ranges of CW. The ranges of CW are fixed for all complicated network conditions, like two nodes or two hundred nodes in the network, which constrains the scalability of the method. The backoff strategy in the DCF method adjusts the size of Contention Window (CW) solely based on the state of the last transmission and does not take the network conditions into account, which can result in a performance decrease in large scale networks.

To decrease the collision probability in large scale networks, many methods are proved in [4] and [5]. In saturation conditions [5], the adaptive access mechanism adjusts the sizes of CW based on the idle slot intervals between two transmissions, which can reflect partly the dynamic states of networks.

In real traffic loads, like E-Mail and voice, it is normal under non-saturation conditions and nodes in the network do not have a packet to transmit all the time. There are many works to analyze the network performance under non-saturated conditions [6] and [7]. Normally, these mechanisms adjust these access parameters based on the channel state information (CSI), like collision probability or channel usage, which increases some network performance. These methods, however, need to judge network states, like saturation or not in the channel, which increases the complication of these methods and decreases network performance, especially in dynamic network conditions.

In this paper, we analyze the channel idle slots between two data transmissions by using the method in [5] and propose a Non-Saturated Access Mechanism (NSAM) that adjusts the CW based on the idle slot intervals between two data transmissions. The new access mechanism doesn't need to judge whether nodes are saturated or not and adjusts access parameters according to the dynamic channel idle slots, which can decrease the collision probability even in dynamic network conditions. The new mechanism can save the energy consumption and improve the network performance, simultaneously.

The rest of the paper is organized as follows. We introduce the backoff algorithm of IEEE 802.11 in the following section. In Section III, we develop an adaptive access mechanism to adjust the sizes of CW based on the idle slot intervals. In Section IV, we give the simulation results. Finally, we conclude our paper in Section V.

2 DCF METHOD OF IEEE 802.11

Even with the development of IEEE 802.11, the DCF method is always the key way to deal with collisions of data transmission in three methods, like DCF, PCF and HCF. We put our interest in the DCF method.

The DCF method is the fundamental access mechanism in IEEE 802.11 MAC [1], in which the adjustment rule of CW is a key way to resolve the collision of data transmission.

The adjustment rule of CW in the DCF scheme, named the Binary Exponential Backoff (BEB) algorithm, is used widely due to its simplicity. The BEB algorithm is divided into two main steps. The first step is to define the size of CW and select backoff time counter uniformly, and then the second is to decrement the counter value while the channel is idle.

As stated in [1], a tagged node using the BEB algorithm will set a low initial size of CW firstly if it desires to transmit data. After selecting the backoff time counter, the tagged node should detect the channel state. If the channel is idle, the tagged node decreases its counter values. When the counter value reaches zero, the tagged node will start a data transmission. The tagged node doubles the size of CW up to the maximum CW (CW_{max}) if its data transmission has a collision. After a successful data transmission, the tagged node will reset CW to the minimal CW (CW_{min}). And the same thing happens with new data transmission.

The detail adjustment rules of CW is as

$$\begin{cases} CW_i = 2 * CW_i - 1, & collision \\ CW_i = CW_{min}, & success \end{cases} \quad (1)$$

As shown in (1), the tagged node doesn't need to sense the status of channel and it adjusts the size of CW solely based on its states of the last data transmission. The reset of CW after a successful transmission is the main default to increase the collision probability and decrease the network performance as analyzed in [2] and [5].

Another problem is the fixed ranges of CW defined in [1] as

$$\begin{cases} CW_{min} = 32 & (slots) \\ CW_{max} = 1024 & (slots) \end{cases} \quad (2)$$

The ranges of CW are used for all network conditions, like two nodes or two hundred nodes in the networks. If there are low traffic loads, any method can be used for data transmission because there is less collision probability in the channel. The IEEE 802.11 can be fit for networks with low traffic loads as any other methods do.

With the increase of traffic loads, however, the channel will be congested by data transmission due to the increase of active nodes in the networks, even if nodes have not been saturated. At this time, the backoff strategies become important to avoid collisions of data transmission, which contains the adjustment rules and definition of ranges of CW.

The adjustment rules of BEB algorithm don't take the channel state into account, which cannot deal with this problem very well as analyzed in [2] and [5]. For a large number of nodes, to increase the sizes of CW and to reduce the collision, the strategies in the BEB algorithm are inefficient and will result in some serious problems, such as the fairness and higher collision probability.

The fixed ranges of CW and adjustment rules in the BEB algorithm are two main limitations to adapt to the dynamic network conditions, which results in poor network performance in large scale networks. The CW_{min} and CW_{max} are PHY-specific and used for any circumstances in the IEEE 802.11 protocols. The adjustment rules of CW that reset CW after successful data transmission are not very appropriate in many scenarios. Both of these aspects are the key points that many works are focusing on.

3 ADJUSTMENT RULES OF ADAPTIVE ACCESS MECHANISM

In non-saturation conditions, nodes start data transmission periodically. If there are no data to transmit, nodes may sleep to save the power between two data transmissions. With the increase of traffic loads, collision probability will increase and network performance will decrease dramatically. To improve performance, access mechanisms need to do two important things: one is to avoid the collisions of data transmission and another is to guarantee the transmission relay of data. The best way is to decrease the collision probability of data transmission in the networks.

We first analyze the relationship between CW and basic network conditions. And then, we further study the proper parameter to calculate sizes of CW with traffic loads.

3.1 Contention window index

With the network delay model in [5], we can obtain the aggregated throughput S as

$$S = \frac{E[P] + F_{ed} \cdot (T_{Data} / T_S) \cdot \lambda \cdot R_{Data}(t)}{E[D]} \quad (3)$$

where $Fed = p_S \cdot T_S \sum_{i=0}^{K-1} \sum_{j=0}^{i} p^i E[U^{(j)}]$ is the time that other nodes transmit successfully except the tagged node during the mean delay time. E[P] and RData(t) are the average packet payload length and the data rate. λ is the ratio of the payload sizes to full packet length with the header messages. Compared with the throughput in [2], the aggregated throughput

S in (3) represents the totally successful transmission of data in the mean network delay time, which includes the tagged node and other active nodes in the network.

The throughput (3) can be further simplified as

$$S = \frac{E[P](1 + p_S \cdot \sum_{i=0}^{K-1} \sum_{j=0}^{i} p^i E[U^{(i)}])}{E[D]} \qquad (4)$$

With $\tau = 2/(CW + 1)$ in [2], the aggregated throughput S in (4) is a function of CW and N. For a given node number N, there is an optimal CW which can achieve the maximum aggregated throughput. We take the derivative of the throughput S in (4) with respect to CW, and let $\partial S / \partial CW = 0$. After ignoring the items which are less than or equal to third order item $1 = CW^3$ and doing some simplifications, we obtain CW as follows

$$CW \approx (\sqrt{\frac{2T_{ECS}}{T_{SLOT}}} - 2) \cdot (N-1) = \theta \cdot (N-1) \qquad (5)$$

where $T_{ECS} = 2T_{EIFS} + 2T_C - 2T_{SLOT}$. T_{EIFS}, T_C and T_{SLOT} mean EIFS (Extended Inter-Frame Space) time, collision time and slot time, respectively. θ is referred to the Contention Window Index (CWI), which is mainly affected by the parameter of T_C. For the convenience of analysis, we select the access mechanism with RTS/CTS to reserve the channel resource. And then, for given a protocol, the optimal CWI (θ_{opt}) is a constant since T_C is a constant and other parameters are PHY character-based. Obviously, the optimal CW is a linear function of the active node number in saturation state. Compared with other adjustment rules of CW, the CW in (5) is associated with the active node number in the network, which can guarantee the high throughput even as the node number changes dynamically. Combined with the optimal CW access mechanism and the main parameters of IEEE 802.11b [1] listed in [5], we can resolve (5) and obtain the optimal CWI $\theta_{opt} \approx 10$.

3.2 CW and traffic loads

To decrease the collision probability, we will introduce an adaptive access mechanism. As described above, although these nodes start data transmissions periodically, the channel will be very busy due to many active nodes in networks. To adapt to the busy state of the channel, the access mechanism has two things to deal with: one is to sense the state of the channel and another is to adjust the access parameters for a better performance of networks. In this paper, we choose idle slot intervals between two data transmissions as the state of channel, which has been analyzed and used in [4] and [5]. According to the measured idle slot intervals, nodes adjust access parameter (CW) to adapt to he dynamic network conditions. We mainly analyze the RTS/CTS/Data/ACK mode of data exchange as done in [2] and [5].

Based on the analysis between sizes of CW and active node numbers in the network, we can obtain the relationship as

$$CW \approx \theta \cdot (N-1) \qquad (6)$$

where N means the number of nodes in the one hop ranges of network. θ is the CW index (CWI) and can be calculated with values of parameters in the access mechanism as definition in [5]. If nodes choose the access mechanism, the value of θ can be calculated in advance. Generally, we take the values of θ as integer and it is larger than zero, which means that nodes need some CW to resolve the collisions of data transmission by using the backoff time counter uniformly selected in CW.

In this paper, we extend the means of (6) and build the relationship between CW and traffic loads. The relationship of (6), however, is derived from the saturated networks, and it can work well with the estimation of active node numbers in the networks.

In non-saturated networks, the nodes don't have data to transmit all alone. In this case, the node number is not a proper parameter to be used as the state of the channel due to the lesser activity of data transmissions. Traffic loads in the channel can be used as the message of channel states. If we do not distinguish the identification of nodes in non-saturated networks, the number of nodes can be replaced by traffic loads in (6) as

$$CW \approx \theta \cdot (N_t - 1) \qquad (7)$$

where N_t means traffic loads in the networks. It means that, with traffic loads sensed in the network, nodes should determine corresponding CW to access the channel. The main problem is to calculate values of traffic loads in the channel.

To simplify calculation of CW, we ignore detail length of any busy states of channel and assume every state as an event, like successful transmission event and collision event. The collided data transmission, however, can be seen as two or more data transmissions happened at the same time, in which nodes have a different start time to begin transmission and the full time length of collision is difficult. By setting every traffic loads as events, nodes can decrease greatly the complicated calculation of full time length of every data transmission. Without paying attention to the full time length of successful data transmissions or collisions in the channel, we just count the number of occurred events, which is enough for the adjustment rules of CW. We have the calculation of N_t as

$$N_t = N_S + N_C \qquad (8)$$

where N_S and N_C mean the total number of successful data transmissions and collisions in the channel, respectively. The total number of N_S and N_C is calculated by nodes during its decrease process of backoff time counter, which can be used in the adjustment of CW for the data transmission. If the tagged node obtains the total number of traffic loads N_t, it adjusts the sizes of CW based on (7).

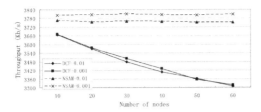

Figure 1. Throughput.

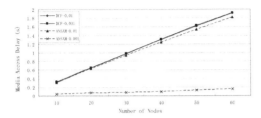

Figure 2. MAC Delay.

4 SIMULATION AND ANALYSIS

In this section, we use the OPNET (version 14.5) modeler to verify the new NSAM mechanism and compare the new mechanism to DCF mechanism [1]. The channel is ideal and there is no noise, which means that collisions come from the simultaneous data transmissions of two or more nodes. All nodes use the RTS/CTS frames to reserve the channel before transmitting the information data. The number of nodes varies from 10 to 60 due to the normal network sceneries.

In the simulation, we adopt the ON-OFF business mode to generate the network traffic, and use the default parameters for the simulation sceneries. The time interval between packets varies from 0.01 to 0.001 second and each simulation runs 20 seconds. Some main parameters can be searched in [5].

The average simulation results of the last ten seconds have been shown in Fig. 1. The aggregated throughput of the proposed NSAM algorithm is higher than that of DCF algorithm. The higher throughput comes from the adaptive adjustment rules of the NSAM algorithm that sets the parameters dynamically. The NSAM algorithm sets the access parameters according to the network states, like idle slot intervals which can be used to measure the states of channel congestion. The MAC delay of NSAM in Fig. 2 shows the better performance than that of DCF.

5 CONCLUSION

In this paper, we propose an adaptive access mechanism NSAM method for IEEE 802.11 WLANs in non-saturation conditions. The NSAM method adjusts access parameters based on the feedback signal of channel (idle slot intervals of channel), which is a main measurement of channel states. Another great contribution of the NSAM method is that, nodes do not need to know or judge whether the network states are in saturation conditions or not, which can simplify the complication of access mechanism greatly and extend the application of the new NSAM method. And then, the network performance of throughput, media access delay and retransmission attempts are analyzed completely by computer simulations. By setting parameters adapting to the dynamic network configuration, the collision probability can be low and the aggregated throughput can reach high. All of these results demonstrate the full characteristics of the adaptive access mechanism proposed in this paper.

ACKNOWLEDGMENT

Thanks to the support by NSFC (No.61362016), NSF of Hainan Province (No.613163, No.613164).

REFERENCES

[1] IEEE Standard for Information Technology-Telecommunications and Information Exchange Between Systems-Local and Metropolitan area Networks-Specific Requirements-Part 11: Wireless LAN Medium Access Control (MAC) and Physical Layer (PHY) Specifications, IEEE Std. 802.11, 2007.

[2] Bianchi G. Performance analysis of the IEEE 802.11 distributed coordination function. Selected Areas in Communications, IEEE Journal on, 2000, 18(3): 535–547.

[3] Sakurai T, Vu H L. MAC access delay of IEEE 802.11 DCF. Wireless Communications, IEEE Transactions on, 2007, 6(5): 1702–1710.

[4] Heusse M, Rousseau F, Guillier R, et al. Idle sense: an optimal access method for high throughput and fairness in rate diverse wireless LANs //ACM SIGCOMM Computer Communication Review. ACM, 2005, 35(4): 121–132.

[5] Chun S, Xianhua D, Pingyuan L, et al. Adaptive access mechanism with optimal contention window based on node number estimation using multiple thresholds. Wireless Communications, IEEE Transactions on, 2012, 11(6): 2046–2055.

[6] Liaw Y S, Dadej A, Jayasuriya A. Performance analysis of IEEE 802.11 DCF under limited load, Communications, 2005 Asia-Pacific Conference on. IEEE, 2005: 759–763.

[7] Zhao Q, Tsang D H K, Sakurai T. Modeling nonsaturated IEEE 802.11 DCF networks utilizing an arbitrary buffer size. Mobile Computing, IEEE Transactions on, 2011, 10(9): 1248–1263.

Future Communication, Information and Computer Science – Zheng (Ed.)
© *2015 Taylor & Francis Group, London, 978-1-138-02653-7*

The construction of outstanding engineers' training-oriented physical chemistry curriculum system

L. Gao, Z. Fang, W. Gao, T. Gu, L. Zhang & Q. Li
School of Chemical Engineering, University of Science and Technology, Anshan, Liaoning, China

ABSTRACT: In the 21st century, how to develop the education of engineering became the common issue in the education field in the world. And how to make the technological university go back to the cradle of engineers has become a universal topic of concern of each university in the world. The article proceeds from the position of physical chemistry course in engineering education, putting forward a curriculum system construction facing the "Outstanding Engineers" training of physical chemistry course, in order to make it meet the requirements of the school's "outstanding engineers" training program.

Keywords: outstanding engineer; physical chemistry; curriculum system; engineering education

1 INTRODUCTION

Many new problems and contradictions of national engineering education have arisen in this new period, and how to develop engineering education becomes the common issue of eastern and western education fields in the twenty-first century. In China, "the Plan for Educating and Training Outstanding Engineers" is not only the major reform project to implement the "Outline of the National Plan for medium and long-term educational reform and development (2010–2020)" and "Outline of the National Plan for medium and long-term talent development (2010–2020)", but also a major initiative to promote Chinese engineering education from big powers to strong powers, aiming at bringing up a large number of various types of engineering and technical talents with innovative ability and high quality meeting the needs of social and economic development, and providing services for the state to take the road of new industrialization, construction of innovative country and the strategy of reinvigorating China through human resource development. It is significant to promote the higher education to cultivate talents orienting towards the social demand and to entirely improve the quality of talents training in engineering education.

"The Plan for Educating and Training Outstanding Engineers" has three characteristics: firstly, the industry and the enterprise participate deeply in the training process; secondly, the schools train the engineering talent according to both the general standard and industry standard; thirdly, it strengthens the cultivation of students' engineering ability and creative ability. The reason why "Outstanding Engineer" is remarkable lies in his abundant professional knowledge, his strong problem-solving skill and his high comprehensive quality.

How to make the university go back to the cradle of engineers has become the universal topic of concern of each university at home and abroad. Since 2010 the Ministry of education started and vigorously promoted "The Plan for Educating and Training Outstanding Engineers", two groups of 194 universities have already become the experimental units of the plan. The Mining engineering, inorganic nonmetal material specialty and metallurgical engineering in the University of Science and Technology Liaoning are right in the second group of the plan. Physical chemistry is the basic course of the three majors. Considering the objectives of "outstanding engineer" – training engineering consciousness, engineering quality and engineering practice ability – physical chemistry curriculum system needs to match the training objectives.

The design and construction of the curriculum system is one of the key tasks to achieve the training goal of university talents. As the guiding ideology of the Plan for Educating and Training Outstanding Engineers, which is the significant innovation and breakthrough of our country's existing engineering education model, the main goal and reform measures are mainly achieved through the formulation and implement of the plan in the professional training program, in which the personnel training curriculum system oriented towards the outstanding engineer reserve and the reform of teaching content are the core contents of participating in the professional training program.

As time advances, the economic globalization, the rise of the knowledge industry and the form of network society make higher education consistently expand in time and space, so there is an urgent need for a brand new knowledge generation mode of transmission and talents cultivation model to match it. The experience of worldwide higher education development shows that knowledge comes from practice, ability comes from practice, and quality needs to be cultivated in practice. Therefore, in many current measures of teaching reform, many colleges and universities deepen and strengthen the reform of practice teaching. Our school's construction of "Outstanding Engineer" training-oriented physical chemistry course system is based on the following aspects.

2 THE CATEGORY AND STATUS OF PHYSICAL CHEMISTRY CURRICULUM

2.1 The category of physical chemistry course

In the twenty-first century, with the progress of a variety of experimental and theoretical research means and methods, chemical research will be much more colorful, and the research level will be further widened, which almost penetrates into the material science and life science study; industrial and defense technology development; life, health, environment and all aspects of research and technology development related to social sustainable development. From the current view of the trend of chemical reactions, chemistry has developed from the tradition that only focuses on the study of the response and change rule of atomic structure and molecular structure to the supramolecular level, multiple molecular aggregation level, and some new classification of chemistry such as supramolecular chemistry, mesoscopic gathered state chemistry, macroscopic aggregation state chemistry and complex system of chemistry. With the cross of chemistry and life science, chemical biology appeared; with the cross of chemistry and material science, nm (material) chemistry appeared; with the cross of chemistry and resource and environment science, green chemistry appeared; and with the cross of chemistry and mathematic informatics and life sciences, chemical informatics appeared. But as a link between physics and chemistry, the character of physical chemistry will not change, but the content will be more abundant, and the role will be greater.

2.2 The status of physical chemistry in interdisciplinary

As the theoretical foundation of chemistry, physical chemistry plays a key role in the development of chemistry and has become the arsenal of assault fortified positions of many other science disciplines. With the closer communication and infiltration of interdisciplinary, physical chemical methods are used widely in other disciplines, and physical chemistry has penetrated into other disciplines, making the major problems of other disciplines be its own study object.

As can be found from the history of more than 100 years of Nobel Prize in chemistry, the issues related to physical chemistry accounted for most of the Chemistry Awards, and physical chemists also accounted for a large proportion. Thus, the physical chemistry discipline itself and the personnel trained in it play a key role not only in the physical discipline but also in the major scientific discoveries.

3 THE THOUGHT AND MEASURES OF THE CURRICULUM SYSTEM CONSTRUCTION

3.1 The thought of the curriculum system construction

(1) Build a teaching team with high level

Build a group of teachers with high level and strong ability, who are good at learning and absorbing the successful experience and outstanding achievement. The team has a reasonable structure, stable staff, higher teaching level, better teaching effect and complete counseling and experimental teachers.

(2) Construct a set of stereoscopic teaching material suitable for our schools' students

Build a set of teaching material suitable for our school's "outstanding engineer" training plan, domestic first-class "three dimensional teaching material of physical chemistry", including paper materials (textbooks and reference books) and multimedia materials (electronic lesson plans, multimedia assisted teaching software, item pool, database and network course).

(3) Construct an experimental and test base

Construct a laboratory and engineering test base matching our "Outstanding Engineer" plan, with a full set of physical chemistry laboratory equipment and venues with comprehensive, design experimental function, which can complete the prescribed confirmatory experiments in the curriculum. It should also have a function to admit the students' engineering practice, research and training and teachers' engineering teaching activity based on the project.

(4) Construct a complete teaching file

Construct a complete, scientific teaching file and teaching archives (including the teaching outline, teaching schedule, teaching plan, teaching cards, as well as the teaching guidance document), to make specific requirements and training scheme about the basic theory knowledge, basic skills of the curriculum and the cultivation of innovative ability according to the outline.

(5) Construct a set of improved course management approaches

Construct a scheme combining variety of teaching methods and evaluation methods. Pay attention to the combination of inspiration, guidance and an interactive teaching method in the process of teaching, which fully embodies the teaching concept of respecting students, inspiring students, student-centered and better teaching effect; at the same time, provide a comprehensive evaluation method, establish an evaluation system of teaching effect combined with grades of theoretical teaching and experimental teaching, which can assess the teaching level and effect of the teachers, it can also test the students' mastery on the contents, practical ability and the level. Reform, improve the teaching methods, and make use of multimedia and other modern teaching means for teaching. Use network and other information technology and resources available to carry out teaching and self-taught tutorship, which include: the survey of the teaching effect, information collection, teaching means and methods reform, and the research, design, inspection, examination system reform research of all the teaching links.

3.2 Measures for the implementation of the curriculum system

(1) Compile three-dimensional teaching material suitable for our school's "outstanding engineer" training objectives of physical chemistry. The theory teaching content should be linked with basic knowledge and professional knowledge, should reduce duplication, and increase the knowledge of physical chemistry in engineering application. Prior to 2013, complete the paper textbook. Before 2014, complete the electronic teaching plan and multimedia teaching software. Before 2015, complete dubbing, and make the multimedia auxiliary teaching software more perfect, more practical and beautiful, which can reflect our highest level.

(2) Reform the experimental teaching contents and methods, increase the comprehensive and designing experiments, which adopt the open type of teaching method. Students can generate their own subjects and also be available for the subject the teachers give. Additionally, they can make open laboratory appointment online about experimental time and content, and complete it independently or with the help of the teachers.

(3) The students can be tested in various ways, including experimental results (20%), interactive classroom achievement (10%), experimental design or design theory subject achievement (10%), the usual performance (10%), theory examination achievement (50%).

(4) Theoretical teaching methods: Theoretical courses are taught by means of multimedia courseware in combination with writing on the blackboard, teachers' teaching in combination with students' interaction in class, lectures in class and answering questions after class and teachers Q&A online; the experimental course adopts the method of explaining the experimental technology combined with doing the experiments independently. It can make students timely understand and master all kinds of physical chemistry experimental research methods and the basic techniques and related skills of system knowledge.

(5) Build a physical chemistry website on campus net

Issue physical chemistry information on the website, for example, students can book the open test time and contents online; teachers publish class contents and methods, making students do the preview and preparation; teachers' Q&A online, so students can solve the problems in time and can enjoy all the teaching resources of physical chemistry in our school at the same time etc.

(6) Train engineering quality through the project

Modern engineering education pays more attention to the cultivation of students' engineering practice ability, and the teaching method based on the project gains more and more recognition. This method is mainly used in a real or simulated project as a case study to train students' cognition of engineering practice and the ability of operation. Bandura thinks that, individual hands-on experience can promote individuals to complete a specific task successfully with confidence, which he defined as self-efficacy belief. High self-efficacy can help individuals to better accomplish specific tasks, and can even influence an individual on the decision of profession and occupation. Therefore, some researchers in the field of engineering education think that, participating in hands-on projects in the study period can enhance students' self-efficacy of engineering practice and their confidence in choosing an engineering occupation, thus the goal of engineering education can be achieved.

(7) Train the innovation ability through scientific research

Excellent engineer training is system engineering; it is not only to foster the elites, but also to combine the elite education with general education; not only textbook knowledge, but also the trinity comprehensive training of the knowledge and skill, process and method, emotion attitude and values. The best way to improve the students' innovative ability is to encourage them to participate in the training of scientific research personally, stimulate research inspiration and desire through learning as well as teaching, then generate the power of new learning and research. In the six sessions of the student research training program of our school, the tutor group guided a total of more than 20 students from freshmen to seniors, the students group included chemical, metallurgical, chemical and biological, applied chemistry and management professional, which fully reflects the advantages of interdisciplinary penetration. The six sessions of the college students' scientific research training provides students with opportunities to be teachers' research assistants, and be involved in the teachers' research directly. Through the training of research assistants, students and teachers have more contact, and students can obtain tacit knowledge which can't be acquired in other places. Through a close professional and practical research task, the students experienced the whole process of scientific research project personally (including research, opening, experimental preparation, experimental operation, experimental data, summary, writing papers and research reports). The students can not only taste the difficulty of scientific research, but also share the joy of success, and they also have the opportunity to experience the process of their own learning progress and the development of scientific research ability; at the same time, college students in specific research can feel their lack of knowledge, and realize the true meaning of the interdisciplinary.

(8) Cultivate "outstanding" ability based on independent engineering practice

Draw lessons from the "engineering workshop" idea of Xi'an Jiao Tong University to establish a communal foundation platform of students engineering practice, closely combine the practice teaching and students' autonomous practice activities within the training program, while strengthening the training program, and the original engineering practice teaching resources of the school can be integrated and the College Students' autonomous practice "DreamWorks" can be strongly promoted, so, driven by the problem and project, the student can carry out "small production, great innovation" of independent practice through independent design, production and invention according to their own interests and dreams. Engineering workshops provide a platform and conditions of practice for students to realize their dreams and cultivate awareness of innovation, innovation ability, practice ability, integrated ability and management ability.

My student has turned to me about it. She found a spring during traveling and brought back some water and wanted to test whether the spring water is suitable for the development of mineral water. The idea put forward by the students is in bad need of a platform such as "engineering imitation" to support the test.

4 SUMMARY

In the process of going from big to powerful, Chinese engineering education must be based on the needs of the country, to achieve a balance between "global" and "local", build a higher engineering education system not only with Chinese characteristics but also with the global development. Colleges and universities should change the educational ideas, reform personnel training mode, adjust the personnel training objectives and adopt new ways of thinking. According to the requirement of "Outstanding Engineer Training Plan", a teaching system of theory and practice should be built. Through the close cooperation between education and industry, universities and enterprises, taking the actual project as background and the engineering as main line to improve the students' engineering consciousness, quality and practice ability, thousands of excellent engineers with innovative ability and suitable for enterprise development will be surely trained.

ACKNOWLEDGEMENTS

This work was financially supported by University of science and technology Liaoning reform and practice of graduate cultivation project (2013YJSCX05); University of science and technology Liaoning graduate education innovation projects (2012YJSCX23); the Liaoning province education science special training mechanism planning subject (lnxwb120241).

REFERENCES

[1] Cheng Guangxu. 2011. H Research in Higher Education of Engineering, 3:14–20.
[2] "Chinese geological education" editorial department statistics. 2012. Chinese Geological Education 1:179–180.
[3] Gao Lijuan. 2011. Education for Chinese After-school 7:29–30.
[4] Gao Lijuan. 2006. Journal of Anshan University of Science and Technology (29):101–102.
[5] Li Shengqiang, Lei Huan, Gao Guohua, et. 2011. Research in Higher Education of Engineering, 3:21–27.
[6] Lin Jian. 2011. Research in Higher Education of Engineering, 5:1–9.
[7] Yin Jie, Yang Jianchao. 2011. Research in Higher Education of Engineering Higher Education of engineering, 3:60–63.

Future Communication, Information and Computer Science – Zheng (Ed.)
© 2015 Taylor & Francis Group, London, 978-1-138-02653-7

Application research of improved adaptive weighted clustering algorithm in the maneuvering communication network

X. Gao, L. Liu & X. Guo
Equipment Academy, Beijing, China

ABSTRACT: In order to ensure the reliable transmission of important service information, that meets the requirements of accessing at any time, this paper puts forward the improved adaptive weighted clustering algorithm. Through simulation and verification, this algorithm can be applied to the maneuvering communication network which can realize the effective supplement and reasonable extension to existing network.

Keywords: maneuvering communication network; clustering algorithm

Communication system is designed to ensure reliable implementation of the communication service, in order to implement the wide area coverage, often using IP network, satellite communications network, and other wireless communication system, but there is a big gap to the goal of complete covering, flexible forming, neutral accessing and communication at any time. In this paper, designs of maneuvering communication network mode based on hierarchical structure will be an effective complement and expand to the existing communication system.

1 MANEUVERING COMMUNICATION NETWORK

In the process of scientific research tasks, in order to transmit and receive the service information of all-round and real-time, and ensure the service information flexible accessing, ground communication node can be a combination of both fixed and maneuvering node based maneuvering communication network mode. In this way, all fixed nodes form a backbone which is responsible for the entire network of information transmission and receiving, mobile nodes as the only access nodes, which is often not as the forwarding nodes on the path.

To ground mobile station, it can change its speed and direction according to the change of task, this network's characteristics is similar to ad-hoc network, so from this article, draw lessons from the research achievements of Ad Hoc network, combined with the characteristics of the scientific research task to study the ground mobile stations maneuvering communication network.

1.1 *The basic structure of maneuvering communication network*

The topology structure of topology variable network includes four basic structures: center control structure, layered center control structure, fully distributed control structure and hierarchical and distributed control structure. To the scientific research task, when performing different tasks, the location of the ground mobile station changes relatively much, when performing the same task, the location of the ground mobile station changes little. Because the node mobility causes the topology to change in the ground mobile network communication based on the mobile station, it is not suitable for use in the center control structure, but can adopt the distributed control structure which is divided into flat structure and hierarchical structure.

1.2 *Topology structure of ground mobile station network*

Considering the diversity and complexity of the scientific research task, more and more ground mobile station is arranged for the communication network, in view of the increasing number and to make a frequent motor on a larger scale, rapid deploy ground mobile station. In case of the flat structure, there are the following disadvantages:

(1) The scale of the flat structure network is limited.When the network scale is enlarged routing maintenance costs will increase.

(2) Because the ground mobile station is fast moving frequently, a different network structure will greatly affect the mobile management costs.

Based on the above reasons, a mobile station communication topology structure should adopt a hierarchical structure.

2 MANEUVERING NETWORK CLUSTERING ALGORITHMS RESEARCH OF GROUND MOBILE STATIONS

The analysis shows that using the hierarchical structure can play a better efficiency of the mobile ground station network in the scientific research task. As a way to build a hierarchical structure, the clustering algorithm which is based on the system requirements and according to certain rules will divide the network into a sub-network which can be interconnected and can cover all nodes of multiple clusters, and the clustering algorithm can update the cluster network structure to maintain the normal function of the network. Good clustering mechanism should try to keep the network topology stable, reduce the number of clustering, optimize the connections within the clusters and between clusters, and take into account the energy level of nodes, network load balancing and the support to channel access protocol [1].

2.1 Requirements of clustering algorithm

Existing typical Clustering algorithm including the lowest (LOWID), the highest – degree of node Clustering algorithm (HIGHD), Weighted Clustering algorithm (WCA), because of the ground mobile stations with high dynamic characteristics in the scientific research tasks, the cluster head of mobile node must meet the following conditions:

(1) Cluster head is responsible for not only the cluster and the cluster communication, is responsible for the fixed nodes' communication with the ground at the same time;

(2) Cluster head must be located in the transmission range of fixed nodes, and the cluster head can realize the communication not only with mobile node, but also with fixed nodes. As shown in figure 1: $d = \sqrt{(x_0 - x_i)^2 + (y_0 - y_i)^2} < r$, d is the distance between the mobile node and fixed node, (x_0, y_0) is the coordinates of fixed node; (x_i, y_i) is the coordinates of a mobile node, $i = 1, 2 \ldots$. Therefore, when considering weight, one needs to increase the transmission range node.

Through analyzing the high dynamic characteristics of the ground mobile stations in scientific research tasks, it puts forward an Improved Adaptive Weighted Clustering Algorithm (IAWCA), the Algorithm adopts the average speed as the standard of measure, to consider comprehensively the best connection degree of mobile node, and factors of distance and transmission range.

2.2 Description of the clustering algorithm

When calculating the node weights, we should consider comprehensively the factors, such as node degree, mobility, distance, the residual energy of nodes, location, and the weight of each factor can be adjusted dynamically according to the requirement of the system and processing power of the node.

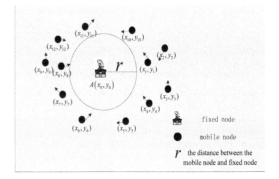

(a)Before the cluster formation

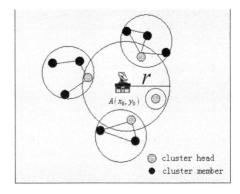

(b)After the cluster formation

Figure 1.   Process diagram of cluster formation.

(1) Search for the neighbor node of each node v, defined d_v is degrees of v, and d_v is the number of neighbor nodes:

$$d_v = |N(v)| = \sum_{v' \in V, v' \neq v} \{dist(v, v')\} < tx_{range} \quad v, v' \text{ not clusters} \quad (1)$$

In the formula, tx_{range} is the transmission range of node v.

(2) For each node, deviation may be calculated between the node degree d_v and the best node degree δ:

$$\Delta_v = |d_v - \delta| \quad (2)$$

In the formula, δ is the ideal number of members of each cluster, $\delta = \frac{W_1}{W_2}\sqrt{N}$, W_1 and W_2 are respectively the bandwidth of communication within the cluster and the bandwidth of communication between different clusters, N is the number of the node.

(3) For each node, to calculate the sum distance D_v which is the node to the all 1 jump neighbor nodes:

$$D_v = \sum_{v' \in |N(v)|} \{dist(v, v')\} \quad (3)$$

(4) Calculate the average speed (M_v) of each node in current cycle T:

$$M_v = \frac{1}{T}\sum_{t=1}^{T}\sqrt{(X_t - X_{t-1})^2 + (Y_t - Y_{t-1})^2} \qquad (4)$$

(X_t, Y_t) and (X_{t-1}, Y_{t-1}) are respectively coordinates of node in t and $(t-1)$ time.

(5) To calculate the distance d of each mobile node $v(x_i, y_i)$ to fixed node $A(x_0, y_0)$ is used to represent the node transmission range T_v:

$$d = \sqrt{(x_0 - x_i)^2 + (y_0 - y_i)^2} < r, i = 1, 2 \ldots\ldots \quad (5)$$

(6) To calculate the weight of each node:

$$W_v = w_1\Delta_v + w_2 D_V + w_3 M_v + w_4 T_v \qquad (6)$$

Among them, w_1, w_2, w_3 and w_4 for the corresponding weights of different parameters, $w_1 + w_2 + w_3 + w_4 = 1$.

(7) Choose the node which is the least W_v of adjacent nodes as the cluster head nodes. If the W_v is equal, choose the smaller ID of the node as the cluster head, 1 jump neighbor nodes of cluster head become the members of the cluster nodes. If the node which belongs to a cluster node cannot serve as cluster head,

(8) Repeat (2)~(7), until there is no new node and node is assigned to a cluster.

3 PERFORMANCE SIMULATION AND ANALYSIS OF GROUND MOBILE STATIONS MANEUVERING NETWORK'S CLUSTERING ALGORITHM

IAWCA algorithm develops from the weighted clustering algorithm, meanwhile considering the LOWID algorithm and HIGHD algorithm are common clustering algorithms, this article mainly simulates and analyzes the LOWID, HIGHD, WCA and IAWCA clustering algorithms, four kinds of clustering algorithms adopt non overlapping clustering strategy [2][3].

3.1 Simulation environment

Twelve ground mobile station nodes are arranged in a certain region within the scope of $80\,km \times 80\,km$, and the node moves at the speed of 5 km per hour in a random direction, w_1, w_2, w_3, w_4 and w_5 are respectively the node's degrees, mobility, transmission power, weight of residual energy and location factors, as shown in table 1.

3.2 Indicators of performance

(1) Number of cluster heads C

As can be seen from figure 2, analysis shows that the number of cluster heads (C) in HIGHD algorithm is the smallest, C is one of the biggest in IAWCA

Table 1. Parameter setting of for algorithms.

Algorithm Parameter	w_1	w_2	w_3	w_4	w_5
LOWID	0	1	0	0	0
HIGHD	1	0	0	0	0
WCA	0.5	0.15	0.15	0.2	0
IAWCA	0.15	0.4	0.15	0.1	0.2

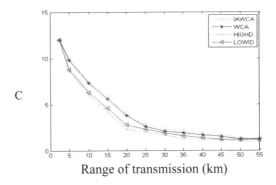

C

Range of transmission (km)

Figure 2. The change of cluster's numbers with the change of transmission's range.

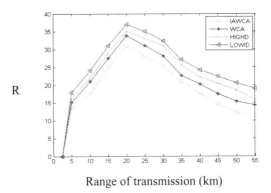

R

Range of transmission (km)

Figure 3. The change of R with the change of transmission's range.

algorithm, this is because the IAWCA algorithm increased the limitation of node position, which has played an important role in regulating C, and distribution of cluster heads is more even, C is slightly more than several other algorithms.

(2) Number of back into the cluster R

From figure 3, with the increase of the transmission's range, the number of back into the cluster (R) is smaller after the first increasing, R reaches a maximum when the range of transmission is 20 km, then R decreases gradually, and it tends to a fixed value finally. In general, R of LOWID algorithm is the maximum, R of IAWCA algorithm is the smallest. This is because the limitation of node position is lower than the mobility of a cluster member. Through comparison, cluster structure of IAWCA algorithm is more stable.

Table 2. LBF's statistical average of four clustering algorithms.

clustering algorithm	LOWID	HIGHD	WCA	IAWCA
LBF's statistical average	0.24	0.09	0.35	0.42

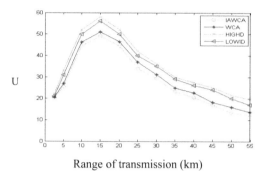

U

Range of transmission (km)

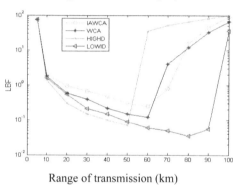

Range of transmission (km)

Figure 4. Update number U Figure 5: the change of LBF with the change of transmission's range.

(3) The update number of node rule set (U)

From figure 4, the conclusion of comparing the values shows the U max of HIGHD algorithm, and U minimum of IAWCA algorithm, the number of cluster heads is more moderate and more uniform in distribution of IAWCA algorithm, and it can adjust adaptively according to the distribution of nodes, the probability of nodes leaves rule domain smaller than other algorithms.

(4) The features of load balance (LBF)

Below are four clustering algorithms under the difference of the load balancing factor (LBF). Table 2 lists the LBF statistical average of each algorithm. The simulation is under the condition of $n = 12$, $r = 100$ km, $v = 5$ km/h.

As can be seen from figure 5, using IAWCA algorithm can improve the stability of the cluster structure, enhance the capacity of the network load balancing, reduce the computation and communication overhead, and improve the network communication bandwidth utilization, effectively prevent communication interruption, ensure the smooth flow of information and communication, can better meet the scientific research task as the requirement of the mobile node of the cluster head, improving the capacity of mobile communication.

4   CONCLUSION

Showed by the simulation analysis, the IAWCA can realize the effective supplement of existing IP network and satellite communications network and a logical extension through the ground mobile stations of maneuvering communication network.

REFERENCES

[1] Zheng Shaoren Wang Haitao etc. AD hoc network technology. People's posts and telecommunications press, 2005:16 to 18, 97, 107, 110.
[2] Inn InnER, WinstonK. G.Seah. Mobility-based d-Hop Clustering Algorithm for Mobile Ad HocNetworks. WCNC 2004/IEEE Communications Society, 2004: 2359–2364.
[3] Zhao jin yuan. Mobile self-organizing network clustering algorithm and routing protocol research. Suzhou: central china normal university, 2007.

Future Communication, Information and Computer Science – Zheng (Ed.)
© 2015 Taylor & Francis Group, London, 978-1-138-02653-7

The research on the change characteristics of Chinese dual economy based on agricultural marginal productivity

J. Ren
Labor Science and Law Institute of Beijing Wuzi University, Beijing, China

A. Li
Landscape Design Research Center of THAD (Architectural Design and Research Institute of Tsinghua University), Beijing, China

ABSTRACT: Changes in the level of agricultural marginal productivity and agricultural wages are important indicators of national economic development stages. Calculation of the agricultural sector through the relevant data of the survey conducted, the marginal productivity of our country over the past decade can be drawn in a rapid increase, in the level of long-term and short-term agricultural workers' wages and agriculture wages both showed an upward state. By comparing the current agriculture marginal productivity and agricultural wages in China, and in accordance with the theory of Dual Economy transition we can educe our current stage of economic development, as well as China's labor policies that should be taken for the next stage.

Keywords: Lewisian turning point; Ranis-Fei model; agricultural marginal productivity; agricultural wages

1 INTRODUCTION

China's labor market presents tremendous changes after the reform and opening. These trends represent in this phenomenon: A large number of rural surplus labor migration to urban labor market, wage growth, capital and non-capital sector departments income gap. In recent years, there was a "labor shortage" rise in China's eastern coastal areas while the agricultural sector income grows and so on.

According to the dual economic theory, during the conversion of the country's economy from the dual economy to unitary economy, some iconic phenomena will occur such as surplus labor absorbed and exhausted, the income gap between skilled and unskilled workers reduced etc.

This paper selects the marginal productivity of the agricultural sector and agricultural wage as the observed objects intended to solve two problems: First, what kind of change occurred to the marginal productivity of the agricultural sector and agricultural wages. Second, what are the theoretical implications of these changes in China's economic development.

2 THE THEORY MODEL AND DATA

2.1 The theory and model

In 1954 British economist Lewis brought forward the dual economic model, which took the viewpoint of classic economics in explaining the process and changes of economic development. The model assumed that at the primary stage of economic development a country or a region's economy was made up of modern industrial and traditional agricultural sectors. In the traditional agricultural sector, constant increase of population reduces the marginal productivity of the agricultural labor force, to a very low level even zero.

Under such condition, there is no effect to the overall production when displacing labor force away from the agricultural sector, the income of the agricultural sector remains at survival level. Meanwhile the capital accumulation in the modern industrial sector promoted its expansion and the demand for labor force. Employment in this sector might exceed population increase, creating a good opportunity for absorbing surplus labor of the agricultural sector.

In the early stage of the economic development the marginal labor force in the agricultural sector is zero, meaning unlimited supply of labor force. With the constant expansion of modern industrial sector, the surplus agricultural labor force is gradually reduced, the labor shortage begins to appear, wages are facing upward pressure, two departments of labor market are becoming integrated.

The Ranis-Fei model enriched this theory, put forward that the transition period is a gradual process. Before the end of the infinite labor the marginal productivity will become positive, but below the survival wage level, and this will form the second turning point.

This model theoretically refined the theories of dual economy transition and became an advanced model of Lewis' theory. This model demonstrates that the shift of labor force in economic development is a gradual process rather than a sudden one.

The new model emphasizes the significance of agriculture in the economic development. During the first stage, the displacing of the labor force will not be hindered because the total output of the agricultural sector does not change.

In the second stage, labor displacement brings the reduction of the agricultural surplus and relative food shortage, together with the price of agricultural products and income level increase, which will affect the displacing of labor force from the agricultural to the industrial sector. If the price increase of agricultural products causes inflation of labor costs in the industrial sector, the capitalization and expansion of the industrial sector will be restricted or even stagnated, then Lewis' second turning point will be postponed.

Therefore keeping a balanced growth between the sectors during the Lewis turning period is very important for economic taking off and structural transition.

The above theory suggests that a country or a region will experience multitude changes in all aspects when it moves from a duality economy to a unitary economy.

According to the Lewisian Turning Point theory, and Ranis-Fei model, this paper intends to analyze changes of Agricultural Productivity and real wages, then investigate the relationship between the two, finally get the judgment on change trends in the non-capital sector labor market.

In the calculation of the marginal productivity of the agricultural sector, applying the Cobb–Douglas agricultural production function to estimate the production flexibility of agricultural labor.

$$Y = Ae^{\lambda t}(Q_N N)^\alpha K^\beta (Q_L L)^\gamma$$

Therein, $Q_N N$ is for the labor, K is for the capital stock, $Q_L L$ is for land area, parameter λ is for the rate of technological progress, the parameters α, β, γ are for the production elasticity of labor, land and capital respectively. Assumed to be linear homogeneous functions, then ($\partial + \beta + \gamma = 1$). After rewriting logarithmic form it can be obtained by the following formula:

$$InY_t = InA + \lambda t + \alpha In(Q_N N)_t + \beta InK_t + \gamma In(Q_L L)_t$$

As the intermediate inputs in agricultural production has great impact on output, in order to fully consider the impact factors of the dependent variables, this paper takes into account the element of intermediate inputs.

For a clearer observation on the agricultural wages in the calculation, the crops had been separated into food crops and cash crops to observe.

2.2 The data and the process of calculation

After appropriate treatment to the above data, the averaged production flexibility can be obtained through regression calculation. And using the agriculture real added-value divided by the total agricultural labor time to draw the average productivity of agriculture.

Finally, multiply the average productivity of agriculture by production elasticity of agriculture labor to get the agricultural marginal productivity. In the calculation of real wages of agricultural workers, the paper applies division calculation between the annual wage of agricultural workers and price index of agricultural products.

In order to reflect the actual situation, we use the household survey data collected by the Ministry of Agriculture from 1995 to 2006. This data come from a continuous tracking survey for 21,000 households, 650 herdsmen, 300 agricultural villages, 15 pastoral villages in the country.

This paper uses agricultural wage as the dependent variable, used the survey statistics of operating income in the food and cash crops from the planting industry in the household operating income. L is for labor date, A is for land area and K for capital stock.

For the agricultural wage calculation, it is mainly obtained by using the three mainly survey data in agricultural workers (day), worker cost and rural price index.

3 THE CHANGES OF MARGINAL PRODUCTIVITY AND AGRICULTURE WAGES IN CHINA'S AGRICULTURE SECTOR

Using the above production function to calculate the level of productivity of food crops and cash crops in

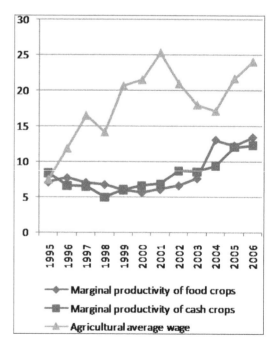

Figure 1. Changes of the agricultural real wage level and marginal productivity.

rural and agricultural wages in our country from 1995 to 2006, the following results were obtained.

As shown in the calculation results, China's agricultural wages have presented an overall upward trend since 1995. Although there was a temporary downturn for the agricultural wage in 2003–2004, this can be considered as a data defect caused by inconsistent standards applied by the new survey method.

The marginal productivity of food crops and cash crops are also showing an increasing trend overall, particularly the growth rate of the marginal productivity showed a big change.

4 CONCLUSION

From the above analysis we can see between 1995–2006 the marginal productivity of the agricultural sector and agricultural wages in China showed the following changes.

First, the average productivity and marginal productivity in the agricultural sector are rapidly increasing, the marginal productivity (average of marginal productivity of food crops and cash crops) reached 8.1% per annum on average. Marginal productivity of food crops and cash crops are showing a rapid rising trend. In particular, the marginal productivity of food crops which presents a more substantial increase, with an average annual growth rate of 8%, with an average annual 4.9 percent for the cash crops marginal productivity.

Second, the wage level in the agricultural sector has been greatly improved with an annual growth rate of 13% in the past decade, with 22.5% growth rate for long-term workers and 11.4% for short-term's.

Third, the gap between the agricultural average daily wage and the average marginal productivity levels is gradually reduced, especially after 2003, the increase of marginal productivity and the decline of agricultural wage growth brought the two closer.

According to the analysis of the above changes using the theory of dual economics, it shows that the marginal productivity of agriculture is infinitely closing to the level of agricultural wages.

We can speculate that, following the further improvement of the agricultural marginal productivity, the future wage level of the agricultural sector will be determined by the level of marginal productivity, while surplus labor in the agricultural sector will become short. Therefore there are reasons to believe that important changes in this period imply the arrival of an economical turning point in the agricultural sector, so our agricultural sector should enhance the labor policy in order to conform to the needs of national economic development as a response to this trend.

REFERENCES

[1] Lewis, W. Author (1954). Economic Development with Unlimited Supplies of Labor, Manchester School of Economic and Social Studies, 22, 2, 139–91.
[2] Ryoshin Minami (1973). The Turning Point in Economic Development: Japan's Experience, KINOKUNIYA BOOKSTORE. 147–167.
[3] Hansen, Bent (1966) Marginal Productivity Wage Theory and Subsistence Wage Theory in Egyptian Agriculture, Journal of Development Studies, Vol. 2, No. 4, 367–405.
[4] CAI Fang (2007) the Lewisian Turning Point of China's Economic Development in Reports on China's Population and Labor No. 8. *By Social Sciences Academic Press (China).*
[5] Wang Dewen and REN Ji (2008) Lewisian Turning Point: International Experiences in Linking up Lewis and Kuznets turning points. By Social Sciences Academic Press (China).

Future Communication, Information and Computer Science – Zheng (Ed.)
© *2015 Taylor & Francis Group, London, 978-1-138-02653-7*

The teachers' teaching ability is the fundamental guarantee for promoting the quality of classroom teaching

Z. Fang, M. Ge & J. Li
School of Chemical Engineering, University of Science and Technology Liaoning, Anshan, Liaoning, China

ABSTRACT: Teaching quality is the lifeline of college, improving it always is the central work of teaching and the key to improving the quality of education. The teachers' teaching ability is the fundamental guarantee for promoting the quality of classroom teaching. This paper suggests some views with regard to applications of the teachers' teaching ability, such as careful lesson preparation, teaching contents, teaching methods, stimulating the enthusiasm of students, students-teacher interaction and multidimensional assessment.

Keywords: the quality of classroom teaching; the teachers' teaching ability; students-teacher interaction word

Teaching quality is life to university, it's the foundation of schools' development. Classroom teaching is not only the palace hall to impart knowledge, but also an important way to cultivate the thinking mode and the basis for other teaching links. Improving the quality of teaching is always the main theme of the school work. How to improve teaching and do a quality education is worthy of careful consideration and discussion.

Currently, some of the problems exist widely in classroom teaching. Firstly, the course hours of most schools are generally reduced after the revisal of teaching plans. Secondly, to supplement the new force, a large number of young teachers, whose quality of teaching needs to be improved because of the lack of teaching experience and job training, stand at the first line of teaching. Thirdly, due to the evaluation of teaching, many young teachers do not focus on teaching but currying favor with students. How to solve these problems and improve the quality of teaching? We believe that a good job of teachers' teaching is the fundamental guarantee of improving the quality of classroom teaching, which is identical to the view of Jianrong Li [1]. We expound the connotation of the teachers' teaching quality from the following five aspects.

1 PREPARE LESSONS CAREFULLY

Preparing lessons well is the premise of giving a good class and the key to guarantee the quality of teaching. And it is also a re-creation process. The teacher not only should prepare knowledge, but also make efforts to find the incisive point, which can stimulate the students' interest in learning. The teacher must understand the teaching materials, and grasp the teaching goals and key points and difficulties. They

should design appropriated questions according to the different ability and quality of the students to guarantee that problems are targeted, and students can learn something during the lesson. Teachers are only good at designing many kinds of teaching plans according to teaching content and students' reality in advance, if they can step into the classroom and start to teach based on the students' thinking ability, and the good teaching effect is received. Combined with blackboard writing which is concise, vivid, creative, and with strong inspiring and appreciate value, it will help students to understand and memorize knowledge, and win the beautiful enjoyment.

2 CONTENT OF TEACHING

First, good teaching quality is in-depth study of the teaching contents, understanding and grasping them, and having the necessary extension and expansion of them. The following needs to be done:

(1) Choosing the teaching contents accurately and logically. The teaching contents should not only be profoundly comprehended, but also be fully grasped. After the revision of the teaching plan, class hours have been reduced. It seems that use of the past requirements is unreasonable, then how to delimit correctly? The Ministry of education course requirements should be taken as the standard. Every teacher should know and comprehend the basic requirements of the course profoundly.

(2) Definition of concepts should be accurate and easy to moderate. We should determine the degree of difficulty of the teaching contents according to different specialty and teaching contents.

(3) Pay attention to emphasis, difficulty and the solving approach. Only set up teachers who have the correct teaching philosophy, have enough teaching wit and understand the art of class-teaching, the class-teaching can be reasonable and with natural decency. The basic, important and difficult knowledge point should be retained in order that students might master the method and learn by analogy, instead of understanding it simply to reduce the content because of the reduction of the class hours. The key point should be concise and easy to master. At the same time, some outdated, inappropriate contents must be removed. To make these, teachers must have a profound study of the teaching contents and careful selection. The class-teaching should focus on the essence of the books.

(4) Focus on building the knowledge structure and logic frame. Taking the course of physical chemistry as an example, the knowledge structure of physical chemistry is a logic network system composed of the strict logical relationship [2]. Through building the thinking logic frame, the students will be able to form a complete, comprehensive impression of the theory of physical chemistry. Such as in the learning process of thermodynamics, teachers can guide students to establish the logical framework of the thermodynamic problems. Thermodynamics is about energy conversion issues during the chemical reaction, including the direction and limitation. These issues can be answered by calculating dU, dH, dG, dA and other physical quantities which are abstract and cannot be measured. Then a key idea to deal with thermodynamic problems is put forward. P, V, T and S, which can be measured, represent the abstract amount by constituting the basic thermodynamic equation between the state functional relationships – the Maxwell relationship. Dealing with related thermodynamic issues will be easier through the Maxwell relationship.

(5) Pay attention to renewing and extending the teaching contents. Students are not interested in the antiquated contents, so it need not be taught. Some new contents which are required should be introduced. Students will not listen to the course if it is boring. Sometimes it is necessary to talk about some history and dynamic development of the subject in order to cultivate students' sense of innovation and scientific spirit.

3 METHODS OF TEACHING

When the contents are determined, appropriate teaching methods should be used to obtain the good teaching effect. According to the characteristics of the physical chemistry subjects, the question type teaching mode is used to lecture the course content. The most important thing of this mode is to set questions guiding the students' thinking, and to give students sufficient space for thinking, and inspiring and guiding them to dare to think and be creative.

The first thing is to set questions ingeniously to activate class-teaching [3]. Questions can contain many aspects, not only from the actual life or production, but also past problems. Generally speaking, questions from our lives can arouse the students' interest, which is the best teacher to them. So it is key to teach such that how to extract interesting questions from life, stimulate the students' desire to solve the problem, improve students' interest and enthusiasm, guide the teaching interaction. For instance, when lecturing the Clausius-Clapeyron equation, there are four questions: (1) Why can't one cook the food in the plateau? (2) Why a pressure cooker cooks the food easier? (3) Why the pressure limiting valve should be buckled up after the steam discharging from the stoma? (4) The pressure inside the pressure cooker is 3 atm, then how many degrees is the pot temperature? Through the above questions, students can form a more systematic and profound impression about the knowledge and enhance the interest in learning.

Secondly, analyzing the nature of the problem and the solving method of the problem. It is necessary for guiding students to analyze the difficulties and how to divide a complex problem into several simple problems. This is not only a kind of learning ability, but also a kind of scientific research ability. Resolving difficulties is a process to guide the students thinking step by step.

Thirdly, solving the problem by combining it with reality. While teaching about the concept of osmotic pressure, teachers may connect it with transfusion in the hospital. As for the penetration theory, teachers may describe the application of the reverse osmosis, such as seawater desalination, wastewater treatment and so on.

Finally, needing discussion and conclusion. Discussion should allow the students to think alone. If there are no conclusions, the students should think about it deeply and independently. In other words, we should have the answers to the questions.

4 AROUSE THE ENTHUSIASM OF STUDENTS' STUDY

To stimulate students' enthusiasm, the primary task is to improve the teaching efficiency in class. If the teachers' lesson is lively, practical, efficient, the students would listen to it carefully. The teaching content must be attractive so that the students will listen to it with interest. These are related to the teachers' sufficient preparation, the proper questions, the concise analysis of the problem, and so on. Only when the teaching is attractive, the students will focus on the course.

While they stand on the podium, teachers with good teaching quality have a responsibility. As a result, teachers can eliminate all the adverse effects on the evaluation of their teaching, instead of feeling ashamed to their students if they do not have a good quality of teaching. He can make students believe that losing lectures are a big loss, and lectures are an appreciation and enjoyment, learning has become an important part of the students' life.

Secondly combined with today's college students' characteristics, teachers should take the multidimensional assessment, which is stressed on evaluating the whole process of student learning, to make sure that students participate in the whole process of teaching. It not only reflects a fair evaluation of a comprehensive ability of students in the whole learning process, and it helps to stimulate the enthusiasm of students who have a good foundation; but also protects the enthusiasm of students who have slightly fallen behind and prompts to change their lazy habits. It is also conducive to cultivate the comprehensive ability of the student and raise the whole quality of classroom teaching comprehensively.

After three years over more than 1000 students practice, the multidimensional assessment has been recognized by students. According to reflection, students listen to the class carefully because of the multidimensional assessment. Since then there is no non-attendance and class skipping. Now it has been an important method for all teachers in the outstanding teaching team of physical chemistry in Liaoning Province to improve the management of class-teaching, stimulate the enthusiasm of students and enhance the quality of class-teaching.

5 THE INTERACTION BETWEEN TEACHERS AND STUDENTS

The teachers should interact with students in classroom teaching. Realizing the interaction between teachers and students is not only the teaching method, but also the teaching idea. The interaction between teachers and students is that the students have a response to the teacher's questions, and students and teachers must have dialogue and communication.

In order to realize the interaction between teachers and students, firstly, one should respect the students' subject position in the teaching activities. If there is no response of the students, the teaching of teachers is passive. It is the high quality teaching effect that realization of the interaction between teachers and students can change from passive effect into active effect, and make students learn to think, reason, compare, identify, analyze, and cultivate their judgment and imagination. Therefore, the task of teachers is not only to impart knowledge, but also to cultivate the students' learning interest, study enthusiasm and the ability of showing themselves.

Secondly, in the process of putting forward problems continuously, we should encourage students not only to answer the questions, the most important thing is to find problems, and put forward problems in an appropriate manner, and make them spurt the largest light and heat and benefit in the learning process.

The problems of success which are set between teachers and students can stimulate the teaching and learning enthusiasm, evoke the 'interaction' between teachers and students, and between students. This 'interaction' can be the discussing type, and also the cooperating type. The discussing type 'interaction': everyone can speak their own opinion freely about questions proposed. The cooperative type 'interaction': students combine freely and establish cooperation group consults mutually, solve some complicated problems. This "interaction" training will cultivate the team spirit of students, and make them become a willingness to cooperate. No matter which kind of "interaction", students will enjoy full autonomy, and become the main body of the classroom. At the same time, the teaching process will be active.

Thirdly, during the discussion and communication, teachers must be good at guiding. If students can not give answers to the problems, the teachers should guide the students to think for themselves. Only in this way can students think in different angles constantly to inspire their intelligence, promote their thinking and deepen their understanding of the key problems. Thus, the role of teachers is not only to impart knowledge, but also to guide the student in thinking.

6 SUMMARY

Teaching is a growth process between teaching and learning, teachers will reorganize and recreate knowledge during teaching. Teachers should not only impart specific professional knowledge, but also should guide students to learn how to grasp the knowledge and acquire knowledge. It is more important to let students know how to learn and solve problems than to learn a specific formula. This is a challenge to teachers. Therefore, excellent teaching quality is vital for teachers. In the teaching process, the teachers should take elastic, flexible, scientific and effective methods to guide the students to be good at learning, and create a good classroom atmosphere by aiming at the contents of teaching, students' interest, knowledge level and understanding, etc. Teachers and students should find out problems, study them and solve them together. Good teaching quality is the basic guarantee to improve the quality of the classroom teaching. So, every teacher must improve their teaching quality unconditionally, for it is related to cultivating quality of talent, and what's more the country's strength and prosperity is closely related.

ACKNOWLEDGEMENTS

This work was financially supported by the Liaoning province education science 'The Twelfth Five Years' planning subject 'The Study on The Classroom

Teaching Evaluation and The Evaluation Criteria of Students '(JG11DB140), University of science and technology Liaoning 2010 landmark achievements construction project 'The exploration of encouraging students' active thinking of the new teaching mode and constructing multi-dimensional evaluation system'(kdjg10-11) and Liaoning province Education Commission (884-4-4).

REFERENCES

[1] Jianrong. Li, China Electronic Power Education, Vol. 134 (2009), p. 113–114.
[2] Yuechuen Zhao, Minghua Li, Advanced agricultural Education, Vol. 8 (2001), p. 55–57.
[3] Liyan Na, Shubiao Zhang, Ruinian Hua, Chemical World, Vol. 11 (2009), p. 703–705.

Future Communication, Information and Computer Science – Zheng (Ed.)
© *2015 Taylor & Francis Group, London, 978-1-138-02653-7*

Electrical Engineering and Electronics course system construction based on outstanding engineers plans

X. Xu

Department of Electronics Engineering, East China Jiao-tong University, Nanchang, P.R. China

ABSTRACT: This paper introduces the present status of "Electrical Engineering and Electronics" course system in our school. On the basis of the project teaching requirements of "Excellent Engineer Education Training Plan", according to the characteristics of non-electric professional, it puts forward the design principle of "Electrical Engineering and Electronics" course system. On the basis of the principles of curriculum system designing for recombination and optimization to construct "Electrical Engineering and Electronics" course system, and strengthen the training of innovation ability, actively explore a variety of teaching methods. Thus, "Electrical Engineering and Electronics" course system reform can comprehensively adapt to the developing requirement of the "Excellent Engineer Education and Training Plan".

Keywords: excellence engineer; electrical engineering and electronics engineering; education reform

1 INTRODUCTION

In June 2010, the Ministry of education put forward the "Excellent Engineer Education and Training Plan", which purpose is to cultivate a large number of various types of high quality engineering technology talent who have good innovative ability and meet the needs of economic and social development, to service the strategy of our country to take the new industrialization development road, and build an innovative and talent powerful country (Jie Zhang et al. 2013). "Electrical Engineering and Electronics" is an important basic course of the non electrical majors in college. It must comply with the teaching requirements of the "Excellent Engineer Education and Training Plan". Optimization and reconstruction of the course system of "Electrical Engineering and Electronics", establishment of the curriculum system as the characteristic of the "thick foundation, wide caliber, good ability, practical, high quality" and the adaption of many departments and many professionals (Changchun Cai et al. 2013). A broad platform of electrical engineering and electronics foundation is built for the students. In this paper, based on the teaching requirements of "outstanding engineers training plan", the "Electrical Engineering and Electronics" course system is optimized and the teaching contents are restructured for non-electric class disciplines.

2 "ELECTRICAL ENGINEERING AND ELECTRONICS" COURSE SYSTEM STATUS

The "Electrical Engineering and Electronics" course as non-electrical major in colleges of engineering is an important basic technical course. It has the characteristics of wide knowledge and strong practice. It acts the important role to meet the students in the high-tech equipment, technology reform, professional technical research on quantitative analysis and system engineering construction (Yuehua Gen. 2013).

At present, "Electrical Engineering and Electronics" course system is divided into three parts which are theory course, experiment course, theory and practice comprehensive course. Theory course content includes the electrician foundation, magnetic circuit, motor and relay contact control, analog electronic technology and digital electronic technology. Main contents include DC circuit and its analysis, sinusoidal alternating circuit and its analysis, the three-phase AC power and load circuit, non- sinusoidal periodic current circuit, circuit transient analysis, magnetic load and an iron core coil, transformer circuit, AC motor, DC motor, control motor and control system, programmable controller and application, power supply and the safety of industrial enterprises, electrical measurement, semiconductor devices and amplifying circuit, power amplifier, integrated operational amplifier and application, sine wave oscillation circuit, DC regulated power supply, combination and sequential logic circuit and its application, memory and programmable logic device, D/A and A/D converter.

For non-engineering majors, such as trade, after, tube, method, language, culture, art, sports and so on. The students master the professional knowledge and basic power knowledge at the same time. To do it like this can achieve the purpose of expanding the students' horizon, enrich their knowledge structure and improve the comprehensive quality ability. We establish a new teaching pattern of "Electrical Engineering and Electronics". Based on electrical technology and electronic technology the goal of the application of electrical engineering and electronics technology is taken. We provide an electrical engineering and electronics teaching system and teaching framework, that is more flexible and adaptable to the requirements of technical development and better adapted to the actual requirements of non- engineering talents. The specific content includes basic electrical technology, electronics technology, power systems, power electronics technology, automation technology, motor and relay contact control, programmable controller, embedded system, microcomputer, programmable logic device, communication and network, consumer electronics, photoelectric display, robots and so on.

The experimental and theoretical courses enrich and complement each other in content. We develop a new experiment to adapt to the related materials. Its contents are verification experiments, designing experiments and comprehensive experiments. According to different professionals specific test items are set. We take the measures of increasing the capital input, adding equipment, exchanging two of a group into one group and strengthening the evaluation of experiment preparation, operation and experiment data processing at the same time. Mechanical major students, not only have experimental courses but also add theory and practice to the comprehensive curriculum. Namely, the opened course designing of "Electrical Engineering and Electronics" will last two weeks. In addition, especially encourage students to sure themselves of experimental items, to take the extracurricular activities of science and technology and electronic competition into the teaching activities. In the open laboratory, classroom learning and out of the classroom learning interaction, integration of classroom teaching and practice teaching are beneficial to open vision and enhance the ability.

With the advance of "Excellent Engineer Education Training Plan", the "Electrical Engineering and Electronics" course system can not adapt to the teaching requirements of "Excellent Engineer Education Training Plan". Therefore, based on the teaching requirements of "Excellent Engineer Education Training Plan" to optimize and adjust the course system of "Electrical Engineering and Electronics", and to satisfy the non-electrical majors demand for electricity has been referred to the agenda (Yan Hou et al. 2013). We take the modular curriculum system, emphasize on introducing that the teaching content is closely related

to the students' professional, pay more attention to the training of students' practical ability and catch the innovative reform on the content of "Electrical Engineering and Electronics" course is the inevitable developing trend.

3 THE COURSE SYSTEM CONSTRUCTION FOR "ELECTRICAL ENGINEERING AND ELECTRONICS"

3.1 The principles of curriculum system designing

"Electrical Engineering and Electronics" involving a wide range of courses includes mathematics, physics, chemistries and other basic courses. It needs a higher mastering degree on basic courses and its theory is more abstract. The students generally believe that this course is not easy to learn. It is not only abstract but also strong in practicality. It has boring and obscure content (Xin Wang. 2013). After many years of teaching practice, the course system design of "Electrical Engineering and Electronics" should abide by the following principle to meet the requirements of "Excellent Engineer Education Training Plan".

(1) We should strengthen the basis teaching of electrical engineering and electronics engineering, pay attention to introduce the basic theory and method and guide students to learn to use the basic theory and method to explain the engineering phenomena and solve practical problems and dilute the pure training of problem-solving skills (Gang Huang et al. 2013).

(2) The introduction of new content should grasp the breadth and depth. According to the actual situation of the teaching object, we should make sure what technology should understand the basic principles, grasp the basic application, the higher level application and development trend (Xia Xu et al. 2013).

(3) We should pay attention to the cultivation of practical ability. According to the teaching content and the various professionals to select some engineering cases to explain and handle the problem, arrange corresponding exercises and thinking exercises.

(4) For non-engineering majors (trade, economy, humanities, etc), the teaching content should emphasize to enlarge the students' knowledge and improve the students' learning interest (Sanyong Jiang et al. 2012). That reflects the characteristics of basic, comprehensive and scientific for non-engineering on electrical engineering course teaching.

3.2 Curriculum system construction

Facing the new era, to reform the "Electrical Engineering and Electronics" course system, to improve the teaching level, teaching quality and teaching efficiency, to optimize the "Electrical Engineering and Electronics" course teaching is an inevitable trend. In order to solve the contradiction between less lessons and a teaching content increase, we must accelerate the reform of the teaching content and the course

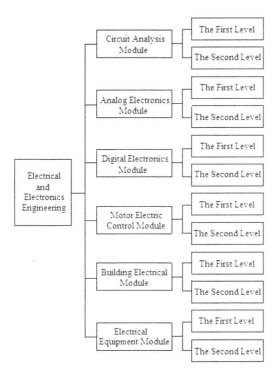

Figure 1. Curriculum system construction frame.

system of "Electrical Engineering and Electronics". Most of the non-electrical engineering demand for electricity is not the same. Our school has the following majors such as mechanics, civil engineering, chemistry, economics and management professional. The professional demand for electrical engineering and electronics teaching is not the same. The module establishment of course teaching of "electronics", respectively the combination of the different module provisions of the teaching content should include electrical engineering curriculum for each professional. Have the same part in module combination and the different part, but the breadth and depth is not the same in the same part. They are divided into the first level and second level, the first level teaching content is deeper than the second level.

Due to the students of non-electrical majors to learn the course of "Electrical Engineering and Electronics" is to understand the electrical engineering and electronics technology applications in the field of non electrical engineering so that they have a wide vision in the future work. So the electrical engineering teaching should pay special attention to the application of non electrical engineering. In view of the present situation and trend of development of our professional engineering technology, we determine the electrical engineering knowledge coverage and take module combination to construct a corresponding professional new outline of "Electrical Engineering and Electronics". The contents of the theory course include the following module parts: circuit analysis module, the analog electronic technology, digital electronic technology module, the

motor electric control module, construction electrical module, the electrical equipment module. In this module, the corresponding teaching outline adapts to the different professional needs.

The curriculum system construction frame is shown in Figure 1.

4 CONCLUSIONS

"Electrical Engineering and Electronics" course system optimization is the need of higher education curriculum reform. "Electrical Engineering and Electronics" course system optimization is the need of "Excellent Engineer Education Training Plan". "Electrical Engineering and Electronics" course system optimization is the need for training high quality innovative talents. We must renew the ideas of education, study the teaching content and improve the teaching level. Teaching is based on the "Electrical Engineering and Electronics" course and pays attention to its engineering application. At the same time, we should strengthen the innovation ability, actively explore a variety of teaching methods and combine various teaching methods. We should make sure the students form the foundation to improve and to cultivate innovation at last and pursuit the purpose of improving the quality of teaching and make the "Electrical Engineering and Electronics" course reform adapt to the requirements of the "excellent engineer education and training plan". Practices have proved that, from the angle of meeting the electricity demand of non electrical majors, taking modular curriculum system of "broad based, application, system", emphasizing the teaching content is closely related to the students' professional, innovation course system on "Electrical Engineering and Electronics" complying with the development direction of the course system construction of "Electrical Engineering and Electronics" is also in line with our goal of culturing students.

REFERENCES

[1] Changchun Cai, Hongxia Miao, Xiong Liu, Zhixiang Deng & Wei Li. 2013. Research on teaching reform of "electrotechnics" for non-electric disciplines, *China Power Education*, (3): 88–89.
[2] Gang Huang, Yongxiang Zhu & Junping Liu. 2013. Electrical engineering theoretical and experimental teaching reform of non electrical majors, *Education Forum* (6): 46–47.
[3] Sanyong Jiang, Jisheng Ding, Cenghuang Qin & Chunying Jin. 2012. In electrical engineering textbooks and teaching the correct handling of four relations—Study on the construction and development of the electrical engineering, *Teaching China University*, (11): 91–20.
[4] Jie Zhang, Qiaoling Tu, Yi Yang & Xia Xu. 2013. Adapt to the non electrical majors needs to optimize the course system of electrotechnics, *Journal of Chongqing University of Technology (Natural Science)*, 27 (12): 149–150.

[5] Xia Xu, Qiaoling Tu & Jie Zhang. 2013. Based on the needs of professional "Electrical Engineering" course overall optimization research, *Electric Power Education China*, (5): 55–56.

[6] Xin Wang. 2013. Electrical engineering curriculum teaching non electric professional university, *Journal of Jimei University*, 14 (1): 119–121.

[7] Yan Hou & Weimin Zhang. 2013. Study on teaching mode of "electrotechnics" based on the experimental, *China Power Education*, (11): 49–50.

[8] Yuehua Gen. 2013. Some electric engineering curriculum optimization of reflection and exploration, *Journal of the Institute of Education of Jilin Province*, 29 (8): 49–51.

Future Communication, Information and Computer Science – Zheng (Ed.)
© 2015 Taylor & Francis Group, London, 978-1-138-02653-7

A time and space efficient algorithm for the tree inclusion problem

Y.J. Chen & Y.B. Chen
University of Winnipeg, Winnipeg, Manitoba, Canada

ABSTRACT: The ordered tree inclusion is an interesting problem, by which we will check whether a pattern tree P can be included in a target tree T, in which the order of siblings is significant. In this paper, we propose an efficient algorithm for this problem. Its time complexity is bounded by $O(|T| \cdot d_P \cdot \log h_P)$, where d_P represents the largest out-degree of a node in P and h_P is the height of P. Up to now the best algorithm for this problem needs quadratic time.

Keywords: tree inclusion; tree matching; ordered labeled trees

1 INTRODUCTION

Let T be a rooted tree. We say that T is ordered and labeled if each node is assigned a symbol from the alphabet Σ and a left-to-right order among siblings in T is specified. A tree T consisting of a specially designated node $root(T) = t$ (called the root of the tree) and a forest $<T_1, \ldots, T_k>$ (where $k \geq 0$) is denoted as $<t; T_1, \ldots, T_k>$. We also call T_j $(1 \leq j \leq k)$ a direct subtree of t.

The preorder of a forest $F = <T_1, \ldots, T_k>$ is the order of the nodes visited during a preorder traversal. A preorder traversal of a forest $<T_1, \ldots, T_k>$ is as follows. Traverse the trees $T_1, \ldots, T_k$ in ascending order of the indices in the preorder. To traverse a tree in preorder, first visit the root and then traverse the forest of its subtrees in preorder. The postorder is defined similarly, except that in a postorder traversal the root is visited after traversing the forest of its subtrees in postorder. We denote the preorder and postorder numbers of a node v by $pre(v)$ and $post(v)$, respectively. Let u, v be two nodes in T. If there is a path from node u to node v, we say, u is an ancestor of v and v is a descendant of u. In this paper, by ancestor (descendant), we mean a proper ancestor (descendant), i.e., $u \neq v$. Using the preorder and postorder numbers, the ancestorship can be easily checked:

> u is an ancestor of v if and only if $pre(u) < pre(v)$ and $post(v) < post(u)$. (See Exercise 2.3.2-20 in (Knuth, 1969), page 347.)

Similarly, u is said to be to the left of v if they are not related by the ancestor-descendant relationship and v follows u when we traverse T in preorder. Then, u is to the left of v if and only if $pre(u) < pre(v)$ and $post(u) < post(v)$.

In the following, we use $\prec$ to represent the left-to-right ordering. Also, $v \preceq v'$ iff $v \prec v'$ or $v = v'$. We will also use $V(T)$ and $E(T)$ to represent the set of nodes and the set of edges in T, respectively.

The following definition is due to (Kilpeläinen and Mannila, 1995).

Definition 1 Let F and G be labeled ordered forests. We define an ordered embedding (φ, G, F) as an injective function $\varphi: V(G) \to V(F)$ such that for all nodes $v, u \in V(G)$,

i) label $(v) =$ label $(\varphi(v))$; (label preservation condition)

ii) v is an ancestor of u iff $\varphi(v)$ is an ancestor of $\varphi(u)$, i.e., $pre(v) < pre(u)$ and $post(u) < post(v)$ iff $pre(\varphi(v)) < pre(\varphi(u))$ and $post(\varphi(u)) < post(\varphi(v))$; (ancestor condition)

iii) v is to the left of u iff $\varphi(v)$ is to the left of $\varphi(u)$, i.e., $pre(v) < pre(u)$ and $post(v) < post(u)$ iff $pre(\varphi(v)) < pre(\varphi(u))$ and $post(\varphi(v)) < post(\varphi(u))$. (Sibling condition) □

If there exists such an injective function from $V(G)$ to $V(F)$, we say, F includes G, F contains G, F covers G, or say, G can be embedded in F.

This problem has been recognized as an important query primitive for XML data and received considerable attention (Kilpeläinen and Mannila, 1995), where a structured document database is considered as a collection of parse trees that represent the structure of the stored texts and the tree inclusion is used as a means of retrieving information from them. Ordered labeled trees also appear in the natural language processing.

Up to now, the best algorithm is due to (Bille and Gørtz, 2011). They got a space-economical algorithm with its space overhead bounded by $O(|T| + |P|)$, but with its time complexity bounded by

$$min = \begin{cases} O(|T| \cdot |\text{leaves}(P)|) \\ O(|\text{leaves}(T)| \cdot |\text{leaves}(P)| \cdot \log\log|\text{leaves}(P)| + |\text{leaves}(P)|) \\ O(|T| \cdot |P|/(\log|T|) + |T| \cdot \log|T|) \end{cases}$$

In this paper, we discuss a new algorithm for this problem. Its space requirement is also bounded by $O(|T| + |P|)$. However, its time complexity is reduced to $O(|T| \cdot d_P \cdot \log h_P)$, where d_P represents the largest out-degree of a node in P and h_P is the depth of P.

Throughout the rest of the paper, we refer to the labeled ordered trees simply as trees.

2 ALGORITHM

Now we begin to describe our algorithm. First, we define some notations in 2.1. Then, in 2.2 and 2.3, we describe our algorithm in great detail.

2.1 Basic notation

Let $T = <t; T_1, \ldots, T_k>$ $(k \geq 0)$ be a tree and $G = <P_1, \ldots, P_q>$ $(q \geq 0)$ be a forest. We will use p_v to represent the virtual parent of $P_1, \ldots, P_q$. Then, in G, every node v, excerpt p_v, has a parent, denoted as $parent(v)$.

Consider a node $v \in V(G) \cup \{p_v\}$ with children $v_1, \ldots, v_r$. We use a pair $<i, v>$ $(i \leq r)$ to represent an ordered forest containing the first i subtrees of v: $<G[v_1], \ldots, G[v_i]>$.

We are interested in a special kind of subtrees, called *left corners*, defined below.

Definition 2 (*Left corners*) A forest $<i, v>$ in G is called a left corner of G if $v = p_v$ or v is a node on the left-most path in P_1.

Clearly, if $v = p_v$, $<i, v>$ stands for a left corner of G, consisting of the first i subtrees in G: $P_1, \ldots, P_i$.

In the following, we will refer to a left corner of G simply as a left corner if no confusion will be caused according to the context.

In addition, we use $\rho(G)$ to represent the left-most leaf node of G. Then, $<i, \rho(G)>$ (with any $i \geq 0$) or $<0, v>$ (with any v in G) stands for an empty left corner.

We also use $\delta(v)$ to represent a link from a node v to the left-most leaf node in $G[v]$, as illustrated in Fig. 1.

Let v' be a leaf node in G. $\delta(v')$ is defined to be a link to v' itself. So in Fig. 1, we have $\delta(v_1) = \delta(v_2) = \delta(v_3) = v_3$. Denote by $\delta^{-1}(v')$ a set of nodes x such that for each $v \in x$ $\delta(v) = v'$. Then, in Fig. 1, We have $\delta^{-1}(v_3) = \{v_1, v_2, v_3\}$, $\delta^{-1}(v_4) = \{v_4\}$, and $\delta^{-1}(v_5) = \{v_5\}$.

Let p_1 be the root of P_1. We have $\rho(G) = \rho(p_1)$. The out-degree of v in a tree is denoted by $d(v)$ while the height of v is denoted by $h(v)$, defined to be the number of edges on the longest downward path from v to a leaf. The height of a leaf node is set to be 0.

As with (Chen, Y. and Chen, Y.B. 2006; Chen, Y. Shi, Y. and Chen, Y.B. 2006), we arrange two functions to check the tree inclusion. However, in (Chen, Y. and Chen, Y.B. 2006), each function returns an integer j, indicating that the first j subtrees in G can be embedded in a target tree or a target forest while in our algorithm each function returns a left corner in G which can be embedded in the target.

Figure 1. A pattern.

If both the target and the pattern are forests, we call a function $A(<T_1, \ldots, T_k>, <P_1, \ldots, P_q>)$. If the target is a tree and the pattern is a forest, we call another function $B(T, <P_1, \ldots, P_q>)$. But during the computation, they will be called from each other.

Let $<i, v>$ be a left corner returned by $A(<T_1, \ldots, T_k>, <P_1, \ldots, P_q>)$ (or by $B(T, <P_1, \ldots, P_q>)$). Then, the following properties are satisfied:

- If $i > 0$ and $v \neq \rho(G)$, it shows that

  - the first i subtrees of $v \in \delta^{-1}(\rho(G))$ can be embedded in $<T_1, \ldots, T_k>$ (or in T).
  - for any $i' > i$, $<i', v>$ cannot be embedded in $<T_1, \ldots, T_k>$ (or in T); and
  - for any v's ancestor $u \in \delta^{-1}(\rho(G)) \cup \{p_v\}$, there exists no $j > 0$ such that $<j, u>$ is able to be embedded in $<T_1, \ldots, T_k>$ (or in T).

- If $i = 0$ or $v = \rho(G)$, it indicates that no left corner of G can be embedded in $<T_1, \ldots, T_k>$ (or in T).

In this sense, we say, $<i, v>$ is the *highest* and *widest* left corner which can be embedded in $<T_1, \ldots, T_k>$ (or in T). We notice that if $v = p_v$ and $i > 0$, it shows that $P_1, \ldots, P_i$ can be included in $<T_1, \ldots, T_k>$ (or in T).

Finally, we say, a left corner $<i, v>$ is higher than a node u (or another left corner $<j, u>$) if v is an ancestor of u.

2.2 A-function

First, let's have a look at a naïve way to evaluate $A(F, G)$, where $F = <T_1, \ldots, T_k>$ and $G = <P_1, \ldots, P_q>$.

1. Two index variables j, l are used to scan $T_1, \ldots, T_k$ and $P_1, \ldots, P_q$, respectively. Initially, j is set to 1, and l is set to 0. (They also indicate that $<P_1, \ldots, P_l>$ has been successfully embedded in $<T_1, \ldots, T_j>$.) In each step, we call $B(T_j, <P_{l+1}, \ldots, P_q>)$.

2. Let $<i_j, v_j>$ be the return value of $B(T_j, <P_{l+1}, \ldots, P_q>)$. If $v_j = parent(p_1)$, set l to be $l + i_j$. Otherwise, l is not changed. Set j to be $j + 1$. Go to (2).

3. The loop terminates when all T_j's or all P_l's are examined.

If $l > 0$ when the loop terminates, $B(T, G)$ returns $<l, parent(p_1)>$, indicating that F contains $P_1, \ldots, P_l$. Otherwise, $l = 0$, indicating that even P_1 alone cannot be embedded in any T_j $(j \in \{1, \ldots, k\})$. However, in this case, we need to continue to search for a highest and widest left corner $<i, v>$ in G, which can be embedded in G. This is done as described below.

194

$F = <T_1, T_2>$: $G = <P_1, P_2>$:

Figure 2.   A target and a pattern forest.

4. Let $<i_1, v_1>, \ldots, <i_k, v_k>$ be the return values of $B(T_1, <P_1, \ldots, P_q>), \ldots, B(T_k, <P_1, \ldots, P_q>)$, respectively. Since $j = 0$, each $v_j \in \delta^{-1}(v')$ $(j = 1, \ldots, k)$, where v' is the left-most leaf in P_1.
5. If each $i_j = 0$, return $<0, \rho(G)>$. Otherwise, there must be some v_j's such that $i_j > 0$. We call such a node a non-zero point. Find the first *non-zero point* v_f with children $w_1, \ldots, w_y$ such that v_f is not a descendant of any other non-zero point. Then, we will check $<T_{f+1}, \ldots, T_k>$ against $<G[w_{i_f+1}], \ldots, G[w_y]>$. Let x $(0 \le x \le y - i_f)$ be a number such that $<G[w_{i_f+1}], \ldots, G[w_x]>$ can be embedded in $<T_{f+1}, \ldots, T_k>$. The return value of $A(F, G)$ should be set to $<i_f + x, v_f>$. □

In the above process, (1)–(3) are referred to as a main checking while (4)–(5) as a supplement checking.

We notice that in the supplement checking, only the first non-zero is utilized for forming the final result while all the other calls of the form $B(T_j, <P_1, \ldots, P_q>)$ (done in the main process) are not used at all. Their efforts for looking for the corresponding return values bring the void, and therefore should be avoided.

For this purpose, we introduce the concept of *cuts* to integrate a kind of control into the above working process.

Definition 3 A cut for a call of the form $A(F, G)$ is a node u $(\neq p_v) \in \delta^{-1}(\rho(G))$, indicating that if the supplement checking in $A(F, G)$ can only bring out a left corner $<i, v>$ not higher than u, the corresponding computation makes no contribution to the final result. □

The following example helps for illustration.

Example 1 Consider a target forest F and a pattern forest G shown in Fig. 2, in which each node in F is identified with t_i, such as t_1, t_2, t_{11}, and so on; and each node in G is identified with p_j. Besides, each subtree rooted at t_i (resp. p_j) is represented by T_i (resp. P_j).

Initially, for $A(<T_1, T_2>, <P_1, P_2>)$, we will set its cut u_0 to be $\rho(G) = p_{111}$, imposing in fact no control on its supplement checking if any. When executing $B(T_1, <P_1, P_2>)$, its cut u_1 is set to be the same as u_0 (i.e., $u_1 = u_0$). It can be seen that T_1 is able to include only $G[p_{11}]$. So the return value of this call should be $<1, p_1>$. Then, for $B(T_2, <P_1, P_2>)$, the cut u_2 for it should be set to p_1, indicating that the supplement checking within $B(T_2, <P_1, P_2>)$ should be cut off if it can only produce a left corner not higher than p_1 since the result will not be used.

Assume that during the execution of $B(T_2, <P_1, P_2>)$, we will call $A(<T_{21}, T_{22}>, <P_1, P_2>)$ and the cut u_{21} for this call is the same as $u_2 = p_1$. Then, after the main checking of $A(<T_{21}, T_{22}>, <P_1, P_2>)$, its supplement checking will be cut off since in its main checking $B(T_{21}, <P_1, P_2>)$ and $B(T_{21}, <P_1, P_2>)$ will return $<1, p_1>$ and $<0, \rho(G)>$, respectively; and the supplement checking will not create a left corner higher p_1. □

With the cuts being considered, the A-function should be changed to take three inputs: $F = <T_1, \ldots, T_k>$, $G = <P_1, \ldots, P_q>$, and $u \in \delta^{-1}(p_1)$. (Initially, u is set to $\rho(G)$.) In the main checking of $A(F, G, u)$, the cut for each B-function call will be dynamically changed as described below.

i) At the very beginning, we will check whether u is higher than p_1, where p_1 is the root of P_1. If it is the case, we simply return $<0, \rho(G)>$ since the computation will not make any contribution to the final result. Otherwise, we will do the following.
ii) For the first B-function call $B(T_1, <P_{l_1}, \ldots, P_q>, u_1)$ (where $l_1 = 1$), set $u_1 = u$. Let $<i_1, v_1>$ be its return value. We will call $B(T_2, <P_{l_2}, \ldots, P_q>, u_2)$ in a next step. If $v_1 = parent(p_1)$, $l_2 = i_1 + 1$) and u_2 is set to be p_{l_2}. If $v_1 \neq parent(p_1)$, $l_2 = l_1 = 1$, and u_2 is set to be v_1.
iii) In general, let $<i_j, v_j>$ be the return value of $B(T_j, <P_{l_i}, \ldots, P_q>, u_j)$ for $j = 1, \ldots, x \le k$, $j_1 = 1$, $j_1 \le j_2 \ldots \le j_x \le q$. Let s be an integer such that $l_1 = \cdots = l_s = 1$, but $l_{s+1} > 1$. Then, for $2 \le j \le s$, we have

$$u_i = \begin{cases} v_{j-1} & \text{if } v_{j-1} \text{ is higher than } u_{j-1} \text{ and } i_{j-1} > 0; \\ u_{i.} & \text{if } v_{j-1} \text{ is not higher than } u_{j-1} \text{ or } i_{j-1} = 0; \end{cases} \qquad (3.1)$$

and for $s + 1 \le j \le x$, we have

$$u_j = p_{l_i} \qquad (3.2)$$

The formula (3.1) shows how the cuts are changed before we find the first T_s which is able to embed some subtrees in G. For all subtrees next to T_s, the cuts are determined in terms of the formula (3.2). Setting u_j to be p_{l_i} will effectively prohibit the supplement checking in the execution of $B(T_j, <P_{l_i}, \ldots, P_q>, u_j)$, which will definitely return left corner not higher than and therefore is useless.

After the main checking, the following checks will be conducted to determine whether a supplement checking will be carried out.

- If $l = q$, we will record the embedding.
- If $j < k$, we will continue to find a next embedding by making a recursive call $A(<T_{j+1}, \ldots, T_k>, <P_1, \ldots, P_q>, p_1)$.
- If there is at least an embedding, return $<q, p_v>$.
- If $0 < l < q$, return $<l, p_v>$.
- If $l = 0$ and $f = 0$, return $<0, \delta(p_1)>$.
- Otherwise, a supplement checking will be conducted.

Let $<i_f, v_f>$ the first non-zero point v_f such that v_f is not a descendant of any other non-zero point. We will make a recursive call $A(<T_{f+1}, \ldots, T_k>, <G[w_{i_f+1}], \ldots, G[w_y]>, w_{i_f+1})$ for this purpose, where $w_1, \ldots, w_y$ are the children of v_f. We notice that the cut for this recursive call is set to be w_{i_f+1} to cut off any possible supplement checking in the execution.

In terms of the above discussion, we give the following algorithm.

function $A(F, G, u)$ (*Initially, $u = \rho(G)$.*)

input: $F = <T_1, \ldots, T_k>$, $G = <P_1, \ldots, P_q>$, u – a cut.

output: $<i, v>$ specified above.

begin

1. **if** p_1 is a descendant of u **then** return $<0, \delta(p_1)>$;
2. $j := 1$; $l := 0$; $v := u$; $f := 0$; $i := 0$;
3. **while** $(l < q$ and $j \leq k)$ **do**   (*main checking*)
4. $\{$ $<i_j, v_j> := B(T_j, <P_{l+1}, \ldots, P_q>, v)$
5.   **if** $(v_j = p_v)$ **then** $\{l := l + i_j; v := p_{l+1};\}$
6.   **else if** $(v_j$ is an ancestor of v and $i_j > 0)$
            then $\{v := v_j; i := i_j; f := j;\}$
7. $j := j + 1$
8. $\}$
9. **if** $l = q$ **then** record the embedding;
10. **if** $j < k$ **then** $\{<i'', v''> := A(<T_{j+1}, \ldots, T_k>, G, p_1);\}$
11. **if** there is at least an embedding **then** return $<q, p_v>$;
12. **if** $l > 0$ **then** return $<l, p_v>$;
13. **if** $f = 0$ **then** return $<0, \delta(p_1)>$;
14. let $w_1, \ldots, w_s$ be the children of v;
15. $j := f + 1$;   (*supplement checking*)
16. $<i', v'> := A(<T_{j+1}, \ldots, T_k>, <G[w_{i+1}], \ldots, G[w_s]>, w_{i+1})$;
17. **if** $(v' = v$ and $i' > 0)$ **then** return $<i + i', v>$;

End

In the above algorithm, we check the cut in two places. One is in line 1, where we check whether p_1 is a descendant of u. If it is the case, we simply return $<0, \delta(p_1)>$ since the left corner produced by the corresponding computation must not be higher than u. The other checking is done in line 13, where we check whether $f = 0$. If $f = 0$, it shows that all the left corners found by the main checking must not be higher than u. In addition, special attention should be paid to the recursive call in line 10 and the supplement checking performed in line 16. For the recursive call, the cut is set to be p_1 while for the supplement checking the cut is set to be w_{i+1}. In this way, in the corresponding computation, a further supplement checking will be prohibited.

Proposition 1 If $B()$ is correct, then the return value of $A(F, G, u)$ $(u = \rho(G))$ must be the highest and widest left corner of G, which can be included in F.

Proof. To prove the proposition, we need to establish the loop invariant for the **while**-loop of lines 3–8:

At the start of each iteration of **while**-loop, $<T_1, \ldots, T_j>$ include $<P_1, \ldots, P_l>$. If $l > 0$, and the current cut is set to be p_{l+1}. If $l = 0$, T_f $(0 \leq f \leq j)$ is the first tree in F, which includes the highest and left corner $<i, v>$ in P_1 among all the left corners included in $T_1, \ldots, T_j$, respectively. In this case, the current cut is set to be v. If $f = 0$, it shows that each of $T_1, \ldots, T_j$

includes an empty left corner and the current cut is set to be u.

For this loop invariant, its initialization, maintenance and termination can be easily specified.

2.3 B-function

In $B(T, G, u)$, we need to distinguish between two cases to use the greedy choice property mentioned before.

Case 1: $G = <P_1>$; or

$G = <P_1, \ldots, P_q>$ $(q > 1)$, but $|T| \leq |P_1| + |P_2|$.

In this case, what we can do is to find whether P_1 or a highest and widest left corner $<i, v>$ in P_1 can be embedded in $T = <t; T_1, \ldots, T_k>$. For this purpose, the following checkings should be conducted:

i) If t is a leaf node, we will check whether label$(t) =$ label$(\delta(p_1))$, where p_1 is the root of P_1. If it is the case, return <1, parent of $\delta(p_1)>$. Otherwise, return $<0, \delta(p_1)>$.

ii) If $|T| > 1$, but $|T| < |P_1|$ or $h(t) < h(p_1)$, we will make a recursive call $B(T, <P_{11}, \ldots, P_{1j}>, u)$, where $<P_{11}, \ldots, P_{1j}>$ is a forest of the subtrees of p_1. The return value of $B(T, <P_{11}, \ldots, P_{1j}>, u)$ is used as the return value of $B(T, G, u)$. It is because in this case, T is not able to include the whole P_1. So what we can do is to check T against $<P_{11}, \ldots, P_{1j}>$.

iii) If $|T| \geq |P_1|$ and $h(t) \geq h(p_1)$ (but $|T| \leq |P_1| + |P_2|$), we further distinguish between two subcases:

-   label$(t) =$ label(p_1). In this case, we will call $A(<T_1, \ldots, T_k>, <P_{11}, \ldots, P_{1j}>, p_{11})$ or $A(<T_1, \ldots, T_k>, <P_{11}, \ldots, P_{1j}>, u)$, depending on whether $u = p_1$. If $u = p_1$, the cut for this call is set to be p_{11} because if the left corner returned by this call is not higher than p_{11}, it will not be used.
-   label$(t) \neq$ label(p_1). In this case, we will call $A(<T_1, \ldots, T_k>, <P_1>)$.

In both cases, assume that the return value of $A()$ is $<i, v>$. We need to do an extra checking:

-   If label$(t) =$ label(v) and $i = d(v)$, the return value of $B(T, G, u)$ is set to be $<1, v\text{'s parent}>$.
-   Otherwise, the return value of $B(T, G, u)$ is the same as $<i, v>$.

Case 2: $G = <P_1, \ldots, P_q>$ $(q > 1)$, and $|T| > |P_1| + |P_2|$. In this case, we will call $A(<T_1, \ldots, T_k>, G, u)$. Assume that the return value of $A(<T_1, \ldots, T_k>, G, u)$ is $<i, v>$. The following checkings will be continually conducted.

iv) If $v \neq p_1$'s parent, check whether label$(t) =$ label(v) and $i = d(v)$. If it is not the case, the return value of $B(T, G, u)$ is the same as $<i, v>$. Otherwise, the return value of $B(T, G, u)$ will be set to $<1, v\text{'s} parent>$.

v) If $v = p_1$'s parent, the return value of $B(T, G, u)$ is the same as $<i, v>$.

In terms of the above discussion, we give the following formal description of the algorithm, in which $B'(\ )$ is used to handle Case 1 and $B''(\ )$ for case 2.

function $B(T, G, u)$ (*Initially, $u = \rho(G)$.*)
input: $T = <t; T_1, \ldots, T_k>$, $G = <P_1, \ldots, P_q>$, u – cut.
output: $<i, v>$ specified above
begin
1. **if** p_1 is a descendant of u **then** return $<0, \delta(p_1)>$;
2. **if** ($q = 1$ or $|T| \leq |P_1| + |P_2|$) **then** return $B'(T, P_1, u)$
3. **else** return $B''(T, G, u)$;
end

function $B'(T, P, u)$ (*Case 1*)
begin
1. let $T = <t; T_1, \ldots, T_k>$; let $P = <p; P_1, \ldots, P_j>$;
2. **if** t is a leaf **then**   (*Case 1 - (i)*)
3.   { let $\delta(p) = v$;
4.   **if** label$(t) =$ label(v) **then** return $<1, v$'s parent$>$
5.   **else** return $<0, v>$;
6.   }
7. **if** ($|T| < |P| \lor h(t) < h(p)$) **then** return $B(T, <P_1, \ldots, P_j>, u)$;
                                      (*Case 1 - (ii)*)
8. **if** label$(t) =$ label(p) (*Case 1 - (iii)*)
9. **then** {
   if p is a leaf **then** $\{v := v$'s parent; $i := 1;\}$
10.   **else**   {
   if $u = p$
        then $<i, v> := A(<T_1, \ldots, T_k>, <P_1, \ldots, P_j>, p_1)$
11.   **else** $<i, v> := A(<T_1, \ldots, T_k>, <P_1, \ldots, P_j>, u)$;
12.   **if** label$(t) =$ label(v) and $i = d(v)$
13.   **then** $\{v := v$'s parent; $i := 1;\}$
14.   }
15.   }
16. **else** $<i, v> := A(<T_1, \ldots, T_k>, <P>, u)$;
                         (*If label$(t) \neq$ label(p), call $A(\ )$.*)
17. return $<i, v>$;
18. }
end

In the above algorithm, we handle Case 1, in which the cut for the B-function is simply propagated to the subfunction calls of $A(\ )$ (see line 7, 11, and 16.) The only change made is in line 10, where the cut set for the call $A(<T_1, \ldots, T_k>, <P_1, \ldots, P_j>, p_1)$ is set to be p_1 to prohibit the supplement checking within the execution of this call.

function $B''(T, G, u)$ (*Case 2*)
begin
1. let $T = <t; T_1, \ldots, T_k>$;
2. $<i, v> := A(<T_1, \ldots, T_k>, G, u)$;
3. **if** $v \neq p_v$ (*Case 2 - (iv)*)
5. **then** { **if** (label$(t) =$ label$(v)) \land i = d(v)$
6.   **then** return $<1, v$'s parent$>$;
7.   }
8. return $<i, v>$;   (*Case 2 - (v)*)
end

This algorithm is for Case 2. Again, the cut for the B-function is directly propagated to the subfunction calls of $A(\ )$ (see line 2.)

Proposition 2 The return value of $B(T, <P_1, \ldots, P_l>, u)$ must be the highest and widest left corner in $<P_1, \ldots, P_l>$), which can be included T and higher than u. $\square$

3 CONCLUSION

In this paper, a new algorithm is proposed to solve the ordered tree inclusion problem. It needs only $O(|T| + |P|)$ space and $O(|T| \cdot d_P \cdot \log h_P)$, where d_P represents the largest out-degree of a node in P and h_P is the depth of P.

REFERENCES

[1] Bille, P. and I.L. Gørtz, L.L. 2011. The Tree Inclusion Problem: In Linear Space and Faster, ACM Transaction on Algorithms, Vol. 7, No. 3, Article 38, pp. 38:1–38:47.
[2] Chen, Y. and Chen, Y.B. 2006. A New Tree Inclusion Algorithm, *Information Processing Letters* 98(2006) 253–262, Elsevier Science B.V.
[3] Chen, Y., Shi, Y. and Chen, Y.B. 2006. Tree Inclusion Algorithm, Signatures and Evaluation of Path-Oriented Queries, *SAC 2006*, ACM, 1020–1025.
[4] Kilpeläinen, P. and Mannila, H. 1995. Ordered and unordered tree inclusion. SIAM J. Comput, 24:340–356.
[5] Knuth, D.E. 1969. The Art of Computer Programming, Vol. 1 (1st edition), Addison-Wesley, Reading, MA.

Future Communication, Information and Computer Science – Zheng (Ed.)
© *2015 Taylor & Francis Group, London, 978-1-138-02653-7*

On the query evaluation in search engines

Y.J. Chen & W. Shen
University of Winnipeg, Winnipeg, Manitoba, Canada

ABSTRACT: In this paper, we discuss an efficient and effective index mechanism for search engines to support both conjunctive and disjunctive queries. The main idea behind it is to decompose an inverted list into a collection of disjoint sub-lists. We will associate each word with an interval sequence, which is created by applying a kind of tree coding to a trie structure constructed over all the word sequences in a database. Then, attach each interval, instead of a word, with an inverted sub-list. In this way, both set intersection and union can be conducted by performing a series of simple interval containment checkings. Experiments have been conducted, which shows that the new index is promising. Also, how to maintain indexes, when inserting or deleting documents, is discussed in great detail.

Keywords: search engine; inverted files; conjunctive queries

1 INTRODUCTION

Indexing the Web for fast keyword search is among the most challenging applications for scalable data management. In the past several decades, different indexing methods have been developed to speed up text search, such as inverted files (Anh and Moffat, 2005), signature files (Faloutsos, 1985; Faloutsos, and Chan, 1988) and signature trees (Chen, 2004; Chen and Chen, 2006) for indexing texts and suffix trees and tries (Knuth, 1875) for string matching. Especially, different variants of inverted files have been used by the Web search engines to find pages satisfying a query (Lumpel and Moran, 2003).

A text database can be roughly viewed as a collection of documents and each document is stored as a list of words. Over the documents, there are two kinds of Boolean queries, that is, queries that can be constructed from query terms by conjunction ($\wedge$) or disjunction ($\vee$). A document D is an answer to a conjunctive query $w_1 \wedge w_2 \wedge \ldots \wedge w_k$ if it contains every w_i for $1 \leq i \leq k$ while D is an answer to a disjunctive query $w_1 \vee w_2 \vee \ldots \vee w_l$ if it contains any w_i for $1 \leq i \leq l$. Conjunction and disjunction can be nested to arbitrary depth, but can always be transformed to a conjunctive normal form:

$$(w_{11} \vee \ldots \vee w_{1l_1}) \wedge \ldots \wedge (w_{k1} \vee \ldots \vee w_{kl_k}).$$

In this paper, we discuss a new method to evaluate conjunctive queries by exploring a new direction to speed up query evaluation by decomposing an inverted list into a collection of disjoint sub-lists, which is substantially different from any existing strategy. To this end, we will

- Represent each document as a word sequence, sorted decreasingly by the word appearance frequency (referred to as a *document word sequence*, or simply a *word sequence*), and then construct a trie structure over all such sequences.
- Associate each word with an interval sequence L, where each interval in L is created by applying a kind of tree encoding over the generated trie structure.
- Associate each interval, instead of a word, with an inverted sub-list. In this sense, an inverted list associated with a word is decomposed. More importantly, such decomposition is always disjoint, which enables us to do both set intersection and union very efficiently by replacing the set intersection for conjunctive queries and the set union for disjunctive queries with the checking of interval containment.

This method will improve the traditional method by an order of magnitude or more.

2 NEW INDEX STRUCTURE

In this section, we mainly discuss our index structure, by which each word with a high frequency will be assigned an interval sequence. We will then associate intervals, instead of words, with inverted sub-lists. To clarify this mechanism, we will first discuss interval sequences for words in 2.1. Then, in 2.2, how to associate inverted lists with intervals will be addressed.

Table 1.   Documents and word sequence.

DocID	words	sorted word sequence
1	c, a, f, m, p	c, f, a, m, p
2	c, f, b, a	c, f, a, b
3	b, a, c, d	c, a, d, b
4	f, d, p, m	f, d, m, p

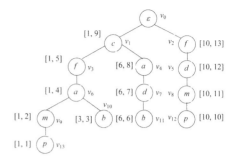

Figure 1.   A trie.

2.1   Interval sequence assigned to words

Let $D = \{D_1, \ldots, D_n\}$ be a set of documents. Let $W_i = \{w_{i1}, \ldots, w_{ij_i}\}$ $(i = 1, \ldots, n)$ be all of the words appearing in D_i, to be indexed. Denote $W = \bigcup_{i=1}^{n} W_i$, called the *vocabulary*. We define the word appearance frequency by the following formula:

$$f(w) = \frac{num.\,of\,documents\,containing\,w}{num.\,of\,documents}, \quad (w \in W).$$

We then define a *frequency threshold* ζ. For any word w with $f(w) < \zeta$, we will associate it with an inverted list in a normal way, denoted as $\delta(w)$, exactly as in the method of inverted files. However, for all those with $f(w) \geq \zeta$, we will create a new index. For this, we will represent each D_i as a sequence containing all those words w with $f(w) \geq \zeta$, decreasingly sorted by $f(w)$. That is, in such a sequence, a word w precedes another w' if w is more frequent than w' in all documents. In addition, for any subset of words that have the same appearance frequency a *global ordering* is defined so that in each sorted word sequence this global ordering is followed. In addition, we maintain a hash table $\mathcal{H}$ that maps each word w to its inverted list $\delta(w)$ or to its new index.

Example 1 In Table 1, we show a set of four documents, their words w with $f(w) \geq \zeta = 0.4$, and the corresponding sorted word sequences, where we use a character to represent a word for simplicity.

Notice that the global order on $\{f, a, c\}$ (with $f(w) = 0.75$) is set to be $c \rightarrow f \rightarrow a$ while the global order on $\{m, b, p, d\}$ (with $f(w) = 0.5$) is $d \rightarrow b \rightarrow m \rightarrow p$.

For each document D_i $(i = 1, \ldots, n)$, we will use s_i to represent its sorted word sequence. Over all such sequences $S = \{s_1, \ldots, s_n\}$, we will construct a digit tree, called a *trie*, as follows.

Assume that $W = \{w_1, \ldots, w_m\}$. If $|S| = 0$, the trie is, of course, empty. For $|S| = 1$, $trie(S)$ is a single node. If $|S| > 1$, S is split into m (possibly empty) subsets $S_1, S_2, \ldots, S_m$ so that a string is in S_j if its first word is w_j $(1 \leq j \leq m)$. The tries $trie(S_1), trie(S_2), \ldots, trie(S_n)$ are constructed in the same way except that at the kth step, the splitting of sets is based on the kth words in the sequences. They are then connected from their respective roots to a single node to create $trie(S)$. In Fig. 1, we show a trie T constructed over the sorted word sequences in Table 1.

In the trie, v_0 is a virtual root, labeled with an *empty* word ε while any other node is labeled with a *real* word. Therefore, all the words on a path from the root to a leaf spell a sorted word sequence for a certain document. For instance, the path from v_0 to v_{13} corresponds to the sequence: c, f, a, m, p. Then, to check whether two words w_1 and w_2 are in the same document, we need only to check whether there exist two nodes v_1 and v_2 such that v_1 is labeled with w_1, v_2 with w_2, and v_1 and v_2 are on the same path. This shows that the *reachability* needs to be checked for this task, by which we ask whether a node v can reach another node u through a path. If it is the case, we denote it as $v \Rightarrow u$; otherwise, we denote it as $v \not\Rightarrow u$.

The reachability problem on tries can be solved very efficiently by using a kind of tree encoding [13, 14, 32, 33], which labels each node v in a trie with an interval $I_v = [\alpha_v, \beta_v]$, where β_v denotes the rank of v in a *post-order* traversal of the trie. Here the ranks are assumed to begin with 1, and all the children of a node are assumed to be ordered and fixed during the traversal. Furthermore, α_v denotes the lowest rank for any node u in $T[v]$ (the subtree rooted at v, including v). Thus, for any node u in $T[v]$, we have $I_u \subseteq I_v$ since the post-order traversal enters a node before all of its children, and leaves after having visited all of its children. In Fig. 1, we also show such a tree encoding on the trie, assuming that the children are ordered from left to right. It is easy to see that by interval containment we can check whether two nodes are on a same path. For example, $v_3 \Rightarrow v_{10}$, since $I_{v_3} = [1, 5], I_{v_{10}} = [3, 3]$, and $[3, 3] \subset [1, 6]$; but $v_2 \not\Rightarrow v_9$, since $I_{v_2} = [10, 13]$, and $I_{v_9} = [1, 2]$, and $[1, 2] \not\subset [10, 13]$.

Let $I = [\alpha, \beta]$ be an interval. We will refer to α and β as $I[1]$ and $I[2]$, respectively.

Lemma 1 For any two intervals I and I' generated for two nodes in a trie, one of four relations holds: $I \subset I', I' \subset I, I[2] < I'[1],$ or $I'[2] < I[1]$.   □

However, more than one node may be labeled with the same word, such as nodes v_9, and v_8 in Fig. 1. Both are labeled with word m. Therefore, a word may be associated with more than one node (or say, more than one node's interval). Thus, to know whether two words are in the same document, multiple checkings may be needed. For example, to check whether p and d are in the same document, we need to check v_{13} and v_{12} each against both v_7 and v_5, by using the node's intervals.

200

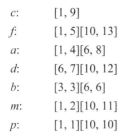

c:	$[1, 9]$
f:	$[1, 5][10, 13]$
a:	$[1, 4][6, 8]$
d:	$[6, 7][10, 12]$
b:	$[3, 3][6, 6]$
m:	$[1, 2][10, 11]$
p:	$[1, 1][10, 10]$

Figure 2.   Sorted interval sequences.

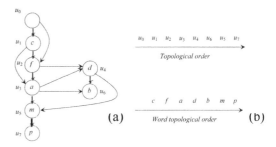

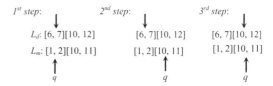

Figure 3.   A transformed DAG.

In order to minimize such checkings, we associate each word w with a word sequence of the form: $L_w = I_w^1, I_w^2, \ldots, I_w^k$, where k is the number of all those nodes labeled with w and each $I_w^i = [I_w^i[1], I_w^i[2]]$ ($1 \leq i \leq k$) is an interval associated with a certain node labeled with w. In addition, we can sort L_w by the interval's first value such that for $1 \leq i < j \leq k$ we have $L_w^i[1] < L_w^j[1]$, which will greatly reduce the time for the reachability checking. We illustrate this in Fig. 2, in which each word in Table 1 is associated with an interval sequence.

From this figure, we can see that for any two intervals I and I' in L_w we must have $I \not\subset I'$, and $I' \not\subset I$ since in any trie no two nodes on a path are labeled with the same word.

As shown below, using such interval sequences, the checking of whether two words are in the same document can be done in a very efficient way.

Definition 1 (*word topological order*) Let $S = \{s_1, s_2, \ldots, s_n\}$ be a set of n sorted word sequences. A word topological order over S is a sequence $\vartheta = w_1, w_2, \ldots, w_m$, which contains all the words appearing in S such that for any two words w and w' if w appears before w' in some s_j ($1 \leq j \leq n$) then w appears before w' in ϑ, denoted as $w \prec w'$.

In Fig. 2, the words are also listed (from top to bottom) in a word topological order with respect to the sorted word sequences given in Table 1. To find a word topological order over $S = \{s_1, s_2, \ldots, s_n\}$ with $W = \{w_1, \ldots, w_m\}$, we will transform the corresponding trie T to an *acyclic directed graph* (DAG) G by splitting the node set of T (except for the virtual root) into m groups such that all the nodes in a group are labeled with the same word, and then collapsing each group g to a single node u. There is an edge in G from u (standing for a group g) to u' (for another group g') if T contains (x, y) with $x \in g$ and $y \in g'$. For example, the trie shown in Fig. 1 will be transformed to a DAG shown in Fig. 3(a).

Using a hash function H on the words in W, the transformation can be done in $O(|W|)$ time, by which all those nodes labeled with the same word w will be mapped to a single node identified by $H(w)$.

Let $G(V, E)$ be such a DAG. It is well known that only $O(|V| + |E|)$ time is required to find a *topological order* of G, which is a linear ordering of all its nodes such that if $u \to v \in E$, then u appears before v in the ordering. Replacing each node in the ordering with the

Figure 4.   Illustration of two-word checking.

corresponding word, we will obtain a word topological sequence, as illustrated in Fig. 3(b). Now we consider two words w, w' with $w \prec w'$. It is easy to see that any interval in L_w cannot be contained in any interval in $L_{w'}$. Thus, to check whether w and w' are in the same document, we need only to check whether there exist $I \in L_w$ and $I' \in L_{w'}$ such that $I \supset I'$. This checking can be efficiently conducted as follows.

- Assume that $w \prec w'$. Let $L_w = I_w^1, I_w^2, \ldots, I_w^k$. Let $L_{w'} = I_{w'}^1, I_{w'}^2, \ldots, I_{w'}^{k'}$.
- Step through L_w and $L_{w'}$ from left to right. Let I_w^p and $I_{w'}^q$ be the intervals currently encountered. We will do one of the following operations:

(1) If $I_w^p \supset I_{w'}^q$ report that w and w' are in the same document. Stop.

(2) If $I_w^p[2] < I_{w'}^q[1]$, move to I_w^{p+1} if $p < k$ (then, in a next step, we will check I_w^{p+1} against $I_{w'}^q$).

(3) If $I_w^p[1] > I_{w'}^q[2]$, move to $I_{w'}^{q+1}$ if $q < k'$ (then, in a next step, we will check I_w^p against $I_{w'}^{q+1}$).

(4) If $I_w^p \not\subset I_{w'}^q$, and $i = k$ or $j = k'$, report that w and w' are not in the same document. Stop.

The above process is referred to as a *two-word checking*, in which each interval in L_w and $L_{w'}$ is accessed only once. So only $O(|L_w| + |L_{w'}|)$ time is required. In Fig. 4, we illustrate the working process to check whether two words d and m are in a same document shown in Table 1.

In Fig. 4, we first notice that $L_d = [6, 7][10, 12]$ and $L_m = [1, 2][10, 11]$. In the 1st step, we will check $L_d^1 = [6, 7]$ against $L_m^1 = [1, 2]$. Since $L_d^1[1] = 6 > L_m^1[2] = 2$, we will check L_d^1 against $L_m^2 = [10, 11]$ in a next step, and find $L_d^1[2] = 7 < L_m^2[1]$. So we will have to do the third step, in which we will check $L_d^2 = [10, 12]$ against L_m^2. Since $L_d^2 \supset L_m^2$, we get to know that d and m are in the same document.

201

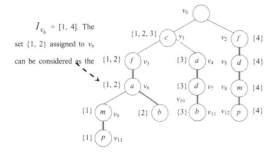

$I_{v_6} = [1, 4]$. The
set $\{1, 2\}$ assigned to v_6
can be considered as the $\{1, 2\}$

Figure 5. Illustration for assignment of document identifiers.

What we want is to extend this process to check whether a set of words is in the same document, based on which an efficient evaluation of conjunctive queries can be achieved. We will address this issue in Section 3.

2.2 Assignment of docID to intervals

Another important component of our index is to assign document identifiers to intervals. An interval I can be considered as a representative of some words, i.e., all those words appearing on a *prefix* in the trie, which is a path P from the root to a certain node that is labeled with I. Then, the document identifiers assigned to I should be those containing all the words on P. For example, the words appearing on the prefix: $v_1 \rightarrow v_3 \rightarrow v_6$ in the trie shown in Fig. 1 are words c, f, and a, represented by the interval $[1, 4]$ associated with v_6. So, the document identifiers assigned to $[1, 4]$ should be $\{1, 2\}$, indicating that both documents D_1 and D_2 contain those three words. See the trie shown in Fig. 5 for illustration, in which each node v is assigned a set of document identifiers that is also considered to be the set assigned to the interval associated with v.

Let v be the ending node of a prefix P, labeled with I. We will use $\delta(I)$, interchangeably $\delta(v)$, to represent the set of document identifiers containing the words appearing on P. Thus, we have $\delta(v_6) = \delta([1, 4]) = \{1, 2\}$.

Lemma 2 Let u and v be two nodes in a trie T. If u and v are not on the same path in T, then $\delta(u)$ and $\delta(v)$ are disjoint, i.e., $\delta(u) \cap \delta(v) = \Phi$.

Proposition 1 Assume that $v_1, v_2, \ldots, v_j$ are all the nodes labeled with the same word w in T. Then, $\delta(w)$, the inverted list of w (i.e., the list of all the documents identifiers containing w) is equal to $\delta(v_1) \uplus \delta(v_2) \uplus \cdots \uplus \delta(v_j)$, where $\uplus$ represents *disjoint union* over disjoint sets that have no elements in common.

Proof. Obviously, $\delta(w)$ is equal to $\delta(v_1) \cup \delta(v_2) \cup \cdots \cup \delta(v_j)$. Since $v_1, v_2, \ldots, v_j$ are labeled with the same word, they definitely appear on different paths as no nodes on a path are labeled with the same word. According to Lemma 2, $\delta(v_1) \cup \delta(v_2) \cup \cdots \cup \delta(v_j)$ is equal to $\delta(v_1) \uplus (v_2) \uplus \cdots \uplus \delta(v_j)$.

As an example, see the nodes v_2 and v_3 in Fig. 5. Both are labeled with word f. So the inverted list of f is $\delta(v_2) \uplus \delta(v_3) = \{4\} \uplus \{1, 2\} = \{1, 2, 4\}$.

3 QUERY EVALUATION

Based on the interval sequences associated with words and the lists of document identifiers with intervals, we design our algorithm for evaluating conjunctive queries.

Let $Q = \{w_1, w_2, \ldots, w_l\}$ be a set of words. Without loss of generality, assume that $w_1 \prec w_2 \prec \cdots \prec w_l$. We will check whether $w_1, w_2, \ldots, w_l$ are in the same document. For this purpose, we need to check whether there exists an interval sequence $I = I_1, I_2, \ldots, I_l$ such that $I_j \in L_{w_j}$ and $I_j \supset I_{j+1} (1 \leq j \leq l)$, where $I_{l+1} = \phi$, representing an empty interval. If I exists, then, the document identifiers associated with I_l is part of the answer

We call I a *containment sequence*.

Lemma 3 Let $Q = \{w_1, w_2, \ldots, w_l\}$ with $w_1 \prec w_2 \prec \cdots \prec w_l$. Denote by I_j an interval in $L_{w_j} (1 \leq j \leq l)$. If for some $1 \leq i < j \leq l$ we have $I_i \supset I_l$ and $I_j \supset I_l$, then $I_i \supset I_j$.

As an example, consider $Q = \{f, a, p\}$ with $f \prec a \prec p$. From Fig. 2, we can see that $L_f = [1, 5][10, 13]$, $L_a = [1, 4][6, 8]$, and $L_p = [1, 1][10, 10]$. Obviously, $I_f^1 = [1, 5] \supset I_p^1 = [1, 1]$, and $I_a^1 = [1, 4] \supset I_p^1 = [1, 1]$. Then, we must have $I_f^1 \supset I_a^1$.

According to the above lemma, the checking of $I_{j+1} \subset I_j$ can be replaced by checking whether we have $I_{j+1} \supset I_l$ if we know $I_j \supset I_l$. Thus, the task to find a containment sequence can be done by slightly modifying step (1) in the two-word checking discussed in 2.1. That is, each time we find $p, q (1 \leq p \leq |L_{w_{l-1}}|, 1 \leq q \leq |L_{w_l}|)$ such that $I_{w_{l-1}}^p \supset I_{w_l}^q$, we need only to further check whether there exist $l - 2$ other intervals $I_1, I_2, \ldots, I_{l-2}$ such that each I_j is in L_{w_j} and $I_j \subset I_{w_l}^q$ for $1 \leq j \leq l - 2$. This will greatly simplify the process for finding a containment sequence.

For this purpose, we define an operation $con(w, I, j)$ to check whether an interval I is contained in some interval between jth and the last interval in L_w. If I is contained in an ith interval in L_w, the return value of $con(w, I, j)$ is a pair $(true, i)$; otherwise, the return value is $(false, i')$, where i' is the least number such that $I_w^{i'}[1] > I[2]$. In addition, to simplify the control process, we place a *sentinel* at the end of L_w, whose value is set to be $[\infty, \infty]$ so that whenever we reach the sentinel of L_w, $con(w, I, j)$ returns $(false, |L_w| + 1)$.

This operation will be used in the following algorithm, by which we will check, for a set $Q = \{w_1, w_2, \ldots, w_l\}$ with $w_1 \prec w_2 \prec \cdots \prec w_l$, whether each $L_{w_j} (1 \leq j \leq l)$ possesses an interval which contains a given interval I.

The input of this algorithm is a triplet (Q, I, b), where b is an integer array of length $|Q|$ with each $b[j]$ indicating the starting position to check $L_{w_j} (1 \leq j \leq l)$.

202

For example, if $b[i] = 2$ for some i, we will check L_{w_i} starting from $I_{w_i}^2$. Initially, each entry in b is set to be 1. We also store Q as an array. Then, $Q[i]$ refers to w_i for $1 \leq i \leq l$.

ALGORITHM *interval-check*(Q, I, b)

---

begin
1. *mark* := *true*; $j := |Q|$; assume that $Q[1] \prec Q[2] \prec \cdots \prec Q[l]$;
2. **while** (*mark* = *true* and $j \geq 1$) **do** {
3. $(x, y) := con(Q[j], I, b[j])$; $b[j] := y$; /*$Q[j] = w_j$*/
4.   **if** $(x = true)$ **then** $j := j - 1$
5.   **else** {*mark* := *false*; }
6. }
7. **if** (*mark* = *true*) **then**   return (*true*, b)
8. **else** return (*false*, b);
end

The output of the algorithm is a pair (t, b'). If in each $L_{w_j} (1 \leq j \leq l)$ we can find an interval that contains I, t is *true*; otherwise, t is *false*. b' is an array satisfying the following properties:

(i) If t is *true*, each $b'[j]$ is an integer i showing that it is the ith interval in L_{w_j} that contains I.

(ii) If t is *false*, there exists j dividing b into three parts: $b_2[1 \ldots j - 1]$, $b'[j]$, and $b'[j + 1 \ldots l]$ such that for any index k,

  1. If $j + 1 \leq k \leq l$, then $b'[k]$ is an integer i such that ith interval in L_{w_k} contains I.
  2. If $k = j$, then in L_{w_k} no interval is able to contain I and $b'[k]$ is $|L_{w_k}| + 1$ or a least number i such that $I_{w_k}^i[1] > I[2]$.
  3. If $1 \leq k \leq j - 1$, then $b'[k]$ is the same as $b[k]$ (see line 5; the execution of this line will enforce the control to get out of the **while**-loop, and leave $b[1 \ldots j - 1]$ not updated.

Based on the above algorithm, we design a process to find a containment sequence, as shown below.

By repeatedly applying the algorithm, we can get all the possible containment sequences, and then the whole answer to a conjunctive query.

4 CONCLUSION

In this paper, a new method is discussed to evaluate conjunctive queries. The main idea is to transform

---

ALGORITHM *containment*(Q, B)

---

begin
2. let $|Q| = l$; assume that $Q[1] \prec Q[2] \prec \cdots \prec Q[l]$;
3. $R := \{Q[1], \ldots, Q[l - 2]\}$;
3. $p := b[l - 1]$; $q := b[l]$;
4. **while** $(p \leq |L_{Q[l-1]}|)$ and $q \leq |L_{Q[l]}|)$ **do** {
5.   **if** $L_{Q[l-1]}^p \supset L_{Q[l]}^q$ **then** {
6. $(x, b) := interval\text{-}check(R, L_{w[l]}^q, b)$;
7.   **if** $(x = true)$ **then** return (*true*, b);}
8.   **else** {$q := q + 1$; $b[l] := q$;}
9.   }
10.   **else** {
11.   **if** $(L_{Q[l-1]}^p[2] < L_{Q[l]}^q[1])$ **then** {$p := p + 1$; $b[l - 1] := p$; }
12.   **else** {$q := q + 1$; $b[l] := q$;}
13.   }
14. }
15. return (*false*, b);
end

---

an evaluation of queries to a series of reachability checkings, which improves the traditional method by a magnitude of order or more.

REFERENCES

[1] Anh, V.N. and A. Moffat, A, 2005. Inverted index compression using word-alinged binary codes, *Kluwer Int. Journal of Information Retrieval 8*, 1, pp. 151–166.
[2] Chen, Y. and Chen, Y.B. 2006. On the Signature Tree Construction and Analysis, *IEEE TKDE*, Vol. 18, No. 9, pp. 1207–1224.
[3] Chen, Y. 2004. Building Signature Trees into OODBs, *Journal of Information Science and Engineering*, 20, 275–304 (2004).
[4] Faloutsos, C. 1985. Access Methods for Text, *ACM Computing Surveys*, vol. 17, no. 1, pp. 49–74.
[5] Faloutsos, C. and Chan, R. 1988. Fast Text Access Methods for Optical and Large Magnetic Disks: Designs and Performance Comparison, *Proc. 14th Int'l Conf. Very Large Data Bases*, pp. 280–293.
[6] Knuth, D.E. 1975. *The Art of Computer Programming*, Vol. 3, Massachusetts, Addison-Wesley Publish Com., 1975.
[7] Lempel, R. and Moran, S. 2003. Predictive caching and prefetching of query results in search engines, in *Proc. the World Wide Web Conf.*, Budapest, Hungary, ACM, 19–28, 20.

Future Communication, Information and Computer Science – Zheng (Ed.)
© *2015 Taylor & Francis Group, London, 978-1-138-02653-7*

Phylogenetic tree constructing algorithm based on K-means quantum clustering

T. Chen

Hunan mobile communication company, Yongzhou, China

ABSTRACT: Phylogenetic tree is the key expression of the evolutionary relationships among species. Clustering analysis is a significant method to construct phylogenetic trees based on gene expression data. In this paper, we propose a K-means clustering algorithm based on quantum mechanics, namely K-means quantum clustering (KQC) algorithm. This method measures different species based on the species' features extracted from their DNA sequence statistical data, and utilizes KQC algorithm to cluster the species' DNA sequence features, constructs phylogenetic tree by dynamic clustering process. Experimental results demonstrate that the phylogenetic tree constructing process based on KQC is more accurate, robust and efficient.

Keywords: phylogenetic tree; K-means clustering; quantum clustering; KQC clustering

1 INTRODUCTION

With the completion of the Human Genome Project, molecular biology and computational biology have got a blooming development. Biology species can be traced back to the collective ancestors and their classification relationships can be expressed by a tree diagram; thus the phylogenetic tree is commonly used to represent evolutionary relationships among species. As the basis of the computational analysis way on biology data, phylogenetic analysis has become an important issue in bioinformatics. Phylogenetic trees can be used in diverse applications such as multiple sequence alignment [1], gene function prediction [2] and drug design [3]. It can explain the evolutionary mechanisms and histories legibly, and biomedicine can also benefit from phylogenetic trees based on DNA or protein sequences. With respect to SARS virus, phylogenetic tree played an important role in finding its origin and prevent the SARS virus [4].

According to the type of processed data, the method for obtaining the phylogenetic tree can be roughly divided into two categories:

The first category is based on distance measurement. In the following, we will discuss the most significant methods belonging to this category; UPGMA [5] is one of the most popular algorithms; it is based on the assumption that the mutation rate, of all branches along the tree, is equal. Therefore, this method always draws the wrong topology when there is a large different evolutionary rate among the different branches or the parallel evolution of homologous sequences.

The second category uses the species' characteristics data expressed by discrete states, such as specific loci of nucleotides in DNA sequence. Methods belonging to this category analyze the evolutionary relationships of taxonomical unit or each feature among the sequence (such as nucleotide loci) emphatically. Maximum parsimony method [6] is one of the typical methods, but this method is likely to yield a wrong topology tree when the number of nucleotide substitutions in the sequence unit loci is relatively large.

At present, the construction of phylogenetic tree mainly depends on the characteristics of the data and that demands a massive data transformation and analysis computation. Thus, it is significant to design a general and efficient method to deal with biology species processing, clustering analysis and construction of the phylogenetic tree. In this paper, we extract the data characteristics by using a statistical characteristic method, and propose a K-means clustering algorithm based on quantum mechanism to construct and analyze the phylogenetic tree by handling species' characteristics data.

2 RELATED WORK

Clustering analysis is an unsupervised pattern recognition technique which divides unmarked models into several subsets by some kind of criterium. It requires the similar samples to be classified in the same category, and it is also known as unsupervised classification.

2.1 *Quantum mechanics*

Volatility of microscopic particles in Quantum mechanics reflects its movement statistical laws, this kind of volatility is called matter wave, also known as probability wave. Microscopic particles or particle system state often referred to as quantum state, the space-time movement state of the microscopic particles described by a nonlinear complex function $\varphi(x)$, called wave function. The wave function is a description of the particle quantum state, and the Schrodinger equation, at equation 2.1, solves the potential constrained wave function.

$$H\varphi = (-\frac{\sigma^2}{2m}\nabla^2 + V(x))\varphi(x) = E \qquad (2.1)$$

Here, wave function $\varphi(x)$ is an equation solution, H is the Hamilton operator, m is particle mass, ∇^2 is the Laplace operator, the former half of H is the kinetic energy of the particle, the latter part of it is potential energy V, Equation (2.1) is the eigen equation of the wave function, and E is the Energy eigenvalues of H.

Wave function portrays the motion and the character of distribution of the microscopic particles in the energy field. The potential energy in the energy field is the key factor that determines this distribution. This potential energy function ultimately determines the spatial and temporal distribution of the particles. From equation (2.1), it is obvious that the same potential fields are the same as the distribution state of the particles are. When the spatial distribution of the particles shrinks and changes to the one-dimensional infinite potential well, the particles gather in a zero potential energy and certain width potential well.

According to the theory of quantum mechanics, for one-dimensional and single sample, the quantum potential energy of the particles $V = (x - x_1)^2/2\sigma^2$, the minimum eigenvalue of equation (2.1) is $E = 1/2$, obviously, for d dimensional the minimum eigenvalue is $E = d/2$.

For a general situation, according to $\varphi(x) = e^{-(x-x_i)^2/2\sigma^2}$, we can learn that:

$$V(x) = E + \frac{\frac{\sigma^2}{2}\nabla^2\varphi}{\varphi} = E - \frac{d}{2} + \frac{1}{2\sigma^2\varphi}\sum_i (x-x_i)^2 e^{-\frac{(x-x_i)^2}{2\sigma^2}} \quad (2.2)$$

In general, the potential energy of any particle potential energy field can be positive, and the minimum value is 0, by specifying zero potential surface. We can get equation (2.3) from equation (2.2).

$$E = -\min(\frac{\sigma^2/2\nabla^2\varphi}{\varphi}) \qquad (2.3)$$

Hence, for a set of sample collections that have d dimensional scale space $x = \{x_1, x_2, \ldots, x_i, \ldots x_m\} \subset R^d, x_i = \{x_{i1}, x_{i2}, \ldots, x_{ij}, \ldots x_{id}\} \subset R^d$, clustering problem is transformed into using the Gaussian wave packet, which width is σ, to describe the distribution of sample points.

David put forward a clustering algorithm (QC) that is based on the quantum mechanism, using gradient descent method, solving particle potential energy through some learning rate, taking C minimum values (C for clustering number) of quantum potential

as clustering center, then using a fixed metrics for samples division.

Before the samples division, each subschema (classification) clustering center must be found. Therefore, David's algorithm is not sensitive to the initial data but robust. However, the algorithm will fall into local optimum easily for using a gradient descent method. The computation efficiency of the algorithm is low.

2.2 K-means clustering algorithm

In 1967, MacQueen firstly proposed K-means clustering algorithm. The main idea is to find C clustering centers $c_1, c_2, \ldots c_v, \ldots c_C$, making the sum of squared distance between each data point x_i and its nearest clustering center c_v minimum.

It is an effective classification method for a large dataset, faster than hierarchical clustering algorithms. Nevertheless, the algorithm is only suitable for numerical data clustering and sensitive to the initial clustering center, it lacks robustness.

Since the K-means algorithm is sensitive to the initial clustering center, combining with the characteristic that QC is insensitive to initial data, we propose a K-means clustering algorithm based on quantum mechanism.

3 K-MEANS QUANTUM CLUSTERING ALGORITHM

3.1 The proposed algorithm main process

Here, we propose a novel clustering algorithm and apply it to construct a phylogenetic tree. Distance is utilized to measure the similarity between two samples, as the sample units of each dimension is not uniform, in order to eliminate the impact of dimension, this paper adopted Mahalanobis distance to measure the similarity. Mahalanobis distance is defined as follows:

$$d(x_i, x_j) = \sqrt{(x_i - x_j)^T S^{-1}(x_i - x_j)}, (i \neq j, 1 \leq i, j \leq m) \quad (3.1)$$

Here, S is the covariance matrix of the sample.

The KQC algorithm is listed in algorithm 3.1.

Two parameters need to be adjusted in the algorithm: σ and ε, the parameter Gaussian kernel bandwidth is an empirical value that needs to be tested many times, meet $\sigma \in [0, 2]$, ε is an adjustable parameter for precision.

3.2 Analysis of the KQC algorithm's efficiency and accuracy

This paper adopted Iris as a test dataset; it is a commonly used and well known dataset. Its clustering results are reliable, and the classification is consistent.

Iris contains 3 classes 4-dimensional 150 samples: setosa, versicolor and virginica three types, each of them contains 50 samples, and each category represents a type of irises. One class with the other two

Algorithm 3.1.   KQC algorithm.

KQC(data, σ, ε, C)

Input:	*data* – original sample *σ* – controlled parameter *ε* – precision *C* – clustering number
Output:	Clustering results
Step1:	Preprocessing of original sample, Clustering pointer $c = 0$;
Step2:	Calculate the potential energy of the sample according to (3.1), (2.2),
Step3:	Clustering pointer $c = c + 1$, if $c \geq C$, turn to step6;
Step4:	According to the potential energy of the sample, find the smallest potential point in samples which are not clustered and use it as cluster centroid G_c of class c;
Step5:	Select non clustered sample x_i, calculate the distance between it and clustering centroid, if $d(x_i, G_j) < \varepsilon$, $1 \leq i \leq m$, $1 \leq j \leq c$, (here , c is for all existing centroid), and it is the minimum distance, then, cluster it into class j;
Step6:	Repeat Step5 until the samples did not meet the conditions, turn to Step3;
Step7:	Utilize K-means method recalculating the clustering centers, getting C clustering centers: $G_1, G_2, \ldots, G_C$.
Step8:	For each sample, calculate the distance between the sample and new clustering centers, if $d(x_i, G_j) < \frac{\varepsilon}{2}$, cluster x_i into G_j class until the sample sets are empty;
Step9:	Repeat Step7-Step8 until centers no longer changes. End

---

Table 1.   Comparison of three algorithms.

Algorithm	Average accuracy of running 30 cycles (%)	Average running time(S)
KQC	82.67	0.2609
QC	82.00	2.3750
K-mean	76.87	0.0037

types are linearly separable, while the other two types are not.

Introducing a concept of clustering accuracy for measurement to verify the effect of the proposed algorithm, the clustering accuracy μ is referred to the ratio of the right clustering samples with the total number of samples set. As shown in equation (3.2).

$$\mu = \frac{\sum_{i=1}^{C} rightNo_i}{m} \times 100\% \qquad (3.2)$$

Here, $rightNo_i$ is the right clustering number of the samples in class i; m is the total number of samples. We compared KQC with the QC algorithm and K-means algorithm to illustrate the performance of KQC algorithm.

Before clustering, we utilized SVD (singular value decomposition) and Normalization to process the original samples. Then, we used three algorithms QC, K-means, and KQC to make 30 time tests for this dataset. The experiment platform consists of MATLAB7.0, computer configuration is Pentium 4 CPU 3.00 GHz, memory is 512 M and Microsoft windows XP. The results are shown in Table 1.

It is evident from the above experiments that QC algorithm has better clustering results than K-means for a good clustering structure of the Iris dataset, but the running time is much longer than both K-means and KQC. The running time of K-means algorithm

is short, but the clustering result is unsatisfactory. It's clear that the KQC algorithm combines the advantages of QC and K-means. KQC shows a good performance in both clustering results and running time.

K-means clustering algorithm has a good computing speed. The efficiency of the algorithm is high but the result is not stable. The running time of QC algorithm is longer than K-means algorithm. KQC algorithm integrated the advantages of the above two algorithms; it has a good adaptability to the initial value of clustering center, and the computing speed is faster than QC algorithm.

4   PHYLOGENETIC TREE CONSTRUCTION METHOD BASED ON KQC

4.1   Statistical characteristics method to obtain 10-dimension dataset

This method can be divided into two phases: Firstly, we calculate the characteristics of the four nucleotides from the whole DNA sequence. Secondly, we divide DNA sequence into several segments; we obtain the eigenvalue from each subsequence, and get a 10-dimensional vector from these eigenvalues.

Provided a length L of DNA sequence, the specific steps are as follows:

Step1:	Count the occurrence frequency of four bases in the entire sequence **A,T,G,C**, and marked for P_A, P_T, P_G, P_C
Step2:	Divide the DNA sequence into n subsequences, each section contains a $= (L/n)$ bases.
Step3:	Calculate the occurrence probabilities of the four bases in the n subsequences, and marked for $P_i^A, P_i^T, P_i^G, P_i^C$ respectively, where i represents the i-subsequences, **A,T,G,C** respectively represents the corresponding four bases.
Step4:	Calculate the related factor of the four bases in the n subsequences.

Step 4 computation formula is as follows

$$
\begin{cases}
\theta_1 = \dfrac{1}{n} \sum\limits_{i=1}^{n} \left\| P_i^{\ A} - P_i^{\ G} \right\| \\[8pt]
\theta_2 = \dfrac{1}{n} \sum\limits_{i=1}^{n} \left\| P_i^{\ A} - P_i^{\ C} \right\| \\[8pt]
\theta_3 = \dfrac{1}{n} \sum\limits_{i=1}^{n} \left\| P_i^{\ A} - P_i^{\ T} \right\| \\[8pt]
\theta_4 = \dfrac{1}{n} \sum\limits_{i=1}^{n} \left\| P_i^{\ G} - P_i^{\ C} \right\| \\[8pt]
\theta_5 = \dfrac{1}{n} \sum\limits_{i=1}^{n} \left\| P_i^{\ G} - P_i^{\ T} \right\| \\[8pt]
\theta_6 = \dfrac{1}{n} \sum\limits_{i=1}^{n} \left\| P_i^{\ C} - P_i^{\ T} \right\|
\end{cases}
\qquad (4.1)
$$

θ_1 represents a related factor of the **A** and **G**, θ_2 represents a related factor of the **A** and **C**, θ_3 represents a related factor of the **A** and **T**, θ_4 represents a related factor of the **C** and **G**, θ_5 represents a related factor of the **T** and **G**, θ_6 represents a related factor of the **T** and **C**. The related two bases are not in a particular order, and every base's factor associated with its own is constant zero, so there are only six factors. The 6 related factors and the probability of the four bases in the DNA segments P_A, P_T, P_G, P_C, can be composed of a 10-dimensional dataset.

This method adds six bases related factor, contains more sequence impact information than the traditional four-dimensional data.

4.2 Construct the 16-dimensinal data

The information theory was proposed and developed in 1948 by Shannon. It is mainly used to improve the reliability and effectiveness of the information system to achieve system optimization. The theory to measure the amount of information through the uncertainty occurred after communication of the various symbols of the signal source. The information of the signal source has an uncertainty, and the bigger the uncertainty is, the more information it owns. The main formula used to calculate the variables is as follows:

Self-information:

$$I(x) = -\log[p(x)] \qquad (4.2)$$

Conditional entropy:

$$H(X \mid Y) = \sum_{x,y} p(x,y) \log \frac{1}{p(x \mid y)} \qquad (4.3)$$

Mutual information:

$$I(X;Y) = I(X) - I(X \mid Y) \qquad (4.4)$$

This method uses the information theory to analyze the bases and duplex amino acids in DNA. First, we consider the single base and duplex amino acids as event probability. Then, we obtain information entropy of the single base, and use the occurrence probability of the 16 duplex amino acids as the conditional probability we can obtain from the conditional entropy and

mutual information entropy from the operation above. In addition, we get a feature vector value of DNA sequence which is composed of 16 mutual information entropies.

Step1:	Calculate the occurrence probability of 4 kinds of bases in the DNA sequence, and marked for P_A, P_T, P_G, P_C. Continue calculating the occurrence probability of duplex amino acids in DNA sequence, and marked for P_{AA}, P_{AT}, P_{AG}, P_{AC}, P_{TA}, P_{TT}, P_{TG}, P_{TC}, P_{GA}, P_{GT}, P_{GG}, $P_{GC}, P_{CA}, P_{CT}, P_{CG}, P_{CC}$
Step2:	Compute the self-information of 4 bases by formula (4.2);
Step3:	Use the occurrence probability of the 16 duplex amino acids to compute conditional entropy, and calculate the conditional probability of duplex amino acids by formula (4.3).
Step4:	Compute the 16 mutual information entropies of the 4 bases by formula (4.4). For respectively: *I(A;A), I(A;T), I(A;G), I(A;C), I(T;A), I(T;T), I(T;G), I(T;C), I(G;A), I(G;T), I(G;G), I(G;C), I(C;A), I(C;T), I(C;G), I(C;C)*

Algorithm steps are as above:

Thus, we got a 16-dimensional mutual information entropy to express the eigenvector of DNA sequence, and it contained more information.

5 EXPERIMENTAL RESULTS AND DISCUSSION

5.1 The experimental dataset processing

In this data, we selected the mitochondrial DNA sequence of eight species as the research objects. The data are selected from the GenBank database (http://www.ncbi.nlm.nih.gov/), species' names and sequences number are shown in Table 2.

10-dimensional data gained by using the algorithm number were shown in Table 3.

The construction of the 16-dimensional data set by Sequence statistical characteristics method based on information theory, is shown in Table 4.

5.2 Experiment results

According to KQC algorithm, combined with the above data to construct the phylogenetic tree, the output results were obtained as follows:

The output of the program running on 10-d data, as shown in Table 5:

In Table 5, from the second to the eighth column are respectively output results of one clustering center to seven clustering centers.

We can construct a phylogenetic tree by Table 5, as shown in Figure 1.

The output of the program running on 16-d data, as shown in Table 6:

We can construct another phylogenetic tree by Table 6 as shown in Figure 2.

Table 2. Related information of species name and sequences number.

N	Species cientifi	abbrevia	accession	Lengt
1	Papio amadrya	baboon	Y18001	1652
2	Hylobates lar	gibbon	X99256	1647
3	Pongo ygmaeus	oranguta	D38115	1638
4	Gorilla gorilla	gorilla	D38114	1656
5	Pan troglodytes	c.chimp	D38116	1656
6	Pan paniscus	p.chimp	D38113	1655
7	Equus caballus	horse	X79547	1666
8	Ceratotherium	w.rhinoc	Y07726	1683

Table 3. 10-dimensional feature vectors.

speci	P_A	P_T	P_G	P_C	θ_1	θ_2	θ_3	θ_4	θ_5	θ_6
babo	0.3	0.13	0.30	0.2	0.0379	0.0043	0.0084	0.0359	0.0162	0.1
gibb	0.3	0.13	0.31	0.2	0.0326	0.0037	0.0079	0.0378	0.0132	0.1
oran	0.3	0.13	0.32	0.2	0.0336	0.0039	0.0076	0.0430	0.0133	0.1
gorill	0.3	0.13	0.30	0.2	0.0353	0.0036	0.0069	0.0357	0.0170	0.1
c.chi	0.3	0.12	0.30	0.2	0.0381	0.0041	0.0071	0.0379	0.0183	0.1
p.ch	0.3	0.12	0.30	0.2	0.0365	0.0040	0.0071	0.0371	0.0175	0.1
horse	0.3	0.13	0.28	0.2	0.0398	0.0069	0.0074	0.0272	0.0174	0.1
w.rhi	0.3	0.12	0.27	0.2	0.0472	0.0070	0.0106	0.0275	0.0183	0.1

Table 4. 16-dimensional feature vectors.

speci	I(A;A) I(A;T) I(A;G) I(A;C) I(T;A) I(T;T) I(T;G) I(T;C) I(G;A) I(G;T) I(G;G) I(G;C) I(C;A) I(C;T) I(C;G) I(C;C)
baboo	0.291194 0.345633 0.295946 0.315204 0.228830 0.253185 0.226527 0.243337 0.292813 0.349857 0.291369 0.310051 0.280271 0.331522 0.289643
gibbo	0.290425 0.345603 0.291154 0.317505 0.230745 0.260468 0.232643 0.245513 0.294910 0.347382 0.292925 0.315089 0.285311 0.326633 0.282314 0.300203
orang	0.293568 0.343359 0.290988 0.317569 0.229779 0.254863 0.222832 0.246431 0.296859 0.352294 0.284435 0.314644 0.281348 0.326948 0.279639 0.303134
goril	0.292988 0.343870 0.297363 0.313129 0.229209 0.254775 0.226726 0.242752 0.295300 0.350153 0.290037 0.309274 0.282900 0.332404 0.288731 0.304036
c.chi	0.291677 0.345285 0.297579 0.313485 0.224360 0.250661 0.222112 0.238269 0.294597 0.350713 0.289405 0.309457 0.280893 0.334149 0.289905 0.303124
p.chi	0.292402 0.344777 0.297008 0.313457 0.225553 0.252184 0.223472 0.240842 0.295615 0.350585 0.289366 0.309257 0.281361 0.333325 0.289112 0.303586
horse	0.289985 0.344896 0.305603 0.309997 0.226920 0.256833 0.233429 0.243931 0.287827 0.345845 0.296957 0.304343 0.281376 0.334100 0.294377 0.305397
w.rhi	0.286804 0.347664 0.305178 0.310677 0.221391 0.252192 0.228710 0.239921 0.281120 0.344466 0.299286 0.305392 0.275795 0.335244 0.298116 0.303544

The phylogenetic tree constructed by DRAW-GRAM program is shown in Figure 3.

5.3 Discussion

It is clear from table 5 and table 6: Swing phenomenon appeared in the 16-dimensional data. Gibbon and orangutan clustered together first, then separated, but the structure of the phylogenetic tree was very clear. Compared with figure 3, we can find that the trees constructed by our method as shown in figure 1 and figure 2 are very similar to the phylogenetic tree constructed by DRAWGRAM program in the PHYLIP package (http://evolution.genetics.\\washington.edu/phylip.html).

Through the comparative analysis, obviously, the KQC method makes the constructing process of the phylogenetic tree more robust, faster and efficient. The KQC algorithm utilizes a quantum mechanical algorithm to calculate the initial clustering centers, and uses these clustering centers as the initial centers of K-means clustering algorithm. This is to avoid the jitter caused by the stochastic clustering centers of

Table 5. The output of 10-dimensional data.

babo	gibb	oranguta	goril	c.chimpa	p.chimp	ho	w.rhi
2	6	6	1	1	1	5	7
2	6	6	1	1	1	5	5
3	1	1	2	2	2	5	5
3	1	1	2	2	2	4	4
2	1	1	2	2	2	3	3
1	1	1	1	1	1	2	2
1	1	1	1	1	1	1	1

Figure 1. Construction of phylogenetic tree by 10-dimensional data $\sigma = 0.5$, $\varepsilon = 0.1$.

Table 6. The output of 16-dimensional data.

babo	gibb	oranguta	goril	c.chimpa	p.chimp	ho	w.rhi
2	6	6	1	1	1	5	7
2	6	6	1	1	1	5	5
2	4	1	1	1	1	5	5
3	2	1	1	1	1	4	4
2	1	1	2	2	2	3	3
1	1	1	1	1	1	2	2
1	1	1	1	1	1	1	1

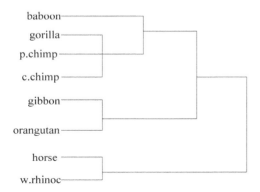

Figure 2. Construction of phylogenetic tree by 16-dimensional data $\sigma = 0.86$, $\varepsilon = 0.5$.

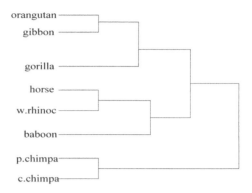

Figure 3. Phylogenetic tree constructed by NEIGHBOR program in the PHYLIP.

K-means clustering algorithm. Meanwhile, it accelerates the speed of the Quantum clustering algorithm.

6 CONCLUSIONS

Firstly, we analyzed the advantages and disadvantages of QC algorithms and K-means clustering algorithms. Then, we proposed a novel algorithm depending on KQC algorithm. The proposed algorithm dealt with the 10-dimensional and 16-dimensional data by using KQC algorithm. Finally, we constructed two phylogenetic trees according to the results, and realized the algorithm's high speed and efficiency for constructing the phylogenetic trees. It is concluded that the thought of using the KQC clustering method to construct the phylogenetic tree is available.

REFERENCES

[1] Löytynoja, A. J. Vilella, and N. Goldman, "Accurate extension of multiple sequence alignments using a phylogeny-aware graph algorithm," *Bioinformatics*, vol. 28, no. 13, pp. 1684–1691, 2012.
[2] H. Mi, A. Muruganujan, and P. D. Thomas, "PANTHER in 2013: modeling the evolution of gene function, and other gene attributes, in the context of phylogenetic trees," *Nucleic acids research*, vol. 41, no. D1, pp. D377–D386, 2013.
[3] E. van der Horst, J. E. Peironcely, et al., "A novel chemogenomics analysis of G protein-coupled receptors (GPCRs) and their ligands: a potential strategy for receptor de-orphanization," *BMC bioinformatics*, vol. 11, no. 1, p. 316, 2010.
[4] E. J. Snijder, P. J. Bredenbeek et al., "Unique and conserved features of genome and proteome of SARS-coronavirus, an early split-off from the coronavirus group 2 lineage," *Journal of molecular biology*, vol. 331, no. 5, pp. 991–1004, 2003.
[5] C. D. Michener, and R. R. Sokal, "A quantitative approach to a problem in classification," *Evolution*, vol. 11, no. 2, pp. 130–162, 1957.
[6] J. Sourdis, and M. Nei, "Relative efficiencies of the maximum parsimony and distance-matrix methods in obtaining the correct phylogenetic tree," *Molecular Biology and Evolution*, vol. 5, no. 3, pp. 298–311, 1988.

Future Communication, Information and Computer Science – Zheng (Ed.)
© 2015 Taylor & Francis Group, London, 978-1-138-02653-7

The multiresolution analysis with radial symmetry elements

A.G. Shoberg & S.V. Sai
Pacific National University, Khabarovsk, Russia

ABSTRACT: Wavelet based multiresolution signal decomposition is effective for analyzing the information of different signals. In this paper we propose radial symmetric scheme wavelet signals decomposition. This scheme allows to restore the signal independent of process direction. It's considered the first transform level and Daubechies basis usage.

Keywords: Wavelet transform, digital signal processing, image coding, symmetry, invariance

1 INTRODUCTION

The transform of a signal (one- or multi-dimensional) is a new representation of that signal. Wavelet model can be easily generalized to any dimension. In many cases we can use serial one-dimensional signal transformation execution (Daubechies, 1992). Wavelet transforms are computationally efficient (Mallat, 1989). Wavelets also usually allow exact reconstruction of the original signal. Wavelet transform and subband coding are two very similar methods. An original signal is usually divided in two sub bands, each having half the size in comparison with the original. The transform result storage consists of a vector of low-frequency and high-frequency components which are written sequentially. A recursive wavelet transform computation is based on pyramidal algorithm use convolutions with quadrature mirror filters (Strang & Nguyen, 1997).

2 SYMMETRY MULTIRESOLUTION ANALYSIS BASICS

The Haar basis is a very simple example of wavelet transform and filter bank implementations. The basic element is a two-point average and difference operation. One channel computes the average and an other the difference of two successive signal samples (Mallat, 1989). The traditional scheme can be described in some matrix multiplications (Welstead, 1999). The $n-1$ transformation level can be presented in matrix form

$$H_n x = \left[\frac{A_n}{D_n}\right] x = \left[\frac{a_{n-1}}{d_{n-1}}\right], \qquad (1)$$

where A_n and D_n = parts square blocks matrix H_n with size $2^n \times 2^n$; a_{n-1} and d_{n-1} = average and difference parts of decomposition column vector, where each

has a length of 2^{n-1}. The H matrix with Daubechies coefficients length 4 (named Db2) has some kind

$$H = \begin{bmatrix} c1 & c2 & c3 & c4 & 0 & 0 & 0 & 0 \\ 0 & 0 & c1 & c2 & c3 & c4 & 0 & 0 \\ 0 & 0 & 0 & 0 & c1 & c2 & c3 & c4 \\ c3 & c4 & 0 & 0 & 0 & 0 & c1 & c2 \\ c4 & -c3 & c2 & -c1 & 0 & 0 & 0 & 0 \\ 0 & 0 & c4 & -c3 & c2 & -c1 & 0 & 0 \\ 0 & 0 & 0 & 0 & c4 & -c3 & c2 & -c1 \\ c2 & -c1 & 0 & 0 & 0 & 0 & c4 & -c3 \end{bmatrix}, \qquad (2)$$

where $c1$, $c2$, $c3$, $c4$ – Daubechies coefficients. It works with 8 numbers x vector.

The average coefficients displacement on the one row vector side (1) does not allow signal recovery to the direct vector order change on the reverse (from d_{n-1} to a_{n-1}). Radial symmetry usage corresponds to division x on two part with equivalent size (Shoberg, 2013). Then we use wavelet signal decomposition

$$Sh_n x = \begin{bmatrix} D1_n \\ A1_n \\ A2_n \\ D2_n \end{bmatrix} x = \begin{bmatrix} d1_{n-1} \\ a1_{n-1} \\ a2_{n-1} \\ d2_{n-1} \end{bmatrix} \qquad (3)$$

where Sh_n = square blocks transformation matrix with $2^n \times 2^n$ size and symmetric placement parts of the center; $D1_n, A1_n, A2_n$ and $D2_n$ = transformation matrix Sh_n parts, each with size $2^n \times 2^{n/2}$; $a1_{n-1}, a2_{n-1}$ and $d1_{n-1}, d2_{n-1}$ = approximate and detail transformation column vector parts with 2^{n-2}. Transformation matrix with center symmetric blocks is used for the wavelet decomposition on the next levels (Shoberg & Shoberg, 2013).

The algorithm application continues next **Sh** matrix' formation (3). This matrix' central part will be two times less contracting to the center. The **Sh**

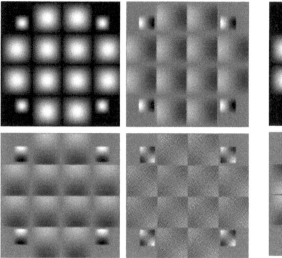

Figure 1.   Test image traditional wavelet decomposition.

Figure 2.   Test image proposed wavelet decomposition.

matrix contains reordered Daubechies coefficients from (2)

$$
\mathbf{Sh} = \begin{bmatrix}
-c3 & c4 & -c1 & c2 & 0 & 0 & 0 & 0 \\
-c1 & c2 & -c3 & c4 & 0 & 0 & 0 & 0 \\
c2 & c1 & c4 & c3 & 0 & 0 & 0 & 0 \\
c4 & c3 & c2 & c1 & 0 & 0 & 0 & 0 \\
0 & 0 & 0 & 0 & c1 & c2 & c3 & c4 \\
0 & 0 & 0 & 0 & c3 & c4 & c1 & c2 \\
0 & 0 & 0 & 0 & c4 & -c3 & c2 & -c1 \\
0 & 0 & 0 & 0 & c2 & -c1 & c4 & -c3
\end{bmatrix}. \qquad (4)
$$

The transformation result $\mathbf{Sw_n}$ is n number $\mathbf{Sh}$ matrixes multiplication formed on described scheme and x vector

$$\mathbf{Sw_n} = \mathbf{Sh_2}\mathbf{Sh_3} \ldots \mathbf{Sh_n}\mathbf{x}. \qquad (5)$$

The average means $a1_1$, $a2_1$ and detail coefficients $d1_1$, $d2_1$ compute on final step. So, the transposed $\mathbf{Sw_n}$ will be row vector with 2^n elements. The formulas set (4-5) allow to produce the signal decomposition and reconstruction in direct and reverse order. It's possible, because we have symmetry in the $\mathbf{Sh}$ matrix and $\mathbf{Sw}$ vector. The reverse reordering of any these multiplication elements reconstructs approximate coefficients in opposite sequence. The two-dimension decomposition result in traditional form used Daubechies basis and is shown in Figure 1. The test image (Gonzalez & Woods, 2002) is decomposed of 4 parts. This decomposition was produced as two one-dimensional transforms applied to rows first and then to columns. Two-dimension test image analysis with proposed algorithm allows to get results with an average part in the center. It allows to reconstruct the source image, if we used rotation on angles multiple 90 degrees. The proposed scheme of image decomposition which is represented in a traditional kind is shown in Figure 2.

3   CONCLUSION

The symmetric multiresolution analysis scheme has a slightly more computational complexity than the traditional scheme. Four frequency bands for one-dimension and sixteen bands for two-dimension symmetric transform can increase the transform coefficients value range in a common case. The symmetric scheme allows to reconstruct the original signal with direction process change. It works with multi dimensional signals. This approach complicates choice wavelet and scaling functions for different applications. The function characteristics are length, symmetry or no symmetry, odd or even. They will have strong influence to reconstruction accuracy in the middle of a signal. This work was supported by "ImPro Technologies Co. Ltd."

REFERENCES

[1]   Daubechies, I. 1992. *Ten Lectures on Wavelets*. Montpelier, Vermont: Capital City Press.
[2]   Gonzalez, R. & Woods R. 2002. *Digital image processing*. Addison-Wesley.
[3]   Mallat, S. 1989. A Theory for Multiresolution Signal Decomposition: The Wavelet Representation. *IEEE Pattern Analysis and Machine Intelligence*, vol. 11, no. 7: 674–693.
[4]   Shoberg, A. 2013. One-dimensional signal multiresolution analysis scheme invariant to processing direction (on Haar basis). *Information science and control system*, no. 2(36): 146–152.
[5]   Shoberg, A. & Shoberg, K. 2013. Multiresolution analysis with symmetric scheme: matrix description. *Bulletin of PNU*. no. 3(30): 49–56.
[6]   Strang, G. & Nguyen, T. 1997. *Wavelets and Filter Banks*. Wellesley-Cambridge Press.
[7]   Welstead, S. 1999. *Fractal and Wavelet Image Compression Techniques*, SPIE, Bellengham, Washington.

Future Communication, Information and Computer Science – Zheng (Ed.)
© *2015 Taylor & Francis Group, London, 978-1-138-02653-7*

Automotive FlexRay communication system implementation

F. Luo, Y. Qu & C. Liu
Clean Energy Automotive Engineering Center, School of Automotive Studies, Tongji University, Shanghai, China

ABSTRACT: To assist in the development of FlexRay applications, an Intelligent FlexRay Communication Validation Platform has been developed on which the whole FlexRay-based simulation and communication process have been implemented in this paper. Network modeling, communication cluster parameters design and configuration have been realized by using CANoe. FlexRay, then real nodes are developed and a dual-channel bus topology network has been proposed and constructed to validate the communication between three FlexRay nodes. With the aid of the whole system, IFCVP has been applied to achieve vehicular Drive-by-Wire control and network data have been successfully monitored.

Keywords: network communication; FlexRay; simulation; Drive-by-Wire

1 INTRODUCTION

The FlexRay communication protocol is specified as a dependable automotive network. Fault tolerance, high transmission rate, predictable real-time transmission delays make FlexRay suitable for by-wire control in vehicles. Automotive vendors have been attempting to expand the applicability of FlexRay, especially being dedicated to developing X-by-wire systems for the FlexRay network.

To assist in the development of FlexRay applications, researches have been conducted extensively. Some related works merely focused on the FlexRay simulation, but they didn't consider the actual hardware communication realization. A hybrid topology network combined with a hardware platform is employed to realize the FlexRay-based communication, but the work was carried out just in a single channel, thus, the reliability of the system can't be guaranteed. In view of this, an Intelligent FlexRay Communication Validation Platform (IFCVP) has been developed on which the whole FlexRay-based communication process has been implemented in this paper.

The structure of this paper is organized as follows. First, the basic protocol specification of FlexRay is introduced briefly. Second, requirement analysis for realizing Drive-by-Wire control is carried out, and a simulation model for the network is established. Third, hardware nodes are designed and IFCVP is proposed and developed to implement a vehicular Drive-by-Wire system. Fourth, software part including underlying and application layer of the system is given out. Last, experimental results analysis and conclusions are presented.

2 OVERVIEW OF FLEXRAY

In the FlexRay protocol, media access control is based on a recurring communication cycle, which is the fundamental element of the media access scheme and is defined by means of a timing hierarchy. The timing hierarchy consists of four timing hierarchy levels as depicted in Figure 1.

Each communication cycle is partitioned into static (ST) and dynamic (DYN) segments. The ST segment is comprised of fixed quantity of equal-length slots, and a specific task is allowed to send a message only during its allocated slot. In the DYN segment, the duration of communication slots may vary in order to accommodate frames of varying length, and task messages are based on fixed priorities.

FlexRay combines the advantages of time-triggered and event-triggered communication protocols which promotes determinism and flexibility. FlexRay also supports redundant channels for fault-tolerant communication.

3 FLEXRAY SYSTEM DESIGN

3.1 *Requirement analysis and design of the simulation model*

To realize Drive-by-Wire (Steer-by-Wire, Throttle-by-Wire) control based on FlexRay, it is a necessity to build a network consisting of 3 nodes, responsible for throttle control, steering control and driver's intention identification, respectively. In order to detect errors of FlexRay system or communication scheduling, building a simulation model before developing

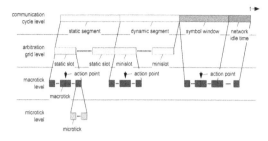

Figure 1. FlexRay timing hierarchy.

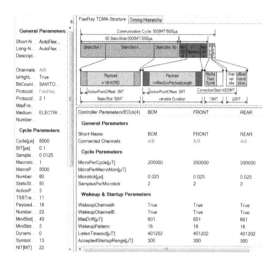

Figure 2. Main parameters of FlexRay system.

Figure 3. Communication scheduling of the entire system.

real nodes is very important. The data transmission is based on slots, which is allocated to each node.

Hence, Vector CANoe is used to set up a simulation model, which is composed of three simulation nodes and two databases (AutoFlexDemo.xml and Env.dbc) to describe the entire network, and then specify bus behavior of individual nodes with CAPL editor. The main parameters in database are shown in Figure 2, and communication scheduling of the whole system viewed in software CANoe FIBEX Explorer is shown in Figure 3.

Data in Trace window reflects the communication of simulation network. It can be seen that setup time from communication cycle start to synchronization is 44 us, and static slot between slot 3 and slot 4 lasts

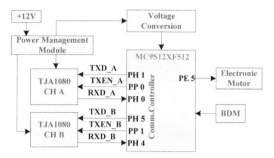

Figure 4. Functional diagram of node Rear.

about 50 us. In data part, the corresponding transmitted data changes in accordance with the change of "current speed" and "steering angle" in the control panel.

From these data, it can be seen that communication timing, data structure, startup time, synchronization time and slot response time are consistent with the protocol specification. Results of simulation can be regarded as reasonable, and the actual communication network can be designed using simulation network slot assignment and cluster parameters.

3.2 FlexRay node design

Based on the established simulation model, hardware and software of nodes for the FlexRay communication cluster are developed. Wherein, the communication protocol parameters of nodes in cluster and actual transmitting and receiving slots allocation should be consistent with that of the simulation model. Thus, with the aid of hardware design and software initialization, synchronization is then set up to realize the accurate communication under the designer's intention.

3.2.1 Hardware structure

The MC9S12XF512 CPU (Freescale) is used as the core microprocessor of all nodes to implement the system functions, and to further complete the FlexRay network of the vehicular Drive-by-Wire system. Its features of built-in FlexRay communication controller with the outbound connection of FlexRay transceiver enable FlexRay communication. The TJA1080 (NXP) is used as FlexRay transceiver in each node and two channels with transmission rate 10Mbps are applied in the FlexRay network. Then, the Intelligent FlexRay Communication Validation Platform (IFCVP) has been developed using FlexRay as the backbone for data communication.

The IFCVP is comprised of Hardware Equipment, Controller, VN8970 FlexRay module and Upper Monitor. Communication between the PC and the platform is routed over a USB cable. There exist three nodes in the system, namely Node BCM, Node Front and Node Rear, and the functional diagram of node Rear is shown in Fig. 4. GPIO of node FRONT and node REAR are utilized to generate PWM to control rotation angle of steering engine and speed of electronic motor separately.

214

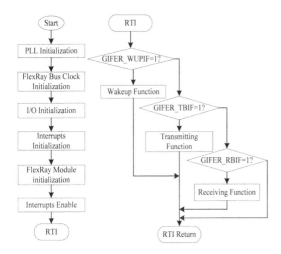

Figure 5. Main program flow diagram of node.

3.2.2 Software design

Regarding the software design of the system, the underlying layer and application layer are concerned. For the former part, main program flow diagram of node is mainly composed of the initialization of CPU and FlexRay module, as is shown in Figure 5. FlexRay frame reception and transmission are achieved by using interrupt service routine. Specifically, on the basis of configuring corresponding FlexRay interrupt registers, when the bus clock runs to the corresponding action point, receiving and transmitting functions are called to complete related FlexRay frame operation.

For the application layer, program flow diagram of implementing vehicular Drive-by-Wire control is shown in Figure 6. Specific slots are allocated to the three nodes and control variables including the steering angle and rolling speed are sent out from control panel on PC to Node BCM, which are transmitted to FlexRay bus in their corresponded frame. These variables are used as input signals to control PWM, and further control the steering angle and rolling speed of the smart car. In case of unpredicted fault, the generated PWM signals for steering engine and electronic motor change depends on consistence of data between reception and transmission nodes.

4 SYSTEM TEST

Drive-by-Wire control system test has been implemented on the basis of self-developed IFCVP. For the purpose of convenient control, a control panel was created for the transmission of real-vehicle control signals necessary for ECU network test, as is shown in Fig 7. When one of the manipulation buttons is pressed, the steering wheel and pedals in the panel change accordingly, and the smart car runs consistently with the driver's intention simultaneously.

Physical FlexRay waveform is not pure frame information, but the combination of frame and some

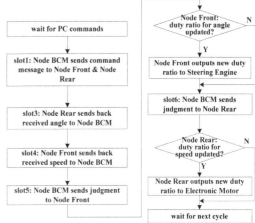

Figure 6. Program flow diagram of application layer.

Figure 7. Control panel.

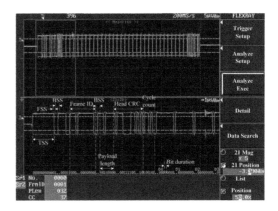

Figure 8. Actual communicational FlexRay frame.

symbolic information with a specific function and significance. Figure 8 displays the actual communicational FlexRay frame including TSS, FSS, BSS, and FlexRay data part.

The waveform in oscilloscope DL7440 shows each division contains 10 bits and its duration is 1us, thus the communication rate is 10 Mbps.

From the Trace window of CANoe.FlexRay as is shown in Figure 9, network data including occupied channels, frame name, transmitting and receiving slots, the payload length and specific content of data have been successfully monitored. In slot 1 of cycle 5,

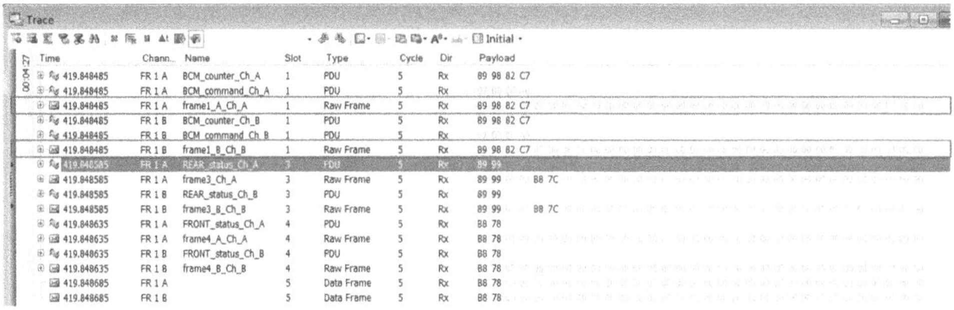

Figure 9. Experiment result.

the message transmitted to Node Front and Node Rear is framed in red. In the payload part, the message (8998) contained in byte 0 and byte 1 represents the rotation angle, and the message (82C7) contained in byte 2 and byte 3 represents speed. It can be seen obviously that there exists no difference between messages transmitted on channel A and channel B.

5 CONCLUSION

In this paper, FlexRay network modelling, communication cluster parameters design have been realized, and configuration of the simulation is performed, then real nodes are developed and a dual-channel bus topology network has been proposed and constructed. Furthermore, an Intelligent FlexRay Communication Validation Platform has been developed on which the whole FlexRay-based simulation and communication process to achieve Drive-by-Wire control have been successfully implemented. In the future, this platform is a useful reference to develop complicated FlexRay network and it can be flexibly integrated in different vehicular test systems for further research on vehicular X-by-Wire applications.

REFERENCES

[1] Yang, J. S., Lee, S., Lee, K. C., & Kim, M. H. (2011, October). Design of FlexRay-CAN gateway using node mapping method for in-vehicle networking systems. In *Control, Automation and Systems (ICCAS), 2011 11th International Conference on* (pp. 146–148). IEEE.

[2] Schenkelaars, T., Vermeulen, B., & Goossens, K. (2011, March). Optimal scheduling of switched flexray networks. In *Design, Automation & Test in Europe Conference & Exhibition (DATE), 2011* (pp. 1–6). IEEE.

[3] Longo, S., Su, T., Herrmann, G., Barber, P., & Gerlinger, U. (2011, September). Scheduling of the FlexRay static segment for robust controller integration. In *Control Applications (CCA), 2011 IEEE International Conference on* (pp. 1487–1492). IEEE.

[4] Sheu, Y. H., & Ku, C. M. (2012, July). The Intelligent FlexRay Safety Monitoring Platform Based on the Automotive Hybrid Topology Network. In *Intelligent Information Hiding and Multimedia Signal Processing (IIH-MSP), 2012 Eighth International Conference on* (pp. 289–292). IEEE.

[5] FlexRay Communications System Protocol Specification Version 3.0.1, FlexRay Consortium, 2010.

[6] MC9S12XF512 Reference Manual, Freescale Semiconductor Inc.

Future Communication, Information and Computer Science – Zheng (Ed.)
© *2015 Taylor & Francis Group, London, 978-1-138-02653-7*

Counts vs. Weights: From traditional to untraditional complexity analysis of computer algorithms

N.K. Singh & D.K. Mallick
Department of Computer Science & Engineering, Birla Institute of Technology Mesra, India

S. Chakraborty
Department of Applied Mathematics, Birla Institute of Technology Mesra, India

ABSTRACT: Although very useful, the power of mathematical techniques is inflicted by its own limitations. This is especially true for average case analysis. Algorithms cannot be compared merely with the complexity measures unless one has an idea about the associated constants which sometimes are non-trivial to obtain (are they constants?). To get a realistic estimate of these constants one has to opt for empirical analysis. Computer scientists did not realize that when they are directly working on program run time they are actually estimating a *statistical weight based bound*. Their main focus is on *verifying the results*, through empirical study, already known theoretically. They should have realized that theoretical complexity analysis is a count based science and consequently the bounds are operation specific. There is no concept of mixing operations of different types and then finding the bound. But in real implementation, the operations perform collectively. So to make complexity analysis realistic there must be a concept of assigning weights to operations so that operations of different types can be mixed. This gives rise to a weight based statistical bound. This research paper puts forward the philosophy behind a relatively novel concept of 'statistical bounds' and empirical-O which we feel are inherent (in this form or some other) to the idea of empirical analysis.

Keywords: algorithm; math bounds; computer experiment; statistical bounds; empirical-O; expectation

1 INTRODUCTION

We are accustomed to the idea of using the notations O, Ω, and Θ to describe the asymptotic running time of an algorithm. Knuth gets the credit of popularizing these notations through his famous 1976 paper Knuth (1976). Knuth traces the origin of the O-notation in P. Bachmann's 1894 book Bachmann (1894). The notation O(f(n)) stands for any function whose magnitude is upper-bounded by a constant time f(n), for all large n. Hardy and Littlewood introduced Ω in their classic 1914 memoir, calling it a "new" notation, Hardy & Littlewood (1914). They defined this notation as a negation of o(f(n)). Later Knuth changed this definition to bring it to the form in which we (computer scientists) use it today. These two notations are further complemented by the Θ-notation which was suggested to Knuth independently by Bob Tarjan and Mike Paterson, Knuth (1976).

The replacement of the phrase "infinitely many n" by "all large n" in the early definitions of these notations is due to Knuth. He defended this change as: "For all the applications I have seen so far in computer science, a stronger requirement (replacing "infinitely many n" by "all large n") is much more appropriate. After discussing this problem with people for several years, I have come to the conclusion that these new definitions will prove to be most useful for computer scientists, Knuth (1976)."

Although very useful, the power of mathematical techniques is inflicted by its own limitations (Graham et al. 1994), Green & Knuth (1982). This is especially true for average case analysis. In the worst case, the math bounds have a guarantee but can also be conservative. A certificate on the level of conservativeness is necessary and we prefer empirical research to get it. In an average case, apart from identifying the dominant operation for taking expectation, the probability distribution over which expectation is taken has to be realistic for the problem domain. Alternatively one can derive one result using uniform inputs but must cross check it for several non-uniform inputs, both discrete and continuous. Otherwise it is risky and misleading to say "my algorithm has a bad worst case but performs better on the average". If the result for uniform inputs does not hold for non-uniform inputs one must see whether these non-uniform inputs are realistic for the problem domain. If they are, the average complexity measure is not robust.

In fact, even some seemingly simple algorithms are found to be very difficult to analyze with mathematical precision and certainty. Empirical analysis is the principal alternative, which is capable of complimenting and supplementing the findings of its theoretical counterpart. Algorithms cannot be compared merely with the complexity measures unless one has an idea about the associated constants (are they constants?). To get a realistic estimate of these constants (?) one has to opt for empirical analysis. This is a very strong argument in favor of empirical research in algorithms. Recently there has been an upswing in interest in experimental work in the theoretical computer science community, because of the growing recognition that theoretical results cannot tell the full story about real-world algorithmic performance (Johnson 2001). It should be noted that the average itself is a statistical term and that the most popular and useful average (arithmetic mean) is itself only a special case of weighted mean where the weights are frequencies.

The ultimate choice of an algorithm for a given problem depends not only on the size of input supplied to it and their representation in the storage, but very strongly on the way in which the records can be expected to be ordered in the input data set itself. The performance of an algorithm can change significantly from one possible ordering to another and hence the analysis of algorithms from the standpoint of probability and statistics is of great importance. The particular distribution of the orderings of the incoming lists is therefore an important consideration in choosing an algorithm (along with others this is especially true for sorting algorithms).

This research paper puts forward the philosophy behind a relatively novel concept of '*statistical bounds*' and '*empirical-O*' which we feel are inherent (in this form or some other) to the idea of empirical analysis.

2 STATISTICAL BOUND AND ITS EMPIRICAL ESTIMATE

Computer scientists did not realize that when they are directly working on program run time they are actually estimating a *statistical weight based bound*. Their main focus is on *verifying the results*, through empirical study, already known theoretically. They should have realized that theoretical complexity analysis is a count based science and consequently the bounds are operation specific. For example, in sorting there would be separate bounds on comparisons and interchanges. There is no concept of mixing operations of different types and then finding the bound. But in real implementation, the operations perform collectively. So to make complexity analysis realistic there must be a concept of assigning weights to operations so that operations of different types can be mixed. This gives rise to a weight based statistical bound. For our purpose the time (CPU time) of an operation is taken as its weight and accordingly the program run time is interpreted as a weighted sum of all the computing operations of different types (including their repetitions) where the weights are the corresponding times

being consumed. This gives rise to a deterministic T that is deterministic for a fixed input. But algorithmic complexity is expressed not for a fixed input but for a fixed input size. The question is: if T is replaced by y as a function of input size only, rather than a fixed input, will y be stochastic? There are two contrasting schools of thought for defending stochastic modeling of program run time–traditional (which algorithm writers mention) and untraditional (which algorithm writers have missed). The traditional argument says that program run time which is deterministic for a fixed input may be regarded as stochastic for a fixed input size and randomly varying input elements, e.g. sorting algorithms, (Mahmoud 2000). The untraditional argument permits us to fit a stochastic model even for non-random data to achieve cheap and efficient prediction (Sacks et al. 1989) and holds for those algorithms where fixing the input size fixes all the computing operations, e.g. n-by-n classic matrix multiplication algorithm.

Statistical bound (non-probabilistic): If w_{ij} is the weight of (a computing) operation of type i in the j-th repetition (generally time is taken as weight) and y is a "stochastic realization" (which may not be stochastic) of the deterministic $T = \sum 1.w_{ij}$ where we count one for each operation repetition irrespective of the type, the statistical bound of the algorithm is the asymptotic least upper bound of y expressed as a function of n where n is the input parameter characterizing the algorithm's input size

For parallel computing, summation should be replaced by maximum.

Empirical-O: We define an empirical-O as "the O that corresponds to the simplest model fitted to (time) complexity data that is neither an under-fit nor an over-fit". More specifically, it corresponds to the leading term of an empirical model, given that the chosen model does not invite the serious problem of ill-conditioning nor does it lead to a loss of predictive power, Chakraborty & Sourabh (2010).

Empirical-O is a *bound estimate* and not itself a bound. The reason is that all bounds are asymptotic while it holds only over a finite range and more importantly because it emerges from an empirical model which is affected by the statisticia's subjective personal opinion and a bound cannot be subjective, Sourabh & Chakraborty (2007).

The weight based statistical complexity bound defined here is different from Ferrari's statistical bound, Ferrari (1990) which is a probability bound being a generalization of the deterministic bounds (which are special cases of probability bounds with unit probability). The reason for calling it statistical is that it is weight based and takes all the operations collectively into a conceptual bound. This non-probabilistic statistical bound is a "separation" (rather than a generalization) from traditional research in that we deviate from counts, bounds, and asymptotes to weights, finite range (in a computer experiment) and estimates (whose credibility can be increased through a proper design and analysis of our computer experiment (Sacks et al. 1989)). For a deeper insight into these concepts, see Chakraborty & Sourabh (2010)

The use of a *separate notation for bound estimate* (empirical O, written as O with a subscript emp) over a finite range is due to Sourabh & Chakraborty (2007). Theoretical-O is a mathematical bound. In a computer experiment, the range is always finite. A bound estimate technically differs from a bound and hence it demands for a separate notation as well. For example, in big-oh, what is inside the O is what matters not the mathematical statement that led to it. Whether we got O(nlogn) from a statement like "a + bnlogn" or a statement like "a + bn + cnlogn" does not matter. But in empirical O, the empirical model, whose leading term is in the O, is important as it is this model through which we can predict the run time even for huge input size for which it is computationally cumbersome to run the code. And this "cheap prediction" is the motive in a computer experiment.

We have also proposed several applications of empirical-O. The statistical bound has a system invariant property, Chakraborty & Sourabh (2010). In an average case complexity, if the dominant operation, or dominant region in the code, is wrongly selected for applying mathematical expectation, the math bound and the stat bound will not be the same. As an example, in Amir Schoor's algorithm for matrix multiplication, Schoor has taken multiplication to be dominant and obtained one result expressed as the expected number of multiplications Schoor (1982). But working on weights, Chakraborty & Sourabh (2007) have found that comparisons are dominant (if there is a sufficient number of zeroes) and they obtained quite a different result. [There has been a debate between Chakraborty & Knuth over the possibility of an empirical $O(n^2)$ complexity in n-by-n matrix multiplication. The interested reader might be referred to chapter 2 of Chakraborty & Sourabh (2010) who have claimed that the proposition is certainly obtainable for two dense matrices.]

The notation O(f(n)) stands for any function whose magnitude is upper-bounded by a constant time f(n), for all large n. It should be pointed out that the asymptotic behavior of y(n) may not hold over a finite range or in other words, f(n) can differ from g(n). Let y(n) be of the order g(n) for all $n_1 \leq n < n_2$ and f(n) for $n_2 \leq n < n_3$ for some reasonable real positive constants. If the range (n_1, n_2) is reasonably large and we are likely to have a large percentage of data sets practically from this finite but feasible range then complexity can better be described as: $y(n) = O_{emp}(g(n))$ for the given range. As a specific case if $n_3 \to \infty$, y(n) is more appropriately O(f(n)) as expected from its mathematical definition. Strictly speaking, for a number of reasons, in an empirical paradigm these thresholds refer to a range of integers and not specific values.

3 CONCLUSIONS

We would like to make the following concluding remarks:

1. Mathematical bounds are ideal for analyzing worst case behavior. Worst case mathematical analysis is a strong science as it in some sense gives a guarantee. At the same time this guarantee can also be conservative. In such a situation a certificate on the level of conservativeness is necessary and the concept of statistical bounds is ideal to get it.

2. In an average case, apart from identifying the dominant operation(s) for taking expectation, the probability distribution over which expectation is taken has to be realistic for the problem domain. An analytically obtained complexity over uniform distribution inputs must also be verified for several other common non-uniform realistic inputs, both discrete & continuous. If the results do not match, the average complexity measure is not robust.

3. The concepts of statistical bounds also can be useful in nullifying the tall mathematical claims in a best case analysis.

4. Asymptotic analysis is based on an idealized sequential RAM-model. Actual computer architecture is not as simple as this RAM-model. In parallel computing, apart from others, the complexity of an algorithm depends heavily on factors such as type and number of processors and computational models. Finding the math bound for a parallel program amounts to fixing a constant depending on the processor speed or other factors. If a factor such as a processor is changed, it is only the weight of the operation that changes. So when the bound itself is based on weights, it is the bound that should be relevant. These constraints make the asymptotic analysis fragile in parallel algorithmic complexity computing paradigm.

Although ideal for average case analysis, the usefulness of statistical bounds should also be investigated for worst and best case analyses. Although it seems not very obvious, we still can hope for some breakthroughs going beyond the worst case theoretical barriers, given that the statistical bound works for finite range (should be big enough for the target application) of data! Apart from estimating a statistical bound, the idea of empirical-O can be extended to estimate the mathematical bound. These estimates are count based and operation specific.

The computer scientists should address the question of selecting weight in an experiment, as it is very crucial in the analysis. Should time be the weight or should it be a weighted combination of time and space? As a final comment, traditional research in algorithm and complexity places it as a branch of theoretical computer science. However, if Sacks' untraditional school of thought is religiously followed it is also a branch of applied statistics (Sacks et al. 1989)!

Author's Profile:

Niraj Kumar Singh is a Teaching cum Research Fellow; Dheeresh Kumar Mallick is an Associate professor; Soubhik Chakraborty is an Associate Professor.

REFERENCES

[1] Bachmann, P. 1894. *Die Analytische Zahlentheorie. Zahlentheorie. Leipzig: B. G. Teubner.*

[2] Chakraborty, S & Sourabh, S.K. 2007. On why an algorithmic time complexity measure can be system invariant rather than system independent. *Applied Mathematics and Computation*, Vol. 190, issue 1, 195–204.

[3] Chakraborty, S. & Sourabh, S.K. 2010. *A Computer Experiment Oriented Approach to Algorithmic Complexity.* LAP.

[4] Ferrari, D. 1990. *Client Requirements for Real Time Communication Sevices.* UC Berkeley, Nov 1990, RFC 1193.

[5] Hardy, G.H. & Littlewood, J.E. 1914. Some problems of diophantine approximation. *Acta Mathematica*. 155–238.

[6] Graham, R.L. et al. 1994. *Concrete Mathematics: A Foundation for Computer Science.* 2nd ed. Addison-Wesley, Reading, MA.

[7] Green D.H. & Knuth, D.E. 1982. *Mathematics for analysis of Algorithms.* Boston, 2nd edition, Birkhauser.

[8] Johnson, D.S. 2001. A Theoretician's Guide to the Experimental Analysis of Algorithms, http://www.researchatt.com/~dsj/, Nov 2001.

[9] Knuth, D.E. 1976. Big Omicron and Big Omega and Big Theta. *SIGACT News*. 18–24.

[10] Mahmoud, H. 2000. *Sorting: A Distribution Theory*, John Wiley & Sons.

[11] Sacks, J. et al. 1989. Design and Analysis of Computer Experiments. *Statistical Science*. Vol. 4, No. 4, 409–423.

[12] Schoor, A. 1982. Fast Algorithm for Sparse Matrix Multiplication. *Information Processing Letters*, Vol. 15, No. 2, 87–89.

Future Communication, Information and Computer Science – Zheng (Ed.)
© *2015 Taylor & Francis Group, London, 978-1-138-02653-7*

Research on teaching reform methods of data structure

Z. Dong, Y. Tang & P. Cao
Beijing Key Laboratory of Signal and Information Processing for High-end Printing Equipments, Beijing Institute of Graphic Communication, Beijing

ABSTRACT: Data structure is an important fundamental course of Computer Science in colleges and universities. There are some problems existing in the teaching process. Some teaching reform methods are proposed to solve the problems, mainly from the following aspects, teaching contents, teaching methods, practical teaching and examination system. The reform methods focus on leading curriculum review, carefully organizing teaching contents, using a variety of teaching methods, improving practical contents and implementation modes, increasing the proportion of practical teaching. These feasible reform methods can help students understand and master teaching contents, and improve their practical ability.

Keywords: data structure; teaching method; practical teaching; examination system

1 INTRODUCTION

Data structure is the academic foundation and software design technology foundation in computer and related disciplines. The research ideas and methods have been widely used in many related research fields. It lays a solid foundation for students to engage in theoretical research, application development and technical management. The course mainly discusses the data logical structure, storage structure and the implementation of various operations defined on these structures. The course mainly describes the basic data structure, the related algorithms, a variety of searching and sorting algorithms, and the algorithm analysis and evaluation technology [1]. The basic data structures include linear structure, tree structure and graph structure. Through the studying of this course, students can make a reasonable choice of data storage structure and effectively design algorithm in the process of software development, so as to improve the overall quality of software.

2 PROBLEMS IN TEACHING PROCESS

The data structure course has many concepts and abstract theories, has strong practicality, and involves a wide range of knowledge [2]. In the process of teaching, there are some problems in the following aspects.

(1) No connection of data structure and C language
Data structure is a course that combines theory and practice. It is very practical. Students should have a basic knowledge of programming in C language.

Indicators and structures are the most frequently used words in a data structure course. They are the foundation for learning data structure. Pointers related to the physical storage of data are difficult to master because of its complex concept, flexible and error-prone usage [3]. During the C language course, students pay attention to the grammar and examination questions for reaching the requirements of the national computer rank examination. Therefore, the key usage of C language can't be mastered. When learning data structure, they can't transfer algorithm to program without a thorough understanding of C language.

(2) Monotonous teaching mode
The data structure course involves lots of concepts and algorithms. The theory is strong and abstract. The multimedia courseware can't completely describe and demonstrate the contents and execution processes of algorithms. Moreover, there is no effective interaction between the teacher and students. So students don't fully understand what the teacher told. On the other hand, computer practice can't play an important role because the students' C language foundation is weak. With less absorption during the course, students lose interest in learning it [4]. Therefore, the traditional teaching model of "transmission-acceptance" is no longer suited to the characteristics of the course.

(3) Practice teaching incomplete
Most computer practical trainings are verification experiments. Students only realize the related algorithms that can be found in textbooks on the computer. The source codes of the experiments also can be found in practice teaching materials or on the network. So some students can't meet the learning requirements. There are few comprehensive experiments. Therefore,

students can't achieve deeper understanding of related algorithms and cannot apply them to practice.

3 TEACHING REFORM METHODS

3.1 Reform in teaching method

(1) Reform of teaching contents

Leading curriculum review should be laid emphasis on. Strong leading knowledge is the precondition to learn a course. C language is the most important leading course of data structure. Students' mastery of C language is a key factor that determines the effectiveness of teaching. Therefore, whether the students study the C language well or not, leading them to review the related knowledge is necessary. The review contents mainly include: 1) Type definition; 2) The difference between passing by value and passing by address of the function's arguments; 3) The definition of the array, pointer and structure; 4) Program development environment.

Teaching contents are well-organized. All contents are based on the teaching syllabus and satisfy the specified materials requirements. The dross is discarded and the essential is retained. Each chapter's knowledge system is refined. Three-level data structures, which are logical structure, storage structure and operations, are highlighted. The teaching contents are carefully organized in accordance with the easier issues first thoughts. Linear list, stack and queue are linear structures which students firstly learn [5]. The sequential storage structures of the just mentioned three linear structures are simple, but their chain storage structures are relatively difficult. Therefore, the logical structures, sequential storage structures and chain storage structures are rearranged and divided into three teaching units. The order of traditional materials is changed and the contents are taught in a three period teaching. Through constantly comparative analyzing and summarizing, students can consolidate what they have learned, but also earn new knowledge.

(2) Reform of teaching means

The multi orientation teaching means, for example, multimedia electronic teaching plan, algorithms' dynamic demonstration system, and curriculum websites, are used. Making full use of modern educational technology, the important, difficult and abstract contents are demonstrated by animation display. It's easier for students to understand and master in this way. Students are required to review related reference books and compare learned knowledge points. On the one hand they can absorb the knowledge from different ways, and on the other hand they can find problems and bring in corresponding solutions. Students are also required to read the relevant basic and applying literatures from curriculum websites. Thus their views not only can be widened but their mastery and application of deeper knowledge are improved.

(3) Reform of teaching methods

The data structure course is a core essential course with a strong practical and theoretical. It is often thought of having many difficulties in teaching. So the reform of teaching methods is to seek better ways to let the students participate, and mobilize students' enthusiasm and initiative. Specific practices include: 1) Summarizing teaching rules. For example, when explaining the linear list, emphasizing stacks, queues and strings are special linear lists, emphasizing their specificity, so that students learn more with a feeling of learning one. 2) Paying attention to teaching skills. According to the degree of difficulty of different knowledge, appropriate adjustments are done to the teaching schedules based on the past student's mastery of different teaching contents. For example, the binary tree traversal is abstract, especially restoring original binary tree from its traversal sequence. We can increase the examples and infer other things from one fact so as to enable students to concentrate on the teaching contents. 3) Combining a variety of teaching methods. Interactive, lecturing, demonstration and other methods are combined in the teaching process to focus students' attention and arouse their enthusiasm [6]. Students' mastery of teaching contents is the final purpose.

3.2 Reform in practice teaching

(1) Reform of practice contents

Practice teaching is the further digestion and absorption of the theory teaching. The implementation of practice can better review and consolidate the basic concepts, principles and methods of learning. It is more important to select the data structure, design and realize corresponding storage structure, and finally solve the problem according to the actual situation. Detailed reform contents include: 1) Refining experiment contents in class and scientifically designing the practice teaching contents. The simple authentication type subject on the textbook is reserved for extracurricular prep lab assignments. Class lab projects based on actual teaching, combined with the life practice and algorithms are very useful to improve students' perceptions of the algorithm. Students can correctly understand the idea and essence of algorithms, and gradually train the computational thinking ability and the ability to independently analyze and solve problems. 2) For top students, encouraging them to participate in the practice of ACM contest questions. It can widen their view and improve their programming skills. 3) Optimizing curriculum design, introducing industry interview questions. These can raise students' interests in learning.

(2) Reform of implementation

The quality of the teaching effect has a direct relationship with the teaching practice, in order to ensure the effective implementation of practice teaching. The following several methods can be used in the teaching process: 1) Increasing the proportion of practice score and leading students to pay attention to practice. According to the curriculum requirements, the course syllabus is updated and the experimental result proportion is increased from 10% to 20%. Increasing practice proportion is aiming to urge students to actively participate in practice, to program, to write and analyze reports, and improve students' practical programming ability and the ability to write documents. 2) Grouping to enhance students' comprehensive ability. In the course design process, in accordance with the principle of voluntary, three to four students were divided into a group and a person is appointed as the leader. The leader is responsible for the implementation of the system, the division of responsibility for the design task arrangement. Everyone has different tasks in one system, and each person's task can't be repeated. Team members work together to accomplish their teamwork. Students are guided according to the actual situation of the project to design their own data structures, form algorithm ideas, and finally realize the program.

3.3 Reform in examination system

In addition to the final written examination, an improved examination system has a stage examination. The final grade is composed of normal records (20%), stage test results (30%) and final examination results (50%). The normal records include learning attitude, lectures effect, knowledge level, and learning ability. They are evaluated by class attendance, class performance, class status, and consult reference materials. Stage test results are composed of homework and a unit test. The test times of each stage are three or more. The test items are students' honesty, knowledge level, degree of innovation, practical ability and comprehensive ability. The total test results are given according to the comprehensive assessment of each stage. The final examination results are composed of program defense and an exam on the computer. The exam contents on the computer include the basis of local knowledge and basic skills that cover the whole the course. It contributes 25%. The program defense mainly checks students' comprehensive ability of applying knowledge and language expression. Its contents include the knowledge to master, spoken, comprehensive capabilities and the ability to apply theory to practice. It also contributes 25%.

4 CONCLUSIONS

According to the characteristics of the data structure course, this paper analyzes the problems of the course in the teaching process, and solves these problems separately from teaching contents, teaching methods, practice teaching and examination system. Various feasible reform approaches are explored. There a variety of modern teaching models used and tried to reach the best combination to complete the task of teaching. So that students learn to analyze the characteristics of data objects, select the appropriate data structure, storage structure and the processing algorithm, and then be able to apply the basic data structure to complex programming. Enabling students to understand and master teaching contents, improving their ability of solving practical problems, and providing the necessary theoretical and practical foundation for the professional follow-up courses are the purposes of teaching the data structure course.

ACKNOWLEDGEMENTS

The research work was supported by BIGC projects of China under Grant No. 22150114002, No. 22150114041 and No. 22150114040.

REFERENCES

[1] Suli Zhang, Xin Pan, Hua Zhang. Reformation and exploration of Data Structure experiment teaching. Modern computer, 2013, (9): 47–49.
[2] Pingyu Qin, Jingshan Ma. Data Structure—C language version (The second edition). Beijing: Qinghua university press, 2012.
[3] Manyin Shi. Teaching reform of Data Structure based upon cultivating application-oriented personnel. Journal of Ningde Normal University (Natural Science) 2013, 25(3): 327–329.
[4] Hua Jiang, Jiahua Lin, Wanfu Zhou. Improvement of Data Structure course in colleges and universities. Journal of Chuxiong Normal University, 2013, 28(9): 14–16.
[5] Weimin Yan, Weimin Wu. Data Structure: C language version. Beijing: Qinghua university press, 2003.
[6] Rihua Xin. Discussion and study on teaching method of Data Structure. Journal of Inner Mongolia Agricultural University (Social Science Edition), 2013, 15(72): 59–61.

Future Communication, Information and Computer Science – Zheng (Ed.)
© 2015 Taylor & Francis Group, London, 978-1-138-02653-7

Reliable design for Earth-Mars transfer trajectory

Y. Yu
School of Computer Science, China University of Geosciences (Wuhan), China

W. Lin
Faculty of Information Engineering, China University of Geosciences (Wuhan), China

G. Dai & Q. Yuan
School of Computer Science, China University of Geosciences (Wuhan), China

ABSTRACT: In the global optimization design of Earth-Mars transfer orbit, the reliability of the solution is hardly considered. However, in the basins of attraction of some minimum of the Earth-Mars transfer orbit, the singularities of which the fuel cost function varies sharply in a short period of time hamper the evaluation of reliability. This paper proposes a method of judging singularities and uses it to analyze the reliability of the solutions. Experiments show that a reliable global optimum can be chosen through the proposed method.

Keywords: Earth-Mars transfer; singularities; reliability

1 INTRODUCTION

Exploration of Mars has become the hot spot of deep space exploration of which the preliminary trajectory design has become an important topic. In the preliminary trajectory design of Earth to Mars transfer by means of two impulsive maneuvers, the designer has to obtain a "low-cost" mission by global optimization methods [1,2]. But when the undesirable interference of uncertainties is considered, the global optima may be suboptimal or infeasible. So the reliability of the solutions has to be ensured.

In preliminary trajectory design of Earth to Mars transfer, there is some local minimum with very large fuel cost function in the basins of attraction. Some authors call them singularities. If there are singularities in the uncertainty set of the global optimum, the solution can not be used for the design of preliminary orbit, as it may be infeasible or risky. This paper analyzes the singularities in detail and applies it to obtain a reliable global optimum.

2 GETTING STARTED

In the bi-impulsive model, as shown in Fig. 1, there are two control variables t_0 and Δt. Here t_0 is the departure date of the spacecraft, expressed in Modified Julian Dates 2000 (number of days from 1st January 2000, MJD2000 in what follows), and Δt is the flight time from Earth to Mars. Given t_0, and $t_0 + \Delta t$ (the arrival date at Mars), the position and

velocity of Earth and Mars at t_0 and $t_0 + \Delta t$ respectively can be computed through analytical ephemeris (i.e., analytical formulas giving the position and the velocity vectors of a celestial body as a function of time). Then, given the mission time Δt from Earth to Mars, the transfer orbit can be calculated by solving Lambert's problem. The solution of the Lambert's problem provides the heliocentric velocities $\vec{v}_1$ at the beginning and $\vec{v}_2$ at the end of the transfer arc. Since the spacecraft initially moves along the orbit of Earth with the same velocity $\vec{v}_E$, the initial $\Delta \vec{v}_1$ is:

$$\Delta \vec{v}_1 = \vec{v}_1 - \vec{v}_E \qquad (1)$$

And then the second in order to move from the transfer orbit to Mars is as follows:

$$\Delta \vec{v}_2 = \vec{v}_2 - \vec{v}_M \qquad (2)$$

The total fuel is as follows:

$$\Delta v = \|\Delta \vec{v}_1\| + \|\Delta \vec{v}_2\| = \Delta v_1 + \Delta v_2 \qquad (3)$$

The modules $\Delta v_1 = \|\Delta \vec{v}_1\|$ and $\Delta v_2 = \|\Delta \vec{v}_2\|$ are the two contributions to the total $\Delta v = \Delta v_1 + \Delta v_2$. So, the total fuel (including initial fuel and braking fuel) is a function of the departure time from Earth and the flight time:

$$\Delta v = f(t_0, \Delta t) \qquad (4)$$

If f is plotted with respect to t_0 and Δt, the result can be seen in Fig. 2. The launch date is from January 1,

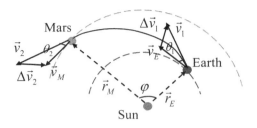

Figure 1.   The bi-impulsive model of Earth to Mars transfer.

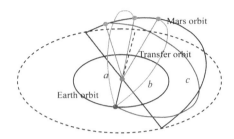

Figure 4.   Transfer orbit when φ close to π.

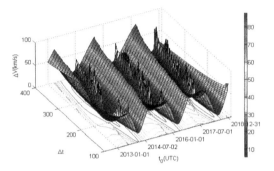

Figure 2.   Distribution of the total Δv problem.

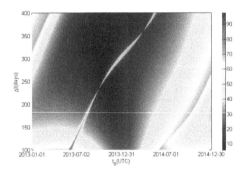

Figure 3.   A cycle of fuel function.

2013 to December 31, 2018 and the range for the flight time is from 100 days to 400 days. It can be shown that f is a quasi-periodic function. The period is equal to the Earth-Mars synodic cycle, approximately 26 months or 780 days. In each period, there are some singularities, near the minimum fuel points, with extremely high error sensitivities.

3   ANALYSIS OF SINGULARITIES

Because of the cyclical changes of the objective function of fuel, only a cycle of f is analyzed and other cycles have the similar case, as shown in Fig. 3.

Singularities exist in an interval where t_0 is from June 2013 to May 2014. There are no singularities in other intervals. We just analyze the region where singularities exist. If t_0 keeps constant, it can be noted f changes from small to very large and then to small where singularities exist in a short period of time. This

process corresponds to a so small change in Δt that the larger is several days and the smaller is less than one second. In such a short period of time, there are no abrupt variations with $\|\vec{v}_E\|$, $\|\vec{v}_M\|$, $\|\vec{v}_1\|$ and $\|\vec{v}_2\|$, according to the fuel function, f changes dramatically attributes to the large changes of the angle between $\|\vec{v}_E\|$ and $\|\vec{v}_1\|$ called θ_1, and the angle between $\|\vec{v}_M\|$ and $\|\vec{v}_2\|$ called θ_2. In the process, f varies dramatically in a short time, the value of the two angles, θ_1 and θ_2, changes from small to large (close to $\pi/2$) and then changes to small. As the spacecraft will not retrograde to the Earth at the launch time, θ_1 is generally not more than $\pi/2$.

3.1   The first factor: value of φ

In a short time, with the variation of θ_1 and θ_2 which changes from small to large and then to small, the angle between $\vec{r}_E$ and $\vec{r}_M$ which is called φ also varies, as shown in Fig. 4. When φ is close to π, the orbital plane inclination of the transfer trajectories reaches a maximum, such as transfer orbit a in which θ_1 and θ_2 are very large at the moment, so the fuel function reaches a maximum value. Now suppose t_0 remains unchanged, when Δt has a small positive or negative change, φ will decrease and the orbital plane inclination of the transfer trajectories tends to decrease too. Consequently, the value of θ_1 and θ_2 will decrease, so that the fuel function will decrease. The same is the case with the orbit of b and c. In order to deal with the pro-grade solutions of the Lambert's problem, orbit c has an opposite direction of motion to orbit a and orbit b.

The lasting time of discontinuity varies with changed position of Earth and Mars. In some cases, fuel function has a significant change when the variation range of Δt is a few days. This is because if the variation of Δt takes a small value, such as just several seconds, the orbit plane inclination of transfer trajectory will not change significantly. However, when φ almost equals π, the transfer orbit, orbit a, is nearly perpendicular to ecliptic if the Lambert's problem is computed by Battin's algorithm [3]. The value of θ_1 and θ_2 is very large at the time, as shown in Fig. 5. Then let Δt add or subtract a very small variable, for instance a few seconds, the transfer orbital plane will close to the orbital plane of Mars, look at orbit b and orbit c. The orbital plane inclination of the transfer trajectories will be very small, so the value of θ_1 and θ_2 are small

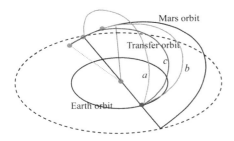

Figure 5. Transfer orbit when φ nearly equals π.

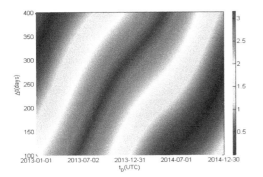

Figure 6. The value of φ corresponding of Fig. 3.

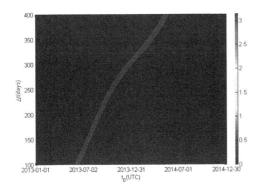

Figure 7. The value of φ when φ is larger than 175 degrees.

too. It can be inferred that the value of fuel function changes from very small to very large and then to very small during a few seconds.

Consequently, we can conclude there is a relationship between the singularities and φ. Fig. 6 shows the value of φ corresponding to Fig. 3.

As can be seen from Fig. 6, the time when φ is close to π is the moment singularities exist. Fig. 7 and Fig. 8 show the value of φ (φ is larger than 175 degrees due to the angle between the ecliptic and the orbital plane of Mars of approximately 5 degrees) and the corresponding value of f.

3.2 *The other factor: changes of direction of motion*

It can be seen from Fig. 7 and Fig. 8, there is a corresponding relationship between the singularities and φ.

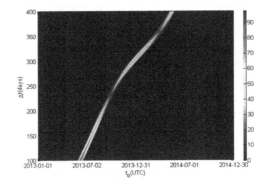

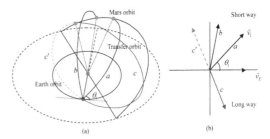

Figure 8. The value of f when φ is larger than 175 degrees.

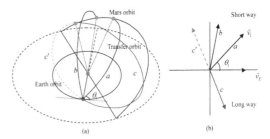

Figure 9. Geometric overview of the transition from the short way to long way solution. (b) is the two-dimensional plan of (a).

Singularities exist where φ is close to π. But it is insufficient to determine whether there are singularities in a region only according to the variation of φ. For example, if the values of φ of two points are close to π, it is hard to determine whether there are larger values of φ between the two points, which may lead to singularities. But if there are singularities between the two points, the value of φ must change from small to large and then to small. In this process, θ_1 changes from acute to abtuse. But θ_1 can not be abtuse according to celestial knowledge, so once θ_1 is larger than $\pi/2$, it must be updated to $\pi - \theta_1$ by means of changing the direction of motion of the spacecraft that is transiting from the "long way" to the "short way" or from the "short way" to the "long way", as shown in Fig. 9. Therefore, along with φ changing from a small value to the maximum and then to a small value again, the direction of motion of the spacecraft changes, whereas in some region where no singularities exist, there may be changes of the direction of motion, so singularities can be determined according to the value of φ and the variation of the direction of motion.

3.3 *Judgment of singularities*

In the reliable design of Earth to Mars transfer, the uncertainty of decision parameters is generally several seconds. We can judge whether there is variation of the direction of motion of the four vertices of a rectangle modeled by the uncertainty set. It can be seen

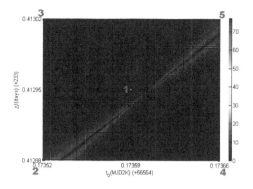

Figure 10. The determination of singularities in sub-domain.

that from section 3.1, the cases of φ being larger than 175 degrees almost contain all cases where singularities exist. So singularities can be determined according to whether φ is larger than 175 degrees and whether there are variations of the direction of motion of the four vertices of the rectangular uncertainty set.

A sub-domain where the value of f varies dramatically in a short period of time is chosen, as shown in Fig. 10. The robustness of point 1 will be calculated. The decision parameters are $t_0 = 56564.17359$, $\Delta t = 233.41295$ respectively and the corresponding value of f is 9.136 km/s. Point 2, 3, 4 and 5 are the four vertices of the rectangular uncertainty sets. The decision parameters are $(t_0 - 6\sigma_1, \Delta t - 6\sigma_2)$, $(t_0 - 6\sigma_1, \Delta t + 6\sigma_2)$, $(t_0 + 6\sigma_1, \Delta t - 6\sigma_2)$, $(t_0 + 6\sigma_1, \Delta t + 6\sigma_2)$, $\sigma_1 = \sigma_2 = 1s$. When the decision variables take value in the rectangular uncertainty set, the maximum of the objective function can reach over 77 km/s. The calculated value of φ of point 1 is 3.1415925 radian. The direction of motion of point 2, 3, 4, and 5 is long-way, long-way, short-way and long-way respectively. It can be determined that there are singularities in the neighborhood of point 1 according to the value of φ which is larger than 175 degrees and the variations of direction of motion of four vertices of the rectangular uncertainty set. The results are conform the actual situation.

4 RELIABLE GLOBAL OPTIMUM

In the design of bi-impulsive Earth-Mars transfer, we expect to find a launch window between 2017 and 2018, and the flight time is between 100 and 400, with the reliable optimal fuel cost. There are two local optimums without considering uncertainties which are shown in Table 1.

Three numerical experiments were performed with uncertainties $\Delta_1 = \Delta_2 = 6$seconds, $\Delta_1 = \Delta_2 = 6$minutes, and $\Delta_1 = \Delta_2 = 6$hours respectively.

The situation of solution 1 and solution 2 are analyzed in Table 2. For solution 1, the value of φ is larger than 175 degrees, at the same time, the direction of motion of uncertainty set changes. As a consequence, it can be known that there are singularities in the vicinity of the solution. According to the same theory, it can

Table 1. Local optimal solutions.

	t_0	Δt	f
Solution	MJD2K	days	km/s
1	6704.401850	249.785210	6.262388
2	6706.550234	204.474837	5.750624

Table 2. Determination of singularities.

		Direction of motion of the uncertainty		
Solution	φ degrees	6seconds	6minutes	6hours
1	179.999959	Long-way	Short-way	Long-way
		Long-way	Long-way	Long-way
		Short-way	Long-way	Short-way
		Long-way	Long-way	Long-way
2	152.891955	Short-way	Short-way	Short-way
		Short-way	Short-way	Short-way
		Short-way	Short-way	Short-way
		Short-way	Short-way	Short-way

Table 3. Neighborhood of solution 2.

Uncertainty Set seconds	Step size seconds	Maximum km/s	Standard deviation km/s	Average km/s
6	0.01	5.750624	0.000000	5.750624
360	0.01	5.750624	0.91×10^{-8}	5.750624
21600	1	5.750756	0.33×10^{-4}	5.750668

be inferred that there are no singularities in the uncertainties of solution 2. So the reliable optimum moves toward a better direction which points to solution 2.

Table 3 shows the neighborhood of solution 2 where the maximum, the standard deviation and the average are computed respectively. All of the standard deviations are so small that they can be neglected. The maximum of the uncertainties nearly equals the solution. It verifies that solution 2 is a global optimum with good reliability.

5 CONCLUSION

The classical optimization design of Earth to Mars transfer generally neglect the tolerance of the decision variables, thus the solution may be suboptimal or even infeasible. Reliable optimization ensures that all implementation of uncertain input has a good performance solution. But in the design of bi-impulsive Earth-Mars transfer, it is every hard to obtain the reliable optimum due to the singularities. As a result, to determine whether there will be singularities in the neighborhood of the solution is necessary. The determination of singularities can be used in any interval of Earth-Mars transfer orbit.

ACKNOWLEDGEMENTS

The research work was supported by the "twelfth five" Civil Aerospace Technology pre research project No. 61103144, the National Natural Science Foundation of China under Grant No. 60873107.

REFERENCES

[1] Vasile M., Locatelli M. A hybrid multiagent approach for global trajectory optimization. Journal of Global Optimization, 44(4), pp. 461–479, 2009.

[2] Myatt D.R., Becerra V.M., Nasuto S.J., et al. Advanced global optimisation for mission analysis and design. Final Report. Ariadna id, 3: 4101, 2004.

[3] Shen H., Tsiotras P. Using Battin's method to obtain multiple-revolution Lambert's solutions. Advances in the Astronautical Sciences, 116, pp. 1–18, 2004.

Future Communication, Information and Computer Science – Zheng (Ed.)
© *2015 Taylor & Francis Group, London, 978-1-138-02653-7*

Data mining application in bank credit risk analysis

Y. Wang & C. Guo
Beijing Jiaotong University SEM, Beijing, China

D. Ma
School of Information Engineering, Shenyang Broadcasting TV University, Shenyang, Liaoning, China

ABSTRACT: This article describes the prediction and assessment of bank credit risk, and proposes and analyzes two methods: distributed data mining method and data mining method based on accounting information. These two methods are carried out on prediction and assessment. At last, the author proposes a new model to control the bank credit risk on a high level.

Keywords: data mining; information entropy; credit risk; fuzzy neural network

1   INTRODUCTION

Different areas have different explanations on credit risk, however, bank credit risk is more commonly used in the following two situations, first one: due to borrowers' unfulfilled promise, bank assets suffer a risk of certain losses; second: in the process of bank credit to clients, a risk of customers non-payment on the appointed time. The nature of these two arguments are the same, bank credit risk means that the customer does not fulfill his promises and causes a risk of loss to the bank. The main reason of bank failures worldwide is the unreasonable operation of credit risk and fund liquidity crisis caused by it.

Data mining technology began to emerge in the late 1980s and it is a technique that digs out useful information from a large number of miscellaneous actual data. After the rapid development of last century, this technology has been widely used in many fields. In order to reduce credit risk and provide guidance for construction and management in other industries, banks need to predict and assess the credit risk. Credit risk assessment occupies an important part of credit risk management and it is an integrated assessment method using credit-related information. The method can directly analyze and evaluate the credit status, solvency and credibility of trustees in their credit relations. This article starts from the theory of bank credit risk assessment and information entropy and proposes two methods of data mining: distributed data mining and a data mining method based on accounting information.

2   DISTRIBUTED DATA MINING TECHNIQUES AND PRINCIPLES

Distributed data mining techniques is the process of extracting information from distributed data sets and a database [1]. A distributed database is the data source of distributed data mining.

The current typical distributed data mining technique is the copying of a string type algorithm and following the main steps: each processor handles the data set simultaneously and takes partial data analysis, then combines local data models on different nodes, so that we can get the global data model (global knowledge) [2]. For general distributed data mining algorithms, it uses the same data mining algorithms on different distribution data nodes, and then combines local knowledge to compose the global data model.

3   APPLICATION EXAMPLES OF DISTRIBUTED DATA MINING TECHNIQUES ON BANK CREDIT RISK ASSESSMENT

Take credit risk for example: using fuzzy neural network on various branches site of the bank to classify credit risk, the neural networks can extract the main features of the relationship implied by the sample itself after a certain amount of training samples with noise, besides, it can also interpolate and extrapolate data under a new situation to infer its properties. In addition, fuzzy logic technology has the logical reasoning

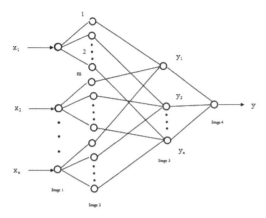

Figure 1.   The structure of fuzzy neural network.

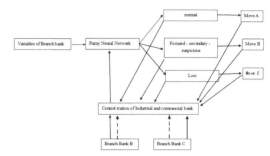

Figure 2.   Bank credit risk based on distributed data mining.

function that can imitate human thinking. To combine neural networks and fuzzy systems together, is what we call a fuzzy neural network. It can foster strengths and circumvent weaknesses and make the risk assessment system with self-learning function and fuzzy logic.

Established fuzzy neural network structure is a single-output and multiple-input system shown in figure 1, and this network structure can be divided into four levels in total.

Input node layer is the first layer in figure 1, input of this layer is a risk indicator, the main factors affecting the bank credit risk are the borrower's credit, the borrower's ability to repay debt, cash flows and trends and other aspects. Operating capacity, liabilities, cash flow, profitability and mortgage case provided by borrowers, etc. of this enterprise are all included in the factors. Quantized non-financial analysis after the screening is also included, in this layer, the input of each node is equal to the output in this figure, and accordingly, the number of nodes in this layer is equal to the entered index number n.

Fuzzy layer is the second layer in figure 1, according to all of its input indicators, the number of divided fuzzy subsets is five. Assume that Mij is the J-th fuzzy subset of section I indicators, and $j = 1, 2, \ldots, n, j = 1$ (normal), 2 (attention), 3 (secondary), 4 (suspicious), 5 (loss), as a result, the number of nodes in the layer is 5n. The input number of this layer is the output number of the first layer, n, while output of this layer indeed inputs membership μ ij of fuzzy subset Mij of indicators.

Fuzzy inference layer is the third layer in figure 1, node number is assumed to be 5. Following are the fuzzy rules we set:

R1:IF x_1 is M_{11} and x_i is M_{i1} and x_n is M_{n1} then y is y

R2:IF x_1 is M_{12} and x_i is M_{i2} and x_n is M_{n2} then y is y

R3:IF x_1 is M_{13} and x_i is M_{i3} and x_n is M_{n3} then y is y

R4:IF x_1 is M_{14} and x_i is M_{i4} and x_n is M_{n4} then y is y

R5:IF x_1 is M_{15} and x_i is M_{i5} and x_n is M_{n5} then y is y.

y is the output variable and y i is a back piece of real number. We can know that the input of this layer is the output of the last layer. Output of the j-th node in this layer is the fitness for fuzzy rules R j. For a given input risk indicators $X = (x1, x2, \ldots, xn)$, output of node j in this layer is: $\mu j = \mu 1j(x1) \times \wedge \times \mu n j(x n)$ (j = 1, 2, 3, 4, 5).

Anti-blur layer (output layer) is the fourth layer in figure 1, and the input of this layer is the output of the last layer. We can calculate output variables of centroid method:

$$y = \frac{\sum \mu_j \times \gamma_j}{\sum \mu_j}$$

output in this case represents the credit risk index. During the training spinning stage of fuzzy neural network, expected output of samples can be given by historical data or other evaluation methods; it can be used to solve practical problems of risk assessment after a successful training model.

From Figure 2 we can see: if the branch site A was in the whole process of decision based on data mining, both head control site and branch site (as branch sites B, C, etc.) will provide corresponding knowledge support. In addition, head control site will also provide a corresponding constraint (knowledge flow from head control site to branch site).

Through the above examples we can conclude that on the basis of the distributed data mining, head control site and branch site can not only always be consistent strategically, but also realize decisions to optimize and share through creative knowledge of each site.

4   THE ROLE OF DATA MINING METHOD BASED ON ACCOUNTING INFORMATION IN BANK CREDIT RISK PREDICTION

Data mining method based on accounting information is based on the information entropy theory, and sets bank credit risk assessment as its target. The linear model of credit risk prediction proposed by such a method is showing its good predictability.

4.1 Measurement method of data mining method based on accounting information

Data mining method based on accounting information has gained more and more attention in recent years. In such data mining method, the study of the index (which can be used as predicted variables of corporate credit risk assessment) has become a key issue. So the study of it is particularly important. Proceeding from information entropy theory and data mining techniques, this paper proposes a method for measurement of the metric information value, specifically as follows [4].

Some indicators for loan application corporate are represented by $x1, x2, \cdots xn, y$, in which y indicates whether the enterprise breach the contract, xi indicates various financial indicators of enterprises. Credit risk is to predict y through xi. The breach of contract is uncertain, but once the value of a certain financial indicator xi is determined, the uncertainty will be reduced. According to the information theory and statistical knowledge, information entropy can measure the uncertainty of random variables, thus the discrete and continuous variables are:

$$I(y) = -\sum_{i=1}^{N} PiLogPi$$

$$I(y) = -\int_{-\infty}^{+\infty} f(x)Logf(x)dx$$

In which, Pi is the different probability value obtained in discrete variables, and $f(x)$ is the density function of continuous variables. Therefore, the corporate default uncertainty in the credit risk prediction can be presented as $I(y) = -(1 - P1)\log(1 - P1) - P1\log P1$, in which $P1$ indicates the corporate credit default probability.

Determining a critical point $x0$ and dividing corporations into two groups, $q1, q2$ are representing respectively the proportion of two groups of companies, and the information entropy of the two groups are $I1(y), I2(y)$, in this case, the uncertainty of y can be measured by

Gain(xi)=$I(y)$-$q1I1(y)$-$q2I2(y)$

It is defined that the critical point when Gain(xi) reaches its maximum is called Threshold, that is the information gain when xi is used to predict y. the degree of information gain indicates the value of xi in credit risk prediction, that is, the bigger xi, the greater the value of y. With the development of data mining technology in recent years, threshold can be easily calculated with a number of data mining software.

The range of information gain is in $[0,I(y)]$, when the value is $I(y)$, $I1(y) = I2(y) = 0$, At this point indicators are at the greatest value for credit risk prediction: Y are no longer an uncertainty after being grouped, one group are all defaults, the other group are not, therefore, using the index, you can establish rigorous prediction rules. On this basis, a prediction method

for measurement information content of financial indicators is established, specifically as follows.

When xi threshold value is a positive or negative indicator, the system should determine whether a company will default, which can be seen as financial indicators issuing a warning to the system. If indicator xi sends a right alert, it can be regarded as xi signal, if the alert is wrong, then it is noise. Define the indicator xi signal ratio and noise ratio of indicator xi: signal ratio is the ratio of the number of default corporate and total number of default corporate when xi sends alert; noise ratio is the ratio of the number of indefault corporate and total number of indefault corporate when xi sends an alert.

When the signal ratio of a financial indicator is large (close to 1), while the noise ratio is small (close to 0), banks will need to increase attention to the alert from the indicators. Thus the measurement method to predict the information content of financial indicators proposed by this paper is: predict the information content of financial indicators = signal ratio – noise ratio.

The role of the predicition of credit risk is proportional to the difference between signal and noise of the financial indicator. When the difference between signal and noise of the financial indicator is close to 1, this indicator is the best credit risk predictor. And when the difference between signal and noise of the financial indicator is close to 0, the financial indicator cannot provide any information to predict credit risk.

4.2 Construction method of credit risk prediction model

The primary method to construct a credit risk prediction model can be divided into two categories: The first is the credit risk prediction model based on corporate accounting data [5]. The second is the credit risk prediction model based on the Morton option pricing theory. Since the basic assumption of the second model is: The stock price of a company can reflect the intrinsic value of the company, but many Chinese shares of listed companies cannot achieve all of the circulation, therefore, stock price of domestic company does not adequately reflect the intrinsic value of the company, as a consequence, the basic assumption of a predictive model based on the Morton option pricing theory does not work in China.

Therefore, the establishment of the domestic credit risk prediction model can only be based on the credit risk prediction model of corporate accounting data. There can be many possible ways to construct a credit risk assessment model, including traditional linear discriminant analysis, Probit model, artificial neural networks based on data mining, Bayesian classification and so on. In order to illustrate the role of signal to noise difference method in bank credit risk prediction more clearly and directly, the following paper proposes a new method to construct a credit risk prediction model: linear model for credit risk prediction [6]. The linear prediction model is widely used in bank credit risk assessment; here we are going to introduce a kind of linear prediction model: the Altman Z-score model. The key point of the construct linear model is to determine the weight of each indicator, Z-score model uses discriminant analysis to determine the index weight, since some initial conditions are not met, this method has often been criticized by some scholars.

From the above analysis ideas, a linear model of constructing method can be presented in this paper. During credit risk assessment, the greater the predictive information content of an indicator, the more the additional importance it should pay to the information, and assign it to find the right play. The foregoing analysis can give us a linear prediction function of structure credit risk:

$$Z=(\lambda_1\rho_1x_1+\lambda_2\rho_2x_2+\ldots\lambda_k\rho_kx_k)/w$$

In which ρ i is the i-th index signal noise difference, $w = \Sigma\rho i$, and when x_i is positive indicator, $\lambda_1 = 1$, when x_i is negative indicator, $\lambda_1 = -1$.

Obviously relative to any single indicator, defined comprehensive index Z can get more information in predicting corporate credit risk, thus the linear prediction model for corporate credit risk can be established: select the threshold Z_0, when the corporate value is greater than Z_0, the loan of the corporate will not default; when the corporate value is less than Z_0, the loan of the corporate will be sentenced default.

5 CONCLUSION

Two methods of data mining method, distributed data mining method and data mining method based on accounting information are used in this paper for prediction and assessment of bank credit risk respectively; finally, a prediction model is proposed and established according to the actual situation in the domestic to provide decision support to reduce credit risk. Due to their limitations, both methods and models have their inevitable inadequacies in the practical application, and need to be continually improved and enhanced.

REFERENCES

[1] Wu Sen, Gao Xue Dong, Data Warehouse and Data Mining. Beijing: Metallurgical Industry Press, 2003.
[2] Xiao Rong, Hu Keqin. Research on Distributed Power Customer Service System. Computer Engineering, 2005, 31(8): 186–188.
[3] Yang Hui Yao, Chen Xuehua. Construction of commercial bank creditrisk classification Warning Model. Guangzhou University Journal (Social Science). 2004(5): 53–54.
[4] Kirkos S, Spathis C, Manolopoulos Y. Data Mining Techniques forthe Detection of Fraudulent Financial Statements. Expert Systems with Applications, 2007(32): 995–1003.
[5] Kaminsky Graciela, Mohsin K. Currency and Banking Crisis: The Early Warning of Distress: The Carly Warning of Distress. IMF Working Paper, 1999: 1–38.
[6] Michael B Gordy. A Comparative Anatomy of Credit Risk Models. Journal of Banking & Finance, 2000(24): 119–149.

Future Communication, Information and Computer Science – Zheng (Ed.)
© *2015 Taylor & Francis Group, London, 978-1-138-02653-7*

Research on embedded system effective teaching method

J. Wang & L. Yang
Institute of Information Engineering, Beijing Institute of Graphic Communication, Beijing, China

ABSTRACT: Embedded system has become an important direction of computer applications with the rapid development of embedded computing technology. And accordingly, the embedded system course has become one of the most important computer science courses. But there are many problems in traditional education on such a practical course. In order to adapt to the new era of change, researches on teaching methods are proposed. Reform on teaching methods includes taking the course in a computer classroom, task-based teaching model based on autonomous learning, and so on. The students' interests and innovation ability has been improved through the teaching reform.

Keywords: embedded system; teaching reform; effective teaching

1 INTRODUCTION

Along with the fusion of the software industry and manufacturing industry, the embedded technique is making rapid progress by giant necessities of intelligent equipments. Embedded technique has been applied in customized electronics, automotive electronics, communication engineering, medical equipment, intelligent home and industry automation. Currently the research and design on embedded system technique has become one focus field in the Information Technology industry. The course of "Embedded System" is one core module of Electronics Engineering. This course combines electronics and computer science, and concentrates on the technique application. The target of this course is to help students to be able to analyze the project requirements and master the basic solution of the embedded the system design.

The classroom teaching and lab experience are the main teaching methods of this course in currently. However, the classical teaching procedure based on the classroom cannot adapt to the requirement of a large amount of experience.

2 THE PROBLEM OF CLASSICAL TEACHING METHOD

The classroom teaching and the lab experience are individual in the classical teaching procedure. In the classical procedure, the thought of teaching is according to the textbook or lecture notes. The contents include the foundation of the embedded system, design method of embedded system, embedded Linux system, design of application program, design of driver

program and application examples. However, embedded system is one practical course. In a classroom, it's difficult and less efficient to describe the reality operation by language, which results in the lower teaching procedure progress and less information. In this typical classroom teaching, students cannot find emphasis easily, and the learning effect is not good enough. From another side, the knowledge about embedded system hardware is difficult, complex and bare. Due to all these causes students lose interest in learning embedded system, and the teaching effect is worse. This classroom teaching procedure only provides the basic concept and theory, which is invalid teaching.

In typical experience teaching procedure, the target is to improve the student's practical ability of the embedded system. However, in the actual experience teaching procedure, students practice is according to the experience textbook in a regulated period. The practice contents are mostly the same as the examples in the textbook, which loads the student with typing, compiling and debugging. Based on this procedure, students only need to be familiar with the design environment and some tips on debugging. However, the most important thing is programming, which is exactly ignored in this typical experience teaching procedure. And the practice effect is discounted.

In Chinese universities, some courses are separated from the actual requirements. There are lots of organizations coming out to train embedded system engineers and driver engineers. The target of these organizations is to provide the specialized and experienced engineer. All these training courses are taken in labs, the classroom teaching and experience teaching are combined, and the teaching contents are regulated according to the requirement of the employing unit. Although the fee of these training course is expensive, there are still lots of students attending these courses to try to find good jobs.

The huge difference between universities and training organizations should be considered by educators. The teaching method of training organizations could be introduced to the embedded system course teaching in universities. In our university, the teaching method of goal-orientation is applied in the embedded system course innovation, and the reaction of students is positive.

3 THE INNOVATION OF EMBEDDED SYSTEM EFFECTIVE TEACHING

3.1 Interests exciting

Interest is the best teacher. To realize the effective teaching it should concentrate on the student's interests. The course of embedded system is open for third year students majoring in electronics engineering. For these students, the future career selection is the most important thing. In this case, at the beginning of this course, visiting research and design companies related with the embedded system, and communicating with embedded the system engineers and HRs could help student planning career direction. After relative visiting and communicating, students can also know the talents which companies really requires and the knowledge that a relative career requires. Based on this, students know the knowledge and skills on embedded system could help their finding a job, and as the salary of relative jobs is higher, the passion of learning this course is also improved.

On the other hand, some interesting cases are introduced into the teaching procedure. The discussion and heuristic teaching methods are also introduced. The learning enthusiasm is improved; the students' ability of analyzing and resolving a problem is also raised. Senior research and design engineers are also invited to the class, their experience and knowledge about this course can cultivate the engineering consciousness of students, and also expand their sight. Besides these teaching methods, the class is divided into two levels. The first level is the foundation part, which consists of some replication experiences. The second level is the professional part, which is also called course design, and aims to train the comprehensive design ability of students [1].

3.2 Teaching in labs

To achieve effective teaching, the traditional teaching method should be changed. Students are the dominant role in the class, and efficient interaction is the way to overcome the traditional teaching disadvantages. The teaching method of training organization can be introduced. Both theory teaching and experience are taken in the lab. In this case, students can practice as soon as possible. The teaching contents are also full of lots of addition operation. The relative knowledge is blended in the experience and subject taught, and students can understand the knowledge points directly, and the teaching effect is improved. For example, during the part of registers in ARM, all 37 registers are introduced and explained in traditional teaching. This is just a simple indoctrination. Students might forget all things after the class. On the contrary, in lab teaching, from the assembler programming, students can operate one example program with sun-program, and then the command of STMFD and LDMFD can be introduced. Because these commands include the SP, LR and PC registers, the registers can be introduced naturally. Students can understand relative knowledge more smoothly. Another example, after one mini embedded system project finished, the design method and procedure can be explained according to this mini project. Then, students can understand the importance of the project strategy. Teachers can emphasize the design strategy and procedure based on their research experience [2]. From the lab teaching method, the concentration of students is focused on the contents of the class, the interaction and passion of students are also improved.

3.3 Task orientation training

The replication experience is the foundation part. After this, a more complex and difficult task can be ordered, which can improve the active thinking of students. In the example of registers, after students understanding the sub-program calling and re-calling, the task of an adder from 1 to N based on the method of sub-program can be introduced. Although this program is simple, the sense of achievement and passion of learning can be inspired. From this way, the fear of programming can be reduced smoothly. In the experience of horse race lamp, students can master how to compile the BIN file in the ADS environment, and how to download a program to the KIT. Another task can be introduced, which is to change the lamp lighting sequence and effect. This task drives students to read and understand the code, they need to understand the functions for the light controller and sequence controller. This is the procedure of active learning, and the learning effect is better than usual. In this case, the discussion in class is increased, and the teacher becomes the support role to help student resolve the questions.

3.4 Multiform examination

Examination is the way to inspire a student. In traditional class, the exam score is the main part of the final score. However, this cannot improve the practical ability of students. Thus, raising the ratio of experience in the final score can drive students in emphasizing the class study and avoid the adventure before the exam. The presentation can be applied in the experience exam, during the presentation and questionnaire one can test the stratagem of design and the ability of resolving questions.

4 CONCLUSION

The teaching method of embedded system course is continued to be explored and innovated, which is aimed at improving the learning passion and help students master the relative knowledge required. In our university, based on this teaching method, students provide more interests in this course, and more aggression in experience teaching, and also more activity in research and learning. From this kind of teaching innovation, effective teaching is achieved.

ACKNOWLEDGEMENTS

This work was financially supported by the Scientific Research Common Program of Beijing Municipal Commission of Education (Grant No. KM18190112005, 22150114002), Beijing Key Laboratory of Signal and Information Processing for High-end Printing Equipments, Beijing Institute of Graphic Communication.

REFERENCES

[1] Martinez D.R., Bond R.A., Vai M.M., 2008, *High Performance Embedded Computing Handbook – A Systems Perspective*. USA: CRC Press.
[2] Wang S.F., Tang Y.H., Lu H.Y., 2011, Embedded System's Exploration of Teaching Methods. *Computer Education*, vol. 20.

Future Communication, Information and Computer Science – Zheng (Ed.)
© 2015 Taylor & Francis Group, London, 978-1-138-02653-7

An ontology of water systems to support BIM content library

Y. Liu
School of Computer Science and Technology, Tianjin University, China

J. Zhang & R. Liu
College of Economics and Management, Tianjin University, China

ABSTRACT: The fast development of Building Information Modeling (BIM) requires content libraries to organize digital models of building components in an efficient way. This paper proposed a semantic approach to encapsulate the knowledge about building water systems in an ontology. This ontology is used to support BIM content libraries in this targeted domain for standardized organization. Furthermore, this ontology empowers the semantic search for desired objects with the help of the Jena reasoning engine.

Keywords: ontology; water system; BIM; content library

1 INTRODUCTION

1.1 *Knowledge model and ontology*

A knowledge model is a computer interpretable body of knowledge about a specific domain. It is a standard specification expressed in some knowledge representation language or data structure that enables the knowledge to be interpreted by software and to be stored in a database or data exchange file.

An ontology is the formal conceptualization of knowledge in a certain domain. As a commonly agreed-upon standard, ontologies provide Web content (to be shared and searched) with both contextual and structural information (Gruber 1993). Different knowledge representation paradigms have different components to describe domain knowledge. Broadly speaking, four basic components are required for representing a domain of knowledge (Gómez-Pérez *et al.*, 2004): *Classes* to represent major domain concepts, *Relationships* to represent associations between *Classes* using binary relations, *Axioms* to regulate the behaviors of *Classes* and *Relationships* by sentences that are always true, and *Instances* to model individuals of domain concepts.

1.2 *Building Information Modeling (BIM)*

Along with the advancement of computer software and hardware technology, the Architectural, Engineering, and Construction (AEC) industry is experiencing a revolution which is empowered by Building Information Modeling (BIM). BIM is an improved planning, design, construction, operation, and maintenance process using a standardized machine-readable information model for a facility which contains all appropriate information created or gathered about that facility in a format useable by all throughout its life-cycle (Isikdag et al. 2007). A building model is a digital representation of physical and functional characteristics of a facility, and serves as a shared knowledge resource for information to support decision-making during its life-cycle from inception onward.

1.3 *Building Object Model (BOM)*

A Building Object Model (BOM) is the digital model of a building component. As the carrier of those geometric and analytical data, BOMs are data-rich and play an essential role in supporting BIM applications.

There are two types of BOMs – generic BOMs and manufacture-specific BOMs. BOMs incorporated in BIM authoring tools, such as Autodesk Revit, Graphisoft ArchiCAD, and Bentley AECOsim Building Designer, are typically generic BOMs which combine fixed and parametric geometry. A manufacturer-specific BOM is the digital model of a particular building product and embeds as much physical and commercial information as possible.

1.4 *BIM Content Library (BCL)*

A BIM content library refers to a system that organizes a set of BOMs in a structured approach to provide easy access, exploration, management, search, and visualization of models. BIM content libraries provide designers with a large number available design options, mainly manufacturer-specific BOMs and some generic BOMs. If BOMs in a library are authored and maintained by their manufacturers, they have the most accurate and up-to-date information. Maintaining a manufacturer-specific BOM also supports the reuse of product information. Designers of different projects using the same building product can use the same BOM instead of respectively creating a generic object in BIM authoring tools then specifying its properties based on the product catalog.

2 THE CHALLENGES

2.1 Information in BOMs

Many BIM content libraries became available to the AEC industry in the last few years. Most libraries host a variety of BOMs which are different types of building components from many manufacturers.

It has been widely discussed what information should or should not be modeled for an object or a project (Hooper & Ekholm 2011; Guttman 2011). For the purpose of supporting analysis and facilitating work processes, it requires different information content at different levels for specific application scenarios. For example, cost information is not required in mechanical analysis but needed for estimation. Intuitively, a BOM needs to include as much information as possible at all levels to fulfill any possible application in the whole life cycle of a construction project. Unfortunately, this will make the entire project model too big and slow in operation for most personal computers and workstations.

In order to balance the usefulness and operability, the current practice is that software vendors integrate information content based on the intended use of their products. Partially due to the different parameter sets between BIM tools and partially due to the interoperability issue, one can see that BIM content libraries contain many types of models. The real issue in this topic is that there is no single consistent format to define and maintain all parameters of a building object to be re-used on demand by multiple BIM applications.

2.2 Object search and validation

Most BIM content libraries support the search function but only by keywords. Some libraries provide a 3D preview of a model but quantitative and qualitative data are not able to be obtained before the model is actually imported into a BIM authoring tool, for example, the size of window frame or the material of a table top. It is very desirable to have a strong search function in a BOM library allowing semantic search for each individual parameter, for example windows that have a heat transfer coefficient not greater than 1 W.m-2K-1.

Along with the need for parameter-level search, there is also a need for validating a model against parameters before it is accepted and stored in the library. Many BIM content libraries welcome user contributed models but either there is no mechanism to check uploaded models or it depends on a manual check to ensure that all necessary parameters are filled

and the value is in a reasonable range. The issue related to this topic is that currently BIM content libraries do not support parameter validation.

3 SCOPE AND METHODOLOGY

3.1 Scope

This paper proposed a knowledge model expressed in the form of ontology to represent the knowledge required in BOMs about the building water system. There are many types of systems in a building and each system includes many types of components to be collected by a BIM content library. However, due to limited time and resources of this research, this paper only covers building water systems, i.e. domestic water supply system, drainage system, and drenching system.

3.2 Methodology

This paper suggests an ontological approach to support the standardization of attributes in representing BOMs of building water systems in a BIM content library.

Knowledge management has been used in many domains to improve the search accuracy and reduce the responding time because of its advantage over traditional syntactic-based (key works) search. A building water system ontology has two key points to address the issues in this research:

- The standardized and structured concepts to organize water system components and their attributes.
- The capability of inferring new knowledge helps the BIM content library improve its search function and validate the model based on the axioms defined in the ontology.

4 BUILDING WATER SYSTEM ONTOLOGY

4.1 Knowledge encapsulation

Building water system ontology captures the knowledge about water systems in a building in terms of their system composition and attributes of all components. For example, a *Valve* is a subclass of *Water Supply and Drainage System Component*, which indicates the relationship of components and systems/sub-systems. This ontology defines a property *has_Price* which is a data property indicating any valve will have a price. The price of a valve should be a positive number which can be defined by an axiom of *Valve* by which the BIM content library could reject the population of a valve instance when the price of that valve is a negative number or a string.

The building water system ontology proposed in this research is developed by Protégé which is a free, open source ontology editor and a knowledge acquisition system (Gašević 2009). Protégé is being

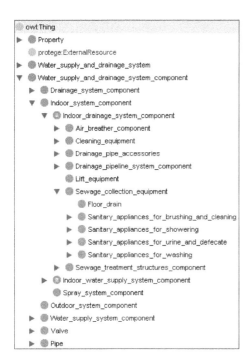

Figure 1.   Ontology Classes.

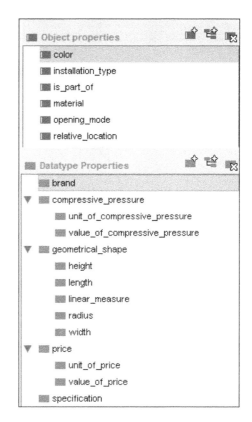

Figure 2.   Ontology Relationships.

Figure 3.   Ontology Axioms.

developed at Stanford University in collaboration with the University of Manchester. This application is written in Java and heavily uses Swing to create the rather complex user interface.

4.2   Classes

The classes encoded in the ontology are systems, system components, and component properties, as shown in Fig. 1. The systems are organized based on two dimensions: indoor/outdoor and water supply/drainage. Components are related to systems by a relationship is_part_of. Properties are used to define attributes of each specific component, such as size, color, shape, pressure, etc.

4.3   Relationship

Relationships between classes are defined as either data properties or object properties in Protégé (Fig. 2). Data properties defines a relation whose range is a type of data, for example, the value of relationship *has_length* should be a number and the value of relationship *has_name* should be a string. Object property defines a relation whose range is a class, for example, the value of relationship *is_part_of* should be a type of system.

4.4   Axioms

Fig. 3 shows an example of defining axioms for the class Hydrovalve. It says, for any hydrovalve, it has one and only one brand and it has at least one color and one type of material.

4.5   Instances

Fig. 4 shows an instance of a lavatory pan which has an ID *Squallting_Pan_8351A*. Based on the definition of this class, the instance is populated with the following attributes defined: the color is *White*, the material is *Ceramics*, the length is *560 mm*, the width is *450 mm*, the height is *230 mm*, and the brand is *FANJU*.

5   REASONING AND SEMANTIC SEARCH

The objective of developing an ontology for a BCL is to conduct semantic search powered by reasoning

241

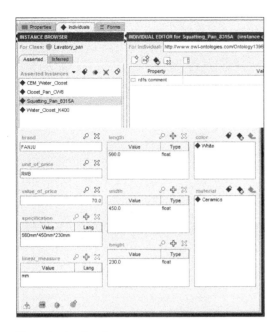

Figure 4.   Ontology Instances.

```
OntModel text_ontmodel = ModelFactory.createOntologyModel();
OntDocumentManager dm = text_ontmodel.getDocumentManager();
dm.addAltEntry("hhttp://www.owl-ontologies.com/Ontology1396285973.owl#","file:" +
                        "Water_supply_and_drainage_system.owl");
text_ontmodel.read("file:C:/test/Water_supply_and_drainage_system.owl");
String prefix = "PREFIX owl: <http://www.w3.org/2002/07/owl#>"+
        "PREFIX rdf: <http://www.w3.org/1999/02/22-rdf-syntax-ns#>"+
        "PREFIX rdfs: <http://www.w3.org/2000/01/rdf-schema#> " +
        "PREFIX base:<http://www.owl-ontologies.com/Ontology1396285973.owl#> ";
String select = "SELECT ?Goods ?Color ?ValueOfPrice ";
String where = "WHERE{?Goods rdf:type base:Goods_of_low_price ;"
        + "            rdf:type base:Sanitary_appliances_for_urine_and_defecate ;"
        + "            base:color base:White ."
        + "        OPTIONAL{ ?Goods base:color ?Color }"
        + "        OPTIONAL{ ?Goods base:value_of_price ?ValueOfPrice } }";
String k = "http://www.owl-ontologies.com/Ontology1396285973.owl#";
String rules = "[Rule1:(?x "+k+"value_of_price ?y) lessThan(?y,300)->(?x rdf:type "
        +k+"Goods_of_low_price)]"+
        "[Rule2:(?x "+k+"value_of_price ?y) ge(?y,300)->(?x rdf:type "
        +k+"Goods_of_high_price)]";
Reasoner reasoner = new GenericRuleReasoner(Rule.parseRules(rules));
Query query = QueryFactory.create(prefix + select + where);
InfModel inf = ModelFactory.createInfModel(reasoner, text_ontmodel);
System. out. println(" query = " + query) ;
QueryExecution qe = QueryExecutionFactory.create(query,inf);
ResultSet results = qe.execSelect();
ResultSetFormatter.out(System.out,results,query);
qe.close();
```

Figure 5.   Code for Rule-based Reasoning.

```
| Goods                   | Color      | ValueOfPrice                                        |
=====================================================================================
| base:Squatting_Pan_8315A | base:White | "70.0"^^<http://www.w3.org/2001/XMLSchema#float>  |
| base:Induction_Urinal_SE-6677 | base:White | "270.0"^^<http://www.w3.org/2001/XMLSchema#float> |
| base:Pedestal_Urinal_CJ | base:White | "145.0"^^<http://www.w3.org/2001/XMLSchema#float> |
```

Figure 6.   Search Result.

engines. This research employed Jena as reasoning framework. Jena can use its own reasoners or import a third party reasoned such as Racer, Pellet, or FaCT++ (Carroll et al. 2004). With the help of a reasoner, the BCL will be able to check the consistency of an instance against axioms defined in the ontology, as stated in section 4.1.

More important, Jena supports rule-based reasoning. For example, we want to find all instances which belong to the type *Sanitary Appliances For Urine And Defecate*, which have a price lower than 300, and which have a white color. Two rules can be defined as follows, where it says if y is the price of x and y is less than 300 then x belongs to a type *Goods of Low Price*:

- (?x value_of_price ?y) lessThan(?y,300) -> (?x rdf:type Goods_of_low_price)
- (?x value_of_price ?y) ge(?y,300) -> (?x rdf:type Goods_of_high_price)

With the help of SPARQL language, the code shown in Fig. 5 is executed in ECLIPSE to query the ontology and its populated instances and the result is shown in Fig. 6 where three instances satisfied the query conditions found.

6   CONCLUSION

This paper proposed an ontology to encapsulate the knowledge of building water systems. This ontology is used to support the organization of BOMs in a BIM content library and also empower the semantic search of desired objects. The proposed ontology represents major building systems, their components, and the attributes of components. The semantic search is validated by incorporating Jena to enable rule-based reasoning.

REFERENCES

[1] Carroll, J. J., Dickinson, I., Dollin, C., Reynolds, D., Seaborne, A., & Wilkinson, K. 2004. Jena: implementing the semantic web recommendations. In *Proceedings of the 13th international World Wide Web conference on Alternate track papers & posters*, pp. 74–83.

[2] Gašević, D., Djuric, D., & Devedžic, V. 2009. *Model driven engineering and ontology development* (Vol. 2). Springer, Berlin.

[3] Gruber, T. R. 1993. A translation approach to portable ontology specifications. *Knowledge acquisition*, Vol. 5(2), pp. 199–220.

[4] Guttman, M. 2011. The information content of BIM: An Information Theory Analysis of Building Information Model (BIM) Content. *Perkins and Will Research Journal*, Vol. 3(2), pp. 28–41.

[5] Hooper, M. & Ekholm, A. 2011. A definition of model information content for strategic BIM implementation. *Proceedings of the CIB W78-W102 2011: International Conference*, Sophia Antipolis, France.

[6] Isikdag, U., Aouad, G., Underwood, J., & Wu, S. 2007. Building information models: a review on storage and exchange mechanisms. In *CIB W78 International Conference on Information Technology on Information Technology*, Vol. 24, pp. 135–144.

Future Communication, Information and Computer Science – Zheng (Ed.)
© 2015 Taylor & Francis Group, London, 978-1-138-02653-7

Application and value of tourism culture

S. Zhang

Department of Humanities and Social Sciences, Sichuan Vocational and Technical College of Communications
Chengdu, Sichuan, China

ABSTRACT: Tourism culture booms due to the mutual promotion between tourism and culture. It helps to spread out culture and develop the tourism business. Besides, it promotes tourism-related business to progress, making tourism experiences more civilized, distinctive and plentiful. Moreover, it brings value to the society and economy. The work analyzed the development and application of tourism culture by making clear the relationship between tourism and culture. The value of tourism culture in tourism industries is also revealed.

Keywords: tourism; culture; application; value

1 INTRODUCTION

With the development of material life after Liberation, people are more eager to tour. People have, since the introduction of reform and opening-up, required more of spiritual life, which promotes the development of a tourism culture. Tourism-related industries sprang up like mushrooms. Besides, the establishment of cultural tourism brands makes the industry chain more plentiful and prosperous. Therefore, it is obvious that culture is the soul and motivation of tourism. Because of tourism culture, tourists can enrich their knowledge and spiritual taste. Besides, people crave more touring and buy tourism products [1]. Moreover, the experiences of different culture help to transmit and develop culture. Therefore, the application of a tourism culture is of great value and significance.

2 TOURISM AND CULTURE

2.1 *Interdependent relationship between tourism and culture*

Tourism and culture are interdependent. Firstly, the absence of culture makes tourism activity meaningless, tiring and boring. For example, the cultural elements in mountain-climbing, such as legend, fairy tale, history and religion, make the climbing more distinctive, meaningful and unforgettable. Without cultural elements, mountain-climbing is just a body-building activity, failing to attract more tourists. Secondly, culture promotes the development of tourism products. For example, the products of village tourism and folklore tourism, which are based on ethnic culture, cannot do without cultural elements. Mo Yan's

winning the Nobel Prize in literature makes his hometown famous. Without such cultural deposits, Mo Yan's former residence will just be an ordinary house [2]. Thirdly, tourism promotes the development of culture. During the journey, tourists become familiar with local and historic culture. On one hand, tourists' curiosity makes local culture expand and develop. On the other hand, through tourism experience, some distinct tourism culture takes shape and enriches the content of local culture, increasing tourists' consumption. Therefore, tourism and culture are interdependent and mutually primitive [3].

2.2 *Inseparable relationship between tourism and culture*

Tourism culture is a new pattern of tourism business. It elevates the level of tourism and meets tourists' needs. Tourism helps to spread culture. In return, culture supplies a bass for the development and innovation of tourism. Therefore, they are inseparable from each other. On one hand, local culture brings tourists a special experience. For example, Shanghai features the prosperity of a metropolis and civilization of a modern city. While Mountain Hua is remembered for its beauty and legend, and Mountain Li is famous for the Tang culture, Taoism culture and the Xi'an Incident. On the other hand, tourism is essential in cultural transmission. For example, Baochai Wang and Wenji Cai are both legendary women, but Baochai Wang's hometown is better exploited than Wenji Cai's tomb. The former, south of Xi'an, is one of the most famous scenic spots. Tourists are so hooked on Baochai Wang's story that they swarm to the place where she has lived. However, fewer people visit the tomb of Wenji Cai, hindering the development of the related culture. Today, with the development of tourism, people have known more about the inseparable relationship between tourism and culture. Tourism culture is not just a combination of culture and tourism, but an inseparable union established on tourism.

3 APPLICATION OF TOURISM CULTURE

3.1 *Features of tourism culture*

National and regional characters of tourism culture are indispensible in the development of tourism products. Firstly, folklore tourism – say the Water-sprinkling Festival or Dragon Boat Festival – is rich in national and regional characters, which arouses tourists' interest and consuming desires. Secondly, catering culture stands out for its regionalism. For example, people associate enjoying Roast Whole Lamb with touring Xinjiang and think of the northeast of China on hearing Braised Pork with Vermicelli. In other words, people will think of different foods when they come to different places. Likewise, having local potherbs and wines is regarded as part of the tourism experience, which is popular among tourists. Thirdly, tourism coupled with intangible culture becomes more prominent. On one hand, region's cultural characteristics abound in China, such as Shanxi opera, Henan opera and the northeast song-and-dance duet. They differ in costume and style. On the other hand, geographical conditions differ from region to region. For example, the south of China is ocean-centered and abundant in rivers, while the northwest part is characterized by plateaus and deserts. Fourthly, national culture is interlinked in essence. Taking the New Year Festival and shadow puppet as examples, they differ in form but coincide in cultural contents. People all take Yangko, stilts, dragon dance and other forms to show joyousness. Although the forms of artistic presentation are different, the activities are similar in essence. The similarities, therefore, make it possible for the tourism culture to blend and develop.

3.2 *Exploitation of tourism culture*

There are many kinds of ways to exploit tourism culture. Firstly, custom culture, such as the Water-sprinkling Festival and Dragon Boat Festival, spreads with the development of tourism. Secondly, the historic culture is well exploited and developed by tourism. For instance, Martyrs Memorial Hall is a good way of applying tourism culture by publicizing the stories and items of martyrs. The exploitation of Red Cliff and Yellow Crane Tower is also based on historic stories and ancient works. Thirdly, folk legends are well developed with the exploitation of tourism. The folk stories of Mount Hua are alluring, increasing the fun of tourism and expanding the tourists' imagination. In addition, some other cultures have been widely used, such as the catering culture, traditional Chinese opera culture and performance culture.

4 VALUE OF TOURISM CULTURE

4.1 *Tourism culture promoting development of tourism*

Culture makes tourism more meaningful and interesting. Firstly, tourism culture promotes the exploitation of tourism products. With the development of tourism, traditional tourism can not meet people's needs. Innovation becomes the key to developing tourism sustainability, which cannot be made without tourism culture. Secondly, tourism culture enriches the traveling experience and expands the tourists' spiritual world. Thirdly, tourism culture is the core of tourism business. Some tourism products will become meaningless without tourism culture. For example, without local catering culture, going to a Food Festival will be the same as going to a restaurant for a meal. What's more, tourism products are updating fast, which creates a better environment for tourists and promotes the development of tourism.

4.2 *Tourism culture spreading and developing culture*

Culture can not develop without tourism. Firstly, tourism is the symbol of local culture. When traveling, people strengthen their knowledge and curiosity, helping to develop local culture. Secondly, tourism makes culture spread further, for which Chinese tourism products abroad set a good example.

5 OPPORTUNITIES AND CHALLENGES IN APPLICATION OF TOURISM CULTURE

The application of tourism culture promotes the development of tourism and local culture. In the information age, internet and multi-media give people a wider and deeper touch with tourism culture. On one hand, the market of tourism business is expanded. On the other hand, the economic benefit of culture is explored, which is an important opportunity for the development of tourism culture. However, improper usage of tourism culture influences tourism negatively. For example, some tourism products are named as "devils into the village", which damages the image of tourists. Therefore, there are still challenges in the exploitation of tourism culture, which should be dealt with scientifically and reasonably.

6 CONCLUSIONS

The application of tourism culture brings much potential to tourism. Influenced by international tourism, Chinese tourism culture has a wide space to be explored. Chinese tourism will succeed through discarding the dross and selecting the essence of culture. And then, tourism culture will enrich social and economic values.

REFERENCES

[1] Shi Qingfang, *Discriminating between Urban Culture and Tourism Culture*, Journal of Heihe University, 2013(03), 31–33.

[2] He Yougui, *Deep Exploitation of Culture Contents for the Sustainable Development of Tourism*, Knowledge Economy, 2013(13), 113–113.

[3] Kang Jiangfeng, *Thinking of Deep Exploitation of Tourism Resources in Zhou and Qin Dynasty*, Modern Enterprise, 2013(05), 55–56.

Future Communication, Information and Computer Science – Zheng (Ed.)
© 2015 Taylor & Francis Group, London, 978-1-138-02653-7

Network management in education system

Y. Qi

Binjiang College of Nanjing University of Information Science & Technology, Nanjing, Jiangsu, China

ABSTRACT: The work mainly analyzed the function of network management in the education system. The advantages of network management were analyzed firstly, and then the Chinese education system was explored. Finally, based on the mentioned two points, the work analyzed, in depth, the function of network management in the education system. It is hoped that people have a further comprehension of the education system through network management.

Keywords: network management; education system; function

The strategy of developing the country through science and education is an eternal theme in China. And the reform and development of the education system has an obvious effect on the teaching achievement. Chinese teaching work presents an orderly development trend by deepening the education system reform. Especially in modern society characterized by a market economy, there is a severe shortfall in supply of qualified people. Whether the direction of the education system is right or not determines the direction of teaching and teaching achievement accordingly [1]. Therefore, it's very important to analyze the function of network management in the education system.

1 ADVANTAGES OF NETWORK MANAGEMENT

With the development of information technology, network management is known and accepted gradually. Especially in modern society, information technology is widely applied, and network management becomes an important management method [2]. The reason why people focus on network management is that it possesses a unique advantage and can play an active role in the education system. Specifically, the advantages of network management are as follows.

1.1 *Reasonable mode of communication*

Nowadays, the methods of communication among people have become various. Besides SMS and phone calls, some advanced communication tools (e.g. QQ, MSN, and even Wechat – the most popular chat tool at present) are springing up. Based on these communication tools, network management can make communication more efficient [3]. Obviously, communication by these tools is a unique advantage of network management. In the process of the education system reform, some management staff and objects involved can communicate efficiently by these network communication tools.

1.2 *Timeliness in network management*

The education system varies in objects or environment. The developing society and the changing education environment are double-edged swords to the education system. On the one hand, these changes can drive the reform of the education system. On the other hand, the education system which has to change according to the education environment will take a hit. Thus, education system integrated into network management can find its feet better and respond to the changing information timely. By analyzing the data reasonably, it can supply better evidence, thereby avoiding the blindness of the education system. In summary, the timeliness and emergency of network management can relieve the pressure of the education system and exploit new space.

1.3 *Universality in network management*

All developments of society, from national development to household affairs, have to rely on network management these days when information technology develops rapidly. Network management has become a hero of management field in modern society. Likewise, just because of its universality and power, network management can promote the efficiency of management. And management of different levels can communicate or learn by network (e.g., management communicate with each other).

2 ANALYSIS ON CHINESE EDUCATION SYSTEM

The education system should regulate the development of educational institutions reasonably and ensure our education working smoothly. In term of the Chinese present education system, education offices in different regions or ranks become a vital part of the education system and promotes its development in various ways. At present, Chinese education can be divided into the following periods.

2.1 Preschool education

In preschool education, educated persons, in general, are children from 3 to 5 years old. These children learn scientific and cultural knowledge and a good lifestyle in order to get a better growth. To say the least, the preschool education period is the base of the whole education stage and can lay a good foundation for other education periods.

2.2 Elementary education

Elementary education is a continuation of preschool education. Both the age of educated persons (e.g., pupils) and their comprehensive qualities have been improved highly. At present, the pupils, from 6 to 11 years old, can know well basic scientific and cultural knowledge and own initial cognition and perception about something.

2.3 Elementary education

Nowadays, education attracts more peoples' attention, and elementary education becomes a necessity. Not only in the city but also in rural areas, nine-year compulsory education has become a basic requirement and even been broken through in some regions. Many students graduate from elementary school and then enter senior school or technical secondary school, which promotes the comprehensive qualities of educated persons from a broader sense. In general, the age of students in secondary education is from 12 to 17 years old.

2.4 The higher education

The higher education emerges when the Chinese education system has developed to a certain period. After secondary education, many students begin their higher education. In this period, students can construct a knowledge or capability system and give full play to their own strengths. And our educational structure will be optimized accordingly.

2.5 Continuing education

Continuing education is mainly aimed at the adult. It can wipe out illiteracy and popularize the educational level. Especially when social demand for qualified persons is rising constantly, continuing education can play an important role and improve the quality of our citizens. Furthermore, it's the continuation of a unified admission system.

3 FUNCTION OF NETWORK MANAGEMENT IN THE EDUCATION SYSTEM

With the development and popularization of network management, the influence on the education system has gradually emerged. Especially in today's education system, the application of network management is an incentive to the education system. And when it has been popularized in a larger range, network management promotes the education reform and impels our education system to an unprecedented development.

3.1 Providing evidence for data analysis from the education system

The education system, in essence, is very complex and combined with a variety of elements at any stage. In the process of education system reform, the effect of a traditional or existing education system should be analyzed on the basis of real data. And effective education reform approaches will be summarized consequently. Network management can coordinate data well and make a reasonable analysis and synthesis to them accordingly. And it can expose the merits and demerits of our present education system in data and provide strong data support for the reform of the education system. The continuous development and expansive scope of network management can guide people to develop and optimize the education system more flexibly and diversely. Thus, the scientific data analysis and sorting could provide an impetus to improve the education system. Moreover, although the information about the education system is more and more complex, network management can still classify, summarize and integrate the same attribute information efficiently. In the process of optimizing the educational system, this information has been combined with the actual situation and can ensure timeliness and realness.

3.2 Promoting the combination of the education system and modern society

Modern teaching methods and modes need a corresponding education system, only when the two are united tightly, education can be developed more effectively. In modern society, science and technology is everywhere and plays an important role not only in economy development but in the education reform. Informationization, technicalization and networking have become an inevitable tendency nowadays. Unfortunately, the Chinese education system, in general, has not yet combined the modern technological society absolutely. Network management, as a new thing or media, can promote the combination of education system and social development. And it also can break traditional teaching media and find the motivational power of educational development from the network. All these can increase the scientificity and sustainability of the education system and promote the reform of the education system actively. Therefore, in the highflying information society, the network will become a bridge between the Chinese education system and other new things. And it's also an important heritage by which our education system could obtain a breakthrough.

3.3 Providing a variety of elements for the reform of education system

Chinese education presents a diversified trend by networking. It has changed to some extent not only in teaching methods but also in the theories guiding teaching. These changes are positive, progressive, and provide an impetus to the development of modern society. However, compared with teaching methods, modes or ideas, the Chinese education system could not follow the pace of reform. Network management, not only in guidance ideas but in staff support, supplies many thoughts about our education system. Therefore, it promotes our education reform in a larger range. Meanwhile, network management can combine the education system with the teaching practice and promote the comprehensive development in the field of education.

3.4 Providing an impetus for promoting the quality of management staff in education system

Nowadays, the informationization has become mainstream. And the staff in any field should master certain technology (e.g. skills in the network operation). Equally, the staffs that take part in the Chinese education system reform need to possess higher comprehensive knowledge. But at the present stage, neither the management staff nor the general staff have this higher capability. Thus, in order to adapt the society, all participants should, by all means, show great initiative in enhancing their ability. Network management, in fact, just encourages these staffs to improve their own ability. For example, it needs the management or general staff to observe new things and learn new skills actively. And consequently, the requirements of the education system reform will be reached.

4 CONCLUSIONS

Network management provides a scientific basis for improving our education system, and through which the combination of education system and modern society can be realized effectively. Meanwhile, it also guarantees the timeliness of the education system and pushes forward consummation and development of the education system by the introduction of advanced elements. Furthermore, network management can expand the Chinese education system and promote its internationalization.

ACKNOWLEDGEMENT

The work is the initial results of "college students psychological quality training" project supported by Zhejiang province moral education excellent optional course. Project Number: Z0122613001. And the initial results of "Research on the practice path of 'example in life' about college students' value training" – supported by The Ministry of Education Planning Fund. Project Number: 13YJA710009.

REFERENCES

[1] Yang Liu, Liu Jiajun. Networked Teaching Methods Based on Modern Information Technology. Journal of Chengde Petroleum College, 2013(1):39–42.
[2] Han Zhihua, An Junchao. Promoting the Scientific Development of Adult Education by Changing Ideas and Deepening the Reform. Journal of Agricultural University of Hebei (Agriculture and Forestry Education Edition), 2013(1):1–4.
[3] Pan Mingying, Zhao Hongming. Research on the Architecture of Campus Network Based Instructional Infrastructure Platform of Physical Education. Journal of Jiangsu Teachers University of Technology, 2012(6):114–117.

Future Communication, Information and Computer Science – Zheng (Ed.)
© *2015 Taylor & Francis Group, London, 978-1-138-02653-7*

Role of rational training management in physical education

G. He

Hainan Vocational College of Political Science and Law Haikou, Hainan, China

ABSTRACT: This work analyzed the role of rational training management in physical education. Firstly, it revealed the dynamic role of rational training management by comparing the increasing input and rational scientific training methods. Then the difference of the role of unitary management and dualistic management in training was analyzed. Finally, from two aspects – the evolution of rational training management and its roles and its contribution to the cultivation of athletes' abilities, the scientific structure and role of rational training management were analyzed. It is hoped to be helpful to physical education.

Keywords: rational training management; scientific structure; dualistic training

1 INTRODUCTION

Sport is inseparable from teaching and everyday training. Physical education in China, although started later than in Europe and the United States, emphasizes the training in everyday life [1]. After hosting the Olympic Games successfully, China's competitive sports have made a greater impact on all aspects around the world. Physical education is one of the most important parts of national sports. In order to achieve more outstanding performances in the sports world, it needs to put more emphasis on daily rational training [2]. A good performance of rational training depends on a good administrative management. In essence, a good administrative role of physical education departments will make an important and positive influence on athletes training.

2 ESSENCE OF PHYSICAL EDUCATION TRAINING MANAGEMENT: SCIENTIFIC TRAINING METHODS

Modern sports development has a history of over 100 years, and the reason why sports have influenced the world positively, especially for its significant progress around the world, are mostly the achievements in physical education. After the founding of P.R.C., the success in sports, especially in competitive sports, attributes to the achievements of physical education training management [3]. In essence, the core of physical education training management in China consists of two parts – the continuous investment in resources and manpower and the continuous improvement on the training methods.

2.1 *Comparison on increasing input and making training methods more scientific and rational*

Experiences from the process of competitive sports development show there are two important factors, training methods improvement and inputs increase, which determine the strength of a country's sports and ultimate result of competition. If the inputs are inadequate, especially for the manpower and resources, it will have some negative influences on sports. However, ultimately, from the history of New China's sports development, it can be found there were still breakthroughs even in the period short of resources. Therefore, in essence, the key factor to physical education management is still training management. A scientific and advanced training method can promote a fundamental development in sports. It can also be regarded that the improvement of the training method is a determining factor in sports performance.

2.2 *Dynamic role of rational training management*

Training method has a certain initiative. Under the circumstances that trainers are in equivalent physical conditions, the improvement of the training method can help trainers improve their performances, and then achieve brilliant scores. Sports industry in China, for some small scale sports, has more advantages because they can make achievements with the strong supports of administrative power. But in some sports with high-gold content, although money is invested equivalently, Chinese athletes' performances fall behind those of western countries because the weak points in the training methods restrict China's sports development. In essence, it is caused by the limitations in daily training management. Therefore, Rational training management plays a positive role on physical education.

3 DIFFERENCE OF THE ROLE OF UNITARY MANAGEMENT AND DUALISTIC MANAGEMENT

At present, theory guides practice in physical education training around the world. In the process of the world's sports development, most countries respect practice and they will adjust the teaching theory according to the results of practice compared with few countries adoring theory. There is no excessive constraint. However, governments participate more widely in China's sports management. Therefore, a series of training methods and theories, recognized by administrative power, will be applied to daily training, thus restricting China's physical education development.

In fact, during the practice of physical education and training around the world, a comprehensive dualistic management theory is produced, that is, techniques and the athlete's own physical condition together constitute an overall athletic performance. Athletes' own techniques in the sport, as well as their physical fitness, are the essence of sport, and they cannot randomly be split up. This unity is regarded the concept of unitary management. The role of unitary management and dualistic management relatively differs a lot.

3.1 *Difference of unitary management and dualistic management in physical education training*

Currently, the leading and managing team in the international sports world and China's sports industry agree with dualistic training management. But from the late 1950's when China's sports industry gradually developed, China has gradually accepted a concept of unitary training management. This concept, on an international scale, is still relatively rare.

The unitary training management is a training with a relatively-specific sport technique. The training of the athletes' own physical fitness trends to technique training and its skillful use, which is an important feature of unitary training management. Although this feature can help athletes gain outstanding achievements in games, its potential is still insufficient, because it ignores the foundation of the athletes' physical fitness which is helpful to athletes' improvement. So relatively, the positive role and impact of the unitary training mode is slightly inferior than those of the dualistic training mode.

3.2 *Influence of unitary training management on China's sports industry*

It shows that the training mode of daily physical education in China differs greatly from the internationally accepted training concept. Such difference relatively causes little efficiency of China's sports industry

in some higher gold content competitive events. It also leads to a deep rooted weakness in China's sports training; therefore, China's sports industry imports a great amount of training experts and coaches from Western countries. But the introduction didn't change the concept of unitary training management, and these specialists are required to obey the administrative management in China. So in this way it can reduce the weak points of China's sports training rather than change the backward situation.

4 SCIENTIFIC STRUCTURE OF RATIONAL TRAINING MANAGEMENT AND ANALYSIS ON ITS ROLE

Either unitary or dualistic training management, in their essences, are for the athletes' overall development. The overall development requires scientific and rational training management, which includes 3 parts – recognition, specific practice and feedback. Actually, the feedback is a re-recognition of performances during training. Such process is a circle of re-recognition, so lastly scientific training management consists of two parts, recognition and training.

4.1 *Rational training management's evolution and its respective role*

During the development of physical education, training varies in 3 different forms. These forms are necessary in the evolution process of rational training management mode and play their own roles.

Firstly, the first rational training management mode is training experience, in which athletes are trained by repeating their experience from the first activity to the second activity. The key method of training experience is that the teacher leads the student to complete a sport mode. In this way, it will bring a strong sense of presence to students, help inexperienced students learn more in a short time and shorten training time. However, students may ignore their own feelings and tend to imitate.

The second rational training management mode is experience training. Compared with the first one, it makes progress. In this mode, teachers pass all of their experiences in sports to students. The unsuccessful experiences will be improved and, during the improvement, an imitation will be made. Such imitation mostly focuses on accumulation of experience and lessons. The continuous accumulation makes improvement. This kind of teaching mode can collect and correct failures made in the past and promote the training process. But there are some obvious disadvantages, e.g., it can only make progress at the cost of more failures and takes a long time.

The third rational training management mode is scientific research training. Scientific research is with certain rules and rational modes, such as controls on variables, so it is forward-looking. Such kind of forward-looking training mode can promote the breakthrough of training methods and innovate the acknowledgement of both teachers and students and improve their training method and skills. Therefore, it is more competitive in sports.

4.2 *Cultivation of athlete's abilities in rational training management*

The result of rational training management finally affects the main body of sports, such as athletes. The physical fitness developed during training actually is produced by the athletes' own body functions. In essence, it is a kind of intelligent physical fitness. It can not only enhance the athletes' reactions to outside, but also improve and adjust their inner environment. The features of intelligent physical fitness are not only found in teamwork games such as football or volleyball, but also in individual sports such as weightlifting and racing.

The development of intelligent physical fitness is contributed to the influence of rational training management. Even the movement of outstretching or shrinking an arm is a complicated system. E.g., changes on the forms of human's organic structures are made by influences on biochemical and physiological structure affected by time and space.

Therefore, rational and scientific sports training management is helpful to cultivate the athlete's comprehensive abilities during competition. The skillful coordination of athletes during rational training brings them a skill, which enables them to use their techniques more flexibly. The skill is guided by many kinds of wisdom and knowledge, especially for a combination of scientific knowledge of sports and physical fitness and philosophic thinking. Then with the integration in multiple levels and factors, the athletes' tolerance and technique are upgraded, and then sports are more adaptable to humans.

5 CONCLUSION

Sports training cannot be separated from rational teaching and management. Only by sticking to rational training management during daily sports training and physical education can the levels of athletes' skills and techniques be improved to achieve the best performances.

REFERENCES

[1] Jin Wanzhang, Modes about Physical Education Training Management, New Sports, 2009(2): 574–575.
[2] Wang Jin. Study on Solutions to Physical Education Problems, Sports Space, 2008(4): 291–292.
[3] Wang Ying. Reflections on School's Part-time Physical Training Situation, Science and Education, 2011(3): 275–276.

Future Communication, Information and Computer Science – Zheng (Ed.)
© 2015 Taylor & Francis Group, London, 978-1-138-02653-7

Consideration of human resource management in competitive sports

G. Li
College of Physical Education, Huaiyin Normal University Huaian, Jiangsu, China

ABSTRACT: The current issues of the human resource management in competitive sports were analyzed. Some optimization tips were proposed based on the general awareness of human resource management. These suggestions are hoped to promote the resource management in Chinese competitive sports.

Keywords: competitive sports; human resource; management

1 INTRODUCTION

China, with the success of the 2008 Beijing Olympic Games, has attached great importance to sports and formulated specific developing plans for competitive sports. Current competitive sports of China have a broad space for development. HRM (human resource management) of competitive sports has drawn great attention with the development of sports. Optimized HRM will promote the development of competitive sports [1]. And Chinese competitive sports will make a comprehensive progress with the support of management personnel.

2 HRM IN COMPETITIVE SPORTS

With the development of competitive sports, some related issues (e.g., HRM) are drawing people's attention. People, especially the fans and management staff, have been familiar with the concept and characteristics of competitive sports [2]. Therefore, clear consideration is primary for HRM of competitive sports.

2.1 *Analysis on HRM*

HRM refers to the comprehensive management and planning of human resources within a certain field. Talents are recruited to the corresponding department for their targeted management. So far, China has completed the mechanism of human resource management. Moreover, both enterprises and institutions have updated their consideration about HRM. Some social departments have established corresponding mechanisms for HRM such as personnel screening, recruiting mechanism establishment, staff training and performance assessment. Most of the Chinese enterprises and social departments have established perfect salary and welfare management. HRM is

becoming a key part for the development of China [3]. It is mainly carried out in two aspects: Firstly, HRM is based on the actual needs. Talents and employers are closely connected to better perform the role of HRM and the function of talents. Secondly, some managing methods and modes of HRM can be utilized to perform its functions. For example, the idealistic change in the management of talents can enhance their efficiency and initiative, which combines personnel and their department. Both actual needs and managing methods are the key factors in HRM and should be combined.

2.2 *Characteristics of HRM in competitive sports*

HRM of competitive sports has a generality with other human resources management – they are both the effective management of human resources. However, the HRM of competitive sports has its own characteristics – pertinence and professional property. The objects of competitive sports in HRM are athletes and managers who have their advantages in competitive sports.

Firstly, the HRM of competitive sports has similarities with other HRM. It should adapt to the overall development of HRM in China to achieve good managing results. From the viewpoint of similarity, the HRM of competitive sports should be a strategic, well-planned, scientific and rational management.

Secondly, the HRM objects of competitive sports have their own characteristics – manager and objects should be associated with sports, which are their professional properties. They need to have professional knowledge and skills on sports. Therefore, sports HRM, from the viewpoint of characteristics, needs professional management of team and objects. Its professional properties make it different from other types of human resources management.

3 HRM ISSUES OF COMPETITIVE SPORTS

In modern society, HRM has been developing and progressing. The HRM issues of competitive sports have hindered its development in China and affected the effect of management. Currently, the HRM issues of competitive sports mainly lie in several aspects as follows.

3.1 Unprofessional mechanism

Chinese HRM of competitive sports needs to be systematized. The HRM of competitive sports needs a scientific and sound management mechanism. However, the incorrect perceptions of managers and rigid management led to the tradition of "appointment" which is contrary to the requirements of the competitive market. It is not conducive to human resources management.

3.2 Unsoundness of talent concept

Talent concept is the supporting concept for HRM of competitive sports. Both the talent introduction and appointment require an advanced talent concept to improve the effect of human resources management. However, this requirement is ignored in the current HRM of Chinese competitive sports. Managers have not established an effective mechanism for the selection, training and development of talents. Unsound HRM will lead to the problem of talent outflow. Some managers ignore the problems of talent fault because they concentrate on the cultivation of talents from some level.

3.3 Unsound incentive mechanism

The competitive-sports HRM had a monotonous incentive mechanism – material incentive or mental incentive, rather than their combination. This monotonous incentive mechanism could not satisfy the management objects. Meanwhile, the unsound allocation mechanism also affected the HRM effectiveness and efficiency of competitive sports. Therefore, unsound incentive and allocation mechanisms suppressed the enthusiasm and initiative of managing objects.

4 OPTIMIZATION OF HRM IN COMPETITIVE SPORTS

The HRM of competitive sports should be optimized according to social development. Targeted HRM should adapt to the essence of human resources. Moreover, the HRM of competitive sports can obtain better results if the specialization requirements are met. Then, the HRM of competitive sports in China will develop scientifically in the long term.

4.1 Improvement of the HRM mechanism

The effectiveness of HRM is related to whether its management mechanism is sound. The traditional mechanism "appointment" in sports HRM is contrary to the requirements of the market economy and needs to be reformed. For example, free competition mechanism, which is a competition mechanism for recruitment, should be established to create a fair competitive environment and select excellent talents. Meanwhile, in terms of scientificity, sports HRM should be developed based on modern HRM to improve its management mechanism.

4.2 Establishment of comprehensive talent concept

A comprehensive concept of talent can avoid talent outflow and retention and effectively promote the development of competitive sports HRM in China. Firstly, humans should be regarded as the objects of HRM in the establishment of a comprehensive talent concept because they are the subject, and their needs should be satisfied. The HRM of Chinese competitive sports can achieve good results only under the guidance of a "human-oriented" concept. Secondly, the development of human resource needs to be focused – talents who adapt to competitive sports will be cultivated. They will receive targeted cultivation based on the demand for talents and the development of competitive sports. Therefore, the development of talents, the HRM and development of competitive sports should be combined to improve the HRM of competitive sports.

4.3 Optimization of incentive mechanism

Optimization of incentive mechanism is the key to promote the long-term development of sports HRM. The HRM of competitive sports, with the development of competitive sports, needs a diversified incentive mechanism and fair distribution mechanism. Therefore, through the optimization, a perfect and scientific incentive mechanism will meet the demands of competitive sports HRM.

In conclusion, according to the management issues and characteristics of competitive sports, the HRM of competitive sports should be reformed and improved to achieve its comprehensive development. Thus the competitive sports of China can be promoted and confirm her place in the sports field of the world.

REFERENCES

[1] Li Ying, Sun Jianpeng. *Strategic Human Resource Management and the Construction of Univeristy Faculty Exploration.* Science-Technology and Management, 2013 (2): 119–122.
[2] Lu Xingtie. *Some Thoughts for Human Resource Management.* Heilongjiang Science and Technology Information, 2013 (6): 127–127.
[3] Wang Hong *Research for the Development Situation of Chinese Competitive Sports.* Success: Education, 2012 (12): 277–277.

Future Communication, Information and Computer Science – Zheng (Ed.)
© 2015 Taylor & Francis Group, London, 978-1-138-02653-7

Financing mode and risk control of supply chain

R. Huang
Guangdong University of Technology, Guangzhou, China

W. Huang
Guangdong University of Finance, Guangzhou, China

ABSTRACT: The work analyzed the financing mode and risk control of supply chain. It first analyzed the current Supply Chain Finance (SCF) and explained the connotation and current mode of SCF. Then it analyzed the risk and control of SCF in detail from the aspects of SCF market, credit and transaction. It is hoped that the analysis and research in the work will be helpful for SCF financing mode and risk control.

Keywords: Supply chain; financial risk; enterprise credit; bank-enterprise relationship

1 INTRODUCTION

Supply chain contains the economic factors of logistics, capital and information flow. Logistics flow from upstream enterprises to the downstream firms as the final result of production. Information flow spreads from the whole supply chain to different directions. Capital flow transfers from downstream to upstream. SCF is generated in the interactive process. The nature and content of this kind of financial activity are different from the previous ones. As the production between up-and-downstream companies, the supply chain is an important organization mechanism maintaining the modern market. This mechanism helps companies integrate production activities into the whole industry. And the relationship between different companies transforms from product exchange to interdependent cooperation. SCF has become an important power influencing the long-term business of most companies. This work presents the analysis and investigation on the connotation, mode and risk controlling of SCF.

2 SCF AND SCF MODE

SCF generated in the operation of supply chain is an important medium of monetary integration between different companies. It is also an advanced form of interaction between the bank and entity economy, which has an epoch-making significance to the bank and production.

2.1 Connotation of SCF

SCF is an important means of funds adjustment to the internal of supply chain. It reduces credit risk and increases the profits of banks. SCF is composed of a monetary settlement and credit between companies. With the implementation of the banking reform in China, the commercial banks are more dependent on modern companies. The banks hope to get the support of dominant manufacturers in supply chain to reduce credit risk and expand off-balance-sheet activity. Accordingly, SCF embodies the banking penetration along supply chain in a sense.

Monetary settlement between companies depends more on banking and should be carried out via banks by law in China. Due to the frequent and close fund exchange and lots of write-offs, it is difficult for banks to expand business to the chain. Banks need to establish business with the dominant enterprises to complete the off-balance-sheet activity.

Credit behavior in the supply chain is always carried out between the dominant enterprises and downstream firms. It is possible for the dominant to provide funds for the downstream because of their strong economic strength and large amount of monetary funds. Meanwhile, the dominant enterprises can discount by negotiable instruments for their strong social influence and easily take steps to freeze the funds if dishonor occurs. During the course of business, the companies in the supply chain can provide guidance for each other and enhance mutual competitiveness.

2.2 Current SCF mode

The current SCF mode is composed of business settlement and credit. As the two parts have different business contents, the operation modes are discussed respectively.

2.2.1 Operation mode of supply chain settlement

Supply chain settlement is the monetary payment between companies within the supply chain. It is usually conducted by the banks which extend the business into the chain via dominant enterprises. The banks need to cooperate with the dominant enterprises and establish the specialized settlement mechanism. As supply chain includes information and capital flow, the enterprises in the chain should integrate the information to compete with the external enterprises. In addition, the capital flow contains information transfer. Thus, the mechanism of mutual trust should be established between the banks and dominant enterprises in the settlement.

2.2.2 Operation mode of supply chain credit

Credit within the supply chain is mainly transmitted from the dominant to the downstream. The capital flow in supply chain is usually from the downstream to the upstream, which leads to a great amount of funds handled by the dominant enterprises. However, upstream enterprises will help the downstream firms with deficient funds get the credit because they determine the completion of the capital circulation.

Upstream enterprises can afford the downstream firms corporation credit and the current credit mode is carried out by discount. Though the interest rate of the discount is higher than that of a bank, downstream firms should accept the discount to withdraw capital because of a smaller scale of fixed assets and restricted credit scale. Upstream enterprises can also guarantee for the downstream with their own credits to get a secured loan and reduce financial risks.

3 FINANCIAL RISK AND CONTROL IN SUPPLY CHAIN

The risk of SCF, including market, credit and transaction risk, is a crucial factor affecting SCF stability. And it should be controlled and not be ignored.

3.1 Market risk and control in supply chain

SCF relies on capital flow and the supply chain is the organic integration of production in different enterprises. Entity economy impacts SCF. The risk of SCF is the losses brought by the changes of the supply and demand relationship. When the market is volatile, it will be tested whether the downstream products achieve their values. When the downstream firms run into trouble in sales and capital withdrawal, it is hard for the banks to call in a loan, which results in financial problems.

The control of the market risk of the supply chain needs close cooperation between the dominant enterprises and banks. As distribution institutions of funds, banks can grasp the basic information of macro economy. The profitability of bank loans will be reduced when the market production declines and demand drops. The decrease of loan demands leads to the reduction in equilibrium interest rate and growth in discount, which reflects the sales difficulties. Banks can feed the information back to the enterprises and help them adjust the production scale and slow the expansion. The dominant enterprises in the chain can alter sales strategies to guide downstream firms with the change of supply and demand.

Another measure for controlling the market risk of the supply chain is to establish the warehouse financial service. This mechanism can help downstream firms get the loan and the upstream enterprises to withdraw funds in time when sales difficulties occur. The warehouse finance depends on the internal logistics management. A lot of stock is usually hoarded in the logistics. Banks can negotiate with enterprises to mortgage the stock. The biggest problem of mortgage loans from current assets is the transportability and instability. Thus, enterprises should inspect and check the collateral carefully to avoid deterioration and unnecessary losses.

3.2 Credit risk and control in supply chain

Credit risk in supply chain is the loss brought by the default of downstream firms. In a supply chain with weak management, upstream enterprises take a lower control of the downstream firms which can quit any time. The financial loans guaranteed by upstream enterprises will be at risk. Enterprises and banks can unify the supply chain and choose downstream firms by credit investigation to avoid dishonest firms. Both upstream enterprises and banks pay more attention to funds return, which relies on trustworthy firms.

There are two important ways in controlling downstream firms – direct acquisition and branding control. Direct acquisition leads to changes of property rights and increases of transaction costs. And it is mainly depended on the weighing between the financial risk and transaction fee whether the acquisition is carried out. On the contrary, the branding control costs less and brand is an important measure of product specialization. High homogenization weakens the ability of price control. After branding and specialization, the enterprises have more control over price and the downstream firms can integrate their products into the enterprises'. Branding is often related to the construction of a complex relationship between rights and obligations, which requires the upstream to take thorough selection from the downstream.

3.3 Transaction risk and control in supply chain

Transaction risk is the most difficult to control in SCF. In the process of development, some enterprises become large-scaled which leads to excessive management levels and high principal-agent costs.

Principal-agent is a big reason for transaction cost. And the principal-agent problems caused by the shortcomings of the managers are mainly manifested as opportunism and bounded rationality etc. Opportunism is quite obvious in the supply chain. Because upstream enterprises and banks do not learn the operation of downstream markets, managers can make use of asymmetric information for fraud. Accordingly, the enterprises should strengthen the management of the supply chain and improve the incentive and supervision mechanisms to avoid transaction risks. It is also a good way to solve the problem by issuing shares for the staff.

ACKNOWLEDGEMENT

This work is funded by Humanities and Social Sciences Youth Fund of the Ministry of Education "Study on finance-supporting model of strategic industries based on risk investment" (13YJCZH055).

REFERENCES

[1] Xuhong Bao. *The Research of SME's Financing Channels Innovation Based on Supply Chain Finace*. Science Technology and Industry. Vol. 9, No. 1, (2009), pp. 74–77.

[2] Shuqi Yang. *Related Game Analysis on Supply Chain Finance*. Journal of Chongqing University of Science and Technology (Social Sciences Edition). No. 4 (2011), pp. 115–117.

[3] Dongxue Ma and Yifei Zhao. *Risk Measurement of Supply Chain Finance in Third-Party Logistics Company*. Logistics Sci-Tech. No. 2 (2011), pp. 54–56.

Future Communication, Information and Computer Science – Zheng (Ed.)
© 2015 Taylor & Francis Group, London, 978-1-138-02653-7

Effect of corporate culture on corporate performance management

X. Yang
Management College, Hankou University, Wuhan, China

ABSTRACT: The work researched the effect of corporate culture on corporate performance management. The function of corporate culture and performance management in enterprise development and the relationship between them were discussed firstly. And then the effect of corporate culture on corporate performance management was analyzed. Corporate culture is a core competence in enterprise development, and performance management is an essential measure for this competence. The work will contribute to the long-term development of enterprises.

Keywords: corporate culture; performance management; research on relationship

1 FUNCTION AND RELATIONSHIP OF CORPORATE CULTURE AND PERFORMANCE MANAGEMENT IN ENTERPRISE DEVELOPMENT

1.1 *Function of corporate culture and performance management in enterprise development*

Corporate culture is a macro concept. It's a culture management pattern with a unique attribute which forms gradually in enterprise development. It's also the soul of an enterprise and an impetus ensuring that all employees have the same values. Corporate culture can infiltrate all aspects of the enterprise (e.g. rules and regulations, employee behavior, organization structure) and promote enterprise development. Corporate culture is seen as the soul of enterprises and the lasting power or key factor of development in enterprise management theory. Professor Cottle, Harvard University, believed that a successful management pattern should be different in different countries or even just in different industries of one country instead of being consistent [2]. The reason for this is the diversity of culture in different countries or enterprises. Management patterns carried out by different cultures are usually stamped with these culture brands, and variations will happen.

Corporate performance management, on the contrary, can promote core competence of enterprises from the micro level. And it's an important way to achieve the strategic goals of enterprises. According to the relevant survey of *Business Week*, the key factor of an enterprises' success is a perfect performance management system. Performance management is a systemic, dynamic management process, and its value is improving the business management. Therefore, optimizing performance management can actually promote core competence of enterprises and guarantee enterprises to achieve strategic goals. Moreover, performance management is a significant method to examine the performance of individuals or teams, and it can also ensure individuals, teams or organizations and the same strategic goals [3]. Scientific performance management can promote employees' enthusiasm and enhance the controllability of management. It also improves the level of modern management and sustainable development of enterprises.

1.2 *Relationship between corporate culture and performance management*

The developments of enterprises are so various that corporate cultures have distinctive features. That is to say, different enterprises have different corporate cultures. Corporate performance management can be improved continuously under the influence of corporate culture. Correspondingly, the scientific performance management system and value distribution system can be formed. Corporate culture and performance management should be complementary to each other. Specific introductions are as follows:

Firstly, corporate culture is a soft environment for developing performance management. It reflects the value of enterprises and is a kind of intangible assets. When it's applied in different management models, corporate culture is a common code of conduct and a basis for the judgment of enterprises. Therefore, as an important method of enterprise development, corporate performance management can be influenced inevitably by corporate culture. And it's also a management means guided by corporate culture.

Secondly, performance management contributes to the development of corporate culture. Performance management is a micro operation, and it is specified in culture values and business philosophy of enterprises.

Employees can feel their existence, and performance management can also restrain or encourge employees' actions and be accepted by them. Thus, performance management can enrich corporate culture in the actual execution process. A good performance management system can stimulate the vitality or potential of enterprises and establish their common values. Therefore, performance management and corporate culture should be complementary to each other, and both are indispensable in the development of enterprises.

2 EFFECT OF CORPORATE CULTURE ON CORPORATE PERFORMANCE MANAGEMENT

2.1 Existential problems of performance management breaking away from corporate culture

Corporate culture directs the enterprise development from the macro level, and corporate performance management can realize this direction from the micro level. Thus, scientific operation of performance management needs a fine corporate culture. Meanwhile, corporate culture influences performance management deeply. The research of Professor Cottle shows that different corporate cultures can bring different performance managements. Therefore, business managers with modern management awareness should emphasize the unity of corporate culture and performance managements. Alibaba, China's largest online retailer, advocates corporate culture and performance management thus obtaining some achievements. Seen from the outside, Alibaba is supervised by its strict performance regulations. In fact, more important is its corporate culture in the process of its success. Unfortunately, many enterprises don't realize this complementariness to each other in China. And in practice, corporate culture and performance management are separated from each other in many the enterprises. Many problems in the enterprises management are as follows:

Firstly, the constraint of corporate culture to performance management is often a formality. Performance evaluation replaces performance management simply in many enterprises management. Performance management is an essential part of the human resources department. And as a method of the enterprise salary system, performance management can be different between salary and bonus distribution. Although some enterprises can distinguish between performance evaluation and performance management, they often take a perfunctory way in performance evaluation of employees without a good corporate culture. Many business managers favor a happy medium in management. For instance, they usually like seniority or egalitarianism and adopt a compromise deal in performance evaluation which becomes meaningless.

Secondly, a derailed corporate culture can cause the differentiation of internal performance targets. Some enterprises are not based on the strategic development of the enterprise when drafting performance targets. They just emphasize the core interests of their own department instead of the common goals of enterprise development. Namely the production, engineering or business department are more concerned about their own performance when making performance management targets. For example, the business department doesn't pay attention to the quality of orders, in contrast, it produces defective orders in order to achieve sales goals. Obviously, the goals are reached but loss is incurred. Such departments disagree with performance management. And they also ignore strategic goals and the whole performance of the enterprise. Reasons leading to this differentiation of internal performance targets are that many problems exist in the development of corporate culture. And there are no common values in enterprises.

Thirdly, performance management will become a bubble performance without the support of corporate culture. Correct corporate culture brings correct ideas of performance management and value. If performance management is drafted on the basis of the monotonous work of employees, employees or departments will scramble to do that which has a visible performance. Then bubble performance incurs. It seems that the production efficiency or profitability is very high in a short time. However, in fact, the long-term development capability of enterprises will be discarded because of over production or uneven development.

Lastly, if corporate culture can't be applied to performance management, the potentiality of enterprises will be affected negatively. Many enterprises take performance management simply as a routine index of employees. In this process, corporate culture can't realize the application or translation. Different values between enterprises and employees will happen because of ignoring employees' sense of identity to values of enterprises or sense of belonging and honor. Enterprises and employees can't stand together whatever happens. In this situation, performance management can only be considered as a means of squeezing employees together rather than promoting enterprise development. It eventually has a negative impact on the potentiality of enterprise development.

2.2 Applying corporate culture to improving performance management

Corporate performance management is just a tool. Only when it is accommodated to corporate culture first can performance management affect it gradually. Performance management needs to integrate with corporate culture and can be improved by corporate culture simultaneously.

Firstly, corporate culture should be introduced into performance management. Corporate culture (e.g. entrepreneurial spirit, values) should be introduced into its every link when designing performance management. Only by this means can enterprise strategic goals be transformed into habits of each employee, and employees will work according to the rules administrators expect them to. Performance management will become a rule all employees observe together. Paying attention to the introduction and construction of corporate culture can integrate corporate culture into performance management effectively.

Secondly, management should put people first. If the profits of an enterprise are based on squeezing employees, the enterprise is doomed to failure. The modern enterprise should put people first. Employees can get a just reward and make a plan for their careers at work. It's the win-win cooperation for enterprise and employees.

Thirdly, corporate culture and performance management should mutually reinforce one another. In practice, corporate culture affects performance management, and performance management reacts to it accordingly. Performance management is innovation of corporate culture in one sense. Control of performance management and influence of corporate culture can be motivated when the two mutually reinforce one another. A unity of the two can realize the objects of the enterprise and employees and the unity of values. Therefore, the integration of corporate culture and performance management is not just integrated simply.

3 CONCLUSIONS

Corporate culture and performance management, two carriages of enterprise development, can promote enterprise development if mutually reinforcing. Corporate culture and performance management are two core concepts and also the most difficult to be solved in modern enterprise management. Integration of them can help administrators to familiarize the relationship of corporate culture and performance management. The modernization of enterprise management at last will come true.

REFERENCES

[1] Niu Yiqing. Construction of Corporate Culture in Performance Management. Modern Business Trade Industry. 2010(17): 178–179.

[2] Kang Hong. Relationship between Corporate Culture and Performance Management. Science & Technology Information (Scientific Research). 2007(34): 152–152.

[3] Wang Yanli. Strengthening Corporate Culture and Performance Management. China Urban Economy. 2011(12): 38–42.

Future Communication, Information and Computer Science – Zheng (Ed.)
© 2015 Taylor & Francis Group, London, 978-1-138-02653-7

Research on the development of regional tourism culture

L. Yuan
Tourism College, Hankou University, Wuhan, China

ABSTRACT: With the rapid economic development, tourism has become one of the core industries in China and expands with each passing year. During the development of tourism, China has developed an abundant regional tourism culture due to the exploitation by virtue of local features. China enjoys a long history and many non-material cultural heritages. Through the protection of non-material cultural heritages, China can achieve a great development of regional tourism culture in view of local features. As a consequence, the economy in China will flourish.

Keywords: region; tourism culture; exploitation; development

1 INTRODUCTION

China enjoys a long history with 5000 years of civilization. From the past to the present, there are many cultural relics for people to exploit and develop. The economy-oriented strategy emphasizes the development of the third industry, which needs the exploitation of non-material industries for the growth of the national economy. Therefore, tourism has attracted much attention in recent years and made some achievements. With the specific development of tourism, various patterns of economic development emerged with great influence [1]. The local feature-based pattern of tourism has gradually been used and played an important role in developing non-material cultural heritages. This work analyzed the development of local feature-based tourism.

2 STAGES OF THE DEVELOPMENT OF TOURISM IN CHINA

Since the beginning of the reform and opening up, China has been emphasizing the construction of its economy and developing the socialist market economy. This helps facilitate the development of the economy and an affluent society. Therefore, the third industry began its rapid development in recent years, especially regarding tourism. Local tourism in certain regions a can not only support the development of its economy but also help the spread and communication of local cultures [2]. Therefore, tourism in China got the support from the economic system and the nation.

2.1 *Foundations of the development of tourism in China*

In order to quicken the development of its economy and comprehensive socialist market economy, China has drawn up plans for the development of the third industry. More local areas can take advantage of their natural landscape and human culture to attract tourists and thus facilitate the economic development [3]. Therefore, these plans encourage regions with natural landscape and historic culture to construct a tourism culture for the economic development. The foundation of the development of tourism in China is derived from the perfection of the economic system. Since the beginning of the reform and opening up, China has emphasized the development of its national economy and achieved comparatively well-off through the economic constrution. However, some special regions can not run industries due to the lack of human resource or material conditions. Therefore, in order to change these conditions and increase people's incomes in these areas, China encourages the construction of a tourism culture by virtue of the resources. Finally, the developed tourism will create jobs for local people and improve their incomes.

2.2 *Booming development of tourism in China*

Over the past years, with the increasing development of people's lives, people tend to pursue spiritual lives after the material lives were largely satisfied. Tourism is a long-term and stable industry and pursuit for the spiritual world. People can cultivate their temperament and improve the quality of their spiritual lives. Therefore, tourism is becoming a more and more important industry and contributes a lot in facilitating the development of the local economy. China should pay more attention to the development of tourism under the existing results. Many main tourism attractions are open to foreigners and thus enhance the import of foreign resources. In addition, to manipulate the development of tourism, China has enacted many regulations to guide tourism and perfect its long development. Therefore, tourism in China has flourished in recent years. The output value of tourism, especially in golden week, has increased many times.

3 INFLUENCE OF NON-MATERIAL CULTURAL HERITAGES ON TOURISM CULTURE IN CHINA

China enjoys 5000 years of history with a brilliant civilization and is abundant in historical relics. Therefore, many old capitals and regions are potential civilization relics and have received more and more attentions. The inherited historical culture is recognized as non-material cultural heritages. Many ancient civilizations in China were passed down to be the symbols of non-material cultural heritages. The poetry of the Tang Dynasty and Song iambic verse are non-material cultural heritages that can not be valued. These non-material cultural heritages, such as the Qinhuai River and Confucius Temple, have become a great wealth in modern society and are cherished by people. Before the university entrance exam, many students in Suzhou and Hangzhou come to visit the Confucius Temple and hope that they will be admitted into the universities. Therefore, these non-material cultural heritages are popular in tourism and facilitate the development of tourism in China and the form of tourism culture. Examples are given below:

Firstly, non-material cultural heritages play a conclusive role in the development of a tourism culture. The emergence of tourism is relatively easy but the construction of a tourism culture is tough. The construction needs some leading elements. Non-material cultural heritages are traditional distinguishing features. They can inspire the people's interests and form a relatively complete form of culture expressed by tourism culture. Many tourists can understand and even touch such a culture. In fact, many non-material cultural heritages in China are open to tourism and their combination is gradually developed. This combination will definitely create a potential tourism culture with more attractiveness.

Secondly, non-material culture is the foundation of tourism culture. Traditional tourism attached much attention to the national landscape. However, the natural landscape is closely related to seasons and needs to be carefully protected. As a potential culture source, non-material cultural heritages can be local features and the nurture of human spirits. They can help to create a special tourism culture. The Li Town in Yibin, an old-world town that has undergone the anti-Japanese War, reflects the Chinese culture and natural spirits with its folk activities. Grass dragon dance, dragon lanterns, beautiful boats, bull lanterns and Sichuan opera are attracting tourists both at home and from abroad. Therefore, it is necessary for non-material cultural heritages to be the foundation of the construction of tourism culture.

Thirdly, the reasonable use of non-material cultural heritages has strengthened the development of tourism. If tourism is expected to be a long-term development, potential cultural deposits are needed to support its sustainable development. The Palace Museum in Beijing is a window for tourists to be affected by the resplendence of the last feudatorial dynasty. However, the Beijing Opera is a kind of non-material cultural heritage and an inheritance of the spiritual world. People can learn the Chinese ancient culture from the Beijing Opera. Likewise, Chinese tourists can learn about their own country through culture and thus love their motherland. Therefore, non-material cultural heritages are supposed to be inherited through tourism. At the same time, the construction of a tourism culture needs the support from non-material culture to ensure its long-term development.

4 EXPLOITATION OF LOCAL TOURISM CULTURE

The development of a local tourism culture refers to analyzing local non-material cultural heritages and folk custom, constructing a tourism culture with local features and attracting more tourists. Many scholars from the regions south of the Yangtze River wrote down famous poetry on the Qinhuai River in Nanjing. Therefore, the Qinhuai River can attract more tourists who love ancient poetry by constructing a tourism culture. It is advisable to develop corresponding cultural industries according to the different interests of tourists, thus achieving the prosperity of the local cultural industry and tourism culture.

The development of tourist attractions should be based on the interests of tourists and thus achieve the reasonable development of a local tourism culture. Therefore, following action needs to be taken:

Firstly, try to command the local cultural atmosphere and advantages. During the combination of culture and tourism, it is necessary to propagate culture with tales and unofficial history.

Secondly, make efforts to learn about the interests and tendency of tourists. A good knowledge of the market orientation can help to achieve a positive development of tourism. The prosperity of tourism relies on not only tourism culture but also the extension of local cultural features. The tourism culture can be better promoted by analyzing target customers and their demands.

Thirdly, recruit professional culture-promotion staff. Tour guides knowing the local culture should be easy to find in tourist attractions. These tour guides can help to propagate the local culture and make the culture a potential power to attract tourists. The existing tourism culture will be expanded.

5 DEVELOPMENT OF LOCAL TOURISM CULTURE

Local tourism culture in China has developed a lot in recent years. The propaganda combined with local cultural features realized the promotion and development of tourism. So, how can the government further expand tourism under the premise of not damaging the local culture and ensure its positive development?

Firstly, ensure that non-material cultural heritages are free from damage and maintain their tradition. Nowadays, many tourism enterprises changed local cultural features to develop and expand tourism. The changed culture deceived tourists and caused serious results. Therefore, do not intend to propagate a mendacious culture and try to fully command local tourism features to avoid negative effects.

Secondly, strengthen the construction and protection of the local culture. The development of a tourism culture relies on culture not tourism. People go to visit tourist attractions because of their unique culture. Therefore, make sure to maintain the completeness of cultural construction and do not aim at short-term benefits. Many tourist attractions have ignored the long-term benefits and thus brought catastrophe to tourism. The construction of Fang Palace did not realize its real purpose and turned out a failure although it cost hundreds of millions RMB.

Lastly, make plans according to national tourism strategies. Developing tourist attractions at will can cause an abnormal development of tourism and is not good for the long-term economic benefits.

6 CONCLUSIONS

It can be seen that it is important to construct a tourism culture after the analysis of the economic system of China and the prospect of tourism. The combination of the local cultural background and local non-material cultural heritages can promote the spread of culture. More people can learn about the Chinese culture and history through travel. The inherited culture lays the solid foundation for the long-term development of tourism. Combined with local features, culture can also facilitate the construction of local tourism.

ACKNOWLEDGEMENT

The work was supported by 2012 Humanities and Social Sciences project of Hubei Provincial Department of Education, "Training Mode and Platform of Innovative Tourism Engineer in Hubei", Project Number: 2012G14.

REFERENCES

[1] Shi Qingfang, *Discriminating between Urban Culture and Tourism Culture,* Journal of Heihe University, 2013(03): 31–33.

[2] Wang Yuanlong, *Several Theoretical Problems on Financial Security*, Studies of International Finance, 2004(06): 13–15.

[3] Kang Jiangfeng, *Considerations on In-depth Development of Zhou and Qin Tourism Culture Resource*, Modern Enterprise, 2013(05): 55–56.

Future Communication, Information and Computer Science – Zheng (Ed.)
© 2015 Taylor & Francis Group, London, 978-1-138-02653-7

Promotion of humanized management for education

F. Li

Sichuan Huaxin Modern Vocational College, Chengdu, China

ABSTRACT: The work focused on the promoting effect of humanized management on education. Humanized management attracts plenty of attention in the knowledge era. The scientific development of human potential has become an important subject of modern management. Reflections were primarily made on the humanized management. Subsequently, a discussion was presented about the importance and promoting effects of humanized management on education. On this base, concrete measures were proposed to enhance humanized management.

Keywords: humanized management; promoting effects; measures

1 REFLECTIONS ON HUMANIZED MANAGEMENT IN EDUCATION

In the information age, all people have to keep on learning new knowledge to win a place in society. Education gradually becomes an indispensable part of social life, because it is the most effective way for people to improve their quality. However, education does not mean merely feeding knowledge to students. It should also integrate all kinds of resources by management to guide the comprehensive development of the subject-human. Humanized management is undoubtedly the core idea of management [1].

Humanized management focuses on human factors and undertakes the responsibilities of realizing human value, exploiting human potential and developing human individuality. The contents of humanized management include praise of humanity, appropriate material reward, sufficient moral encouragement, and so on [2]. Besides, the subject-human is provided with many opportunities for growth. Education is for people who want to learn. The ultimate target of humanized management is to help these people achieve their goals.

2 PROMOTING EFFECTS OF HUMANIZED MANAGEMENT ON EDUCATION

There are two definitions of education. Education is narrowly defined as a educational form with special organizations, such as full-time school, cram school, night school and correspondence school. Its essence is that children and teenagers of school age should be trained in knowledge, and it is aimed at training the new generations to step into social activities. At the same time, the generalized definition of education is any social activity that affects the knowledge and moral character of humans. In the broad sense,

education updates knowledge and skills and promotes ideology [3]. Therefore, the promoting effects of humanized management are also divided into two types – promoting effects on school education and social education.

2.1 Promoting effects on school education

Humanized management focuses on the coordinated relationship between school and individuals of teachers and students. It embodies humanistic care and a loving environment. More respects are shown to the will of a human. Besides, school administrators and "teachers & students" are in equal positions. As a result, school education greatly improves with the realization of two-way communications between school and "teachers & students".

Humanized management promotes the teachers' work. With the implementation of humanized management, teachers are provided with a good working environment. This improves the teachers' efficiency. Teachers have direct contact with students in school education, so their efficiency is closely related to the quality of teaching. School is supposed to maximize the teachers' efficiency through humanized management. In this way, teachers will be fully devoted to the education of talents. Humanized management motivates teachers to take part in the school work consciously and develops their potential. Furthermore, teachers regard the imparting of knowledge as their personal fulfillment. They take a positive attitude instead of passive one and try to be the masters of the school.

Humanized management also promotes the study of students. School emphasizes the importance of civilized education and moral education. Besides, the school makes rules and teaching methods according to the developing characteristics of students, because students represent the teaching quality. Humanized management centered on students can shorten the distance between students and school administrators. Students' suggestions about school construction and teaching mode will reach the administrators without delay. Besides, humanized management contributes to the construction of an academic atmosphere in school, because the atmosphere has an impact on students. In a good academic atmosphere, students tend to study consciously and make rapid progress in academic performance. Apart from the promoting effects on the work of teachers and study of students, humanized management emphasizes the cultivation of the students' nature. Students should be polite, obliging and selfless in daily life.

2.2 Promoting effects on social education

In china, there are different institutions for social education, such as cultural centers, Children's Palaces, libraries, museums, memorials, radio stations and TV stations. Among these institutions, libraries top the list, because they house a large number of books on different fields. Social education is broad, various and complicated. It includes all kinds of knowledge or skills and a variety of educational forms or thoughts.

A library is a non-profit and academic institution, and the collection of books are intricate. So it is necessary to apply humanized management to the library work. Considering that the library is a holy place of spirit, administrators ought to motivate librarians with moral encouragement rather than material reward. For the readers, humanized management improves the environment for reading and simplifies the process of book lending.

People usually receive a social education in their spare time. They learn new knowledge and new things after work or school. Social education not only satisfies people's demands at work, but also enriches their spiritual world. Meanwhile, with the conduction of humanized management, workers in institutions of social education strive to be more dutiful. Their efforts make it more convenient for people to study and ensure that the goal of education is achieved. For example, workers of memorial halls should provide guidance to visitors so that people can be deeply impressed at the exhibition. Furthermore, people longing for knowledge are allowed by humanized management to receive education whenever they want to. They also take part in the construction of social educational institutions, such as by making suggestions on the improvement of radio programs.

3 MEASURES OF IMPLEMENTING HUMANIZED MANAGEMENT IN EDUCATION

3.1 Heart-to-heart talks

In daily teaching, teachers meet with troubles inevitably. If these troubles are not solved by administrators in time, teachers will misunderstand the school administration. Therefore, school leaders are supposed to have zero distance communications with teachers to understand the troubles they meet at work. The leaders should try to stand in the teachers' shoes and solve the problems from a teachers' prospective. By doing this, school leaders manage to change from a passive role to an active one and succeed to stir up the enthusiasm of teachers. Compared with formal communication, a relaxed conversation at the table is more likely to pay off. In the informal atmosphere, teachers are willing to express their dissatisfaction and accept comfort from school leaders. After the conversation, school administrators, in turn, can realize the blind spots and deficiencies of their humanized management. They can propose targeted solutions for teachers to further optimize the system of humanized management.

3.2 Creating opportunities for teachers' self-improvement

The students' learning efficiency is directly related to the teaching ability of teachers. However, teachers often fail to give full play to their abilities in daily teaching. School administrators should encourage teachers to set up a lofty goal and arouse their enthusiasm. Consequently, teachers can set up an ambition and gain the sense of fulfillment. Besides, school administrators should create opportunities for teachers to show their talent, such as by holding public classes, assessing teaching plans and holding skill competitions. Moreover, backbone teachers and promising teachers ought to be encouraged to pursue higher degrees, or be sent abroad for advanced training. As a result, the development of the teachers' ability will bring great benefits to students.

3.3 Enhancing the emotional communication between teachers and students

Teachers are the first-line managers of student management. They not only participate in the humanized management of students, but also convey students' feedback on the management to school administrators. Unlike animals, humans have emotions. Emotional communications can shorten the distance between people. When the emotions of students are aroused properly, they tend to study hard and correct their bad behaviors. Teachers can communicate with students either in class or at recess. Communication in class usually focuses on study. If students ask any questions in class, teachers will try to answer them patiently so that students can solve their problems on the spot. This is different from the traditional teaching mode where students only listen but cannot ask questions. Communication at recess usually focuses on the students' feelings. Students' assessment is the best feedback on humanized management.

3.4 Strengthening the humanized management in institutions of social education

The personnel management in institutions of social education is looser than that in schools. So it's more necessary to implement humanized management in these institutions. In these non-profit institutions of social education, a democratic atmosphere empowers the staff to take part in the decision-making process. The staff are fully respected and allowed to pronounce on the management and operation, so that they can have a better understanding of the policy. This measure guarantees the effective implementation of management in these institutions as well as the combination of individual interest and collective interest. Social educational institutions have a strong cultural atmosphere, thus self-management of the staff should be implemented. Self-management is an upgrade of democratic management. The staff should make plans for their own development according to the characters of institutions. Only then can they improve the management of the whole institution.

3.5 Stabilizing the structure of management team

The implementation of humanized management depends on the management team. Necessary measures should be taken to attract talents to stabilize the structure of the management team. Meanwhile, available talents ought to be treated with good welfare in case of the brain drain. Besides, the team should invite experts on humanized management regularly to give lectures. In such lectures, experts analyze successful cases and existing deficiencies of humanized management. They also offer solutions as examples to follow. However, members of the management team should be punished if they violate the principle of humanization. Every one of the team ought to raise the awareness of humanized management and make efforts to do a good job.

4 CONCLUSIONS

Humanized management in education emphasizes the important role of humans. Both teachers and students can achieve their goals in education. This is the promoting effects of humanized management on education. Besides, the mentioned measures are expected to promote further research on humanized management.

REFERENCES

[1] Wan Yi. Humanized Management in Higher Education—Organization and Implementation. *Continue Education Research*, 2013(01): 98–99.
[2] Liang Wei. Reflections on School Humanized Management. *Journal of Educational Development*, 2006(10): 11–14.
[3] Ma Qiaoxia. Reflections on Humanized Management in Higher Education. *Journal of Shaanxi Normal University (Philosophy and Social Sciences Edition)*, 2004(S1): 131–132.

Future Communication, Information and Computer Science – Zheng (Ed.)
© *2015 Taylor & Francis Group, London, 978-1-138-02653-7*

Influences of trade barriers on international economic situation

M. Bai
Xinzhou Teachers University, Xinzhou, China

ABSTRACT: International trade has become more frequent with economic globalization, but trade barriers and frictions have occasionally occurred. The work analyzed the influences of trade barriers on the current international economic situation. Based on the causes of international trade barriers, the influences were discussed in two aspects – tariff barrier and non-tariff barrier. It's hoped to help the development of import and export trade in Chinese enterprises.

Keywords: trade barriers; international economy; influences; analysis

1 INTRODUCTION

According to the types of influences on trade, trade barriers can be generally divided into tariff and non-tariff barriers. Everything appears for certain factors, so does a trade barrier. It's an inevitable phenomenon in the constant development of the international community and economic situations of various countries. Trade barrier reflects the self-protection of a country to a certain degree. The protection has a huge impact on the international economic situation, deserving deep study [1]. The influences of trade barrier on the international economic situation were analyzed as follows.

2 CAUSES OF TRADE BARRIERS

2.1 Causes of conventional trade barriers

The conventional trade barriers are tariff barrier, licensing barrier and quota barrier, among which the tariff barrier is the most central. They're not produced by economic and social developments, sometimes exerting great negative effects to international economy instead of positive ones. The weak economical growth in some developed and capitalist countries is the reason. Therefore, protectionism is possible to rise. The Sino-US case of tires, for example, truly shows trade barriers. In this case, the development of American tire enterprises was greatly affected due to relatively cheap tires from China [2]. Therefore, on June 29, 2009, the United States International Trade Commission proposed special tariffs for China in three consecutive years – respectively 55%, 45% and 35%. The tire exports of China were significantly influenced, and the cooperation between the two countries became fissured.

2.2 Causes of new trade barriers

People's requirements for the health and safety of products has become increasingly stringent with the progress of the economy and living standards. Therefore, based on safety, hygiene and health, the new trade barriers have become greater in international trade [3]. With reasonable causes, the new trade barrier has a role in promoting the harmonious development of the international economy. Thus it's seldom criticized, for having a good space for development. The new trade barriers include green barriers, social barriers and technical barriers.

A world-wide sustainable development has been promoted with the enhancement of environmental awareness. People become more concerned about the environment of the earth. Thus with corresponding requirements, the products and production processes in any international trade are forbidden to pollute the environment. Meanwhile, the health of the workers can't be sacrificed, and their surplus value can't be exploited. Consequently, green barriers and social barriers come into being.

3 INFLUENCES OF TARIFF BARRIERS ON THE INTERNATIONAL ECONOMIC SITUATION

In the import of goods and labor resources, the tariff barrier is formulated by the national government. It's a trade barrier caused by the tariff that the customs authorities collect from the export enterprises. Based on the collecting purpose, tariffs can be divided into two kinds – fiscal tariffs and protective tariffs.

Fiscal tariffs, also known as revenue tariffs, are aimed at increasing the state revenue. On the contrary, as high tariffs, protective tariffs are to protect domestic economic industry. Therefore, the price of imported products can be the same as or higher than domestic products. The market demand for the domestic products can be stimulated to protect the domestic industry. And the higher the protective tariff is, the bigger the protection will be. The establishment of a tariff is even affected by the prohibition of import.

In the implementation of fiscal tariffs, two factors require consideration. One is to collect the import duties with low tax rate. Then the massive imports of foreign products can be simulated to increase the state revenue in import tariff. The other is to collect the export tax. Collecting extra tariffs on high-tech products that holds the monopoly can increase government revenue in export tariff. Collecting fiscal tariffs can inhibit domestic development to a certain degree, but the influence is small since the restrained enterprises are mostly high-tech and monopoly companies. However, the reduction of import tariffs can greatly promote the international economic development. Low tariffs strengthen the circulation of products among various countries and promote the development of industries related to imports and exports. What's more, the reduction of the world's import and export tax rates is good to the mutually beneficial development of multiple countries. Although the specific import tariff of the country has declined, the total import tariffs have increased. And the tariffs of the country in other countries show a downward trend, forming a win-win situation. Once the fiscal tariffs violate the two factors, the same consequence as protective tariffs can be caused. Then it's difficult for the international economy to develop.

Protective tariffs usually appear in developed and capitalist countries, used as the means for the monopoly of international markets and profiteering. Sometimes, they appear in developing countries to oppose the economic aggressions from capitalist countries and safeguard national economic development. Protective tariffs protect the operation of the company and domestic economic development, but a high tariff has a great negative impact on the international economic development. If a country raises the import tariffs, so will other countries. Afterwards, a malicious competition can be triggered to harm the export and import of various countries, especially for developing countries.

4 INFLUENCES OF NON-TARIFF BARRIERS ON INTERNATIONAL ECONOMIC SITUATION

4.1 *Influences of technical barriers on international economic situation*

Technical barriers are set up based on technologies. They have numerous categories and flexible performances in various countries and regions, such as technical regulations, standards and certification systems. The standard of practice for the relevant technologies even seriously goes against common sense. Using technology as a disguise, most of these barriers have become the most covert and intricate non-tariff barriers among trade barriers.

Contrary to the liberal and global principles of the world economy, technical barriers hinder the free development of international trade. They're seriously disadvantageous for the free circulation and optimal allocation of international resources. For example, the United States has serious strict standards for most imports of electronic products. The imported meat and food are also faced with various and repeating test items. Thus the international trade encounters much inconvenience. Meanwhile, the costs of export products from other countries are improved, going against the rational operation and development of international trade. With technical barriers, the benefits of international trade tend to chronically go to capitalist countries. Since most constitutors of the current international systems and standards are developed countries, developing countries have to accept them and disburse the expensive cost of exports. With high a scientific and technological level, the standards formulated by developed countries are harsh and even difficult to reach when compared with developing countries. Developed countries often treat the standards as a means of protecting their economies. Then they can occupy the dominant position in the international market and inhibit the economic development of developing countries. In 1996, failing to reach the quarantine standard of the EU, the frozen chicken from China was rejected to enter the European market until 2001. But it was allowed only a handful of companies back then. As a result, the annual loss of the frozen chicken industry in China is hundreds of millions of dollars in recent years. According to the number of technology patents around the world, the share of developed countries is almost 90%. The distance between developed and developing countries won't disappear immediately. Thus the developing countries will remain unfavorable in the trade for a long time.

4.2 *Green Barriers*

Created by import countries, green barriers (GBs) are to protect the natural ecology and health of the populations in the country. Through a variety of environmental regulations and standards, the export countries are restricted or even prohibited to export the products into the country. Most of the regulations or standards are harsh. They are either higher than the internationally accepted environmental requirements or unacceptable to most countries. As such, the economy of import countries can be protected.

With the change of the international economic society, the countries are struggling to develop the competitiveness of domestic enterprises to improve the overall competitiveness. Thus it's natural to see the endless trade protection policies. However, with the operations of the World Trade Organization and the General Agreement on Tariff and Trade, the tariffs continually decline. Thus the non-tariff barriers are greatly affected. On this basis, various countries and regions begin to search the new trade barrier. Consequently, green barriers come into being with the fastest development. Therefore, green barrier is a part of protectionism, utilized by the developed countries like technical barriers. With the increasing demands for environmental protection in the present world, the countries have increased the efforts on environmental protection. However, it can't be used as an excuse for some countries to deliberately set up trade barriers. Dressed in the cloak of environmental protection, these countries protect the domestic enterprises with weak competitiveness and suppress the trade of other countries. For example, America rejected the gasoline imports of Venezuela for excessive lead content. The EU prohibited Canadian leather products from entering the European market because of the Canadian hunters' excessive capture of wild animals. In the 1990s, due to the destruction to the ozone layer by Freon, the sales of Chinese refrigerators were forbidden in Europe. Consequently, the export rate of Chinese refrigerators declined by 59%, and the economic development was negatively affected.

4.3 Social barriers

In recent years, several developed countries have tried to cover for their behaviors of this new protectionism, making it more justifiable. The labor issues and economic development have been obstinately combined to form more complex and formidable social barriers.

Social barriers are composed of the social contracts or regulations in various countries and regions, including labor compensation, social insurance, labor responsibilities, powers and technological standards. Such as the SA8000 standard, the main manifestation of social barriers is that the employers reduce or limit the labor conditions and wages of the employees. A social barrier is the metamorphous use of a social labor contract. The original guarantee for a win-win situation between the employers and employees has transformed into a means to restrict the economic development of developing countries. The labor-intensive products from developing countries are the main object. The scientific and technological level of some developing countries is far less than that of the developed countries, so labor-intensive products are their main export commodities. Once the developed countries implement the social barriers, the exports of labor-intensive products from developing countries will dramatically decline on the short term. The survey of the American Trade Organization shows that once the SA8000 standard is implemented, more than 50% of the companies in America will consider resigning the contract. Several non-certified companies will lose orders and development opportunities. In the long run, trade barriers will lead to the rising costs of the export products from developing countries. Thus they will lose the only competitive advantage – price. Afterwards, reduced investment and idle labor can be caused, influencing the industrial structure of the entire enterprise or even the industry. In addition, the long-term idle labor can cause a series of potential impacts, seriously restricting the stable development of the international economy.

5 CONCLUSIONS

There is no doubt that the emergence of trade barriers has a great impact on the international economic situation, especially for developing countries. The expansion of the exports and the diminishment of the distance with developed countries are obstructed. And the security of the state economy is brought with instability factors. Therefore, breaking through trade barriers is a problem that deserves studies from global scholars.

ACKNOWLEDGEMENT

The work was supported by Xinzhou Teachers University project "Culture and Tourism Development of Yanmen", Project Number: ZT201319.

REFERENCES

[1] Li Jinning. Influences of Technical Barriers on International Trade and Strategies. *Guangxi Social Sciences*, 2004(2):75–77.
[2] Xiong Furong. Strategies for Green Barriers in International Trade. *Productivity Research*, 2012(2):158–160.
[3] Wu Hansong. Strategies for Trade Barriers of International Trade in China. *Special Zone Economy*, 2004(11):178–179.

.

Reasonable application and management analysis on tourism culture

J. Dong & H. Li
Shijiazhuang Institute of Railway Technology, Shijiazhuang, Hebei, China

ABSTRACT: To further enhance the impact and expand the scale of the tourism industry, managers of China's tourism industry have made several adjustments and reforms in the industrial structure. There are many important factors in tourism industry development besides its structure. From a neglected perspective of tourism culture, the work analyzes reasonable application in tourism culture, discusses its management, and thus extends the influence of tourism culture and pushes forward the development of the tourism industry.

Keywords: tourism culture, reasonable application, management

1 STRUCTURE AND CONNOTATION OF TOURISM CULTURE

Tourism culture is a kind of special tourism resource which combines the behavior of tourism with the artistry of culture. It makes up two parts, spiritual resource and material resources, and can not be treated as being abstract according to its definition. The material resources include tourists and their property, managers and their premises, equipments. The spiritual resources cover cultural tradition, folklore and social customs [1]. In addition, the latter has an impact on the former, such as how to play and buy tourists, and how to develop resources for managers.

When it comes to its connotation, tourism culture contains a natural factor, human factor and other factors. The natural factor, such as topography, landscape, architecture and climate, is the foundation of tourism culture because beautiful natural resources are a direct consumption for tourists [2]. The human factor, including historical events, folkways, traditional program, religion, music and diet, becomes the core of tourism culture for tourists are the theme of tourism. Tourists can better integrate into the scenic spot and enjoy its unique charm through the human factor. Besides the natural and human factor, other factors, such as government policy, business opportunity and development planning, are a supplement to tourism culture. They can help to establish a more perfect and standardized scenic spot.

Tourist culture plays an important role in the tourist industry. On one hand, tourist culture, which the carries local humanistic culture of a scenic spot, can not only increase the attraction of a scenic spot, but also influence people's thoughts. Taking the Wuhou Temple in Sichuan for example, there are always lots of tourists every year because Zhu Geliang's story is known far and near. It is a good place to have fun and also to experience his royalism called "being loyal and devoted to the last" [3]. On the other hand, tourist culture is the medium of cultural transmission. Tourists can get a better understanding through the explanation of the local customs from the tour guide or through mass media such as television broadcasting.

2 RESONABLE APPLICATION OF TOURISM CULTURE

2.1 *Encouraging people to travel*

With the rapid development of modern society and the economy, for one thing, people's standards of living are getting higher and higher, and the city's construction process is becoming more and more quickly. For another, smaller living space and heavier living pressure come shortly. Therefore, people hope to find a good way to relax after becoming rich. Tourism has been one of the leisure methods ever since. It is not only a way to ease people's tension, but also a way to exercise themselves during traveling. What's more, with the popularization of education and the great progress in cognition, tourists are more eager to relax and sublimate their minds and spirit instead of having fun. As a result, tourism culture is an effective way to attract people. In order to entice people to travel, developers and managers in the tourism industry should increase the scenic spot's popularity through media publicity, such as a video on TV, a brief introduction on the radio, and so on. Tourism culture should be covered in the whole process during traveling to avoid tourists' disappointments and to increase the chance of being introduced. For example, a text label should be put in some important tourist areas, and scene simulation of historical events and traditional food can also be showed to tourists.

2.2 Improving the brand force of scenic spot

A bigger and better tourism industry needs the establishment of brand awareness and brand influence. To set up their own brand, many attractions are applying for the national natural protection zones or key cultural relics protection units at present. The leading principal idea that culture is the core of enhancing brand force is the same even though attractions images shaped by a brand are various and the expressions and disseminations of cultural thought also differ. To further improve the brand force of attractions by tourism culture, three points should be taken into consideration. Firstly, have a profound understanding of spiritual culture and material culture that the attractions include, and establish a brand image in accordance with the cultural characteristics. For example, a foreign friendly brand image should not be built in a scenic spot that is full of ethnic struggle. Secondly, make a survey on the tourism market, obtain some information about what tourists really think, and then conduct a scientific and rational analysis on them. Finally, combine the information above with the attraction features so that people can join in tourism heart and soul.

2.3 Helping the tourists improve their ideological level and mind

Lots of scenic spots have their own special historical background and cultural implication. The cognition level of tourists can be improved and sublimated through the promotion of tourism culture. For example, the red tour, which has been long advocated in China, is a typical application of this approach. The attractions of a red tour, most of which are former sites of war and memorial buildings, can visually reproduce a scene of years to tourists. By watching the historical relics and listening to the scenic spots introduction, people seem to experience the life of a war. Thus, emotional resonance can be produced, and the understanding of patriotism and devotion can be deepened.

Moreover, some scenic spots form their own tourism culture, not for their profound culture, but for a beautiful landscape, a unique folk custom or fresh air. Although there is no strong educational significance, the tourism culture can help people form a soft state of mind, which is a natural and relaxed attitude towards challenges and difficulties. For instance, when traveling in some minority residences, tourists will pay special attention to rest and health preserving in the increasing fast-paced city life by knowing and joining the local life style called "going out to work at sunrise and going back home at sunset."

2.4 Enriching the culture constitutes of China

The culture of China consists of many parts, including humanistic, natural and social culture. Theirs components are uneven and formed by precipitation of a long history or accumulation of social change. The tourism culture, a kind of special culture, includes humanistic culture and natural culture. Moreover, it brings two cultures together, and finds a balance point in order to promote the common development of both sides. People usually believe that tourism is only a way of leisure, and don't expect to learn useful knowledge during traveling when the tourism industry has not been defined as a service industry yet. Since the importances of tourism culture have not been realized in China, tourists consider some folk customs and historical events as entertainment and don't have any further understanding of its existence reason and value. The tourism culture, brought into the development of the tourism industry by developers, can not only play an indelible role in promoting the industry, but also enrich the culture constitutes of China. With the help of tourism, a medium easily accepted by people, tourism culture has eliminated dullness and passivity from traditional cultural dissemination.

3 STRATEGY OF TOURISM CULTURE MANAGEMENT

3.1 Paying attention to the development and use of national culture resources

China, a country with a history of 5,000 years and one of the four ancient civilizations in the world, is very rich in natural resources and cultural resources. There are not only many Provincial and National Historical and Cultural Heritage Protection Units, but also a dozen World Cultural Heritages listed in the World Cultural and Natural Heritage nationwide. What's more, the unique national culture – such as language, character, food, clothing, folk custom, and religion – affected by climate, region and various social factors, shows a variety of human civilization. We shall fully develop and make advantages of national culture resources. On one hand, we should enhance the market competitiveness of scenic spots, and attract millions of tourists from home and abroad. On the other hand, the impact of national culture should be strengthened so that the patriotism of people can be improved. Scenic spots should be built on places rich in national culture resources, and open up a unique folk custom tourism route and a show multi angle, multi cultural presentation according to the characteristics of local culture and planning. Meanwhile, scenic spots have to minimize the chance of urbanization and westernization during the development.

3.2 Developing actively tourist commodity with local features

The tourism cultures between scenic spots are different from each other because of regional differences even though some tourism cultures belong to one people. For example, both northeast China and Guangdong Province are places where the distribution of the Han People is highly concentrated, but the tourism culture in the two areas is different. The souvenir with local characteristics, including various works of art, sculptures, foods and national costumes, can highlight the characteristic and connotation of the local culture. These souvenirs, deeply marked by culture, are used as memorial, collection and presentation. Furthermore, to avoid being put aside, souvenirs should be practical and have a space in daily life besides just a distinctive local characteristic, scenic spot feature and national custom. Taking embroidery, a traditional art and craft of the Li People in China as an example, in the past, the major embroidery products were clothes and handkerchiefs, which were not very commonly used in daily life. Nowadays, developers consciously create belts, bags and headdress. As a result, the souvenirs are of great practicability and their connotations have been improved to a new level because of the stone forest features on embroidery.

3.3 *More efforts into tourism culture publicity*

For a scenic spot, publicity is the most effective way to attract the public. Without good publicity, a tourism attraction would never be known to people even if it is full of unique tourism culture. Although there is vast territory and large number of provinces and cities in China, only a few tourism attractions can be known. When coming to the City of Ice and Snow, people associate it with the city of Harbin even if more than one city in China has snow in winter. The reason is that publicity in other cities and tourist areas is inadequate. Publicity has certain methods. A blind publicity not only wastes a lot of human, financial and material resources, but also has a negative effect. Publicity possesses great pertinence and innovation. Pertinence means that different landscapes should be targeted at various groups. For instance, to stimulate the interest of Guangdong tourists in travel, local customs and natural landscapes from different places, instead of seascape, should be shown to them. Innovation refers to a publicity method that can't be too rigid and traditional. There should be creative thinking and a way of publicity. Otherwise, it will generate aesthetical fatigue. For example, some interesting and scene promotional video online, rather than a promotional video with simple language, should be produced to grab people's attention.

3.4 *More efforts to improve the cultural accomplishment of employees in the tourism industry*

Although it is an objective existence, the tourism culture of scenic spot can not get an effective promotion and application without tourism staff. The education degree of managers should not be too low because the tourism industry is a high-cultural industry. Firstly, for a better publicity, managers should have a systematic understanding of cultural resources in the scenic spot. Secondly, managers should also have a higher cultivation of history and ideological cognition in some scenic spot rich in history and culture, so they can combine humanity and nature better, and introduce it to tourists in appropriate language under good aesthetic ability. Finally, systematic training – such as learning and aesthetic education – for the staff of a scenic spot should be conducted before taking their posts because both their regions and degrees of culture are different. In order to master the change of social demand and make any necessary adjustment, regular lectures are also needed.

4 CONCLUSIONS

As one of the core components in the tourism industry, tourism culture is an important embodiment of the charm and attractiveness of tourism. For a better development and more participation, people should combine the application of tourism culture, strengthen the management of tourism culture and improve the competitiveness of the tourism industry according to the actual demand of society.

REFERENCES

[1] Zhong Jiwei, Tourism Culture in China and Its Strategy Management, Youth Literator, 2010(11):227.
[2] Zhou Bentao, Xie Chunshan, New Theory of Tourism Culture, Journal of Beijing International Studies University, 2009(11):20–24.
[3] Jiang Jie, A Study on the Development and Utilization of the Tourism Culture Resources, Journal of Liaoning Administration Institute, 2008(11):163–165.

Future Communication, Information and Computer Science – Zheng (Ed.)
© 2015 Taylor & Francis Group, London, 978-1-138-02653-7

Development of interactive computer-aided teaching mode of foreign languages

G. Chen

Foreign Language Department of Qiqihar Medical University, Qiqihar, Heilongjiang, China

ABSTRACT: Foreign language teaching plays an important role in the exchanges and co operations of global economies and politics. Nowadays, China is advocating educational reform to cultivate talents who meet the demand of social development of this new era, which requires innovation on foreign language teaching. The computer-aided foreign language teaching is one of the commendable methods of innovation. This study focuses on the current situation of modern computer-aided foreign language teaching and then proposes the student-oriented principle of foreign language teaching. Several strategies for the interactive computer-aided teaching mode of foreign languages are put forward to demonstrate that it is significant to use computer-aided technology to enhance the quality of foreign language teaching.

Keywords: computer; interactive; foreign language teaching

1 INTRODUCTION

The interactive computer-aided teaching mode of foreign languages refers to using advanced computers technology to assist foreign language teaching and enrich teaching resources, which makes foreign language teaching to be more varied and more dynamic, stimulating students' learning interests of foreign languages and enhancing the corresponding teaching quality. The focus of the traditional teaching mode of a foreign language is teachers, textbooks and classes, emphasizing the explanation of foreign language knowledge. Accordingly, students can only passively receive some knowledge, which contributes to a lower learning efficiency and initiative. However, the traditional teaching mode has some advantages because students can directly exchange with their teachers in class to solve their problems in time. Yet, modern educational ideas and technologies in China always change with the constant changing demand of our times, which requires a timely update of corresponding teaching methods so as to keep pace with the times. The fast developing computer technology has been applied to various kinds of industries. Therefore, when conducting foreign language teaching, it is also advisable to scientifically and appropriately use computer technology to enhance the flexibility and vitality of foreign language teaching, thus promoting the teaching quality to cultivate new-type talents who meet the demand of social development.

2 DEVELOPING TREND OF MODERN FOREIGN LANGUAGE TEACHING MODE

Modern foreign language teaching in China focuses on practicality and pays attention to the cultivation of language proficiency, which aims at enabling students to apply what they have learned. Under the background of the new educational reform, the main purpose of foreign language teaching is not to enable students to acquire fixed language and incondite grammar knowledge but to enable students to command good foreign language skills. Nowadays, many universities in China have not yet deeply realized it. Their foreign language teaching mode mainly followed traditional teaching, excessively emphasizing grammar knowledge; the teaching of audio-lingual skills like pronunciation and intonation, the most important part of language learning, has not been placed in a vital position. Students cultivated by traditional teaching mode can only pass different kinds of examinations, but they can not fluently exchange with others in a foreign language. However, it is a good start that some teachers have already used this technology to conduct an appropriate teaching reform, like using multimedia to strengthen the training of listening, speaking, reading and writing due to the equipment of multimedia facilities in many universities.

Modern foreign language teaching in China should also pay attention to the foreign culture, emphasizing the cultivation of the students' intercultural communication competence. Since language is often closely related to the culture of the country, students ought to have a cross-culture recognition in order to better command and use the foreign language. Only in this way can misunderstandings due to a wrong interpretation of the language be avoided. For example, many foreigners have little knowledge of Chinese culture. When they first hear a Chinese sentence "Talk of the devil, and he is sure to appear", they can have various interpretations for this sentence. However, many Chinese people will immediately know what the speaker wants to express when they hear this sentence because they are familiar with their own cultural background. Exchanges among different countries are becoming more and more common owing to the free international trade and coordination. Under such a circumstance, foreign language teaching in China should connect culture with language education and strengthen the students' communication skills based on the cultivation of their comprehensive capabilities and qualities, which is beneficial to China's rapid economic development and its close cooperation with the world.

Modern foreign language teaching emphasizes student-oriented study. The traditional foreign language teaching mode has always put teachers on the core position, but modern foreign language teaching needs to change this situation, regarding students as the core of the teaching. Students of different ages have their unique characteristics and some of them can be different from others in cognitive ability, learning method and motivation. In foreign language teaching, these characteristics should be carefully studied and analyzed so as to conduct targeted teaching, enhancing the learning quality of students.

3 PRINCIPLES OF THE INTERACTIVE COMPUTER-AIDED FOREIGN LANGUAGE TEACHING MODE

Traditional foreign language teaching in China laid much stress on examinations. It only cared about the students' examination performance instead of their language using abilities and the smooth communication skills. Nowadays, one of the major problems in foreign language teaching in China is that students have poor listening and speaking abilities. The computer-aided foreign language teaching can commendably improve the traditional teaching mode. Many universities require that their teachers use multimedia to promote the vitality and dynamic of the class, but there are still some basic principles in conducting foreign language teaching with multimedia, which can help promote the sound development of foreign language teaching.

3.1 Choosing appropriate multimedia materials according to in-class requirements to conduct foreign language teaching

When conducting foreign language teaching, teachers can not regard using multimedia technology as a goal but should regard it as a tool to choose appropriate internet resources to assist the teaching according to the requirements for knowledge explanation, which aims at enhancing the students' comprehensibility and their learning efficiency. For example, in a foreign culture class, teachers can use classic movies, videos and pictures of a certain period to intuitively demonstrate the settings, allowing students to be closer to and easier accept the cultural background. Moreover, these multimedia resources can also add some interests to the teaching and attract students' attention, which will increase their initiative to actively learn and explore. For example, *Forrest Gump* is an excellent inspirational American movie. Teachers can play some classic clips of this film to students to practice their oral English and listening because the actors' and actresses' pronunciation in this film is very standardized. Likewise, students can spend after-class-time to enjoy the whole movie and enhance their listening and speaking abilities by repeatedly watching this movie or imitating the pronunciation along with the movie dialogues.

3.2 Choosing scientific and appropriate Internet resources for students to conduct foreign language learning

Many foreign language learning the students, especially those high school students, are often lost in the abundant foreign language learning materials on the Internet because they can not identify the quality of these Internet resources and massive data. Therefore, teachers should guide their students in conducting online learning by recommending some high-quality foreign language learning resources on the Internet. For example, there are many kinds of foreign language learning websites today. If you input *foreign language learning* in Baidu, you will get numerous interlink ages of foreign language learning websites. Many students may have no idea which website to choose since there are various kinds of foreign language learning materials on each website. In this case, it is necessary for them to get their teachers' guidance.

3.3 Choosing computer resources close to real life for teachers in foreign language teaching

The resources close to the students' life should be introduced naturally and reasonably when teachers explain the knowledge of foreign language textbooks. But not for the application of computers, it is unreasonable for teachers to introduce teaching videos and materials mechanically not in accordance with practical theory or spend a great deal of time playing video and browse images, thus deviating from topics in the class. It is advisable that teachers maintain the right balance in the classroom, for the purpose of introducing materials close to life – explaining the knowledge better, thus greatly enhancing the effects of foreign language teaching.

It is easier to arouse psychological resonance of students when the materials introduced are close to real life, so as to promote their thinking and explanation. The American TV drama *Friends*, for example, is a good resource for learning English because its plot is interesting and close to real life and its lines have a strong applicability, conducive to learning English. Some video screenshots conforming to the classroom subject should appropriately be chosen by teachers in the class to assist the teaching. In this way, the learning ability of students can be promoted, thus improving the quality of teaching.

3.4 *Being student-oriented in the process of the interactive computer teaching of foreign languages*

The student-oriented principle should be adhered to the process of foreign language teaching, because the main purpose of foreign language teaching is to cultivate the foreign language ability of students in the foreign language teaching. Therefore, it is necessary that teachers know the characteristics of their students at this age in daily life – mental and physical development and cognition, and individualize the students' learning. Only in this way can the better teaching effect be gained.

4 COUNTERMEASURES OF THE INTERACTIVE COMPUTER TEACHING MODE OF FOREIGN LANGUAGES

The quality of foreign language teaching can be improved and new vigor and vitality added by computer interactive teaching of foreign languages, which conforms to the requirements and expectation of education reform in China. At present, the computer multimedia technology has been applied to the foreign language teaching, but there are disadvantages in the application of modern computer teaching of foreign languages because the computer teaching of foreign languages started relatively late. Some countermeasures are put forward aiming at the improvement of the interactive computer teaching of foreign languages in this work.

4.1 *Building the interactive mode of student-oriented teaching*

The interactive computer teaching mode of foreign languages is aimed at the interaction between students and computers, and reflection of students learning a foreign language by the interactive features of the multimedia computer itself. The traditional teaching mode, a teacher-oriented one, has its disadvantages and deficiencies, but the mode can provide an opportunity for students – face-to-face communication with teachers, which contributes to communicating with teachers and solving problems better. This is the advantage of the traditional teaching mode. As a result, it is advisable that disadvantages be improved by application of the computer technology when the advantage is

maintained. Nowadays, a lot of people sign up for some foreign language courses on the internet and there is no need for them to go out having classes if they have computer access to the internet at home. In this way, it is more convenient for people to learn, but there is a problem that the main position of students cannot be presented very well because teachers have less communication with students, and cannot master their learning state and attitudes. Teachers, therefore, should regard students as the main part in the teaching of foreign languages, exchange ideas with students properly and apply the computer to their teaching in accordance with the characteristics of students. For the student-oriented purpose and result of teaching, for example, it is advisable that teachers inquire what learning materials the students are interested in, according to the content scheduled for the next class, which helps teachers find some materials of foreign languages in much accordance with the students' interest on the internet. In this way, the students' interest can be aroused by the knowledge and atmosphere in the classroom.

4.2 *Improvement of teachers and students in computer technology laying a foundation for the interactive computer teaching of foreign languages*

It is imperative that teachers are skilled with the technology of computer-aided foreign language teaching, optimizing the use of computer multimedia technology. Thus this can improve the quality of the foreign language teaching class and produce a good effect in foreign language teaching, because they need to apply the computer technology to the foreign language teaching. Students, moreover, as the main part in learning a foreign language, should master the skill and basic means of computer-aided language learning. Only in this way can they actively use computers to learn English outside class. Nowadays, for example, there is a lot of auxiliary software to help students memorize words, such as youdao dictionary, foreign language learning software widely used in China. Software, meanwhile, is with this function – providing vivid explanations of words with interesting pictures, which makes word learning not so boring. At the same time, it is with another function – providing example sentences of each word. Students can learn the word and master its usage better when the word is placed in the context. Only in this way can skills of foreign languages be improved. The students, therefore, should be equipped with basic computer skills, in order to obtain resources from the network and install and use the software in computers.

4.3 *Cultivation and introduction of advanced computer technicians providing a favorable supporting technology for foreign language teaching*

It is necessary to cultivate and introduce computer technicians with advanced science in each university, building technical teams to support classroom teaching. Good computer technicians, meanwhile, lay the important foundation for the school computer-aided teaching because the computer teaching is not limited to the foreign language teaching, and has been gradually into the different curriculum teaching. During the class, for example, if there are multimedia problems delaying the course, it is needed that computer technicians repair them quickly and timely so as to ensure the original schedule and normal teaching activities.

5 SIGNIFICANCE OF THE INTERACTIVE COMPUTER-AIDED TEACHING MODE OF FOREIGN LANGUAGES FOR PROMOTING THE QUALITY OF FOREIGN LANGUAGE TEACHING

The national demands for talents are increasingly high with the enhancement of the economic level and state strength, especially the ability of reading, writing, listening and speaking for communication in foreign languages. The examination-oriented shortcomings in traditional teaching, therefore, should be changed in order to cultivate new-type talents with a comprehensive ability who contribute to the socialist construction in China. The launch of the interactive computer-aided teaching mode, moreover, according to the demand exactly, teachers can apply rich internet resources to foreign language teaching by computer aid, gaining many more materials of audio and video, which provides a better opportunity for students to train listening, speaking, reading and writing. In this way, the two aspects can be promoted, the comprehensive ability of students to communicate in a foreign language and the quality of foreign language teaching.

6 CONCLUSIONS

The interactive computer-aided teaching mode of foreign languages is conducive to the improvement of China's traditional foreign language teaching, which can make the foreign language teaching meet the demands in modern times. At present, teachers should adhere to the student-oriented principle and introduce the corresponding materials of computer foreign language teaching reasonably in order to create a lively classroom atmosphere for the computer foreign language teaching is not mature enough. Effective computer-aided measurements should be taken in terms of foreign language teaching, in order to promote the students' active learning, the quality of foreign language teaching, which can contribute to the rapid development in the cause of foreign language teaching.

REFERENCES

[1] Cheng Xue, Wu Zhe, *Computer-assisted Foreign Language Teaching under Reform of the University English Language Teaching*, Higher Education Research, 2011(2):111–111.
[2] Ren Ningning, Zhang Wang, *Application of computer aided methods in language teaching*, Teaching Reform, 2013(7):67–68.
[3] Shao Wanhong, Problems and Countermeasures of Computer-assisted Foreign Language Teaching, Teaching and Management, 2013(11):146–148.

Future Communication, Information and Computer Science – Zheng (Ed.)
© *2015 Taylor & Francis Group, London, 978-1-138-02653-7*

Detection and common repair strategy applied to automobile knock

J. Zhang
School of Mechanical Engineering, Chongqing Vocational Institute of Engineering, Chongqing, China

ABSTRACT: The performance of the car can be improved by engine the automobile slightly. To control knock intensity within a certain range, the monitoring of the automobile knock is important. Research has focused on the detection methods and the common repair strategies in the work. Based on the concept and characteristics of the automobile knock, some common repair strategies are proposed. These strategies are applied to the sensor, knock control equipment, the engine itself, etc., combined with the importance of knock detection and the knock signal selection. Based on the application to the traditional automobile knock, the effect of electronic equipments to the automobile knock is analyzed.

Keywords: Automobile; knock; monitoring; repair

1 INTRODUCTION

With the development of the Chinese economy, the automobile vehicle has been widely used in China. A slight engine knock can improve the performance of the automobile engine, but the serious knock can damage the automobile components. So it is important to monitor the automobile knock. Once the unreasonable engine knock occurs, measures will be taken in time. Because of the special historical factors, economy, science and technology started late in China, causing a large gap in the automobile manufacturing process compared to some western developed countries. So the well-known automobiles at present are mostly produced by foreign companies, limiting the development of the Chinese automobile industry. In order to improve the automobile manufacturing in China, the technology related to the automobile should be analyzed deeply and be applied to practical production by starting with details like the automobile knock [1].

2 BRIEF TO AUTOMOBILE KNOCK

2.1 Concept of automobile knock

The engine support power for the automobile to run. The energy source of most engines is a fuel like gasoline, which will explode at a critical point when mixed with air. Then the high-pressure gas, brought about by explosion, pushes the piston of the engine and converts chemical energy into mechanical energy. The utilization of the fuel for the automobile is low in the limit of technology, generally only about thirty percent, resulting in a waste of resources. Therefore, how to improve the utilization efficiency of the fuel

has become the problem many experts and scholars research: some scholars proposed that the existing engines should be improved; some put forward some new way of combustion. Engine knock usually happens in the engine working process [2]. Normally, the explosion in the engine only brings about one spark. In special, two sparks will trigger the explosion reaction, known as engine knock. Massive research indicates that a correct use of the knock can improve the efficiency of fuel use and increase the performance of the engine. It is important for the automobile production enterprises and promotion of the idea of green environmental protection. For this purpose, detection and corresponding control devices are installed to keep the automobile knock in a reasonable range, achieving distinct results.

2.2 Application of automobile knock

The analysis above indicated that if the knock is controlled within a certain range and reasonably applied, the performance of the engine will be improved. The knock often occurs based on the working mode of the engine. And the automobile components are damaged because of the high intensity of the knock. So how to decrease the knock has been researched for a long time [3]. Obviously, the direction is wrong till nowadays. Reasonable application of the knock, combined with the actual condition of the working engine, can not only control the influence of the knock to the components like the engine, but also improve performance of the automobile. It is important to the actual manufacture. The idea is proposed and attracts people's attention. Many automobile factories have made some improvement to the automobile based on the actual conditions of themselves. The knock can be controlled within a reasonable range via adding the equipment like ECU, detecting the knock and controlling ignition advance angel. Nowadays, The above technique is applied to most of the production of automobiles. The investigation shows that different automobiles have different engines, especially, different use makes the engine different. Some automobiles need an engine with large horsepower, so the corresponding devices and modes of the knock detection should be changed [4].

3 KNOCK DETECTION AND ANALYSIS IN AUTOMOTIVE SYSTEM

3.1 *Importance of knock detection in automotive system*

In order to make use of the knock, the specific parameters are first detected, and then measures are taken to control the knock. The above shows that knock detections are the base of the using of the knock. Although the knock will cause the shaking of the air cylinder, and the knock with enough intensity will lead to the damage of the related components, the importance of the knock is recognized with the development of it. The knock has been found early in the development of the automobile, however, with the limit of the technical level, there are no related equipments for detection, and the knock intensity cannot be particularly detected. Usually, it is considered that the knock will influence the normal use of the automobile. So many measures have been taken to release the knock, however, with no obvious effect. The automobile knock has been re-recognized with the development of science and technology. Many intelligent devices are applied to the automobile, converting it from traditional mechanization to electronization. The current automobiles are integrated by lots of intelligent devices, improving the performance. So the knock can be detected and controlled within a reasonable range, and the performance of the engine can be improved obviously.

3.2 *Knock signal detection in automotive system*

The knock occurs with the air cylinder shaking, however, the investigation shows that many reactions occur while the engine is working, and there are many factors leading to the shaking of the air cylinder. Obviously, detection of the shaking of the air cylinder cannot determine the knock intensity, so the determination of the knock signal is important. The characteristics of the knock indicate that the knock often occurs between the two top dead centers of the stroke and the knock can be detected in the interval. The knock generates a certain voltage, making a change, so the output voltage of the engine can be detected to determine whether the knock occurs. To take advantage of the knock, the knock should be divided into different levels based on the change of the voltage. Then the optimal knock point will be found by particular experiments, and some

special measures will be taken to control the knock within a certain range, making the knock beneficial. In general, the knock is controlled by changing the ignition advance angle. The actual controlling method will change as different engines have different performances. In order to make the controlling work well, it is necessary to understand the performance of the engine first of all and to draw a curve of the relationship between the knock and the engine performance. Then the optimal ignition advance angle will be found in this curve.

3.3 *Application of knock detection in automotive system*

The purpose of knock detection in the automotive system is mainly to make use of the knock reasonably. Early in the development of the automobile, some problems came about in the automobile when it was running, making the automobile useless. The actual investigation shows that the main factor causing the problems is the knock. Especially, because of the poor quality of the fuel and the engine, the knock can easily damage the components. After years of development, the materials and technology of the automobile production have made great progress. The engine, as a core component of the automobile, has been widely improved in performance. The knock cannot lead to the damage of the components, but can influence the use of the automobile. The decrease of the ignition advance angle can decrease the occurrence of the knock to a large extent, but for modern automobiles, the improvement of the performance of the engine is very important. A reasonable use of the knock can achieve this purpose. Especially after years of research, the concept of pulse detonation engine has been put forward based on using the knock to the largest extent. And all the applications of the knock are based on the detection. Only when the concrete intensity of the knock and the influence on the performance of the engine are detected based on the actual condition of the engine, can measures be taken to improve the performance of the automobile.

4 COMMON REPAIR STRATEGY APPLIED TO AUTOMOBILE KNOCK

4.1 *Repair of sensors*

With the rapid development of electronic and information technology, massive electronic equipments are integrated in modern automobiles. In a sense, the modern automobile has been changed from mechanical equipment to electronic equipment. Many automobiles even have the function of automatic drive. The above shows that there are a lot of electronic equipments in the automobile, running on a basis of sensors, especially the knock detection. In actual knock detection, the corresponding sensors are installed to detect the shaking of the air cylinder and the output voltage of the engine. If the sensor is in trouble, it will cause the interruption of the signal and the concrete condition of the knock will not be detected. The actual investigation shows that the knock problem nowadays is mainly caused by the trouble of the sensors. So in the concrete repair process, if the problem caused by the knock is found, the sensor will be detected. The sensor repair is simple and widely used. At present, many manufactures have produced different types of sensors on the basis of the actual use, with a low price. The new sensors will be replaced when the sensors are damaged. If allowed, the sensors with better performances should be selected to ensure the accuracy of the detection of the automobile.

4.2 Repair of knock control devices

If the automobile components are damaged because of the engine knock, the knock control link can be in trouble. If the devices don't control the knock intensity within a reasonable range, then the knock intensity will be larger, leading to the damage of the related components. However, lots of practices show that the probability that the knock control devices meet with problems has been low. Especially, with the rapid development of the information technology, the improvement of the production process can ensure the quality of the devices. If the problem occurs by detection of the knock control devices, it will be difficult to solve. Considering the complication of the control devices, it is confirmed that either the hardware or the software is in trouble: if the hardware is damaged, then replace it; if the software, then write new software without replacing the device. Because the price of the knock control device is high, many car owners choose the cheaper in the current market considering the costs when the car is out of the maintenance period. These devices can be damaged again after being used for a period of time, even the surrounding components will be damaged. Consequently, to choose some devices produced by the original factory can ensure the performance and compatibility with other devices.

4.3 Repair of engine fault

The engine, as the core component of the automobile, is used to improve the motive power for the automobile. So if the engine faces a problem, the automobile will not carry on running. Early in the development of the automobile, the mechanism and the structure of the engine were simple. Although the principle of the engine has not changed for several years of development, a large number of electrical devices have made the engine more complicated. So in the actual working process, lots of reasons lead to the problem of the engine. If the engine itself comes across

the problem, the automobile knock detection will not be in progress obviously, and the knock will not be controlled within a reasonable range. Considering the importance of the engine for the automobile, the maintenance personnel will detect the engine and ensure whether it works normal when the automobile faces a serious problem. If the engine comes across a problem, it will be difficult to repair. Especially, different automobiles have different engines. In order to have the engine perfectly repaired, different engines must be understood. This demands that the maintenance personnel have enough experiences and professional quality. Usually, the engine can be repaired to keep working and changing the engine is rare.

5 CONCLUSIONS

The analysis of the work shows that the engine knock is a normal phenomenon when the automobile is running. Early in the development of the automobile, the knock often caused the damage of the automobile's components. So measures like decreasing the ignition advanced angle has been taken to decrease the knock. However, with the development of the automobile industy, the improvement of the performance of the engine has been an important problem. The reasonable using of the knock can improve the performance of the engine. In this context, massive sensors are used to detect the automobile knock, and the related equipments are used to control the ignition advanced angle. So the knock intensity can be controlled within a reasonable range and the performance of the engine can be improved. When the knock detection system comes across a problem, firstly the concrete link of the problem will be determined and the repair strategy will be taken pertinently.

REFERENCES

[1] Liu Yiming, Chen Bin, Teng Qin, Zuo Chengji. Detection and Control of Coalbed gas engine knock. Internal Combustion Engine & Power Plant. 2007(02): 1–5.

[2] Wu Pingyou, Huanghe, Chen Qin. Analysis and Control of Gasoline Engine knock. Drive System Technique. 2003(03): 36–38.

[3] Gao Qing, Jin Yingai, Li Ming, Fang Ying. Control and Characteristics of Spark Ignition Engine Critical Knock. Jounral of Jilin University (Engineering). 2003(04): 7–11.

[4] Jianhuan Wei, Huafeng Sun, Shengjie Yang. The cycle measurement of sulfur blank value with the CS-444 infrared ray carbon sulfur analyzer. Advanced Materials Research, 2012: 2173–2176.

Future Communication, Information and Computer Science – Zheng (Ed.)
© 2015 Taylor & Francis Group, London, 978-1-138-02653-7

Analysis on the manifestation of the green environmental protection concept in architectural engineering survey

C. Shi

Faculty of Engineering, Shandong Yingcai College, Jinan, Shandong, China

ABSTRACT: With the development of industry, the environmental issues start to attract more and more attention. The boom of the real estate market in recent years results in an increase of survey works in architectural engineering. This work mainly analyzed the manifestation of the green environmental protection concept in the architectural engineering survey. The concepts and characteristics of the architectural engineering survey, along with the combination of the importance of the green environmental protection concept, lead to the conclusion that the green environmental protection concept can be demonstrated well only if three aspects are assured: the application of advanced measuring equipment, personnel with high quality and essential post processing. Among them, post processing is a novel method.

Keywords: architectural engineering; survey; green environmental protection

1 INTRODUCTION

After decades of development under the reform and opening-up, the overall economic level in China has significantly improved, which makes it the second largest economy in the world. However, in the early stage of the development, limited by the technology at that time, our country had to sacrifice resources for economic growth, which exerted a heavy influence on the environment. Now in China a pollution issue occurs in many regions such as the long-term hazy weather in Beijing-Tianjin-Tangshan region. With the appearance of the environment problem in recent years, how to minimize the influence on the environment during the actual construction and production process becomes a hot issue among the experts and scholars. Afterwards, the concept of sustainable development was proposed on the basis of their studies [1]. Recently, the popularity of the construction industry in China brings about a large amount of construction projects. During the actual construction, the environment will be polluted by the use of many traditional materials and technology.

2 CURRENT SITUATION OF ARCHITECTURAL ENGINEERING SURVEY IN CHINA

2.1 *The concept of architectural engineering survey*

Recently the construction industry is so popular in China that an increase in construction projects is observed. In addition to the residential buildings, there are many other buildings with different functions. As the building type varies, the materials for construction and the technology have to differ accordingly. Thus before the actual construction, a plan on every detail during the construction is required. Planning requires plenty of reference data, which has to be obtained via survey [2]. Thus architectural engineering survey plays an important role in construction planning. As a traditional subject, people started to design different kinds of buildings for different purposes a long time ago. However, limited by the technology at that time, only a simple geological survey could be made during the engineering measurement. Therefore, the information acquired from the measurement was limited, which could hardly meet the needs for construction. After years of development, architectural engineering survey, becoming an independent subject, has improved a lot. With the growth of modern science, the equipment and technology in the architectural engineering survey have been highly upgraded, which leads to some changes to the concept of architectural engineering survey. During modern measurement, besides an essential geological survey, soil in the location of the construction will also be sampled [3].

2.2 *Function of architectural engineering survey*

As mentioned above, planning is very important to construction in architectural engineering. The process of which requires plenty of reference data, including the geological conditions in the area of construction. Only with the specific data can the plan be made accordingly, and that is the importance of an architectural engineering survey. However, according to the investigations, the survey skill in China is relatively backward compared to some developed western countries due to the limited technology. In modern architectural engineering survey, advanced survey equipments are always adopted. With the popularization of the computer, the automation level has been improved to some extent. Measurement can be conducted by just a few technical staff members. After putting samples such as soil into the facilities, the result will come out automatically, highly improving the efficiency of the measurement. Now these advanced facilities are quite popular in the architectural engineering survey in western countries. However, these facilities haven't been adopted yet because of the cost restriction. Although some companies introduced some advanced facilities from foreign countries, highly improving the hardware, the low quality of the technical staff is still a problem. For example, timely repair can hardly be ensured when the facilities don't function properly, which becomes a barrier to the application of them [4].

3 THE IMPORTANCE OF THE CONCEPT OF GREEN ENVIRONMENTAL PROTECTION

3.1 Brief analysis on the concept of green environmental protection

The concept of green environmental protection is gradually formed during the development of technology and economy. In the industrial production, massive trash and exhaust have to be released because of the limited technology, which causes pollution to the environment. Many harsh climates such as the El Nino phenomenon, acid rain and haze, will not only interrupt people's daily life, but also lead to economic loss and physical harm. In this situation, protecting the environment during the development of the industry becomes a crucial issue. That's when the concept of green environmental protection comes into being. In a sense, the main purpose of this concept is to protect the earth, keep the environment good enough for descendants to live in. And a sustainable development of economy and technology has to be reached at the same time. After years of construction, people have realized the importance of environmental protection. Many organizations have been established to advocate green environmental behaviors. Some positive effects have been achieved by them according to the investigation. The application of modern chemicals is an important fact of influence on the environment. Nowadays, new materials are adopted to improve the quality during the construction, but most of them are put into use without detection. These materials can release poisonous gas, such as the gas released from decoration materials, which can do harm to people's health. This problem has attracted wide attention from the society, and many green environmental materials are introduced into the market to solve this problem. These materials can meet construction requirements and minimize pollution at the same time.

3.2 Function of the concept of green environmental protection

After the second industrial revolution, western countries took the lead into the industrial age. Via the application of massive chemicals and machinery, the efficiency and category increased a lot. While the environment was polluted by the trash and exhaust released during the production. There wasn't an environmental concept at that time because of the limitation of technology. And it always takes a long time for pollution to accumulate before the environment goes wrong. That's why people didn't realize the importance of environmental protection until phenomena such as acid rain and El Nino appeared. London's famous name, foggy city, should owe to the pollution during the industry age. And the situation failed to improve a lot after years of effort, from which we can see that environment recovery is a long-term work. The industry in China started late because of some special historical factors. When the western countries had already entered the information age and made efforts to develop light industry, we were still in the heavy industry stage. In order to catch up with the western countries, China had to sacrifice resources for the economy. However, the limited technology led to heavy pollution of the environment. The hazy weather in Beijing-Tianjin-Tangshan region is a result of the pollution because of the lack of the concept of green environmental protection. With more and more attention paid to environmental problems, China proposed some method to recover the environment, which will take a long time before any sign of good news appears.

3.3 Manifestation of the concept of green environmental protection

For many experts and scholars, when they do research on green environmental protection, applying this concept to practice is the main subject. Nowadays there are many international environmental organizations with emphasis based on their purpose in practical work. According to different fields, these organizations will come up with their own manifestation to assist their advocacies for environmental protection. To reveal the importance of environmental protection, proposing a slogan is always a first. And if allowed to, these organizations will send their staff to some place in a harsh environment. The duty of this staff is guiding and training the grass-root staff to learn advanced construction technology, which is a good example of the manifestation of environmental protection. In one sense, during construction, green environmental protection is mainly reflected in the application of material, equipment and processing technology. And among them, environment friendly material should be the first one to be ensured. After years of development, there are different kinds of materials with similar property in the market. But their prices may vary with their composition and processing technology. Thus during the production, considering the cost, many enterprises prefer cheaper material, which is harmful to the environment.

4 MANIFESTATION OF GREEN ENVIRONMENTAL PROTECTION IN ARCHITECTURAL SURVEY

4.1 *Application of advanced survey facility*

In this information age, facilities upgrade very fast. The companies are always willing to introduce the newest facilities, because with every upgrade, higher performance can be achieved to improve the productivity significantly. As for architectural survey, limited by the technology in the old time, only naked eye observation and simple measurements can be done in the survey. Although the environment will take no damage from this method, the results can't be accurate enough to meet the requirement for modern architectural survey. Nowadays, a lot of auxiliary equipments are applied to architectural survey to improve the accuracy and efficiency. According to the research, the traditional equipment will do harm to the environment during working. Recently, reducing pollution has been taken into consideration when designing survey facilities because of the growing attention paid to the environment. These advanced and environmental friendly facilities are a good manifestation of the concept of green environmental protection. That's why they are adopted in the construction survey in western countries. But in China, traditional facilities are still in use because of the limitation of technology. Especially with the rapid development of information technology, facilities are upgrading so fast that companies have to keep importing the newest foreign facilities to ensure progressiveness. In order to control the cost, most companies won't replace the facilities until they don't function properly. This fact weakens the manifestation of environmental protection in Chinese architectural projects.

4.2 *Staff with high quality*

In a construction survey, all the survey equipments have to be operated by the staff, disregarding their types. And the quality of the staff plays an important role in the manifestation of environmental protection. Improper operation may do harm to the environment by staff with low quality and limited environmental awareness in their work. To prevent this, staff with high quality is required. Strict standards will be helpful in recruitment, and if possible, the companies should hire people with a rich experience. Experienced workers can be skilled enough to handle survey facilities and prevent pollution to the environment. Thus staff with high quality is very important to the manifestation of environmental protection. For the existing staff, regular trainings should be held for them to learn the newest facility, technology, and proper method about how to prevent pollution during a survey. This reflects the concept of environmental protection, and a positive effect has been achieved by many companies in China who adopt this method to ensure the quality of the staff.

4.3 *Essential post processing*

However, in a practical survey, even hi-tech facilities and qualified staff can't promise zero damage to the environment. And slight pollution is still unforgivable in achieving green environmental protection. So as a remedy, essential post processing after survey work is required to eliminate any pollution to the environment. Good post processing can not only erase the pollution from a survey, but recover the environment as well. In foreign countries, post processing will always be conducted after a survey. According to the actual situation, the corresponding recovery work will be undertaken to minimize the damage to the environment. Plenty of practice proved this method to be very effective. After years of development, architectural survey has become an independent subject by improving the facility and method in practice. Nowadays many colleges and universities have offered this course and trained many qualified personnel in survey. According to the research, the Chinese architectural survey lacks post processing, which weakens the manifestation of environmental protection.

5 CONCLUSION

According to the analysis in this work, architectural engineering is a traditional industry. There were plenty of construction projects in ancient times for residential purpose and other functions. But limited by the technology at that time, only a simple measurement could be conducted before the construction. With the development of architecture, material and processing skills improved a lot. Therefore, people came up with higher requirement for architectural engineering, which increased the difficulty of the construction survey. In traditional survey work, the improper operation of facilities and survey method leads to polluting the environment. To avoid this consequence in actual engineering survey, advanced measuring facilities and essential post processing should be ensured for the utmost manifestation of environmental protection.

REFERENCES

[1] Cheng Zhou, Importance of Engineering Survey in Construction Quality Management, Heilongjiang Science and Technology Information, 2012(10):290–290.
[2] Aiping Zhou, Xinghui Xu, Importance of Engineering Survey in Construction Quality Management, Sichuan Building Materials, 2011(02):219–219.
[3] Walter, Shengyang Zhao, Construction Survey in Architectural Engineering, Science and Technology of Private Enterprise.
[4] Jianhuan Wei, Huafeng Sun, Shengjie Yang. The cycle measurement of sulfur blank value with the CS-444 infrared ray carbon sulfur analyzer. Advanced Materials Research, 2012:2173–2176.

Future Communication, Information and Computer Science – Zheng (Ed.)
© 2015 Taylor & Francis Group, London, 978-1-138-02653-7

Influence of supply chain integration on business performance

L. Zhang
Business School of Shandong Yingcai University, Jinan, Shandong, China

ABSTRACT: As an important trend in business management, supply chain integration is the transmission chain of business capital, information and materials. Currently, there are problems such as double marginalization among different companies, delay in a supply chain, misinformation, and diseconomies of scale in integration. Therefore, based on theories of industrial economics and supply chain management, the work analyzed patterns of supply chain integration, and finally proposed reasonable suggestions. The work highlighted the significance of e-commerce and a theory of vertical restraints for the first time.

Keywords: supply chain integration; E-commerce; business performance

1 SUPPLY CHAIN INTEGRATION IMPROVING BUSINESS PERFORMANCE

As an important trend in business management, supply chain integration is drawing attention from managerial force; logistics, information flow and capital flow in supply chain are currently concerned by companies. Previously, companies in the supply chain split each other, resulting in varieties of problems which reduce their economic benefit, counteract their competitive edge, and consequently cause a waste of resources [1]. It is such problems in the supply chain prior to integration that make the integration a necessity and contribute to the wide spread of it.

1.1 *Double marginalization and supply chain integration*

Double marginalization, which is prone to spring from core companies, suppliers and distributors in the supply chain, is the biggest promoter of supply chain integration. If double marginalization occurs, it indicates that, due to their relative solidification in the market featuring the reprising of materials or products by suppliers and downstream companies, suppliers and distributors are likely to exert the market force to increase the price above marginal cost and thus make a profit. However, this sort of reprising will result in a higher price of final products, which, to companies of a high product similarity, greatly frustrates their competitive edge [2].

Double marginalization can also hinder the economies of scale. Companies and their suppliers have technical interfaces with each other, which means raw materials and other production factors in companies are distributed on a particular mixture ratio. This mixture ratio is first of all embodied by the fact that the production function of upstream companies is the demand function of downstream companies. Accordingly, on the principle that marginal benefit equals marginal cost, the use of raw materials in downstream companies is subject to price provided by upstream companies; the market force, exerted by upstream companies, results in a higher price, which partly limits the reserve capacity of companies, especially the ones with greater fixed input.

Supply chain integration, therefore, can greatly strengthen the companies' competitive edge and make a profit through weakening the market force exerted by upstream companies, and selling final products at a lower price. However, double marginalization – especially that in industries with highly diversified products – goes long in the supply chain. This makes it difficult for core companies in the supply chain to have control over their product price [3]. Generally, companies have realized that a control over the supply chain means a better competitive edge.

1.2 *Supply chain integration correcting information flow*

Information flow, an important part in the supply chain, includes supply & demand information, price information, industry prosperity, etc. Before integration of the supply chain, different product levels are less correlated with each other. Therefore, originally resulting from information augmentation, information can be distortedly transferred in the chain of suppliers-core companies-downstream distributors. With the absence of an effective inter-companies communication channel, companies can overreact to factors such as price and quantity demands, causing inventory accumulation or sharp fluctuation of the production plan during different periods of production and operation.

Procter & Gamble Company is the first to have realized this problem: even the downstream companies' single reaction to the market can frequently misguide upstream companies in their strategy-making. Then a blind expansion or production decline ensues. Consequently, the supply & demand condition of final products fluctuates, the production plan of the company changes, and an accumulation of inventory or raw materials occurs. It is to the detriment of both capital and operation of companies.

In supply chain integration, the information monopoly between companies can be broken up. However, random fluctuation in the final product consumption doesn't mean fundamentally changing the supply & demand structure or extent of industry prosperity. This allows companies to make a scientific production plan and marketing plan, lower the cost of production and operation, and save money and business risks. Given that the heavy chemical industry is suffering from severe excess capacity, such integration can ease the companies' competitive stress, as well as materials & energy accumulation to the maximum possible extent, thus diminishing the side-effects of excess capacity, and, indirectly, promoting the modulation in the companies' capacity.

1.3 Supply chain integration ensuring an effective capital turnover

Capital turnover means the repeated rotation of capital; rotation of capital is the continuous transformation of companies' monetary capital and tangible & intangible assets in the form of money capital, productive capital, and commodity capital. Capital turnover and rotation of capital require different forms of capital to remain continuous in time and coexistent in space.

The previous supply chain management pattern, to wit monopoly, blocks the effective communication among companies. This leads to a failure of the prompt supply of raw materials needed, a loss of business opportunity, a delay of realization of product value, or a delay of transformation of commodity capital into money capital. Besides, the velocity of capital turnover can be slowed down. A faster velocity of current assets turnover equals a higher annual profit margin and annual rate of surplus value of companies. Therefore, a slower velocity of capital turnover means a narrower profit space of companies.

After the supply chain is integrated, suppliers or distributors become part of the companies' production and marketing plan. These companies and their production plan are more practical which can relieve stock or inventory accumulation. It is obvious that by doing so the cost of companies can be saved. Continuous speed-up of capital turnover can be maximally added to the organic composition of business capital, which increases the profit margin. The reason is: speeding up capital turnover can shorten the investment payoff period of fixed assets, and then cut investment in all fixed assets for more advanced production equipment.

2 PROBLEMS SUPPLY CHAIN INTEGRATION CAN CAUSE FOR COMPANIES

In supply chain integration, major businesses and the owner's equity in companies can be changed. Then companies can see diseconomies of scale and decreasing returns to scale due to scale expansion.

Firstly, a more difficult management can switch companies from economies of scale to diseconomies of scale. Economies of scale, which means that average cost drops with the increase of production scale, comes from the fixed input diluted by a larger number of products. However, overstaffing of management organization can occur in business expansion, causing potential diseconomies of scale for companies.

Secondly, after the supply chain is integrated, raw materials can be easier accessible, and the material supply can be faster. These changes will force companies to add numerous variable elements, causing the diminishing of marginal returns which again increase the average cost. Once these negative effects outweigh the decrease of the average cost from economies of scale, economies of scale will fail to occur. Besides, with raw materials easier accessible, the technological improvement of companies can be slowed down. This phenomenon is economically defined as path dependence, for elements added to boost the output through available technology is safer than through technological improvement. With the boom of the high-tech industry and continuous application of high-tech products, risks in technological improvement will bring companies fiercer path dependence.

Finally, supply chain integration can monopolize the means of competition among companies. It makes technical upgrades more difficult and slower, which means companies can hardly maintain the competitive edge of product differentiation on a long-term basis. With the absence of such a competitive edge, companies will set price and output as the main means of competition, and manufacturers will loosen their control over the market price. As a result, in dealing with potential competition manufacturers, companies can have difficulty maintaining market share.

3 SCIENTIFIC INTEGRATION TO IMPROVE BUSINESS PERFORMANCE

As analyzed above, supply chain integration can greatly improve business performance, reduce the frequency of double marginalization, reduce information distortion of companies, accelerate capital turnover, and, therefore, effectuate economies of scale. However, such integration can also cause diseconomies of scale, path dependence, and fewer means of competition. Therefore, to improve business performance, the work dealt with scientific integration of the supply chain as follows:

Firstly, vertical restraints can be used to reduce path dependence. This can be achieved through trade agreements – which cover product supply, product quality, product price, etc. – among companies, suppliers and downstream distributors. In these agreements, core companies should maximize their profit whereon the trading capacity and trading price are nailed down. This is called in the industrial organizational theory constraint of rational man. Besides, considering the interests of suppliers and distributors, preferential policies should be formulated, i.e. increasing the supply with a somewhat lower price. This is called incentive compatible constraint. Vertical restraints are dependent upon cooperation agreements between companies; this not only enables companies to promptly get raw materials or market products for a relatively low price, but also reduces the companies' cost of expansion and management for a safer operation. Vertical constraints can be carried out in companies both downstream and upstream.

Secondly, supply chain integration can also be achieved through horizontal integration of core companies. The number of the supply chain is determined by core companies. The Cartel between core companies can intensify control over downstream suppliers and reduce cutthroat competition between manufactures. As for the suppliers, an alliance of core companies can provide more orders in bulk for them to achieve economies of scale in a more effective way, and, as a result, cut production cost for a lower producer price. However, horizontal integration of companies will cause a reputation risk. As the Cartel is formed, a sudden change in competitive strategy of a certain company therein – reducing price or raising purchase price of raw materials, to be exact – will mean a severe loss to companies. Actually, either in a one-shot game or in repeated games, companies can't say for sure that the Cartel agreement will successfully regulate both sides. It is after countless repeated games that companies can get a chance to collaborate with each other. Therefore, manufactures can share both distribution channels and supply channels to reduce risks for each other. Based on that, long-term market cooperation can be achieved.

Finally, e-commerce should not be downplayed in supply chain integration. It can offer via the Internet a platform for companies to exchange supply & demand information more easily and effectively. Highly dependent upon modern electronic information technology and internet technology, e-commerce systems can greatly reduce transaction costs for companies; with information communication achieved, randomness in production and operation can be reduced as much as possible; sudden changes in demand such as an accidental event useless for

production & operation cycle and even the whole industry prosperity, once made accessible to companies, can reduce the blindness and fluctuation caused by market-oriented management. Besides, with the help of e-commerce systems, especially when offers of software or electronical invitation are available on the e-commerce platform, companies can effectively choose suppliers and distributors.

4 CONCLUSIONS

In the final analysis, supply chain integration can improve business performance by reducing double marginalization, correcting information flow, accelerating capital turnover, etc. If not properly done, however, it can also originate diseconomies of scale, path dependence and single means of competition. To solve these problems and scientize the supply chain integration, the work argues that companies should achieve long-term benefits through a variety of resources in the supply chain, e-commerce platform, upstream & downstream sales agreements and horizontal cooperation of core enterprises, etc. Supply chain integration is an important vehicle for companies to sharpen the competitive edge and therefore improve the operation performance. For a long-term development, improving supply chain management and integration should come as the first choice of more companies.

ACKNOWLEDGEMENT

The work was funded by *Research on Cultural Fusion of Supply Chain Management*, a Project of Shandong Province Higher Educational Humanities and Social Science Research Program, program number: J12wf60.

REFERENCES

[1] Zhao Li, Sun Linyan, Li Gang, et al. *Research on Relationship between Supply Chain Integration of Chinese Manufactures and Business Performance*. Journal of Industrial Engineering/Engineering Management, 2011, 25(3):1–9.

[2] Wang Qun, Lu Shenghui. *Research on Exchange Mechanism of Supply Chain Integration inside and outside Based on Embedded Structure*. Commercial Research, 2012, (11):120–125.

[3] Yin Congchun. *Discuss of Problems and Implement of Supply Chain Management*. China Management Informationization, 2014(4):103–104.

Future Communication, Information and Computer Science – Zheng (Ed.)
© 2015 Taylor & Francis Group, London, 978-1-138-02653-7

Influence of computer technology on the development of E-business

Y. Chen
Guangdong Engineering Polytechnic, Guangzhou, Guangdong, China

ABSTRACT: Covering related contents of E-business, the work explained the relationship between E-business and computer technologies. Meanwhile, main problems involved in current E-business activities such as illegally falsifying and destroying trade information were revealed. Therefore, multiple technologies including digital signature, identification technology, data encryption, intelligent firewall and virtual private networks (VPNS) were introduced to E-business. Besides, the key roles computer technology plays in promoting the future development of E-Business were also particularly addressed in the work.

Keywords: computer technology; E-business; development; promote

1 INTRODUCTION

With the development of China's economy and the promotion of reforming and opening, computer network technology has achieved great progress. Therefore, E-business has been quickly accepted and become the most popular mode for business trade at present. E-business refers to a series of business activities by using the network and computer. Merchants can show their goods in network malls or platforms where consumers can freely shop through webpages and achieve online trade. E-business is an emerging trade mode, which enables people to get through a complex business process conveniently at any time, in any place. Built on the support of computer technology, E-business will not be able to work once it lacks the support of computer technology [1]. Therefore, computer technology plays a vital role in promoting the development of E-business. The work expounded a series of current E-business problems in the process of its operation. What's more, it also discussed how relevant technologies of computer technology played a role in promoting E-business.

2 BASIC CONTENT OF E-BUSINESS

Today, E-business generally refers to the process of business activities of some organizations or individuals. With the help of a computer system based communication network, E-business has a huge group of consumers. Platforms associated with E-business provide various trade activities as well as an efficient and convenient service for general consumers. Those platforms are generally built on advanced communication channels with lower marketing costs and efficient marketing methods [2]. The construction of those platforms is inseparable from computer technology; that is to say, E-business cannot lack the support of computer technology. The influence of computer technology is particularly prominent in the related information security technology and network technology. E-business in general is based on the construction of a relevant trade website; thus the maintenance of those websites is of high importance for the normal application of trade information security.

3 RELATIONS BETWEEN E-BUSINESS AND COMPUTER TECHNOLOGY

Today, the application of E-business usually refers to a systematic concept with multiple meanings and covering many industries. Firstly, as the product of the Internet age, E-business generally involves the business activities involving the use of Internet technology and computer systems. Those business activities usually include network trading, network marketing, internal management network of enterprises and other activities. Secondly, E-business usually covers business-related activities based on electronic communications and the computer [3]. It consists of manufacturing computer products, developing computer application technology, researching and applying related products, developing and applying Internet-related technologies and telecommunication information technologies. The delivery of goods after placing an online order, the manufacture of computers and the construction of network platforms are the most common sorts of E-business activities. From the 1960s to 1970s, E-business and related data exchange developed rapidly. Since the emergence of relatively reliable computer technology, computer technology has been widely used in business related fields and is now promoted in the Local Area Network, World Wide Web and Internet as well as E-business. Obviously, the promotion and popularization speed of E-business is far greater than that of any other industry. Meanwhile, the development of E-business also promotes the development of computer component manufacturing, optical fiber manufacturing, electronic communication and other computer related fields. To a certain extent, the development of computer technology and E-business are complementary to each other.

4 SECURITY ISSUES OF CURRENT E-BUSINESS ACTIVITIES

During the rapid development of E-business activities, many problems inevitably appeared in E-business activities due to their online transaction management vulnerabilities and computer technology limitations. The main problems are summarized as follows:

4.1 *Arbitrarily falsifying and destroying transaction information*

Rapid development of computer technology enables more and more people to master this technology, which results in the emergence of hackers. Hackers are those who master computer technology very well and are very familiar with Internet information rules as well as online transaction information. In the process of E-business activities, some hackers make use of their advantages to illegally invade into E-business platforms. Usually, they try to modify, copy or disclose related transaction information to third parties, making the normal trade between two parties or multiple parties fail to be achieved. In addition, some hackers even try to attack E-business platform while copying or modifying an E-business information, leading to the collapse of an E-business platform.

4.2 *Creating fake information to deceive E-business users*

Besides copying and modifying transaction information, those lawbreakers who invade E-business platforms also steal customers' private information and transaction information stored at the backend of an E-business platform. Then they send some fake production information or transmit some phishing site to other remote clients according to the private information they have stolen. As a result, some downstream clients of E-business will be taken in unconsciously and some of them may suffer serious economic losses.

4.3 *Stealing information from E-business platform*

In the process of E-business operation, information stored on E-business platform is an important resource including personal accounts and the password of E-business users, transactions involved, information on money and prices. In order to strengthen their competitive edge, some E-business platforms hire

hackers to attack other E-business platforms to obtain private information of E-business users. Thus, they try to understand the customers' psychology as well as the competitors' business model and weaknesses through analyzing the information so as to win the competition.

5 ROLE COMPUTER TECHNOLOGY PLAYING IN PROMOTING THE DEVELOPMENT OF E-BUSINESS

Widely applied as it is, E-business still faces many serious problems, which makes it increasingly prominent to enhance the security management of an E-business platform. The development of computer technology and network related technologies enable computer technology gradually to go deep into the E-business payment system, website management and construction and safe transaction. With the security of its activities guaranteed, E-business can be greatly improved. The specific roles of computer technology on improving E-business are as follows:

5.1 *Application of digital signature technology in E-business*

The continuous development of computer technology enables digital signature technology to play a vital role in E-business security and secret system. For example, digital signature technology is applied to both source signature service and non-repudiation service. Therefore, with the help of a digital signature technology, the signed parties cannot deny their signatures, and it becomes possible to identify the authenticity in front of a notary public immediately. What's more, some other problems such as other illegal forgery of a signature are also properly solved. In addition, a digital signature helps trading parties to effectively identify their original messages, avoiding trading information to be falsified and denied. Consequently, many difficulties in E-business activities can be effectively solved.

5.2 *Application of data encryption technology in E-business*

As a widely applied security measure in E-business, data encryption technology mainly includes asymmetric encryption and symmetrical encryption. The common public key encryption belongs to asymmetric encryption. At present, PKI technology is adopted to build an encryption system in the process of E-business operation, effectively ensuring the safety of E-business activities. The specific implementation process of PKI technology is as follows. First of all, decompose the secret key into two parts. Then, open any one part of the secret key to others in an unclassified way and save the rest part as a solution to the secret key. By this means, users can make use of a public secret key to encrypt important information and later decrypt the information through the corresponding secret key. Thus, the security of trade information can be guaranteed for both trade parties.

5.3 Application of identification technology in E-business

In the process of E-business activities, identifying accurate information of both trade parties is a very important section. At present, identification technology includes mark identification and password identification. Password identification is the most widely used method in current E-business activities. In general, users should select their passwords based on some basic rules. For example, they should choose the password that is difficult for others to guess or analyze and is easy for them to remember. A password is usually a character string, and a mixed password of special characters, numbers and letters is highly recommended. This kind of mixed password can increase the difficulty to crack it and improve its safety. Today, a magnet is usually used as the medium of mark identification to record private information, which is convenient for a machine to identify. This to some extent improves the development of smart card identification technology, which in return ensures the security and efficiency of identification technology.

5.4 Application of intelligent firewall technology in E-business

Firewall technology is built on the basis of information security technology. The rapid development of network information security and network communication helps to popularize firewall technology in modern public and private networks, which also improves the development of intelligent firewall technology. Intelligent firewall technology in E-business can effectively prevent the spread of a virus and the intrusion of some advanced virus forms. Intelligent firewall technology reflects the main developing trend of modern firewall technology. The main intelligent firewall technologies being adopted in modern E-business include agreement normalization technology, anti-fraud technology, anti-scanning technology, anti-attack and other related technologies.

5.5 Application of virtual private network technology in E-business

As an emerging technology, virtual private networks itself are not a real professional network. In addition, the information transmission medium that virtual private network adopts is also not a reliable public Internet network. Therefore, a virtual private network tries to achieve its functions which are similar to a private network mainly by additional safety tunnel, access control, user authentication and other related technologies. Thus, the transmission security of some important information in E-business activities is further guaranteed.

5.6 Application of virus prevention and illegal intrusion detection technology in E-business

Intrusion detection technology effectively combines control technology and detection technology. This technology can detect the eternal or internal illegal intrusion behaviors of the Internet according to the attack trace left by former attackers, which can effectively defend it against illegal invaders. Intrusion detection technology is the most important technology of computer network security technology and is widely used in E-business activities, providing an important guarantee for the security of E-business activities. In addition, the wide use of network virus prevention technology also provides a powerful guarantee for preventing a computer virus invading and infecting the computer system and related software in E-business activities. For now, network virus prevention technologies involved in E-business activities mainly refer to document monitoring and scanning, anti-virus chip and other related technologies.

6 PROSPECT OF E-BUSINESS'S FUTURE UNDER THE IMPETUS OF COMPUTER TECHNOLOGY

With the development of the economy and the popularity of the computer network, E-business shows a growing influence on people's daily life as a new type of efficient transaction form. The rapid development of E-business is closely related to that of computer technology. Nowadays, the constant update and perfection of computer technology can ensure more efficient and securer E-business activities. In return, the development of E-business enables computer related fields to expand widely, further improving the rapid development and application of computer technology. At present, E-business in China has entered a new period. Besides, electronic network and computer related technologies also have deepened into various industries and fields, greatly accelerating the development of the market economy in China and effectively guaranteeing the global expansion pace of the domestic market in China. Therefore, computer scientific researchers spare no effort to explore and research new methods and technologies to improve computer technology and ensure the steady development of E-business in China. Meanwhile, constant improvement of computer technology progress and update enables computer technology in China and its related fields to show their importance on the international stage. In addition to the guarantee of computer technology, the future development of E-business in China also relies on the support from related laws and regulations. Therefore, the development prospect of E-business in China has an optimistic and bright future.

REFERENCES

[1] Cai Yang, Ding Meiling. Application of Computer Technology in E-business. Silicon Valley, 2012(21): 127–127.

[2] Qin Jingwei. Relationship between Computer Technology and the Development of E-business, Market Modernization, 2012(25):63–63.

[3] Zhang Lei. Roles of Computer Technology Network Consumer Interaction, Value Engineering, 2014(3): 190–191.

Future Communication, Information and Computer Science – Zheng (Ed.)
© *2015 Taylor & Francis Group, London, 978-1-138-02653-7*

Analysis of methods dealing with computer image post processing

G. Feng
Qinghai Nationalities University School of Physics and Electronic Information Engineering, Xining, Qinghai, China

ABSTRACT: A computer is always used to deal with image post processing in the current digital photography and film and television production. But different processing methods will have different influences on the results and efficiency of processing. This work mainly studied which kind of the processing method should be used to obtain the best processing effect. Based on the present situation of the post processing of computer images and in combination with some problems in the process, it was found that only the advanced hardware equipment and software with high-quality staff can guarantee the effect of the image post processing to the largest extent. What kind of equipment should be chosen lies in the actual needs of the work.

Keywords: computer; image; post processing

1 INTRODUCTION

Computers can dispose of tasks automatically to improve the efficiency of the work, thus people attach much importance to the application of computers. Some companies even put forward the concept of the paperless office. Along with the development of the computer software technology in recent years, people have developed the software with corresponding functions specifically according to differentness in different areas. It improves the computer application to a certain extent. In daily life and work, computer applications are everywhere. Affected by the special historical factors, economy, science and technology, the start of China has not been as early as other countries, resulting in a low modern computer technology. Currently, the computer hardware equipment on the market is produced by foreign companies, as well as a great deal of computer software. For example, in order to improve the quality of images in the field of the photograph and film production, a computer will be used in later processing, but computer technology of image post processing is not well applied in China.

present electronic calculators in terms of its size and performance. Along with the application of the transistor and the integrated circuit, computer performance has been improved greatly and its volume became smaller but the tasks it can conduct tend to be more complex. In order to apply computers to more fields, people invented high-level languages such as C language on the basis of computer language. A great deal of software with different functions has been developed to change computers application to a great extent. Nowadays, computers have been widely used after years of development. Some fields are even changing with computers, for example, shooting effect is comparatively poor in traditional photo-taking, because of the limited technology without any post processing work. With the development of the computer, if photos are stored in digital form on computers, photos can be modified by some software to improve the effect. It is very crucial for the actual photo-taking and film-making work. Computer image post processing has become an independent discipline after years of development. A lot of colleges and universities have opened this course and trained a large number of relevant technical personnel.

2 PRESENT SITUATION OF COMPUTER IMAGE POSTPROCESSING IN OUR COUNTRY

2.1 *Development of computer image post processing*

The appearance of computers fundamentally changed the ways to work. However, in the early computer development, the computer could only do simple mathematical calculations and it was not even as good as

2.2 *Situation of computer image post processing in our country*

Compared to some developed western countries, there are still certain differences in the computer application, especially due to the limited technology in China. Plenty of the hardware and software in the market are produced and developed by foreign companies, so companies in China have to import them from abroad at a high cost. It is this kind of phenomenon that furiously affects the photographing and film production in China. China has become the world's second-largest economy, and the overall economic strength has been improved greatly after more than 30 years of reform and opening-up, people begin to computer image post processing work, and spend a great deal of manpower and material resources on study. So image post processing has become common. But, through the actual investigation, it was found that traditional equipment is still used in image post processing work. Especially, the computer hardware upgrades at a high speed, from Moore's law, we can know that computer hardware performance will be doubled every 18 months and the production cost remains unchanged. It shows that frequent equipment replacement is the guarantee of the computer advancement in image post processing work. But computers can work for a long time and it will need a high cost to replace them, if computers are not damaged. What's more, it is inevitable that hardware updates are accompanied by software replacement, so few enterprises can ensure the advancement of computers. They only update them accordingly after a long time or when problems appear.

3 PROBLEMS IN COMPUTER IMAGE POSTPROCESSING

3.1 *Behindhand hardware equipment*

Computer image post processing mainly relies on the hardware and software. Affected by such factors as costs, it is difficult to timely replace hardware devices and promise advanced computers in the work. Image post processing work can be finished by using the relatively backward equipment. On the contrary, they will lead to low efficiency and poor image post processing. The present hardware situation must be improved to settle the problem. Given the reasonable resources use, the most advanced computer may be required, but the computer performance should be ensured to meet actual needs. As the information technology develops in recent years, the image has been improved greatly. For instance, pixel and analogous color are enhanced a great deal. A higher performance is necessary in dealing with work. If a computer processor's operation ability is poor, it can't finish some special processing works in that it will spend a long time waiting. Especially with the development of computer software in recent years, software companies have developed a lot of processing software in order to better complete the image processing work. However, each software has requirements for using hardware, the more advanced the software is, the higher the performance requirements for hardware are.

3.2 *Low staff quality*

No matter what hardware equipment and software technology are adopted, it is sure that operators are needed in the actual processing work. So the professional quality largely affects the efficiency and effect

of image processing. Although people have realized the importance of the image post processing, and many colleges and universities have opened the corresponding courses to train technical personnel in image post processing, because of the limited education, technical personnel trained in our country have a poor practical ability with a good grasp of the theoretical knowledge. Education reform has been carried out in recent years, but students need to adapt to the job after graduation. It is obvious that problems on the education system cannot be solved in a short time, thus it can only be solved from the perspective of the staff. When employees enter jobs, they will be trained to understand the importance of post processing and master methods to use the advanced hardware and software. If the condition allows it, the company can also establish a cooperation relationship with some advanced enterprises to provide a practice environment for the staff. This way, the staff can largely improve their own quality by co-working with some experienced people.

3.3 *Software at allow level*

The computer has been widely applied, but the reason why the computer is able to meet different needs in different fields is mainly due to the role of the software. According to various needs in areas, people developed targeted software with corresponding functions. For instance, there is a lot of software on image post processing on the market. Photoshop and 3Dmax are among them. Various image parameters can be adjusted by software. From the actual survey, it was found that software updates at s high speed. For example, the processing software available on the market is regularly updated. It will also add some new functions in addition to fix bugs, and the function will change a lot. Thus the new software should be used in order to ensure the effect and efficiency of image post processing, costing lots of money to develop the processing software. As a result, money is the premier of using. When the version is updated, it is needed to buy it again. Because of expensive costs, it is not often for the company to use the latest ones, affecting the image post processing work to a large extent. Especially for some advanced software, because there are fewer users in our country, there is no simplified Chinese version. A good command of English is required for using the software skillfully, which also affects the use of the software to a certain extent.

4 METHODS OF COMPUTER IMAGE POST PROCESSING

4.1 *Adopting advanced computers*

It is necessary to adopt the advanced computer hardware equipment to ensure the efficiency and effect of computer image post processing. Especially the image, pixel and color display have developed in recent years.

A very high operation frequency is needed in the processing of the images. Considering the cost, as long as it can ensure the computer hardware performance and satisfy the actual processing need, it is not essential to choose the most advanced computer equipment. If the image post processing technology and software are updated properly after a period of time, the efficiency will be greatly improved. Then we replace the computer hardware. If so, it can not only control the cost, but also ensure the effect of the image post processing at the same time. It is of great importance for the actual computer image post processing work. At present, some enterprises have already started to use it. In our country, the modern computer technology is at a low level. The computer hardware equipment on the market is from foreign companies. We have to spend a lot of money in introducing them from abroad. Our technology must be enhanced and the ability to develop software must be advanced, meanwhile, the hardware production capacity must be improved.

4.2 Improving professional quality of stuff

As the principal part of the image post processing work, the staff's professional quality can largely influence the image processing effect. Present software is highly automatic and as long as corresponding parameters are out in, the computer can complete the processing work by itself, greatly improving the processing efficiency. If people can understand the characteristics of the software and hardware, and master some processing skills, they will be able to use the computer proficiently. Especially after years of development, great progress has been made in computer image post processing, and many processing methods have emerged. However, in the actual work, the concrete treatment method is mainly determined by the staff. For different images should adopt different processing methods to gain enough processing efficiency on the basis of image situation. To guarantee the quality of the staff, the company should try to raise the standards for employees in recruitment. If the condition permits it, some assessment methods can be used to investigate all aspects of employees specifically. It can, to the largest extent, ensure that employees are armed with sufficient professional quality. For the current staff, employers can adopt such appropriate methods as training.

4.3 Using advanced software

Considering that computer software updates fast, each new generation of the software will, to a certain extent, improve the efficiency of the image post processing. To guarantee the effect of computer image processing, the company can only use advanced software. But affected by such factors as cost, enterprises usually adopt some cheap software, leading to poor effects of image post processing. If the company wants to find a good solution to this problem, first of all, it is needed to attach importance to software. Then the actual the needs of the image post processing are combined to specifically select the best software on the basis of cost. In the actual choosing process, the various performances of the software should be compared. Normally, focuses of the work are different, due to different processing areas. Software on the market is integrated Software with a number of processing capabilities. But different software still has some slight differences in detail. If the details can be fully understood, one of the best software can be chosen on the basis of the actual needs of the image processing work. So that even if some cheap software is chosen, it can also meet the actual needs. After years of development, China's software industry has improved a lot. Although compared to some western developed countries, there still is a certain gap, but our country has been able to develop software with a better performance.

5 CONCLUSIONS

From the work, it can be known that the computer has been applied widely. According to different needs of different fields, people have developed targeted software with corresponding functions, meeting actual needs. Image post processing plays a significant role in digital photography and film and television production. The efficiency and effect of image processing can be largely improved by using advanced computer hardware and software. But under the influence of such factors as cost and personnel quality, computer image post processing is at a low level in our country. Advanced hardware equipment should be adopted and high-quality staff should be employed to settle the problem. At last, according to the actual image post processing situation in China, some advanced foreign software should be introduced to improve computer image post processing to the maximum extent.

REFERENCES

[1] Shenghua Xu, Study on Later Stage Editing, AE and DF of Digital Film and Television, Science and Advice (Science Management), 2011(03):78–79.
[2] Liang Zhao, Influences of Post Production of Digital Photos on Photograph Works, Artistic Research, 2010(01):134–135.
[3] Kerong Li, Study on Some Problems about Non-linear Editing System in Post Production, Scientific and Technological Information (Scientific Research), 2008(05):56–56.

Future Communication, Information and Computer Science – Zheng (Ed.)
© *2015 Taylor & Francis Group, London, 978-1-138-02653-7*

Analysis on teaching skills and mode of physical training

X. Zhang
Institute of Physical Education, Sichuan University of Science & Engineering, Zigong, Sichuan, China

ABSTRACT: In the current physical (PE) education teaching, low teaching efficiency and physical fitness of students are the main problems, which were investigated in the work. Results show reforming the teaching mode and exploring new teaching methods actively can stimulate the students' interest for physical exercise and improve the teaching skills of physical training. The innovation of this study is that the information, standardization and psychologization training mode has been introduced to physical education. Therefore, the teaching effect can be intuitively understood, and the corresponding training plans can be adjusted timely.

Keywords: Physical Training; Teaching; Mode

1   IMPORTANCE OF PHYSICAL TRAINING TEACHING

The current PE teaching has many problems: things like sports resources shortage, consideration of the personal safety of students, lack of evaluation pressure for PE. On this account, PE teaching has the following deficiencies, less class time for outdoor training, and outdoor activities are just free activities. These phenomena result in reducing the PE teaching hours to a minimum then physical qualities and sports skills of students cannot be correspondingly improved. Therefore, reform and innovation seem urgent for PE teaching mode. As the organic combination of training and teaching, physical training achieves not only the goal of teaching activity but also the purpose of training [1]. That means it can expand the relevant skills and tactical knowledge of students, also improve their skills, perfect tactic sand skills. Although there are different emphases, once combined the huge effect will appear. Then the efficiency of physical education will greatly increase.

Physical training can not only exert the advantages of teaching and training, but also narrow the gap between students and teachers. Only when the strangeness and fear between students and teachers disappears, can teaching ideas be delivered to the students. Thus students can seriously complete the task assigned by teachers, so as to exercise their physical quality and improve sports skills [2].

2   PROBLEMS IN PHYSICAL TRAINING TEACHING

2.1   *Lacking enthusiasm for physical training from students*

Under the current examination-oriented education background, schools and teachers generally believe that the most important is the academic achievement of students. Only to learn well, can students have a bright and beautiful future. Under the guidance of this concept, students bury themselves in the classics and ignore what is going on beyond their immediate surroundings. That means they only learn the importance of knowledge while ignoring the physical exercise. Consequently, physical education is placed in the edge discipline place, or is even canceled by some schools in the tense final phase. In addition, PE courses generally have no score evaluation system and will not affect the rankings of students. These also increase their contempt for PE. Some students even think PE course is a waste of their learning time and should be canceled. With less time for PE classes and a lacking understanding of sports training, students lose their learning interests of physical education [3]. That is one of the most obvious problems existing in the current physical education and the most fundamental reason for the low efficiency of physical training teaching.

2.2   *Attaching little importance to the training and teaching for PE teachers*

Some teachers think that students are just in the primary stage of sports learning, unlike professional athletes who learn systemically. So they only organize outdoor activities for students in school, without seriously considering the teaching methods and results. In addition, the comprehensive qualities and skills of occupation of PE teachers are relatively weak, compared with other subjects' teachers. And PE teachers usually do not use the textbooks and prepare lessons, so the teaching schedule and content completely rely on the experience of teachers. This mode of teaching activities will not guarantee the progress and quality of teaching, thus sports training and the improvement of physical fitness of students are affected. Paying no attention to the training and teaching, teachers also won't teach students about the role and significance of some training during the class. In addition, even with enough interest for physical training, but with a negative attitude of teachers, students will gradually kill their interests in this negative environment. Then, it is so easy to imagine the final physical training effect.

2.3 Old teaching method of physical training

Some PE teachers work in a relatively closed environment so long that they know little of innovative teaching methods; thus they cannot apply these in class skillfully. Generally, teachers are still inclined to traditional and conventional teaching methods, which have been used for many times and rarely lead to errors. So they can complete the class without spending much experience. But these old methods cannot meet the current teaching environment to a great extent, such as the change of the physical quality of students and the development of sports skills. These changes undoubtedly set new requirements for the methods of training teaching. With the information and multimedia of current sports training teaching, more and more innovative thoughts of teaching methods emerge. Under this background, PE teachers should adapt to the times actively, learn and master new teaching methods, so as to play its due role in the training of teaching.

2.4 Backwardness of comprehensive training model for PE teachers

Innovative training teaching method promotes the reform of teaching mode, which contains the transformation of thought and the improvement of theory. Yet the current training mode is relatively backward. Some teachers have brought innovative ideas into the training teaching, such as game style training, to enhance the learning interest of students. While there are still some loopholes in the entire training system. Training teaching is a kind of learning activity by training. Different from the formal training process, teachers rarely analyze and evaluate the training results after class. Without acquaintance with their training, students will not attach importance to training. Gradually, the effectiveness of training will be affected.

3 ANALYSIS OF THE SKILLS AND MODE IN PHYSICAL TRAINING TEACHING

3.1 Analysis of the skills in physical training teaching

3.1.1 Using game teaching, stimulating the training enthusiasm of students

Games, as the teaching method in game teaching, will relive students from the boring process. Absorbed in physical training, students will be more actively participating in PE classes and challenging the problems they face.

Sports games not only are rich in content and varied in style, but also contain fierce competition and far-reaching cooperation. Therefore, the students can improve their physical fitness and sports skills involved in the process of the game. In the meantime they also learn how to correctly deal with cooperation and competition, how to conduct tactical arrangements and present themselves in the game. In addition, in a relatively independent game, with no compulsory rules, students can adjust the training intensity and manners according to their actual situation during the game. There should be some limitations for the game, not too rigid, to guarantee a harvest for everyone involved in the game. Without constraint, some students with good physical conditions will occupy a large number of game resources, so as to hit the enthusiasm of others.

3.1.2 Reasonable grouping, expanding the role of mutual help group

Grouping can promote mutual assistance among students. Even without the guidance of teachers, students can help others to solve the training problems or finish the project that cannot be carried out independently. For example, during the 100 meters sprint training, students themselves cannot be both running and timing, which requires students of the same group to measure the time. Obviously, grouping brings a great deal of benefits in training teaching. To give full play to the role of grouping, it is necessary to synthesize various factors while organizing teams. Firstly, teachers should have a general understanding of the students' physical fitness and sports skills or something, which is helpful in the distribution balancing of groups. That is: one group should contain both the good physical students and poor ones; one group should contain both the students who understand the physical training and the ones who know nothing; one group should contain both male students and female ones. Thus, with active students driving the inactive ones in learning, all students will progress and develop. In fact, teamwork is necessary in many sports training, such as volleyball, football and basketball, etc. At this time, teachers can invite the outperforming teams in training to show, so that other team members can find their deficiencies and correct it while watching.

3.1.3 Organizing small game in class, inspiring the sense of collective honor among students

Students in a vibrant stage are more intense about the outcome of a competition than those of a different age. Therefore, teachers can hold some small games in class using this feature. In this way, the training enthusiasm of students can be mobilized and the training results of students can be tested. As for the small game, attention should be paid to the time. If it is too long, the enthusiasm of students will be exhausted in repetitive activities; if it is too short, students cannot fully play and show their strength. Thus, the game time should be fixed at 2-3 classes, which can meet the aspirations of students without delaying the progress of future teaching. As the organizer, teachers should maintain normal competition and ensure the results of the competition fair, just and open. After the game, teachers should systematically analyze the information provided by the game to provide a reference for the next PE teaching plan.

3.2 Analysis of the training teaching mode

3.2.1 Strengthening ideological education, penetrating psychological teaching mode in physical education

Thoughts guide the behavior of people. With positive thought, his behavior would be passionate and sunny while with negative thought, his behavior would violate thics or even infringe law. Physical education curriculum is a discipline of motility, unlike Chinese, politics and history that involve the thought of students. Therefore, teachers seldom attach importance to ideological education in physical teaching. Especially in PE teaching, teachers focus on the effect of the training with little communication to students. Strengthening ideological education requires teachers to communicate with students in class or after class. By understanding the students' current views of PE teaching mode, the students' thoughts can be determined. Then for the problems reflected, teachers can carry on with the ideological persuasion of students. For example, when students reflect physical training is too tiring, teachers should help students understand the importance of sports training. In addition, combined with their actual experience, teachers should give psychological intervention to students to understand the significance and role of training teaching.

3.2.2 Using information teaching methods

In the 21st century, the era of information technology, application of computers and all kinds of software makes people's life more regular and normal. Introducing information technology to teaching methods of physical training benefits mainly two aspects: one is that information technology software can help teachers to better analyze sports training; and the other is that using multimedia makes teaching more humane. In physical training, teachers should analyze the situation of the trained students. The complexity and difficulty of this process can be reduced with the emergence of information technology software. Once the basic information and training scores of students were typed into the computer, the software will automatically generate a variety of statistical charts. Under the guidance of this visual information, teachers can correct the problems of training in time.

In physical training and teaching, a multimedia approach is often used as an auxiliary tool. For instance, after completing some physical items of huge consumption, students can be relaxed by the soothing music played by multimedia devices; in the competition, passionate music can allow students to adapt to the game quickly; by using videos to show the training programs, students will have a detailed understanding of every training part.

3.2.3 Following the scientific methods of sports training and teaching

The method of physical training should be scientific and perfect, not blindly viewing the effectiveness of training as the only goal. Firstly, before training, teachers should draft a comprehensive training plan according to the mind and physical condition of students. This plan should meet the needs of the development of students, so as to improve the quality of training. Secondly, on the basis of their experience, teachers should give students some training tips to help them avoid hurting and improve score. However, the training with skills should be done step by step, because hasty training will bring pressure to students. Finally, the implementation of the training program should go step by step, neither too fast nor too slow, so the quality can be guaranteed and time can be saved.

4 CONCLUSIONS

Consequently, based on the project training, physical training and teaching aims at improving the physical fitness and reducing the pressure of students. And it is a PE teaching mode to help students develop and grow healthy. In order to further perform the functions of physical training and teaching, teachers need to analyze the deficiencies in teaching carefully as well as innovate and reform actively. Thus, the right training methods and strategies can be found.

REFERENCES

[1] Kanlian Tang, Related Research on the Teaching of Physical Training of Junior School, Sports World Scholarly, 2013 (8):137–138.
[2] Zhuoyue Tan, Analysis of Physical Education and Training, Scientific and Technological Information, 2012 (12):297–298.
[3] Wei Mou, Brief Analysis of Physical Training Teaching, New Curriculum Research, 2013 (05):22.

Future Communication, Information and Computer Science – Zheng (Ed.)
© 2015 Taylor & Francis Group, London, 978-1-138-02653-7

Current situation of Chinese education management ethics

X. Wang

Vocational and Technical College, China West Normal University Nanchong, China

ABSTRACT: This work analyzed the current situation of education management ethics in China. The education management ethics and its significance were introduced firstly. It discussed the values of Chinese education management ethics and the core issues of this field. And based on the current situation, some suggestions were proposed for the improvement and identification of an ethical framework. Some practical problems of the education management were also discussed. Finally, the work gave advices for the current development of the management ethics.

Keywords: education management; management ethics; organizational behavior; values

1   DEVELOPMENT OF EDUCATION MANAGEMENT ETHICS AND ITS SIGNIFICANCE

Research and application of education management ethics are of ancient origin in China. Confucius and Mencius, who were the educators in the Spring and Autumn Period and the Warring Period, had put forward the thought of "benevolence". They aimed to train politics and excellent educators with the benevolent quality. Since then, Chinese educators have attached great importance to ethical education and education management ethics [1].

1.1   *Specificity of education management cause*

Education, as a process of teaching, is actually aimed at the realization of the self-development for the individual. And education management, as a management behavior responsible for society, is to cultivate talents in society and improve the quality in China [2]. Therefore, deliberation is needed for the organization of educational activities and the formulation of an educational policy. The Chinese education management cause is closely connected with the consideration for the values with the requirement of ethical guidance. Humans are the implementers, influencers and the ultimate goal throughout the education management system. Therefore the education management should have a moral character and guide technical behavior such as education with morality and justice, not just as a tool [3].

1.2   *Stimulatory effect of ethics for the education management*

Ethics is stimulating for education management activities, especially for the effect produced by education management behavior. Basically, the legal restraint is only for the illegal and criminal activities, not much including the threat to public order and interest. Thereby, there is a need for the guidance of morality and values which have a deep influence on different levels of society. In fact, moral values promote personal development and co-operation between people. Education management relies on all the people in the education system, not only on the individual. Ethics of education management will improve moral qualities of education managers, which stabilizes the relationships and public relations of education management. Moreover, education management ethics even benefits effectiveness and increases efficiency of education management.

2   CURRENT SITUATION OF CHINESE EDUCATION MANAGEMENT ETHICS

In the past 30 years of education, the research and practice of education management ethics in China have made valuable progress. Profound studies are conducted especially on the significance and some focus and difficulty problems of education management. The current development of education management ethics in China is summarized as follows.

2.1   *Values of education management ethics*

Currently, most of education management scholars in China have shown an affirmative attitude to the theory and practice of education management ethics. Meanwhile, scholars and the community of practice of education have made broad efforts on the practice of education management ethics. Education management, as a specific practice of the current Chinese education community, is closely connected with the guidance of correct scientific theory – morality and ethical values. Regulations and related system of current education management have gradually formed in China. The educational community is restrained by the values, in addition to the restraint of some education laws and regulations. Particularly, the hidden ethical morality has a positive impact on the development and direction of the entire education community. And in some ways, it can be said that the hidden ethical morality has a more profound influence than the laws and regulations. If the ethical morality forms a consensus, the values will form a cultural atmosphere, which could have a lasting impact on the whole education management community. Therefore, deep explorations of the education management ethics will solve the current problems such as moral anomie and disorder of teaching ideas. Besides, it will promote harmonious scientific sustainable development of education by maintaining the interests of the education practice community and building a communication platform for the interests of stakeholders. Generally, values of education management ethics play an important role in the guidance and improvement of the blank in education management and the ideas of education localization.

2.2 Core issues of education management ethics

The education management ethics have a relatively fast development in China. The scholars have been exploring the function and core issues of education management ethics in the Chinese education community. The explorations have gotten some hot theoretical research achievement with Chinese characteristics. Many current education policies of China also have already taken issues of education ethics into consideration. When making policies, the education administration authorities often consider the influence of the relevant education management policy on the Chinese education community and the ethical values. Updated views have emerged, especially on equality of educational opportunities, educational resources sharing and fairness, justice of policy and the value and free development of educated groups. Additionally, the formation and establishment of the current education policies have been an important part of education management ethics.

Another aspect are the ethical problems in the management of schools. Many scholars have explored ethical issues of the current examination-oriented education management. It should be realized the school management needs to focus more on moral cultivation and the relationship between students and teachers, not just on the performance management. Therefore, the school management should be flexible and hierarchical. Fortunately, besides the establishment of relevant scientific management standards, Chinese school management gradually has formed harmonious relations between teachers and students. Meanwhile, the justice and objective fairness of ethical values

in education management also are incorporated in the improvement of the evaluation system between teachers and students.

3 SUGGESTION OF THE CURRENT CHINESE EDUCATION MANAGEMENT ETHICS

Scientific outlook on development widely spreads the people-oriented concept. The management of educational institutions no longer totally refers to the assembly line operations or industrial model, but on more humanity. The education management ethics produce a significant impact on the education community. However, in spite of a deep understanding of ethics and booming research, the development of education management ethics is still not satisfactory in China. What remains a crucial problem is the application of theory to practice and the solution of debate produced in the practice of education management. Therefore, based on the current situation of Chinese education management ethics, ideas and suggestions are given in this work.

3.1 Improvement and identification of ethical framework in education management

An effort for the further improvement and identification of the ethical framework is the systematization and scientization of education management ethics and the rational guidance of ethical values. The ethical values of current education management in China have not formed a relatively complete and mature discipline system. Therefore, the research department should have clear research objects and missions and classify related core issues and basic concepts. Through these measures, Chinese education management ethics will gradually form a stable and clear framework. Besides, the framework system will be the foundation of deep ethical researches and the basis of the formation and improvement of the discipline.

3.2 Extension for the practice of education management ethics

The extension of ethical values vision and research dimensions particularly relies on the basic way and logic of ethical methodology in the education management. The education management ethics, as an interdiscipline and an emerging discipline, covers a wide range of subjects. In addition to inheriting research methods of the ethics and management, the education management ethics should also have appropriate innovation and abandonment. Meanwhile, based on the theory of pedagogy, the educational management ethics should gradually form an independent discipline with the practice of management and ethics. And the discipline includes the universal research methods of humanities such as social surveys, case studies and the analysis of utilizing tools of organizational behavior. Such efforts will improve the practicality and bring necessary technological support and theoretical resources for education management ethics.

3.3 *Ethical consideration about practical problems of education management*

Education management ethics should focus on the solution of problems in the education process and the ethical values of the subject behavior related to educational behavior. These efforts contribute to the practical value of educational management. Education and management ethics as a humanities, should regard practices and the solution of practical management problems as important objectives of development. Consequently, researchers and practitioners of education management ethics should pay more attention to some practical problems of education management. And the further improvement of the theory will prevent an empty theoretical framework of education management ethics. Besides, the problems education managers face, particularly in education management, should be integrated into the research framework of education management ethics. The moral values of education managers are deeply influenced by the process of solving problems, which finally forms their thinking and behavior patterns with ethical values.

4 CONCLUSIONS

Ethics of education management is an important breakthrough of current Chinese education. A range of studies have been conducted since the founding of China. Actually, current education management ethics in China should include five important aspects – ethics of education management behavior, ethical relationship between subject and object, ethics of policies, ethics of management ideas and the ethical values of regulations. Ethics are a kind of value that identify a certain behavior or idea of education management as valuable. The identification and thought powerfully promote the development of the educational cause in China.

REFERENCES

[1] Zhi Tingjin. Education Management Ethics: a New Field of Research. Journal of East China Normal University (Educational Sciences), 2005 (3):38–42.

[2] Xie Yanlong, Shuai Xiangzhi. Ethical Pursuit of Educational Management. Journal of Xinxiang Teachers College, 2003 (5):74–76.

[3] Zhi Tingjin. Ethical Problem of Education Management System. Journal of East China Normal University (Educational Sciences), 2006 (4):32–37.

Future Communication, Information and Computer Science – Zheng (Ed.)
© 2015 Taylor & Francis Group, London, 978-1-138-02653-7

Application of computer graphics processing technology

Y. Lv
Jiangxi Environmental Engineering Vocational College, Ganzhou, Jiangxi, China

ABSTRACT: Graphics processing technology has gradually evolved into a new discipline widely used in multiple fields with the development of computers. According to the concept and characteristics of computer graphics processing technology and some problems in its application—the quality of operating personnel, software and hardware, the work is aimed at an analysis of its application. A conclusion is drawn that the effects of computer graphics processing can be ensured as high-quality personnel and advanced software are selected. For a practical selection of software, moreover, it is necessary to carry a targeted selection rather than choose the most advanced software, combining it with hardware performance.

Keywords: computer; graphics processing technology; application

1 INTRODUCTION

Work efficiency has been greatly improved with the popularization of the computer and the internet. Some targeted software with corresponding functions, meanwhile, has been developed according to the actual needs in different fields whose emergence promotes the application of the computer to a great extent. Based on computer application, even the concept of paperless business is presented by a lot of enterprises. Compared with some western developed countries, however, the application of computers in China is in poor condition for the development of computer technology is relatively late subject to some special historical factors. Graphics processing technology, one of the important applications of computer technology, needs improving more and more in accord with the development of photography and film in recent years [1]. But the poor technology can not meet the demands of photos and graphics processing. For the practical needs, nowadays, it is advisable to use most of graphics processing software developed by the foreign companies in the market.

2 COMPUTER GRAPHICS PROCESSING TECHNOLOGY

2.1 Concept of computer graphics processing technology

Early in the development of computers, only some simple math can be done because of its relatively low performance and large size, not even as an electronic calculator now. With the use of transistors and integrated circuits, computer performance has been greatly improved, the volume also to a certain amount of control, and able to complete more complicated

tasks. People, on the basis of computer language, have developed a high-level language like C language, which greatly promotes the development of computer software technology. The efficiency of graphics processing is relatively low by the traditional methods of graphics processing – use of artificial drawing and its modification with some instruments, paper and pen. Graphics processing technology, therefore, has gradually formed in this context. With the popularization of the computer, image is transformed into digital form, stored in the computer, and can be freely modified by some application software, which is the technology of graphics processing. At present, there are a lot of graphics processing software, such as the commonly used Photoshop and 3DMAX in the market. The software itself, after years of development, has been very sound, with a variety of functions of graphics processing [2]. The software can do some processing automatically by some plug-ins and tools when appropriate parameters are entered, greatly improving the efficiency of graphics processing. In addition to this, some artificial graphics can be produced by the software [3].

2.2 Characteristics of computer graphics processing technology

In recent years, the computer graphics processing technology has gradually developed into a discipline with the popularization of the computer. Considering the importance of graphics processing technology, nowadays, such courses have been set up and a number of talents have been cultivated in many colleges and universities. The graphics processing technology, on the basis of computer application, is characterized by high efficiency and precision, and especially through the emergence of graphics processing software, a lot of processing can be completed automatically. According to the Moore's Law, showing that the development speed of computer hardware is very fast, the performance of the computer can be doubled every nineteen months, but the production cost is unchanged. The performance of computer hardware has been very high, after years of development, and can meet the demands of graphics processing. For the current processing of high-definition pictures, after entering the appropriate parameters, it can be done almost instantly and the effect can be presented immediately after treatment. With the development of information technology in recent years, people have gained a new understanding, proposing some concepts such as pixel and display color. The deep processing of graphics, meanwhile, can be achieved, by precise adjustment of various elements in graphics. The purpose of graphics processing, normally, is mainly to improve picture quality and the synthesis of a new image. Thus it has very important applications in the field of digital photos and film and television production.

2.3 Development of computer graphics processing technology

Early in the development of the computer, just some simple tasks could be done, for it is difficult to transform image into digital form stored on the computer, influenced by the performance of computer hardware and software level. Thus it was impossible to achieve appropriate graphics processing by computers. The application of the computer has been greatly promoted along with the popularization of the operating system, windows visualization based on which a variety of visualized operating software have been developed, meeting the needs of different fields. The graphical display plays an important role in the visual interface. In the early stage of computer development, the needs of industry drove its application and development, such as CAD, the earliest graphics processing software for industrial design. There is, meanwhile, a sharp increase in the number of graphics processing software along with the computer applied to different fields. According to practical needs, in the graphics processing, it is advisable to select targeted software from a lot in the market. The survey, however, shows that most of the graphics processing software on the market is developed by some foreign software company, subject to the limit of software technology in China. Software needs introducing from abroad if necessary, some of which is less used in China. Thus there is still no Chinese version, which has a great effect on the application of graphics processing technology in China.

3 PROBLEMS OF COMPUTER GRAPHICS PROCESSING TECHNOLOGY

3.1 Operating personnel with low professional quality

After years of development, the processing can be handled voluntarily after entering appropriate parameters by corresponding software because the computer has a high-level automation. But computer operation needs to be completed by professional personnel whose quality can influence the results of graphics processing to a great extent. There is low efficiency of processing if the staff cannot operate the computer skillfully because of poor professional quality. There is a survey showing that the quality of the graphics processing personnel is very poor influenced by the level of education in China. After professional courses, some students can master the theoretical knowledge of graphics processing technology but have a poor practical ability due to which they cannot skillfully use the graphics processing software. According to the actual situation, meanwhile, each enterprise selects different software from a lot in the market, especially each software with its own characteristics, and thus the students cannot be competent in their jobs. The teaching in colleges and universities cannot apparently cover the characteristics of all kinds of graphics processing software but just the software more widely used for some explanation. There is less chance for students to put into practice, influenced by poor education infrastructure and even no computer rooms in some colleges and universities, thus influencing their practical skills to a great degree.

3.2 Computer hardware with low performance

In accordance with the Moore's Law, showing that updating the speed of computer hardware, the performance of the computer can be doubled every nineteen months. The hardware should be replaced frequently to ensure its advance, thus increasing its cost. Given cost, advanced hardware is rarely used in application of graphics processing technology because of expensive computer equipment. Even with the installation completed, hardware will not be replaced after its updating, which influences the application of graphics processing technology to a great degree, subject to low-performance hardware equipment. Especially in recent years the emergence of high-definition image technology, computers need to have the ability of high operation. Due to low-performance hardware, it takes some time to wait during the processing, and some processing can not be completed, which is not in accord with the characteristics of computer graphics processing technology. After many years of computer application, there are different hardware on the current market for the needs of practical application. The survey, however, shows that a lot of hardware equipments produced by foreign companies, are subject to the level of information technology. Thus the equipments need introducing from abroad if necessary in China with a certain cost unavoidable, while multiple enterprises select some cost-effective hardware rather than the latest products.

3.3 Application of computer software in poor condition

Software plays an important role in the practical application of the computer. In a sense, different functions performing of the computer benefit from software application while the fulfillment of actual graphics processing is in need of corresponding graphics processing software. Various graphical parameters can be modified by the software's visual interface, obtaining graphics processing the effectiveness desired. The survey, however, shows that there is a lot of graphics processing software on the current market most of which is developed by foreign companies, such as the commonly used software, Photoshop and 3Dmax. Generally, the fulfillment of actual graphics processing is in need of these graphics processing software characterized by the function of powerful processing, which can meet the needs of different fields. But it is necessary to pay for the use of software because the development of software needs certain cost, and repay for it with the software updating. With the development of information technology in recent years, it is imperative to keep software frequently updated for ensuring its advance since the software version updates faster. In the graphics processing, given the cost, the most advanced software is rarely used in China, because new software is with more perfect functions but needs some time to adapt.

4 APPLICATION OF COMPUTER GRAPHICS PROCESSING

4.1 *Operating personnel with high professional quality*

The processing effectiveness can be improved if the operating personnel are with qualified professional quality and skilled in graphics processing software, since a computer needs to be operated by people. Working experience, therefore, plays an important role in the practical graphics processing. Graduating students have less knowledge of graphics processing software and just gradually complete corresponding operation in accordance with theoretical knowledge because they lack experience of practical graphics processing. In the actual process, some operating skills can be gained if the operating personnel are experienced. Then they can find an optimal solution as soon as possible according to the actual situation of graphics and software when they receive the task, saving processing time and improving processing effectiveness. It is necessary to employ some operating personnel with high quality during personnel recruitment, in order to ensure their competence, and even some experienced personnel in handling can be hired if the conditions allow it. Currently employed personnel, meanwhile, should be trained so that they can understand the latest concept of graphics processing and advanced graphics processing software. Operating personnel can be divided into several groups, choosing an experienced leader, to improve the overall quality of them. In the actual graphics processing, an optimal solution can be found by means of group discussion.

5 ADVANCED SOFTWARE ADOPTION

The advanced software has an effect on the processing because graphics processing is on the basis of software. There is different software on the current market, and each one has its own characteristics, whose use needs certain fees. The more perfect the software functions are, the more expensive the prices are. Given cost, the version of graphics processing software used in China is relatively backward. For the same software, regular updating of the software version is not only to repair bugs and errors, but also to add some new functions and even in the large of the version update, changes of the operation interface happen. In a word, every version update can improve the application of software. Thus efficiency and effectiveness of graphics processing can be improved to a great degree by adopting advanced software. In the actual graphics processing, given cost and other factors, it is not necessary to adopt the most advanced software. According to the actual needs of graphics processing, however, optimal software should be selected for better graphics processing. The achievement of the software function is on the basis of computer hardware, thus combined with the hardware condition. A lot of software functions, even the advanced one, cannot be achieved if the hardware's performance is poor.

6 CONCLUSIONS

The computer graphics processing technology has been applied to a lot of fields. A conclusion, however, is drawn by the above analysis that the graphics processing technology in China is still lower because the quality of operating personnel and performance of hardware are in a poor condition. There are a series of problems in the practical application, and it is imperative that some corresponding measurements be taken – learning from some foreign advanced processing technology and experience for the needs of practical computer graphics processing in China, so as to improve our graphics processing technology. In the particular computer graphics processing, meanwhile, it is necessary to adopt some high-quality operating personnel while ensuring the performance of hardware and advanced software. Only in this way can the efficiency and effectiveness of graphics processing be improved to the maximum extent.

REFERENCES

[1] Chen Yamin, Jin Xudong, *Computer Graphics and Image Processing Technology*, Journal of Changchun University of Science and Technology, 2011(01): 138–139.
[2] Jiang Yuzhen, *Teaching Research on the Application of Matlab in Computer Graphics Animation*, Computer and Technology, 2010(19): 5337–5338.
[3] Mu Qianhua, *Application of Computer Graphics in Practice*, Value Engineering, 2010(09): 22–22.

Future Communication, Information and Computer Science – Zheng (Ed.)
© *2015 Taylor & Francis Group, London, 978-1-138-02653-7*

Security protection measures of wireless sensor network

J. Wang & M. Liu
School of Computer Science & Engineering, Xi'an Technological University, Xi'an, Shaanxi, China

ABSTRACT: The characteristics of Wireless Sensor Network (WSN) make it attract much attention, but the network security is a vital issue in its practical application. This work mainly studied the security protection measures of WSN. Based on the concept and characteristics of WSN and its security issues in application, this work found that only by adopting an advanced transport protocol, installing corresponding prevention software and combining it with good operation habits can network security be guaranteed to the most degree.

Keywords: wireless; sensor; network; security protection

1 INTRODUCTION

The application of computers has greatly enhanced the working efficiency. With the increasing number of computers, the data switching among different computers has been studied by many experts and scholars. In the early phase of the computer, people used specific circuits to connect different computers and then conducted certain data switching by setting up corresponding transport protocol [1]. However, constrained by the technology of those days, the data transmission was too inefficient. After years of development and construction, now the mondial Internet has been set up, and the emergence of optical fibre transmission largely promoted the application of the Internet. Since the access to the Internet requires appropriate connecting circuits, the development of the network has been constrained to some extent. Wireless network has been developed on the basis of cable network with the widespread use of mobile devices in recent years, which places higher demands on the network [2].

2 BRIEF ANALYSIS ON WSN

2.1 Concept of WSN

WSN, on the basis of cable network, is a new technology gradually developed according to the needs of actual use. The emergence of wireless transmission technology was relatively early, and the primary devices like radios, were all based on it. However, that time was still the era of analog signals with the low efficiency of wireless transmission. Along with the emergence of digital signal processing technology, the wireless digital network (WDN) appeared and modern mobile devices like cell phones are operating under WDN environment. After many years of development, the 3rd generation of WDN has been widely used, thus highly promoting the efficiency of network transmission. Base on the TD-SCDMA, a 3rd generation of WDN technology developed with independent intellectual property, China has developed the 4th generation of WDN technology, LTE, which has been in commercial operation in some developed cities in China. The application of LTE indicates that WSN technology in China has played a leading role in the world. As the general cable network, WSN can be divided into two categories, local area network (LAN) and wide area network (WAN). The WAN is a wider coverage network offered by service providers while the LAN is a smaller one set up with sensors like wireless routers. In a sense, the wireless WAN is composed of numerous LANs and the coverage of each signal base station can be regarded as a LAN [3].

2.2 Characteristics of WSN

Compared with traditional cable network transmission, WSN has its distinctive characteristics. The first characteristic is its convenience. WSN has no circuit connection as its main devices serve as a corresponding function, connecting the network by searching wireless signals. When there is no need for network connection, WSN can immediately disconnect the network. After many years of development, the WSN has had a wide coverage of signals. For the most commonly used WSN, WIFI, has already achieved signal coverage in many important public places. Although affected by special history factors, the WSN technology has a certain gap between China and western developed countries, WSN is developing quickly in China since the economic power of China is largely promoted and China has already been the 2nd largest economies through more than 30 years of reform and opening up. Many families in China, according to their practical needs, use devices like wireless routers to set up corresponding wireless family networks. Therefore, due to the absence of circuit connection, many people hold that WSN is the further development trend of network because it reduces the cost of network construction and frees it from circuit maintenance.

2.3 Development of WSN

WSN, on the basis of cable network, is a new technology gradually developed according to the needs of actual use. In the early phase of WSN, the analog signal transmission was of low efficiency and bad quality. The emergence of digital signal processing technology is a revolutionary progress for WSN because the transmission of the signal and operation of hardware devices have been largely improved. Through practical analysis, it can be concluded that modern mobile devices, like cell phones, can all become digital devices, which adopt digital WSN. Compared with traditional analog signal transmission, the performance of digital devices has been enormously promoted and these devices can undertake a more complicated operation, for example, digitalizing pictures and videos to store them to corresponding memory devices, then transmitting these data through WSN. After many years of development, WSN technology has been so mature that the bandwidth of wireless transmission is of MB scale. The 4th WSN transmission standard of WAN also reached such a scale. After a period of commercial operation, the WSN has excellent performance though there are still some security issues that need to be improved in its practical use.

3 SECURITY ISSUES OF WSN

3.1 Security of transport protocol

In substantial network transmission, regardless of wireless transmission or cable transmission, a unified transport protocol is needed to conduct data transmission because different computers may have different storage structures. The design of transport protocol can not only influence actual transmission efficiency, but also have an impact on the security of data transmission. For example, the TCP/IP protocol, now widely used in global cable network transmission, mainly focuses on the efficiency instead of security of data transmission since the design of this protocol is relatively early and constrained by the technology of those days. Therefore, security of data transmission can not be guaranteed under this protocol. On the contrary, the transport protocol of WSN, in consideration of communication and message privacy, takes the security of data transmission as a priority in its design. However, after many years of development, the 3rd and 4th wireless network gradually emphasize the efficiency of data transmission instead of security issues, which greatly influences the network security to some extend. In conclusion, the characteristics of

transport protocol can largely affect the security of the network. In recent years the development of wireless LAN, constrained by the quality of the working staff, the construction of wireless LAN, with some traditional security measures on the basis of wireless devices, has not been integrated with the idea of security protection. Such security measures obviously can not reach its goal and thus family networks are often usurped.

3.2 Damages caused by harm programs

With the popularization of mobile devices, in order to bring better user experience, modern mobile devices are intelligentialized to some extent. For instance, most popular cell phones adopt Android or Mac operating systems. Under these operating systems, users can install many software and manage devices. Meanwhile, the performance of theses mobile devices is greatly enhanced, and some of them have 1.5G basic frequency and quad-core CPUs. In a sense, the performance of some mobile devices is equal to that of some computers. Under such a background, there are many harm programs aimed at mobile devices. If these programs are installed on many devices, data loss or even more serious economic damages can be caused. Moreover, some operating systems adopted open source for the popularization of the mobile operating system. Though open source can promote the improvement of software and the deep revision of operating systems, some harm programs are also easier to be written down. In the face of this, some operating systems opened up their application stores accordingly to conduct some examination on the software, but security issues can not be fundamentally solved. In the early phase of WSN, mobile devices had primary performance, so harm programs were rare and people did not recognize their damages. However, along with the popularization of WSN over the recent years, more and more harm programs have appeared on many people's mobile devices. Some harm programs, like automatic message sending, were installed when people were using their cell phones, which have bought much negative influence on their user experiences.

4 SECURITY PROTECTION MEASURES OF WSN

4.1 Adopting advanced transport protocol

Considering the influence of transport protocol on the network security, a safe transport protocol has to be designed in order to further promote the security level of wireless WSN. Thanks to the development of wireless network technology in recent years, on the basis of practice, many experts and scholars have conducted researches on the wireless transport protocol and put up different kinds of transport protocols. However, constrained by many factors, many protocols have not been applied to practical use, especially in the construction of wireless WAN, which needs to adopt international unified transport protocols in view of some factors like network connection. However, these international protocols, due to their early design, did not fully deal with security factors, and thus more security issues aroused with the development of wireless network technology. Although, people have continuously tried to improve transport protocols in practical use, no prominent effects have been achieved. Therefore, in order to better solve security problems, an advanced transport protocol has to be adopted. Considering the characteristics of WAN, advanced transport protocol can be first used in LAN; after a certain period of test, if the protocol is safe and does not influence the transmission efficiency, then the range of application can be widened. In this way, the security protection level of WSN will be greatly promoted.

4.2 *Installing firewall software*

In order to keep mobile devices free from the damages caused by harm software, viruses and Trojans, antivirus software and firewalls should be installed on these devices. After many years of development, people have already learned the damages of viruses and Trojans in the cable network and almost every computer has installed corresponding antivirus software. However, due to the late development of mobile devices, people have little knowledge about the damages of viruses and Trojan in wireless network even though some mobile devices are intelligentialized to some extent. To better solve this problem, people should first raise their awareness on viruses and Trojans. Practical research showed that most users of mobile devices had installed harm software and suffered damages caused by viruses and Trojans. Due to the restriction of the technology level, firewall software on the market are rare and they can not fully guarantee the security of devices. Moreover, installing these software could occupy large amounts of resources of the operating system, which would lower the performance of these devices. Thus it can be seen that bad performance of hardware devices greatly influenced the installation of firewall software; with the development of WSN, security issues are attracting more and more attention, so the use of firewall software will be an upcoming trend. Some measures have to be taken to protect the security of devices in view of the increasing number of viruses and harm software in the WSN.

4.3 *Forming good operating habits*

Compared with cable network, WSN needs to be operated by people. If operators have low qualities, some improper operations will bring problems to the network security. To solve these problems, good operating habits of operators must be formed although it usually takes a long period. Therefore, it is necessary to establish a standard of security operation. Different WSN, because of different devices and transport protocols adopted, are bound to be different in operating modes. In the construction of the operating standard, the different situations of different WSN should be taken into consideration and the operating standard needs to undergo corresponding adjustments when the network updates. On the premise of scientific and reasonable operation, the network security can be guaranteed and free from the security issues caused by human factors.

5 CONCLUSIONS

As analyzed above, WSN is a new technology gradually developed on the basis of traditional wireless network. It soon gets popularized for the characteristics of flexibility, convenience and high transmission efficiency. Restricted by technology level, the construction of WSN in China is barely satisfactory. In the practical use of WSN, transport protocol and harm software brought some impacts on the network security. To fundamentally solve these problems, advanced transport protocol must be adopted to guarantee the security of data transmission. Then by installing corresponding firewall software, hostile attacks can be avoided. Finally, according to the concrete situation of WSN, an operating standard aimed at regulating operating modest should be established to prevent security issues of network caused by improper operations. Only in these ways can the security level of WSN be promoted to the most extent.

REFERENCES

[1] Qin Boping, Zhou Xianwei, Yang Jun, Song Cunyi, *Secure Routing Research in Wireless Sensor Networks*, Chinese Journal of Sensors and Actuators, 2006(01): 16–19.

[2] Dai Hangyang, Xu Hongbing, *Overview of Security in Wireless Sensor Networks (WSN)*, Application Research of Computers, 2006(07):12–17.

[3] Ren Xiuli, Yu Haibin, *Security Mechanism in Wireless Sensor Networks*, Journal of Chinese Computer Systems, 2006(09):1692–1694.

Future Communication, Information and Computer Science – Zheng (Ed.)
© 2015 Taylor & Francis Group, London, 978-1-138-02653-7

Study on sustainable development based on geography

Z. Ming
Funing Higher Normal School, Yancheng, Jiangsu, China

ABSTRACT: From the perspective of geography, the work explores optimization ways of sustainable development, aims at problems such as a lack of regional specificity that arose in China's sustainable development, and proposes solutions to strengthen the application and awareness popularity of sustainable development. From a series of discussions, it can be concluded that the decision-making and research of sustainable development should be optimized through geography, including geography education and geographic information technology. The work features a joint analysis on the application of geography to sustainable development from multiple levels of geography.

Keywords: geography; geographic information systems; sustainable development; man-land relationship

Researches related to sustainable development have been an important social issue since the beginning of the 21st century. In practical terms studies and discussions on sustainable development are closely linked to specific conditions of regional geography. Geography plays an important reference role in the theory and practice of regional sustainable development. With the rapid development of China's reforming and opening, city development in China has entered high-speed track, but the development model of companies and cities nationwide is still extensive. Great achievement of city development has been made, but huge resource waste and environment damage are also generated due to the neglect of resource and environment. There are the important social premises for promoting sustainable development [1].

1 CURRENT SITUATION AND DILEMMA OF SUSTAINABLE DEVELOPMENT FROM THE PERSPECTIVE OF GEOGRAPHY

China, a country rich in resources, faces tense situation due to resource damage in the development process in recent years. Planning for China's resources is needed in accordance with the construction of socialist society with Chinese characteristics in the 21st century. The plan should be built on geography and regional conditions because both specific resource and environment conditions differ from one area to another [2]. Under the reasonable development, promoting for effective protection of resources is the priority of the whole country's social sustainable development. Therefore, the application of geography to sustainable development is a necessity and urgency of the times [3].

1.1 Integration between sustainable and geography

For sustainable development itself, there is a relationship between application and expansion. More knowledge and technology of geographical sciences are needed to support indicators and scales of sustainable development. Currently, geographical sciences researches on China are rich, while those of specific regions are still relatively blank. Specific regional environment features a specific geographic condition, such as regional resource reserves, population capacity. In a word, geographical sciences research on regional geographical environment and sustainable development linkage between various regions should be strengthened.

1.2 Shortcoming of concepts and research methods in the current sustainable development

Sustainable development is not only a specific strategy and method, but also a combination of spirit and material, and integration of multi-dimensional continuum between geographic time and geographic space in terms of geography. Therefore, it is essential to strengthen the coordination of man-land relationship, to rely on awareness popularity of man-land relationship through geography education, and to promote acceptance of sustainable development by all parties on the spiritual level.

Lots of research works on sustainable development are carried out in a regular way, and too many unnecessary discussions on sustainable development decision-making are conducted because there are shortages of geographic professional guidelines for policy research of current sustainable development. In fact, the research works are difficult to promote practices related to sustainable development, and the conclusions drawn from researches, which lack regional characteristics and specificity, have little significance in promoting the sustainable development in China.

Researches on current sustainable development are often lacking multi-dimension, and not thorough enough in terms of time dimension. Actually, long-term geography detection is not very popular in China. As a result, many researches on sustainable development in China are kind of a short-term research and unable to conduct sustainable observation of the man-land relationship nationwide. In essence, insufficient learning and mastery of knowledge for geography sciences, results in no application of its scientific knowledge to sustainable development, and leads to insufficient scientificalness, lack of continuity and no expansion of multi-dimension in terms of sustainable policy making.

2 IMPACT MODEL OF SUSTAINABLE DEVELOPMENT BY GEOGRAPHY

2.1 Geography being the theory basis of sustainable development

Related to multi-disciplinary knowledge, sustainable development is a comprehensive act which is linked to a lot of relevant scientific knowledge. According to the original idea of sustainable development, its core lies in the contradiction between natural environment and human development. The contact between human and nature is closely related. A relatively stable and huge ecosystem formed by the two and centred around mankind, has a great impact on human survival and development. Therefore, the contact between human and nature, namely man-land relationship, gets more attention in the Post-Industrial Revolution.

Only if man-land relationship is coordinated, are human beings able to gain living space and developmental foundation. Therefore, the coordination between man and land is an important prerequisite and foundation for achieving sustainable development of individuals. Although sustainable development focuses on development, no harmonious man-land relationship and over-prominent contradiction between man and land can easily lead to great pressure and thus have impact on sustainable development of mankind. Modern geography researches mainly pay special attention to the influence of the overall natural environment and resources brought by human behaviors, as well as population, environment and resource conflicts caused by economic activities. The solution of these conflicts is the key point in the entire process of sustainable development. As a result, modern geography, the core of contradiction between man and land, is to coordinate the influence of the overall natural environment and resources brought by human behaviors, to solve a series of conflicts caused by economic activities, and to build a harmonious system between people and land.

In a word, a more theorematic reference to geography can promote sustainable development in the reform and use of natural resources. Therefore, geography is an important theorematic basis of sustainable development.

2.2 Geography science being the pillar science for sustainable development strategy

Being a comprehensive discipline, geography science covers a wide range of areas and focuses on the relationship between the natural environment and mankind. Geography science has proposed many questions worth thinking of and discussion for sustainable development since its beginning. China has paid attention to harmonious development between heaven and mankind ever since the development of ancient geography. It emphasized coexistence, common prosperity and harmonious development between man and nature, which was similar to modern geography science. Meanwhile, scientific knowledge of sustainable development concerns solutions to problems such as the environment, resources and population, resources concentrated on the surface of the earth, which is also similar to geography science. Therefore, for sustainable development itself, it possesses a relatively strong component of geographical science. Regarding the sustainable development issue, geography science can provide some new evaluation scales, including a regional development mechanism and evaluation mechanism, as well as planning methods on regional sustainable development. In conclusion, there is no effective promotion of sustainable development in China without the comprehensive participation of geography science.

From the perspective of regional scale, geography science can not only bring more advanced programs into sustainable development, but also provide more scientific evaluation scale. Sustainable development is a very important goal of human and society development during new periods. The smooth achievement of it lies in a better coordination between mankind and nature, and a scientific evaluation scale of man-land relationship. Multi levels of theory researches and practices on geography science have provided scientific evaluation scale for regional sustainable development, and promoted sustainable development of environment, ecology and resource. The difference of man-land relationship in regional development reflects that geography science should be adjusted in accordance with regional conditions, so as to get richer methods and experiences.

3 THE APPLICATION METHOD OF GEOGRAPHY TO SUSTAINABLE DEVELOPMENT

3.1 Strengthen the application of geography to sustainable development

In fact, the formulation of sustainable development strategy relates to multi-disciplinary, masses of information, and dynamic response and changes between systems. In order to provide more references for scientific decision-making of sustainable development, its emphasis and difficulty are to make adjustments to various information data in a dynamic and quantitative way during the process of sustainable development. On one hand, geographic information technology contained in geography science can strengthen the application of geography to sustainable development. Moreover, geographic information science can integrate all kinds of relevant information and conduct integration and classification for dynamic geographic information so as to provide a more extensive information base and technical support for the sustainable development strategy.

As an important aspect of geography science and combination of computer technology and three-dimensional space technology, geographic information technology can provide reference to various research areas among sustainable development strategy, and offer in-depth technical support for solving large-scale regional problems. Being a comprehensive environment, the earth contains diverse resources and environmental factors. Therefore, geography science should use information integration technology and then provide some more appropriate solutions to relevant questions when facing different resources and environmental problems in different regions. The integration of geographic information science and technology can obtain relevant information data from different levels, strengthen quantitative research on relevant geographic information, precisely conduct specificity of sustainable development strategy and test the effect of the sustainable development strategy carried out in certain areas. As a result, geography science, which can provide timely relevant information and make adjustments to specific strategies and arrangements for sustainable development according to specific conditions of geographic data, plays an important role in promoting the application of sustainable development and scientific decision-making.

3.2 *The popularity of awareness and concept of sustainable development driven by geography*

Sustainable development can not only bring about an effect on practical application, but also raise awareness of national sustainable development in a larger scale. The goal of sustainable development can't be achieved without joint efforts of all Chinese people in the 21st century. On one hand, geography is required to promote national awareness. China's national awareness of commitment for sustainable development is closely related to all social classes, such as farmers, workers, students and women. In order to become a cause of all people, sustainable development needs certain relevant skills and knowledge, which can be largely gotten through the popularity and promotion of geography.

Geography education among geography involves the core areas of sustainable development, including

man-land relationship, protective development of resources and environmental protection. It is unmatched by other disciplines in terms of the richness of data and systematization of the entire education. Geography education can help young students to gradually build correct values towards the man-land relationship, and establish ideas about protective use and development resources. Actually, the fundamental purpose for popularity of geography science is to continually enhance national consciousness of sustainable development, and allow people to correctly tackle the conflicts between man and land. Therefore, the popularity of geography science has a great impact on the utility of sustainable development driven by geography.

Furthermore, geography education can bring up future researchers on geography science among young students. Enhancing the entire nation's basic skills and knowledge of geography science is not the whole countermeasure to sustainable development from the perspective of geography. In fact, cultivation of a batch of experts on geography science is of importance because they are the important power to deal with the man-land relationship in the future. Continuity of the sustainable development strategy can't be guaranteed unless people are armed with professional skills and academies.

4 CONCLUSIONS

Actually, sustainable development is the most complex issue in the history of mankind. To deal with such a problem, more knowledge and methods related to geography are required to promote the scientific decision-making of sustainable development. In the process of sustainable development, balance of development difference between regions, promotion of regional common development in terms of geo-politics, coordination and fairness in geographic resource allocation should be noted. For geography itself, its research paradigm should be in line with features and content of sustainable development, achieve optimal configuration of the man-land relationship development system, and finally realize comprehensive sustainable development.

REFERENCES

[1] Shui Wei, Zhang Qichun, Wang Shanhe, Feng Xianhui. Contemporary American Geography and the Theory and Practice of Sustainable Development—Based on Relationship between Man and Land. Geography and Geo-Information Science. 2004(04):56–60.

[2] Lu Dadao. Several Issues Worth of Considering in the Development of Chinese Geography. Geographical Journal. 2003(01):3–8.

[3] Liu Guobin. Research on Sustainable Development of Geography. Journal of Sichuan Normal University (Natural Science). 2001(06):640–644.

Future Communication, Information and Computer Science – Zheng (Ed.)
© *2015 Taylor & Francis Group, London, 978-1-138-02653-7*

Financing strategy analysis of small and micro businesses

P. Ling
College of Economics and Management, Yibin University, Yibin, Sichuan Province, China

ABSTRACT: At present, most small and micro businesses in China are faced with difficulties in financing, which are rooted in their narrow channels for fund raising, high risk of default and little attention from financing institutions. Combining the fundamental principles of institutional economics and finance and establishing on the basis of its development actuality, the work strives to explore the strategy and method for small and micro businesses to get rid of the financing problem. Innovatively, the work integrates the financing difficulty with the background of China's financial system so as to propose a solution from multiple points of view.

Keywords: small and micro businesses; financing difficulty; bank credit

Small and micro enterprises, as an important part of our national economy, have made a great contribution to the continuous steady growth of the national economy, and have provided a great amount of jobs as well. But, nonetheless, at present, small and micro businesses across China still have trouble with financing, the sources of which lie in the expected improvement in the management and operation mechanism of financial institutions, in addition to the fact that the small and micro enterprises are suffering the deteriorating market environment. The work will combine the institutional economic theory with relevant principles of finance, and probe into a scheme to handle this problem [1].

1 ROOT OF FINANCING PROBLEM IN SMALL AND MICRO BUSINESSES

The financing problem in small and micro businesses has obtained common concern from the national economic management authority as well as business and academic circles. In order to resolve the financing difficulties in small and medium-sized enterprises, the first step is to identify the root, and formulate correlative countermeasures on this basis.

1.1 *Single funding resource*

Funding resources for small and micro businesses are more single than those for large enterprises, which is an important source for the lack of capital elements in small and micro enterprises. Funding resources refer to the ways and methods used to raise money. For enterprises, they include sorts of loans, investment and borrowing from financial institutions [2]. Due to the reality that small and medium-sized enterprises often run into a stone wall when seeking financing by going public through the capital market, they consequently tend to choose to invest through their own capital, in such a way to complete the expanded reproduction and reduce the borrowing from financial institutions. In the mean time, on account that the monetary market in China has not fully developed, a large number of small and medium-sized enterprises encounter obstructions to use funds outside of their own in the form of negotiable instruments circulation [3]. As a result, they can only rely on bank loans, and thus a problem of single funding resources is formed.

A single funding resource not only brings a hindrance for small and micro enterprises to effectively improve their own financing combination, with an aim to reduce the risk and cost of capital; on the other hand, it becomes easier for them to over-rely on the single channel, resulting in a further mismatch between the supply and demand structures of capital in credit markets, and giving rise to the potential trend of a virtually price increase in the capital elements.

1.2 *High risk of default in small and micro businesses*

Apart from the expected profit, the main factor to consider when banks are providing funds credit and loan is the risk of default that the enterprise cannot return on schedule after borrowing. When the responsibility for breach of contract arises again, the bank or other financial institutions tend to recoup their investment by obtaining the ownership of enterprise assets and resorting to auction. But due to the unstable business environment of small and micro businesses themselves, their ability to rein the market is relatively weak, and thus they are likely to land in a predicament of returning the principal and interest regularly.

For banks, the loan risk for small and micro enterprise is relatively high, so they will not be inclined to extend credit to small micro enterprises.

However, small and micro businesses mostly enter into a certain industry for a relatively short period of time and the assets owned by those enterprises are in a relatively small scale. When faced with a responsibility for breach of contract, they can only provide a relatively small quantity of mortgage items. Besides, small and micro enterprises, in most cases at a low ebb in a certain industry, do not acquire advanced technology and equipment or finished goods of high market value. In consequence, financial institutions have to undertake the risk that if the small and micro enterprise is caught in default, it will be difficult to recoup the capital outlay. It is a must for financial institutions to take a considerate account of it, so the credit applications of small and micro businesses are often replied less timely than the large capital requirements of large and medium-sized enterprises because of the high risks, or even been laid aside for a longer time, thereby leading to a delay of opportunities for the development of small and micro enterprises.

1.3 Financial institutions generally paying no attention to investment oriented in small and micro enterprises

Due to historical reasons, China's financial institutions give a priority to state-owned and state holding enterprises that are more sensitive to make responses to the government's policy orientation while their attention on the realistic demand of the market tends to be weaker than private capitals. Based on such a reason, the loan demand of small and micro businesses cannot truly receive attention from large financial institutions.

Since state-owned and state holding financial institutions still don't have a clear property rights boundary of their own as private enterprises, the phenomenon of government intervention still remains. In this context, the objective functions of banks and other financial institutions contain not only the part of profit, but also include such factors as the degree of support from the government's policies, which accordingly affect the financial institutions' management decisions. Under such a background, the relevant financial institutions will be more inclined to grant a loan to investment projects in state-owned enterprises of large and medium-sizes and governments of all levels. The investment needs of small and micro enterprises will confront with the situation of a credit resources shrinkage and credit demand increase.

At the same time, China's current financial market liberalization continues to process ahead. It has existed for quite a long time that the price of real interest rate is much higher than that of nominal interest rate in virtue of abundant public investments. Such will directly affect the small and micro enterprises' ability to obtain credit funds.

2 ESTABLISH MODERN ENTERPRISE SYSTEMS TO BROADEN THE FINANCING CHANNELS

To solve the small and micro enterprises' financing difficulties, the first step is to iron out the problem of single investment and financing channel in small and medium-sized enterprises. An important cause to this problem is that the management systems of those enterprises are relatively too chaotic to have an inherent characteristic of the modern enterprise system, which imposes restrictions on the financing channels of small and micro enterprises.

The modern enterprise system is concentratedly reflected as incorporated enterprise, where ownership and management right are separated; in other words, the board of directors is responsible for making enterprise management decisions, while the enterprise manager team represented by general managers is in charge of the supervision and management on specific business matters. Only in that case will alterations in the condition of capital formation not lead to a significant change in the enterprise's internal management process, which can sequentially embody the continuity and scientificity of enterprise management. Meanwhile, the owner of the enterprise is also not able to produce excessively, or even arbitrarily interfere with the management of enterprise by virtue of their ownership.

On this occasion, the enterprises will meet a downward trend towards their reliance on the single investment channel. This is because under the modern enterprise system, the enterprises will take the limited liability system. Thus, the investor's risk will be greatly reduced, for the enterprise's problems in debts and bankruptcy will not affect the owner's personal property. Compared with the unlimited liability feature of the enterprise management model widespread in small and micro businesses, this will significantly promote the diversification of corporate investors.

3 FURTHER DEVELOP THE MONETARY MARKET

The monetary market, mainly as a place for circulation of all kinds of creditor's rights certificates, is an important channel for the enterprises' short-term capital financing. The currency circulating in the monetary market tends to be able to provide a powerful tool for small and micro businesses' short-term capital operation. For them, settling the drafts and checks of an obligatory relationship in a more convenient way, can help the enterprise to transform its future cash flow into current monetary funds. As for enterprises, loans are obtained through the real commodity business relationship.

The drafts most commonly used in the monetary market can help enterprises to get credit funds and to take advantage of the few resources. As an unconditional pay warrant, the draft itself contains a certain consideration. The drawer often acts as the buyer of commodity, while the drawee generally is the seller. There is a debtor-creditor relationship between the payer and the drawer. Therefore, the draft incarnates a form of deferred payment. As the draft can bring in a certain amount of future cash flows, the draft, for small and micro businesses in need of capital financing within a short term, can serve as a certain amount of assets to mortgage in financial institutions and thus to obtain a circulation fund. Discount business in commercial banks is the main form of transforming the negotiable instruments into loans. Discount itself shows that the bank will convert the net present value of future capital flow into a circulating fund and then transfer it to the applicant for a discount. The discount rate shall be regarded as a direct embodiment of currency prices.

At present, the discounting behavior of various bills is still under the strict national macro control. The discount rates, in fact, often make adjustments in terms of national rediscount rate. So, in essence, China's monetary market, especially the discount business therein, is still subject to the impact by the state planning management model; consequently, the anthropogenic interference on interest rate and discount volume remains massive. The rediscount rates floating based on the rediscount rate formulated by the state, can hardly completely reflect the trend of supply and demand change in market, so it has a certain lag relative to the changes in market.

With relatively few resources, small and micro enterprises can only provide very little pledge goods compared with large and medium-sized ones. All the proprietorship and creditor's rights certificate circulated in bill circulation market are such a form of assets that are more easily for banks to confirm. The asset's intrinsic value is exactly the monetization of final products formed by their own production and business activities. Letting go of the monetary market and encouraging free circulation of all kinds of bills can help small and micro businesses find an important way to improve their operating conditions.

4 BY VIRTUE OF REFORM IN STATE-OWNED FINANCIAL INSTITUTIONS

The degree of marketization reform in state-owned financial institutions remains to be further improved, which is one of the direct reasons why small and micro enterprises cannot obtain sufficient funds credit share. The state-owned commercial banks have played a leading role in the market. They serve as a current main provider of China's capital element, as well as the important implement leverage for China's fiscal and monetary policy at the same time of pursuing for profit maximization as its management goal. Meanwhile, since commercial banks are still largely influenced by all levels of local governments, the flow direction of monetary capital does not entirely exist

based on the market price mechanism and supply-demand mechanism. This in fact boosted the malady in the planned economy period when the budget constraint was softened, to remain in the present economic construction.

Due to the interference from policy factors, the state-owned commercial banks will tend to transfer its own credit resources to state-owned enterprises and public construction projects. And the products required by these projects are frequently not enough to provide for small and micro businesses, therefore, the difficulty for small and micro enterprises to obtain credit funds from state-owned commercial banks is imaginable. In addition, as the demand for monetary capital amplifies, it will in fact result in a rise of real interest rates, and ultimately further prevent the state-owned commercial bank system from putting in credit and loan to small and micro enterprise. At last, small and micro enterprises can only produce and operate in the absence of credit resources. In the actual operating process of state-owned commercial banks, the national interest rate control system has long been an important factor that affects the bank's business model. Once the interest rate under the state control is in deviation to the market demand, it will give rise to a decrease in the banks' will to grant credit and loan, and generate corruption within the commercial banks.

To speed up the reform in state-owned commercial banks, not only a modern enterprise system should be established, but more importantly, a clear property right relationship should be straightened out as well. Besides, the situation where all state-owned commercial banks are directly affected by the local governments of all levels waits for a change, and an independent legal person system, independent accounting, as well as a profit-oriented managemental model also needs to be truly established. In the process of reform, the internal incentive mechanism of state-owned commercial banks should also be transformed: ensuring that the banking institutions of medium and low levels can receive benefits from crediting to small and micro businesses, and ultimately include the loan to small and micro enterprises into their main business income. This can largely degrade the original administrative operation mode, and invest the capital of state-owned commercial banks, in accordance with the entrepreneurial configuration mode, to those small and micro enterprises with high production efficiency and monetary capital requirements.

5 TAKING ADVANTAGE OF THE CREDIT GUARANTEE SYSTEM DOMINATED BY FINANCIAL INSTITUTIONS

Small and micro enterprises have a disadvantage to compete in the market economy. Its assets are relatively narrow, and the anti-risk ability is relatively weak. As a result of its limitation in technology, small and micro enterprises often have to engage in the research and sale of low-end products, so their products are highly homogenized, with poor market competitiveness. No guarantee for safe and timely capital interest repayment is an important hurdle for their financing ability. To settle down this major problem, the relevant insurance and financial institutions can promote some related products, covering the small micro enterprises' requirements for loans into their own products, and on this basis, offer security for small and micro enterprises' financing to ensure that they would be able to obtain all kinds of money smoothly.

Granting security for small and micro enterprise loans is actually more similar to risk investment, so financial institutions can choose to cooperate with venture capital companies to a certain extent, and carry out relevant business by the powerful financial advantage of financial institutions and experiences of risk investment, project evaluation and human resources from venture capital firms. But unlike venture capital and bank credit, the credit insurance provided by financial institutions does not pledge the enterprise's resources, so in order to spread risk – their products shall be promoted in regions with development potential for small and micro enterprises, thus to disperse the enterprise's financial risk, and ultimately achieve earnings.

Small and micro enterprises' financing difficulty is actually generated by their market environment and management system of financial institutions. In order to promote the development of small and micro enterprises, actions ought to be taken from the angles of enterprises and financial institutions to propose an overall reform scheme. Moreover, its growth calls for the joint efforts from multiple subjects to straighten out the market structure and to make sure that the monetary funds can really be configured to the most efficient department.

ACKNOWLEDGEMENT

The work was supported by research-funded project of Sichuan Provincial Department of Education, "Financing Survey and Channel Innovation of Small and Micro Enterprises – A Case Study on Yibin, Sichuan", with Project Number: 13SB0284.

REFERENCES

[1] Li Mingxing, Nelson Amowine, He Di et al., An Innovation Research of Small and Micro Enterprises' Patent Financing Model under the Background of Transformation and Upgrading. Scientific and technological Progress and Countermeasures, 2013, 30(18): 138–142.

[2] Zhu Xinrong, Li Hong ham, Yang Yingjie et al., A New Thinking of Technology-based Small and Micro Enterprises' Financing Strategy under the Background of Inflation—Experience Reference of America and Japan. Scientific and Technological Progress and Countermeasures, 2013, 30(13): 96–101.

[3] Wang Bin, Wang Jian, Tan Qingmei, et al., Available Channels for Technology-based Small and Micro Enterprises' Financing Assurance: Bonding Companies with Corporate Investment. Scientific and Technological Progress and Countermeasures, 2013, 30(18): 128–131.

Future Communication, Information and Computer Science – Zheng (Ed.)
© 2015 Taylor & Francis Group, London, 978-1-138-02653-7

The optimization analysis of the physical education major training program in the sight of innovation education

S. He
Institute of Physical Education, Beihua University, Jilin, China

ABSTRACT: With the development of the times and social progress, innovative talents have increasingly become the focus of attention. The Physical Education major plays an important guiding role in the physical culture in China. Therefore, the level of the students' professionalism and innovation ability is an important factor for the sports development. The main purpose of this work is to put forward effective training programs for innovative education. The approach is to gain enlightenment by analyzing the problems existed in innovative education. The innovation is to combine the students' education with their future vocational development.

Keywords: innovation; physical education; program

1 THE NECESSITY OF INNOVATING THE PHYSICAL EDUCATION SPECIALISTS TRAINING PROGRAM

The main occupation directions of Physical Education Majors are gym teachers, coaches and so on in different stages of the schools. These posts demand strong professionalism, which could enable them to give full play to show their professional knowledge and technical skills. Hence, they are ideal jobs for physical education majors. Besides, to make physical development plans in a better way, related Sports Administrations would need professional physical education talents to participate in the making so that it can match the scientific, reasonable development principles [1]. Due to the unceasing enhancement of people's living standard and the deterioration of the living environment, people become more aware of their own health and pay more and more attention to physical exercise. However, unless physical exercise gets scientific guidance can it achieve the expected goals and avoid the occurrence of self-injury during the workout. Therefore, physical education graduates could play a significant role in these fields. Though they have a wide field of employment, physical education graduates will still be at a disadvantage in the career competition if they don't have a good professional knowledge and quality themselves. What's more, even though they enter the workplace, they still cannot make their own contribution to China's physical culture and people's health. Strengthening the cultivation for Physical Education Majors from the perspective of innovation education could achieve the training purpose remarkably and improve the students' professional level efficiently [2]. Traditional professional physical education fosters the students' just viewing of the students as future sports teachers, whose practical and operational abilities are

also cultivated for adapting to the working environment of schools. The training program restricts the physical education majors' knowledge structuring. And it also goes against the students to adapt to different sports skill requirements for all walks of life. Innovative training program, while adapting to the society's constantly updating requirements, could in turn promote the development of social sports undertakings. Through participating in social practice and having close allocation with the society, the students could gain a wider knowledge of the organization and a higher ability of getting jobs.

2 THE THEORETICAL BASIS OF OPTIMIZING THE PHYSICAL EDUCATION TRAINING PROGRAM IN THE SIGHT OF INNOVATIVE EDUCATION

2.1 *Requirements for physical education talents' innovation in the modern society*

Currently, China is in the bottleneck stage of development. Though it is still in the progressive stage, the pace of progress is obviously slowing down. In order to cope with this situation, China has put forward a strategic planning for innovative development, striving to build an innovation-oriented country. An innovation-oriented nation needs innovative talents to take participation in the building. Education being one of the key development projects in China, the talent quality is a critical factor for career progress [3]. What are innovative talents? The prevailing view seems to be that talents "possessing innovative thinking" or "possessing innovative knowledge skills" are innovative talents. These two points are the basic characteristics wherein talents should be qualified. The innovative talents, different from traditional talents, possess a more alert mind, broader perspective, liberated thinking and richer knowledge. They dare to challenge the obsolescence in common learning and life. And they dare to express their unique ideas and integrate these innovative thoughts into realistic social practice. The training mode to optimize the physical education talents in the view of innovation need to be carried out in the following three aspects: the first is to expand the students' ideological sphere, making their thought walk in the forefront of the times and have extremely big openness and acceptability. The second is to activate the students' mind, making their understanding and addressing of the event extremely characteristical, thereby getting rid of habitual thinking and mind set and having different adaptability to every event. The third is to possess strong sensitivity to the events occurring in the society, capable of perceiving some unique, unusual flash points from those seemingly ordinary things.

2.2 *Basic quality for physical education innovative talents*

Though most of the current physical education majors have an innovative spirit or certain innovative consciousness, only a very small part of the students already had the innovative ability, which means applying their own innovative consciousness and innovative spirit to the common students' and social practice. Therefore, it is urgent for schools to train the students' innovative ability. Besides the innovative ability, physical education majors still need to be qualified with a good occupational accomplishment, solid professional basis and advanced learning and education ideas, thus in a position to undertake the tasks of the schools' physical education, government physical guidance and social sports guidelines. Though educators don't know what specific occupations physical education students will take, they could take the training of basic qualities aimed at several typical occupations. For example, for sports teachers, they should intensively cultivate their scientific research ability and management ability. The scientific research ability could enable them to put forward an innovative research for occupation skill and teaching content while the management ability could guarantee implementing the innovative teaching content smoothly. All in all, physical education majors should be equipped with basic qualities in various aspects, covering profession, life, thought, morality and so on, so that they could adapt to the development and change of the society better.

3 PROBLEMS EXISTING IN THE OPTIMIZATION OF PHYSICAL EDUCATION TRAINING PROGRAM IN THE SIGHT OF INNOVATIVE EDUCATION

3.1 *A lack existing in the construction of professional teacher team*

After making related training plannings for innovative physical education talents, the teachers' teaching level

is one of the most crucial factors for implementing them in the common teaching activities effectively and gaining good effects. Currently, a physical education major is an unusual specialty in China. This is because it is between professional sports and professional education. It can neither provide excellent athletes for the country like professional sports, nor can it provide regulatory occupation skills for schools like the teacher education major. Consequently, schools devote less teaching resources to the physical education major. The low wages and ordinary working environment, therefore, can't attract good teachers. Meanwhile, the current teachers, badly divorced from the society, don't know the employment situation and trend of the physical education major well. Their educational ideas are universally too old-fashioned. They can't understand the schools' training for creative talents, so they will have some resistent feelings when carrying out the training program.

3.2 *Uneven development of general knowledge education course and endorsement course*

General Knowledge Education Course, also referred to as public subjects, is the general name for the subjects involving the cultivation of the students' comprehensive quality, such as Ideological and Political Course, Computer Class, Foreign Language Class and so on. Endorsement Course refers to the particular subjects of a major, which are the foundation of the students' future career development, strongly correlated with the students' academic level and professional skills, such as Humane and Sociological Science of Sports, Exercise and Sports Science and so on.

As far as the current talent training programs are concerned, schools and teachers generally attach little importance on the General Knowledge Education Course. They put a lot of energy and hardware resources into the Endorsement Course, hoping to improve the student's professional quality effectively. Students' mental health level and skills in other aspects are not included in the highlighted teaching scope, which will lead to a students' low moral trait level and social adaptability. Among the General Knowledge Education Courses, most students also prefer the computer class and English class since these two subjects have a wide application range while they are obviously in a passive state when attending classes such as the Ideological Political Class and History class.

4 THE OPTIMIZATION APPROACHES OF PHYSICAL EDUCATION TRAINING PROGRAMS IN THE SIGHT OF INNOVATIVE EDUCATION

4.1 *Optimizing the discipline structural layout*

Innovative education perspective demands the optimization of the physical education major's discipline structure. The discipline structure must follow scientific, reasonable principles so that it can give full play to the mutual function between subjects. For example, when planning theoretical courses and practical courses, practical courses should be put behind the theoretical courses so that the knowledge and skills gained from the class can be applied to the actual environment. Besides, subjects with a strong professionalism and general studies should alternate with each other, thus giving the students' mind to have a relax cycle, which is helpful to improve the learning efficiency. The optimization of the discipline structure layout not only lies in the planning of the schools' teaching tasks and progress. The macroeconomic control of related education departments are also playing an unignorable role. Based on the social sports undertakings' demand on the talents and combined local schools' teaching resources with the teaching's present situation, the related education departments make overall settings and arrangements for professional disciplines and guide schools to optimize the discipline structures to increase the physical education major's teaching effects and improve the students' professional course performance and comprehensive quality. The subjects of physical education majors cannot be limited to some subjects with strong professionalism, such as Sports Statistics and Sports Training. Some subjects closely related with daily life also need to be included. On one hand, these subjects should be easy to understand; on the other hand, they should be easily applied, such as Community Sports Guidances and Leisure Sports.

4.2 Building teaching and research-integrated innovative talent training program

Though the modern pure-indoctrinated education training system could enable students to master more knowledge in limited time, it still can't fully meet the requirements for the cultivation of innovative talents. Apart from a rich knowledge reserve and broad professional perspective, what's more important is to possess strong creativity and ability to find out unusual knowledge during the learning. To improve the students' innovation ability, theory teaching is the foundation and practical teaching is an effective approach. Scientific research teaching is a teaching mode combining theory with practice, which not only has the research for knowledge theory, but also has the chances for hands-on experiments. Therefore, schools should combine teaching and scientific research, both of which promote and develop mutually, so as to improve the building of the teaching training system. It is still a little difficult for students to do scientific research on their own, so teachers need to undertake the guiding role, which asks of them to indicate the direction for students when they meet trouble while not disturbing the students' innovation ideas. When selecting a scientific research subject, there are two basic principles that need to be satisfied: firstly, to not be so complicated that it will be beyond the students' ability; secondly, to match with the society's demand, which would be helpful for the students' future employment. When processing the scientific research, students should think positively and dare to

explore. Don't seek help from teachers as soon as you encounter a bottleneck. When scientific research is over, students should pay attention to make a summary and get a deep understanding of which innovative thoughts get identification and which don't get practices. For those who lack analysis, pay more attention in future scientific researches.

4.3 Enhancing the improvement of teachers' teaching ability

The implementation performance of the training programs under the innovative teaching system has an inalienable relationship with the teachers' capability. Therefore, building a teaching team with an innovative consciousness and ability has a tremendously stimulative effect on optimizing the training program. Firstly, completely transform the teachers' bias on the innovative training program and let them fully realize the importance of innovation on the students' education. Secondly, strengthen the teachers' education innovative consciousness and education innovative ability through various training activities. Once having an education innovative consciousness, they will enhance the analysis and research on the development of methods of teaching activities and actively explore the innovative teaching mode. Education innovative ability is one of the key factors for implementing the innovative thinking and idea into reality. Teachers will take such approaches as applying their own extensive knowledge structure and good professional skills, adopting actual teaching and scientific research teaching and other ways to cultivate the students' innovative consciousness and innovative ability. When practicing innovative teaching, teachers should pay attention to penetrate innovative ideas into the whole teaching activity from the beginning to the end, constantly remind students to learn innovative knowledge and encourage students to come up with their own unique ideas boldly.

4.4 Strengthening the supervision and evaluation of innovative talent training program

Teachers should fully arouse the students' learning interest during the teaching activities, inspire the students' innovative enthusiasm extensively and stereoscopically, guide and inspire innovation activity, respect the students' individual development, thus achieving the goal of innovative training. Nevertheless, besides making efforts to the curriculum setting and teachers' teaching ability, the supervision and evaluation of innovative education should also be attached as being important. Schools should first supervise the teaching quality under the innovative teaching mode by establishing a thorough supervision system, such as making the questionnaire investigation to students regularly, holding different forms of innovative competition activities and so on, trying to get related specific material about the innovation training plan, submit these materials to schools for a serious analysis and find out the insufficiencies existing in the implementation of the plan, thus better serving the innovative education. After getting the feedback information, schools should take a serious analysis of every detail during the implementation process of the training plan, adjust the teaching plan combined with the real teaching effect and give some improvement suggestions for implementors instead of leaving it aside and only focusing on the final results, thereby giving full play to the positive function of the supervision system. Besides education supervision, schools also have to make evaluations on the implementation of the innovative talent training program and send it to the whole school in a written form, letting teachers and students involved know the insufficiences and flash points so that they could improve the teaching methods and learning attitude self-consciously.

5 CONCLUSIONS

In a word, optimizing the physical education talent training programs in the sight of innovation could enhance the students' innovative consciousness and ability, thereby enabling them to adapt to the social development and requirements better after graduation and finally make their own contributions to the physical development in China.

REFERENCES

[1] Sun Yiliang, Wang Bing, Chou Xianjiang, Tang Honggui, Gao Zhi. *The Building and Innovation of Physical Education Talent Training Mode under New Situations and Circumstances.* Journal of Wuhan Institute of Physical Education, 2011(05):61–65.
[2] Wang Bin. *The Thinking of the Reform and Development of Physical Education Major.* Journal of Neijiang Normal University, 2006(02):110–114.
[3] Huang Hanshens, Chen Junqin, Mei Xuexiong, Xu Hongfeng, Lin Quyong, Hong Taitian, Yu Kuikang. *The Research on the Reform of China's Common Colleges' Physical Education Course System toward the 21st Century.* 1997(03):69–72.

Future Communication, Information and Computer Science – Zheng (Ed.)
© 2015 Taylor & Francis Group, London, 978-1-138-02653-7

Research on undergraduate students' ideological and moral core values education

Y. Lu

Shandong Labor Vocational and Technical College, Jinan, Shandong, China

ABSTRACT: Students' ideological and moral core values are significant for characters and personal development. The main purpose is to analyze problems in the students' ideological and moral core values education from perspectives of times, society and undergraduates. Then efficient ways of education are explored under the guidance of established education principles to provide help for constructing the students' ideological and moral core values. The differences between this work and others lies in the combination of undergraduates' values education with national spirit and zeitgeist, revealing how college students adapt to the society and country in a new era.

Keywords: undergraduates; ideological and moral quality; core values

1 IMPORTANCE OF UNDERGRADUATES' IDEOLOGICAL AND MORAL CORE VALUES

Nowadays, college students have become a large scale of talent groups with a higher education, being the reserved army of social construction and the future of nation and country. Undergraduates' ideological and moral core values are not only important to undergraduates, but also to the social morality, national spirit and culture. College students, shouldering historical responsibilities, are the hope of ethnic and national revitalization, therefore their ideological and moral core values need to be guided.

First of all, as the development of the economic globalization and cultural diversification, all sorts of social thoughts, varying greatly, are pouring into our country from abroad [1]. Although some are excellent, there are poor cultural thoughts in front of college students. Only by strengthening their education to teach them how to find a worthy part in the confused minds, the ability to distinguish right from wrong and the insight will be improved. Then their moral quality will be enhanced to participate in the social life with more enthusiasm and a strong sense of a national mission. Secondly, the college students have not yet fully entered the society, thus their inexperience and immaturity will make them to be easily induced and deviate from the correct path. It is needed to raise college students' ideological quality and strengthen the scientific and cultural knowledge and moral quality to establish the correct outlook on life, values and world view, getting away from the worship of money, selfishness and hedonism and establishing patriotism, dedication, ideological doctrine and the spirit of being ready to help others, positive, rigorous and innovative [2]. However, there are some unhealthy factors in

ideological and moral core values education to obstruct their personality shaping, moral cultivating and quality formatting. Wherefore, how to overcome negative factors and establish good values are worth pondering for every educator.

2 STATUS QUO AND PROBLEMS OF IDEOLOGICAL AND MORAL CORE VALUES EDUCATION

2.1 Teaching philosophy

College students' ideological and moral education mainly relies on the political course, therefore the teachers' teaching quality largely affects ideological and moral core values. From the structure of current ideological teachers, it can be known that the most common constitute is the combination of professional and unprofessional teachers. But the teachers major in ideology and morally take up a relatively small proportion, and most teachers also work in other departments of the university, such as work as the leader of the Party Youth League, student counselor or head teacher [3]. They have a heavy workload, and therefore the energy into the ideological and moral education will be greatly reduced, resulting in the neglect of students' core values. In addition, some ideological and moral teachers are also problematic in ideological and moral quality. These shortcomings and deficiencies will be directly displayed in front of students, and the role of the teacher is greatly reduced. The more important is that the students imperceptibly develop a poor quality of thought and behavior, affecting their future development and construction of social atmosphere.

2.2 Connection between environment education and times

With the development of China's modernization, a variety of external social factors merge with internal social factors to promote the development, forming a unique social environment in a new era. On the one hand, new requirements are proposed for college students to adapt to the society; on the other hand, the construction of the undergraduates' ideological and moral core values will also be influenced. For example, a few years ago, our society advocates Good Samaritan acts to enable young people to come forward to curb bad behavior in the face of adverse circumstances. However, as college students have not yet entered the society, this kind of rash actions would threaten their personal security. Therefore, with the increasing social respect for life, Samaritan wisdom has emerged, enhancing college students' rationality to handle the emergencies. As modern mass media and Internet develop rapidly, lifestyle and thinking have changed tremendously. Thanks to the free talk environment on the Internet, more and more people tend to enjoy expressing personal views on life, society and nation. However, due to the inadequate supervision of Internet speech, some lawbreakers wantonly spread false information to lead people to incorrect ideas. It has negative effects on immature college students and their quality and behavior will become vicious.

2.3 Undergraduates' ideological changes and national spirits

National spirit, one of the key aspects of the ideological and moral core values construction, answers what the role and style of college students in the historical trend of China' developments. But with the passage of history, especially in the era of peace and development, national spirit and patriotism have slightly dissipated. They believe that the national spirit and patriotism are just something during the war, not peacetime. This wrong perception will weaken national pride and ideological and moral quality, failing to realize self-struggle and make contribution to the motherland under the guidance of national spirit. This change is very dangerous, so universities need to strengthen the construction of national spirit. A strong sense of shock and pride will be inspired by telling the Communists' hard work in the Anti-Japanese War to lay a solid foundation for ideological and moral education in universities.

3 PRINCIPLES OF IDEOLOGICAL AND MORAL CORE VALUES EDUCATION

3.1 Principle of direction

The so-called principle of direction is to realize the consistence of the students' ideological and moral core values with modernization. That means the students' ideological and moral quality should meet the social requirements for talents, and should not go against social morality. As a result, college students' ideological and moral core values should be built under the guidance of Marxist ideology, adhering to the party's policies and guidance and improving the spirit. When conducting ideological and moral education, universities should not only educate students' behavior, but also instill Marxist and socialist ideas to students to improve college students' ideological and moral quality fundamentally.

3.2 People-oriented principle

College students' ideological and moral core values education, the emotional education, is different from other courses. Although excessive brainpower is not required, tense emotion is needed. Therefore, ideological and moral core values education should be consistent with the people-oriented principle. The way to help them build core values is found through the analysis of each student's personality and ideology to avoid the students' quality being affected by rigid ideas. For the students with low ideological and moral quality, their resentment for learning will be reduced and the relationship with educators will be closer so that conscious self-education and self-evaluation will be achieved.

3.3 Principle of incentive

The principle of incentive is of great importance for the undergraduates' ideological and moral core value education. First of all, incentive-based education can greatly increase the students' enthusiasm and initiative, so that they can consciously pay attention to their demeanors to meet society better. Secondly, the ideological and moral core values are too abstract for college students, but incentive education is able to combine the ideology with the reality, enabling a more intuitive understanding of the importance of the ideological and moral core values. For instance, college students are praised by teachers after they give a hand to others, so he will understand that good quality is welcomed by others.

4 WAYS TO CULTIVATE UNDERGRADUATES' IDEOLOGICAL AND MORAL CORE VALUES

4.1 Coordinating family education, social education and school education

College students are in the development stage, being energetic and able to accept and learn new things. Colleges and universities should grasp this chance to systematically implant ideological and moral core values. Moral consciousness will be established to practice good behaviors in daily life through learning history, patriotism, outlook on life and world view. In addition to school education, family education and social education also play an important role in the undergraduates' ideological and moral core values education. Those constitute the students' growing and living environment. Expected results are based on a healthy, optimistic and positive atmosphere. Although we cannot ensure that the adverse atmosphere will lead to many vicious and reprobate students, "living in the silt but not imbrued" only occurs to a few people. College education is universal, and it should be for a large group of students. Therefore, it is needed for schools, families and society to establish good communication channels. Feedback mechanism is a good example. A favorable environment should be created to train the students' quality and gradually promote students' comprehensive development.

4.2 Innovative and scientific education philosophy

With the development of time, ideological and moral core values show new connotations, so educational philosophy should be updated constantly according to the social changes. Ideological and moral education should embody characteristics of a new era, so that the students' ideological and moral quality will adapt to the society. Students' thoughts have changed so greatly that some traditional education modes are no longer suitable for practical teaching activities. Some of them, rigid and stiff, do not meet the active thoughts, and some teaching methods, traditional and backward, do not comply with the students' physical and psychological characteristics. As a result, universities need to innovate the context of updating educational philosophy. Innovative educational mode includes multiple aspects. The first one is the innovation of education methods. Teaching methods at this stage should be more targeted, so as to enable each student to develop better. But it is necessary to pay attention to the latest educational research to achieve comprehensive and multi-channel applications of new educational methods to educational activities. Followed by the innovation in teaching, though the ideological and moral core values education mainly relies on language teaching, extracurricular practice is also important. On the one hand, it can make the course more interesting. On the other hand, it is also convenient to inspect education results.

4.3 Emphasis on mechanism of model and incentive

Because college students are lively and active, boring language teaching will not guide them to think deeply and fulfill a noble ideological and moral quality in daily life even though it seems that they listen carefully. A fine example has boundless power, thus a model has to be found out in teaching to encourage students to emulate consciously. For example, a national upsurge of learning from Comrade Lei Feng began in the 20th century, cultivating a generation of excellent talents with dedication and thrift.

Incentive is an effective supplementary method for education. For some students with poor ideological and moral quality, incentive rather than criticism is more likely to arouse their enthusiasm and self-confidence. While the students with a higher level, can get satisfaction and a sense of accomplishment. Hence, such positive cultivation methods as incentive mechanism should be infiltrated into the students' ideological and moral core values education to enhance the efficiency of education.

4.4 Developing new teaching models

Although ideological and moral core values education is the education of thoughts, the education effect relates to the students' behaviors. New teaching models require the combination of practical teaching with classroom teaching. It begins with theoretical cognition in classroom teaching and then switches to practice teaching. For example, students are organized to participate in social activities so that they will understand what kind of talents the society needs and arm their own thoughts on the basis of core values. Although all students are the same age, there is still a maximum age gap of 4 years. Naturally, freshmen and graduates may have different ideological and moral core values, the education to the former highlights ideological and moral quality while the latter highlights social thoughts. So that the former will improve ideological and moral quality during four years and the latter will adapt to the work and life.

5 CONCLUSION

All in all, undergraduates' ideological and moral core values education is long-term activity. College students are the future builders of the motherland and successors of socialism, thus universities are expected to devote more time and energy to research ideological and moral education modes to find out the most efficient cultivating method.

REFERENCES

[1] Guangxin Xie. *How to Cultivate Young Students' Ideological Quality*. Journal of Tongren Vocational and Technical College, 2004 (03):15–17.
[2] Jianhua Luo. *Psychological Structure and Content of Ideological Education*. Theory, 2013 (11):303–304.
[3] Zhiming He & Shuo Zhou. *Theoretical Exploration and Practical Innovation of Undergraduates' ideological and Moral Core Values Education*. Jing Guan Wen Yuan, 2013 (02):49–51.

Future Communication, Information and Computer Science – Zheng (Ed.)
© 2015 Taylor & Francis Group, London, 978-1-138-02653-7

Analysis on emotional communication in foreign language exchange

F. Huang
Foreign Language Department, Shanghai Business School, China

ABSTRACT: In people's life and communication, emotional communication is omnipresent, a complex psychological changing process of human communication in language. It is people's unique personality characteristics that make emotional communication become one of the important factors in foreign language communication an obstacle. In addition to language, the most important in foreign language exchange is emotional communication. Only language full of emotion can make people understand each other's true feelings, likes and dislikes, personality characteristics, promoting the further development of people's relationship, and making society more harmonious. However, it's not simple to implement emotional communication in foreign exchange, which is closely related to personality characteristics, education and life background of both sides. This work briefly analyzes the implementation of emotional communication.

Keywords: foreign language communication; communication; promote each other

Affective disorder is most important in the process of language learning and communication, especially facing with the rapid development of society, accelerating pace of economic globalization, increasing mutual cooperation throughout the world, progress and fusion of the language driven by a developing economy. However, due to the difference of the world's culture, people's living and education environment, lifestyle and ideas differ very much, which directly affects the foreign exchange to achieve its effect. Thus, how to solve this problem has become a focus of people's study [1]. To solve a problem is to understand it first, so the first step is to understand the emotional communication obstacle.

1 THE IMPORTANCE OF EMOTIONAL COMMUNICATION

So-called emotion is part of the human attitude. It is consistent with inward feelings and intention and is a complex and stable physiological evaluation and experience of attitude in the physiological.

Humans have a very rich emotional world, and they have different subjective emotions on various things in life. For example, people feel disgusted with what they hate, and fond of or happy with things they like. During the exchange of people, various different emotions from the two sides exist. When the two sides can understand correctly and accept each other's viewpoint, they will reveal more harmonious feelings [2]. If the psychological resonance is reached, the further communication will be promoted, realizing the

ideal objective. However, when both sides can't correctly understand or accept each other's viewpoint, the feelings won't be harmonious and contradiction in conversation will generate more easily. Then serious impact on their further development will appear. In addition, when both sides can't communicate effectively in the same language, the shortcoming can be compensated providing effective emotional communication is built, making both sides understand each other's meaning based on it. So, in order to talk deeply, both parties are required to have a certain knowledge and psychological quality, the respect each other's conversation characteristics, values, life and world views, culture and religious beliefs to communicate sincerely, reaching the maximum meet of the emotional communication in foreign exchange based on the heart and heart [3].

Similar to the above exchange ideas, emotional communication in language communication is of considerable importance. Obviously, people using a foreign language to communicate should come from two countries or two language cultures. The differences of culture, education level, values, and world and life views will easily lead to the situation of both sides finding it difficult to understand each other deeply in the language communication. Thus, it's required for both sides to communicate on the emotional basis, truly show the psychological feeling, and move each other with a sincere heart to promote further exchange. In recent years, with the development of economic globalization, China and the world's economic exchanges are increasingly close, and demand of foreign language talents is becoming bigger and bigger. In order to conduct more effective economic cooperation, the emotional communication with partners is also vital for the economic negotiations in addition to clear expression of professional terminology.

2 THE PROBLEMS OF EMOTIONAL COMMUNICATION IN LANGUAGE EXCHANGE

The first is a personal one. On account of limited education, it's not comprehensively enough and detailed for the individual to master a foreign language or know about the foreign language culture.

Moreover, a unique individual character doesn't make the emotional communication play a proper role in language exchange.

The second is the difference of the world's culture. The global economic integration has been developed, so that cultural exchange between nations all over the world will continue to be deepened and a variety of Cultural Festivals are developed. Moreover, the communication between different cultures is enhanced. However, the unique cultures force people to pay attention to respecting each other's culture. When one doesn't know about the other's culture, it is likely to cause contradiction even conflict in communication.

Finally, the defects of China's education system make English difficult to use. Although China has implemented many reforms in education, achieving great achievements, China's education system still hasn't changed. The 'cramming' method of teaching still widely exists, and the 'cage' of the exam-oriented system is still tightly wrapped around many students' head, leaving their development not free. In learning a foreign language, grammar has the priority. Thus, in order to improve students' performance, schools put the grammar education first, so that the students can accurately use grammar to solve problems. However, the performance of the actual use is unsatisfactory with the effect not obvious.

3 THE METHODS OF EFFECTIVE EMOTIONAL COMMUNICATION IN FOREIGN EXCHANGE

3.1 *Realizing the educational and emotional communication – Guide to the emotional communication in foreign language education*

The rapid development of modern society and the increasingly fierce competition have made increasingly high demands on talents. At the same time, the development of economic globalization has also made China and the world's economic exchanges increasingly close, resulting in the bigger and bigger demand of foreign language talents. In order to conduct more effective economic cooperation, the emotional communication with partners is also vital for the economic negotiations in addition to a clear expression of professional terminology. To enable people to conduct correct emotional communication

in foreign exchange, they should be trained at the beginning of learning a foreign language.

3.1.1 *A good learning attitude and interest is needed in the foreign language teaching*

Learning a foreign language is a very difficult thing, and learning it well is more difficult. Meanwhile, interest is the best teacher. When interested in learning it with persistent perseverance, you will get half success in foreign language learning. So teachers should pay attention to stimulating the students' interest in learning in many ways, improving the enthusiasm of students to participate in teaching activities. Specific methods can be used such as the following.

(1) Grasping the minutes before class. Starting from the social hot topic combined with the practical content of the curriculum, teachers can animate the class atmosphere, arouse students' curiosity, give students the chance to speak actively and discuss facts. Moreover, students can be encouraged to analyze facts in English, understanding the texts' content as well as facts, so as to get touch with society.

(2) Using English songs, movies, and TV series. Teachers can choose some English songs that have an obvious sense of rhythm and easy content to play for learning. A popular English movie or TV play can also be chosen. This artistic teaching method can stimulate the students' interest in learning more easily. Furthermore, when watching the TV plays, students can also understand foreign customs, culture, and communication style and so on. To conclude, it is a very efficient way of teaching a foreign language.

(3) The use of games in class to stimulate the students' interest. Teachers can make their own English games and invite students to join in with participants encouraged to use foreign language to communicate. Other students not participating can comment on the results. This method can regulate the atmosphere of the class, promote exchange between teachers and students, and help students master the vocabulary and sentences.

3.1.2 *Increasing the proportion of foreign culture and custom in foreign language teaching*

Language and culture are closely related to each other, so foreign language learning is also a process of an in-depth understanding of the foreign culture. As a tool of communication, language reflects the connotation of national and regional culture. Meanwhile, understanding the cultural connotation is an essential item in language learning. Only by knowing the cultural background of each other, can one respect each other in communication and better understand the meaning that the other expresses. In concrete teaching and education, suggestions about how to make students learn more about foreign cultures are as follows.

(1) Selecting teaching materials that contains much about the foreign culture. The main source of foreign language learning is teaching materials, so the selection of textbooks is of much importance. In order to allow students to understand a foreign culture more

comprehensively, teachers should choose textbooks containing details of foreign cultures, customs, religious beliefs, local customs and practices, dietary habits, social system, history and geography and other aspects. Not only can students learn the foreign language, they can also know more about the foreign culture, killing two birds with one stone.

(2) Playing documentaries. It's also an approach to arousing the students' interest in learning. With the popularization and development of computer and television, there are more and more documentary recordings all kinds of countries' history and culture, so that the channels of people to understand a foreign culture are increasing. Through television or in film form, documentaries show people foreign history, geography, humanities environment, local customs, practices and so on which they want to understand. Compared to articles on papers, this form of education can stimulate the students' interest more, enhance their memory and make them learn more about the foreign culture.

(3) Focusing on issues as taboo in foreign language communication.

Due to the difference of world culture, people from different language background have different ways, rules of communication, and also many sensitive and privacy issues not involved. These privacy issues reflect the foreign cultural environment, so more attention should be paid to the problems in foreign exchange, focusing on the other's emotional change.

For example, in the Anglo-American countries, when talking to women, you should show your respect to them, and don't ask for their age. When getting on with foreigners, you should get the consent of others about smoking. Don't put your tongue out, because this represents an insult in foreign culture, and so on. In foreign language communication, inadvertently body posture will also impress the other party. If an action offends the other, it's unfavorable for the mutual communication, leading to the other's bad mood, which makes emotional communication difficult to achieve sincerity. On account of being able to learn more about these problems, it requires the teachers and students to discuss and study together, striving to fully know about these problems. However, everyone is unique, has his own characteristics, and issues each person concerns and taboos also vary. So that it demands people to make a concrete analysis of concrete problems in the future exchange, touch each other in emotional exchanges, deepen mutual understanding, know more of the psychological situation of each other. So that good results can be achieved in emotional communication.

3.2 *Abandoning Chinese foreign language in foreign language communication, and using correct expression in accordance with English culture*

Foreign language translated according to the features of Chinese is called Chinese foreign language. Take English for example, Chinese-English has become a very common phenomenon. Resulting from a lack of systematic English learning training, many people will communicate in Chinese-English according to their own understanding of Chinese and English. Because of cultural differences, a lot of foreigners are difficult to understand, which causes people to remain in a difficult and embarrassing situation in foreign language communication. Many students living and studying overseas have summarized this issue that they had found, the grammar taught in Chinese classrooms is of little use in daily English communication. When they use what they learn in the class to talk to foreigners, most of the time they can only communicate on a very shallow level, and emotional communication is few. In addition, some other overseas students have said, in the domestic, as long as sincere communication is built, it's very easy to become friends with others. But in foreign countries, in the absence of a speech problem, it's still very difficult to make long time communication with no substantive topic. After analysis, it's discovered that, the main problems is that, English grammar and sentence emphasized in domestic teaching is not important in actual communication, and the use of speech compatible with the foreign customs in the exchange of communication environment is more prone to narrow the distance between people.

Chinese-English has become the biggest obstacle to people's communication, which has a very big relation to China's 'cramming' examination-oriented education system. Although in recent years China has made a great effort in education reform, no substantive breakthrough has been conducted, leaving China still an examination-oriented education. In the examination-oriented education system, English grammar learning is important in English learning. However, the contact of cultural learning under the background of English language is rare. Those with experience have claimed that the grammar that Chinese students understand is even more than that which students from English-speaking countries know. This phenomenon is a unique one affected by the teaching environment of performance determining everything. Faced with the development of society, the improvement of the English teaching method is urgent.

4 CONCLUSIONS

In foreign language exchange, affective factors actually occupy a large part, and language exchange is emotional contact. Furthermore, correct and reasonable use of foreign exchange is the premise of effective communication, which has a very high demand for language learning. Thus, in the process of daily learning, you should ensure you often have strong power and confidence, persevere on the way of foreign language learning. Meanwhile, you should keep the enthusiasm of learning, so that it can be accomplished with half the effort. In addition, keep yourself in an unsatisfactory state forever, not just confining yourself in the range of usual knowledge learning. You should also have the mentality of the eagerness for knowledge, and struggle to explore the knowledge that you never come across in foreign language learning. At the same time, try to know about the foreign culture, humane environment, history and religious belief and so on. Refuse being the puppet of grammar and illiteracy of culture. Only when the two are combined, can the emotional communication always go through language communication, can you get more support in the effect of foreign language environment. Besides, you can also add a little more competitiveness in the future, and live a wonderful life in the increasingly fierce social environment.

REFERENCES

[1] Zhang Deng Zhi. Foreign Language Teaching and Cultivation of Intercultural Communicative Competence. Dajia. 2011(16).

[2] Shan Jie. Cultivating the ability of outside the foreign language. Chinese translation occupation Exchange Conference in 2010. 2010.

[3] Lu yang. Using of Affective Factors in English Teaching. New curriculum research. 2011(8).

[4] Yang Ming Xia. Emotional Education in English Teaching. Xue Yuan Education. 2011(18).

Future Communication, Information and Computer Science – Zheng (Ed.)
© 2015 Taylor & Francis Group, London, 978-1-138-02653-7

Development and application of computer software technology

X. Zhang
Department of computer science, Mianyang Vocational and Technical College, Sichuan, China

ABSTRACT: With the development of computer technology, the development and application of software technology are constantly updated. To effectively increase the development and application efficiency of computer software technology, software development technology was introduced. Additionally, the exploration was made from three respects: importance and methods of development, as well as its application. As a result, the development and application of software should abide by the development philosophy of being people-oriented, thus promoting usability. The innovative point of this thesis lay in an analysis on performances of different software development techniques from specific applications. Also, the people-oriented importance was analyzed as the core of software technology development and of security.

Keywords: computer; software technology; development; application

1 INTRODUCTION

With the development of computer technology and the arrival of the information era, the network has become an important transmission medium. Also, diverse needs start to emerge from the technology and application of computer software development. Additionally, technical requirements for software development, as well as safety requirements for information, have become increasingly high. Thus, the promoted development and further application of computer software technology will be achieved by its optimization, along with the introduction of advanced security encryption and technology fusion [1].

2 OVERVIEW OF SOFTWARE DEVELOPMENT TECHNOLOGY

2.1 *Analysis on concepts of software*

There are two main types of computer software. One is system software, which is a kind of interface based on a bare computer to connect machine and human. Its function is to manage and maintain all the computer software. The other one is application software, which is developed into various types, such as the Human Resource Management System, Enterprise Resource Planning System and Customer Relationship Management System. The purpose of the two kinds of software is to provide a simple computer work environment and to extend the application of the computer. For users, the essence of using a computer is the application of various software on it. Thus, a lot of software in the 1950s and 1960s were manually written in assembly language. This machine language could not meet the production needs of software due to high error rates or low production efficiency.

With the development of software development technology, large-scale software and its commercialization started to emerge. However, it was difficult to guarantee the reusability and function enrichment of software without clear specification. Therefore, specification documents should be developed to standardize the programming, debugging and operation of programs. Software engineering began to appear in the 1980s, which referred to standardized rules and methods of realizing the computer program functions. And the sum of relevant documents and programs operating on the computer was called computer software. Thereby, the definition of computer software was expanded, and standard processes and methods of software development were further standardized [2]. So a theoretical support was provided to the increasing of software sizes.

Software engineering experienced three ages from nothing, including programming, software and software engineering. To take the development of construction projects as an example, a considerable difference could be seen between the software development process and construction. There were rare traceback problems during construction projects since the construction demand schedule and design drawing were confirmed, because it was difficult to modify the original design during the construction phase. However, every step of design in computer software engineering would undergo several traceback problems to ensure the stability of each link, including numerous modifications, testing and adaptation. Additionally, the convenience of software maintenance and stability of operation should be considered after the software being put into use [3].

2.2 Importance of software development technology

As one of the most core techniques in software engineering, software development technology plays an important role in promoting the development of computer networks and technology. It can promote the innovation of computer network information technology, and achieve a network-based control and support, so that a coexistence of software and internet could be formed. Also, the application could solve the problem of software development behind network needs and promote the software development technology of computer. Meanwhile, the security of computer network and reliability and convenience of software could be promoted. In a sense, the relationship between software development technology and computer technology and network security is mutual interdependence and promotion. Thus it is essential to promote software development technology.

3 METHODS AND APPLICATION ANALYSIS OF SOFTWARE DEVELOPMENT TECHNOLOGY

3.1 Method analysis on computer software development

Currently, there are three methods of computer software development, including the prototyping method, software life-cycle method and automatic form. In the prototyping method, software should be strictly defined and pre illustrated before software development. Meanwhile, overall functionality and needs should be understood deeply and comprehensively by software developers and users. The prototyping method could be adopted when the input and output requirements during data processing could not be described in detail from software requirements and goals given by the users. It is also applicable when the final program algorithm, applicability and interpersonal interface morphology of a system could not be confirmed by software developers.

The software life-cycle method is defined into several stages in chronological order from the software development to the end, including software development, the definition and maintenance. The beginning and end of each stage is standardized by corresponding standards. Generally, there are six stages linked to each other in the life style of software, namely feasibility analysis, requirements analysis, overall design phase, program phase, testing phase and maintenance phase. With the confirmation of one stage being completed, the next stage could be entered. On the contrary, without confirmation, re-optimization should be conducted by tracing back to the last stage. Thus, the stability of the whole software could be achieved.

As the latest method of software development technology, automatic form is a system development methodology. This method adopted the software in the fourth-generation software development as a development tool. And development efficiency could be improved effectively only with functions, contents and goals of software illustrated by software developers. By contrast, how to implement these functions and methods is not necessary. However, for numerous large-scale application software, automatic design and coding according to requirements still cannot be achieved because this method is in the development stage.

3.2 Application of software development technology

Software developers aim to make the computer, an important tool, to make greater contributions to mankind and to better meet the demands of people to related work. Therefore, software development technology attaches great importance to its application value to play its role better. To further explore the application of computer software in development technology, an example is needed to illustrate the important application value of software development technology. For example, two different software development technologies were applied to a human resources information management system. One was the combination of VB and SQLSERVER database in a stand-alone mode. The other one adopted ActiveX and MYSQL database server through WEB and XML.

Then corresponding client-side software was installed based on these two products. If the system developed by the first technique was adopted, the client could directly install the appropriate client software, along with SQLSERVER database service software. The VB runtime environment was already contained in Microsoft systems, while the database environment should be installed later. But it was quite convenient to use the corresponding software due to good compatibility between Microsoft systems and SQLSERVER and numerous database interfaces provided by VB. However, if the other client was installed based on ActiveX, corresponding ActiveX components were also needed. Additionally, security settings of browsers should be changed and Mysql database server be installed. Thereby, one more security step was needed during the installment of the second system.

When two clients are installed successfully, the usage could be monitored after a period of operation. The software developed by the first technique was relatively stable, and users could backup and restore data by themselves due to the simplicity. Meanwhile, there were numerous disadvantages. The speed of the client would significantly decrease with the growing number of clients, increasing service time and data in the database. Moreover, data sharing and extension of software functions were relatively weak due to the lack of B/S structure of the stand-alone client. On the contrary, the system developed by the second technique played a prominent role in function extension and data sharing and usability. But there were still some inconveniences caused by limits on the operation of security settings of the browser and some operating system versions.

From this example, usability of software could be reflected on cognition and usage of development technology through software development. However, there was a significant gap in convenience and functionality. Meanwhile, differences also existed in security of the two software development techniques. Without attention being paid to security encryption technology, the computer would be easily invaded based on the ActiveX technology, thus causing safety problems. With computers becoming more and more important, software developers cannot guarantee the software functions could be well extended, because not all the computer operators are professional. Thus, the software development technology should be brought in from the perspective of usability, security and stability of software. Also, the technology should be innovatively applied based on the development philosophy of being people-oriented. Additionally, the software security should be guaranteed.

3.3 *Application of software security technology*

Application of software security technology contains two respects. One is application management of the key, which refers to the primary encryption means of information of software and the main work object of information privacy and eavesdropping. Generally, the key is private and non-public. However, there are a lot of the same keys in the process of computer technology development. Once the key was obtained by a third-party user, related information would be eavesdropped by a third-party through interaction between the user and others. As a result, the information security of the user would be threatened. Furthermore, the more a key is used, the more likely it will be leaked. So the information should be guarded against illegal eavesdropping to enhance the software security. In other words, the key should be replaced regularly to reduce the eavesdropping according to its attributes. Nowadays, organizations of software development have established distribution centers on the Internet to guarantee software security, which could provide safe and reliable information keys. Each user can only obtain one key to have a dialogue with the distribution centers. In this way, reuse of the key could

be reduced effectively and security requirements of the software will be met.

The other application is quantum cryptography, which could determine whether the computer will be under attack. This technique could transform the traditional cryptosystems into an all-optical network. Also, the key exchange and information encryption could be absorbed into the level of optical fiber. In this way, once there is an illegal invasion, the software will send out related invasion information based on significant changes in the quantum. Then the software could raise the vigilance of users through self-shielding or alerting immediately, or remind users of cleaning up the computer with anti-virus software. So software security could be increased effectively. Therefore, quantum cryptography has been applied by a lot of software, especially those with higher requirements of information security.

4 CONCLUSIONS

Overall, with computer hardware as an important carrier, software is an important approach of acquiring various applications through hardware. As the core of computer thought and application, computer software could process numerous logic and computing information for the maximum promotion of the application value of the computer. Thereby, In order to maximize the functions of the computer, the application approach of computer software should be expanded, as well as development methods. Meanwhile, user-friendly development of computer software should be paid attention to. Because only based on convenience and usability, the software could be applied widely, thus contributing to the rapid development of the software industry.

REFERENCES

[1] Zhao Mingliang, *Development Technology of Computer Application Software*, Heilongjiang Science and Technology Information, 2011 (26): 98.
[2] Chen Bin, *Depth Development and Application of Computer Software*, Guide of Sci-tech Magazine, 2012 (21): 61.
[3] Liu Mo, *Analysis on Importance of Software Development Technology in Computer Network Education*, Management & Technology of SME, 2011 (18): 275.

Future Communication, Information and Computer Science – Zheng (Ed.)
© *2015 Taylor & Francis Group, London, 978-1-138-02653-7*

Relations between regulation of socialist market economy and economic law

H. Wang
Ningxia Justice Police Vocational College, Yinchuan, China

ABSTRACT: The work analyzed the relations between the regulation of socialist market economy and economic law. First of all, a general research was provided on the relations between economic law and market economy. Next, the work analyzed the regulatory functions of economic law on market economy from three aspects. They are special mechanisms of economic law, the duality – economic characteristic and social standard and the effects of economic law on the market economy. Finally, the work analyzed the regulation of economic law on economic subjects based on two aspects – regulation of economic law on enterprise behavior and market subjects. Furthermore, the ideals of making full use of the mechanism of economic law were suggested.

Keywords: economic law; market economy; spontaneity; behavioral agent

1 OVERVIEW ON ECONOMIC LAW AND REGULATION OF SOCIALIST MARKET ECONOMY

With the development of society and the formation of the market economy, the blindness and spontaneity of the market economy are increasingly prominent. Consequently, the state, as public administrative power, needs to intervene in market activities. With the emergence of the socialized mass production and market economy, the state, rather than being a "watcher", should be more dedicated to the macro-control for economic development [1].

It is in a shorter time that the market economy develops in China, relative to that in western countries. However, thanks to the superiority of the socialist system and powerful macro-control, China is building a fair environment of market competition. In the development of the socialist market economy, the disadvantages and drawbacks are comparatively prominent [2]. Therefore, they can possibly cause failure in a self-regulating mechanism. Therefore, it is critical to establish economic law and give full play to its regulatory function.

Economic law is the key part in making full use of macro-control. Possessing national authority and representing the people's will, economic law is the most compulsory, normative and general means of macro-control. It plays a main and fundamental role in the macro-control. Without economic law, the macro-control will be out of order [3]. This potential risk stems from complex economic relations and operation law of the economy.

2 REGULATORY FUNCTION OF ECONOMIC LAW

Economic law is good at regulating the socialist market economy. The regulatory function has many aspects which will be analyzed as follows.

2.1 *Special regulatory function of economic law*

On the one hand, with respect to the socialist market economy, economic law has a regulatory function that is comparatively special. With incomparable binding force, economic law is more general in contrast to traditional regulating methods of a planned economy. Economic law is different from civil law. It is not agreed, but a binding law aiming to regulate economic behaviors in the socialist market economy. Similar to obedience and command in administrative law, the concept of economic law is obedience and management. On the other hand, economic law is more equal in some degree and gives more freedom to economic subjects in private rights. It can be found that economic law shows more respect to private rights, displaying a more normative form.

2.2 *Duality of economic law*

Economic law has a dual nature. Firstly, economic law has economic characteristics. It is established to fix the failings of self-regulation of the market economy. The fails cannot be fixed by traditional regulation or self-consciousness of behavioral agents in the market. Thus, the establishment of economic law is to remedy the limitations of market regulation. Secondly, economic law has the characteristic of a social standard. Economic law is not for protecting the interest of administration and public power. Instead, it is for the normal operating of the market economy. Therefore, economic law is aimed at maintaining steady social functioning. The economic base determines the superstructure, so the stability of the operation mode determines the stability of society. As a result, the social characteristic is actually the social standard of economic law.

2.3 *Impact of economic law on market economy*

The greatest aim of economic law is to regulate and guarantee administrative power in balancing economy. The analysis shows that economic law is for maintaining social stability and normal social functioning. Therefore, the commonwealth of economic law can ensure the stability of social interest. Given the defects in the self-regulation mechanism of the market economy, economic law can regulate and control the market economy appropriately. Therefore, the major function of economic law is to safeguard stable public administration. The major function contains two aspects in fact – economic law safeguards the authority of public administrative power and economic law defining the power of the government. That is because economic law defines the legal processes when public administration intervenes in economic operation. In addition, economic law focuses on the validity of administrative intervention. As a result, economic law, as a defined framework, serves as a reasonable limitation on public power implemented in macro-control.

It can be seen that economic law is closely related to the market economy. For the overall interest of the society, economic law guarantees the whole economic benefits, protecting the rights and interests of an economic entity. These call for regulation on different economic agents with various and flexible means. Hence, economic law can cause considerable influence on the economic operation. It can guide economic development and adjust the direction of the national economy, thus boosting a healthy development of the socialist market economy. Therefore, without the regulation of economic law, there is no socialist economy. The regulation of economic law is the crucial part of the macro-control in the socialist market economy.

3 REGULATION OF ECONOMIC LAW ON MARKET SUBJECTS

In essence, behavioral agents in the market economy are not just companies or consumers. They contain traders in social reproduction, consumers, operators and producers in the market. Besides, states, administrations, companies as well as individuals are also covered. Therefore, adjustment objects of economic law should not contain traders only. As public administrative subjects, behavioral agents take part in the market transaction as well. States or public administrations are not only empowered to supervise and regulate socialist market economy. But they, acting as business entities or consuming subjects, can conduct direct transactions in the market. So, the range of economic law in regulating market subjects is broad. Any subject, which is able to take part in a socialist market transaction, should be restricted by economic law.

3.1 *Regulation of economic law on enterprise behavior*

The enterprise, among different behavioral agents in the market, is a concrete and active behavioral agent with the greatest expansion capacity. On the one hand, the enterprise is the producer in social reproduction. On the other hand, it also takes part in the primary distribution. Besides, the enterprise undertakes the marketing exchange in social reproduction, and it is also the consuming behavioral agent of raw materials. From this perspective, the enterprise is the most valued behavioral agent. The realization and development of the function of market economy rely on the restriction of economic law on the enterprise. Economic law is not the only law that has a great impact on the enterprise – the behavioral agent. The impact is of co-movement.

Economic relations of the enterprise can be grouped in two types – External economic relations and internal economic relations. External economic relations of the enterprise cover external relations of economic cooperation and management. But internal economic relations focus on the internal regulation.

In fact, the effect of economic law on the enterprise is exerted by the regulation and control of economic law. Economic law and civil law are supposed to regulate external economic cooperation of the enterprise. The external administrative relation of the enterprise can directly reflect the will of state administration. Besides, it is the result of national macro-control in regulating the enterprise operation. However, administrative regulations are not mandatory to adjust the external administrative relations. These relations should be chiefly adjusted by the enterprise.

However, external cooperative relation of the enterprise calls for the joint coordination of civil law and economic law. An enterprise will deal with others in all kinds, and it particularly deals with the contractual relationship between enterprises and other economic subjects. Consequently, the enterprise is supposed to be regulated by both economic law and civil law, instead of economic law only. That is because the union of the two laws can cause an effect of restraint.

3.2 *Regulation of economic law on market subject operation*

Operational relations among market subjects refer to the relations among different economic properties in the market. Operational relations still involve all market subjects. But they mainly refer to competitive, joint and cooperative relations formed by market subjects. These relations are also the adjustment objects of economic law. However, they can be distinguished by relations of administrative nature and transactional nature.

As a matter of fact, economic law regulates the relations of market management. As mentioned above, the regulatory function of economic law, which is not a panacea, is influenced by hysteresis, blindness and spontaneity of the market economy. In order to control nonstandard phenomena of management, such as local trade barrier and administrative monopoly, the government is obliged to build open and unified orders of market management. In addition, the government should contribute more to eliminating unfairness in market competition. Meanwhile, limitations of the market economy also require economic law as a state invention for remedy.

Market transactions should be intervened and regulated by civil law. That is because transactional behaviors depend on price mechanism, competition mechanism as well as supply and demand. When there is behavior against market economic criteria, civil law can play a role in safeguard and regulation without intervention of economic law. This is actually the stabilization of the market order.

4 CONCLUSIONS

In conclusion, economic law plays a key role in regulating the socialist market economy. Therefore, the government should make full use of economic law for a better function of macro-control. With the law involved with market management, the government can guide the healthy development of the socialist market economy within the legal framework.

REFERENCES

[1] Liu Yana. Analysis and Restruction on Adjustment Objects of Economic Law. Estate & Science Tribune, 2006(12): 64–65.
[2] Zhang Xuyong. On Independence of Economic Laws. Journal of Zhejiang Normal University (Social Sciences), 2007(02): 25–29.
[3] Wang Zaihang. Research on Effect of Economic Law in Modern Administration. Journal of Xuzhou Education College, 2008(04): 29–30.

Future Communication, Information and Computer Science – Zheng (Ed.)
© *2015 Taylor & Francis Group, London, 978-1-138-02653-7*

Financial security and improvement of financial law system in China

L. Pan
Huanghe S & T College, Zhengzhou, Henan, China

ABSTRACT: Financial security is an important factor affecting the stability of financial markets in China, while the financial law system is a general term for all the laws that manage financial relations. Therefore, only when the financial law system is improved can financial security be effectively protected. Currently, the financial security of China has many risks, and there're deficiencies in the financial law system. Thus this work creatively analyzed the establishment of a law system. It's hoped to strengthen legislative acts to restraint the behaviors threating financial security, ensuring financial security.

Keywords: financial security; financial law system; proposal

1 OVERVIEW AND RELATION OF FINANCIAL SECURITY AND FINANCIAL LAW SYSTEM

Financial security refers to the unrestricted circulation of funds in the economic market and stable operation of the financial system, which is very important to the entire financial market. The structure of the financial market has become increasingly diverse with the rapid and diversified development of world economy. However, the managers of markets and governments can't fully guarantee the perfection of every detail, causing increasing risks in financial security. Generally, financial security is closely integrated with financial risk and crisis: the three of them mutually interact with each other. The situation of financial security can be described and explained by the risk index and conditions of crises. In addition, it can be used to measure and determine the degree of financial risk and crisis. To a degree, the higher the financial safety factor is, the smaller the risk will be. And the probability of a crisis will be smaller [1].

Restrained and guided by financial law, the financial law system is a mode guaranteeing the optimal allocation, harmonious development and safe operation of resources in financial markets. In fact, the financial law system and financial market have never separately existed. Only the organic combination of the two can promote the improvement of economic society. Due to the lure of materials and money, people nevertheless only emphasize the significant role of markets in financial development regardless of the importance of the system in the current stage. The system has two major roles – incentive and constraint. In other words, it can not only combat the misconducts disrupting the order of financial markets but also promotes the progress of financial markets [2].

It's inevitable to generate deviations during the operation of markets. A perfect financial law system can promptly detect this phenomenon, investigating and correcting it based on relevant laws and regulations. Therefore, with the deterrence of law, the people who rightly carried out financial activities will be more careful about their business – the constraint of the system. On the other hand, several legal provisions will provide government supports for financial components in certain special cases. Consequently, financial markets can become better and more diverse, promoting the progress of the entire financial market – the incentive of the system [3].

The above analysis indicates that there is a prodigious link between the security of financial markets and the degree of perfection for a financial law system. Therefore, the construction and improvement of the financial law system should be vigorously conducted in China in order to ensure financial security.

2 ISSUES OF FINANCIAL SECURITY AND DEFECTS IN FINANCIAL LAW SYSTEM IN CURRENT CHINA

2.1 Issues of financial security

2.1.1 Illegal inflows and outflows of capital
The competitiveness of China has continued to improve in international financial markets through the import and export of capital. However, the ensuing problems are prominent. China, as a developing country, doesn't have a perfect enough construction of the financial law system and market system. As a result, lots of funds will secretly circulate at home and abroad without relevant approvals by the government

in order to escape high taxes and strict monitoring. These inflows and outflows of illicit funds disrupt the normal market order and reduce the effectiveness of the currency, which can easily lead to an economic bubble. Subsequently, financial risk and crisis can be induced. Meanwhile, the work of the supervision department is becoming tough. It's difficult to identify the source and use of funds once a financial accident occurred. This is not conducive to the timely and proper proposition of solutions. At present, the inflow and outflow channels of several funds are not very consistent with the relevant provisions of the financial market system. However, the financial law system didn't make specific legislation based on this issue. Consequently, these the illegal funds presumptuously circulate in the market with the vacancy of law. Thus a thorough financial law system should be constructed as soon as possible to prevent the breed and spread of the illegal use of funds.

2.1.2 *Instability of network finance*

Network finance has been formed with the development of the computer and network. Thereby people have realized the desire of completing financial activities without meeting. In fact, networks have freed people in different regions from traveling long distances for negotiations. A face-to-face communication can be achieved via online video calls and other advanced network means. What's more, the constraints of time have been eliminated. Participants at different times are able to have network meetings at a unified time, greatly improving the quality and efficiency of financial activities. However, networks are not completely secure. In recent years, there were many incidents in which several hackers attacked corporate websites and caused a significant loss of business. These facts have given rise to a great deal of vigilance: the force of network protection should be emphasized in network financial activities. Otherwise criminals can be provided with convenience, leading to the leakage of important data. Currently, network finance has rapidly developed in China, including a variety of online money transactions and banking. The payment system of online banking and the settlement system of the credit card have penetrated into every part of financial activities. However, the online financial technology of China is immature on account of a late start. Online financial activities are often attacked by hackers. Therefore, relevant supervision departments of online finance should be established in China to severely strike illegal activities and protect the safety of financial markets. At the same time, the software guaranteeing the security of network finance should be developed with great efforts to block pathways of invasion from hackers.

2.2 *Defects in financial law system*

2.2.1 *Problems of bank law system*

The bank law system is the core of the financial law system since most of the financial activities in China are conducted based on banking. Currently, the major problems of the bank law system in China are the ambiguous legislative basis of subject law and imperfect range of specification of currency law. Such deficiencies have led to many problems: a lack of guidance and restraint in international financial activities, the inconsistence of the currency exchange rate with market trends, non-compliance of various financial services and inadequate management of funds. The ambiguous legislative basis of the subject law is mainly manifested in a low legislative level. Thus there is not a valid legal provision to guide and restraint several special financial services of banks. At present, the currency law of China has not established legislation regarding the nationalization of monetary sovereignty. Therefore, the international financial security of China will inevitably be threatened due to the gradual internationalization of the RMB and the rising trend of the central parity rate of the RMB against the USD. In addition, big banks have been owned by the country for many years since the reform and opening up with little or no case of bank failure. However, some commercial banks have gradually participated into the banking system of China with the diversification of the economy. Some of them are owned by local governments, while some are stock systems or foreign banks. The national government provides little support for them, and customers seldom choose them on account of security. Thus the probability of bankruptcy faced by commercial banks is much higher than state-owned banks. China nonetheless doesn't have a legal system specifically dealing with bank failures. Once a bank is unable to sustain operations, the customers are bound to suffer significant losses without the guidance of law. Meanwhile, turmoil in the financial markets can be triggered.

2.2.2 *Problems of securities law system*

At present, the securities law system of China is based on stock exchange, which can meet the needs of social and financial development for now. However, it can't be ignored that the structure of the stock market is not perfect without counter and international markets. Several foreign companies are forbidden to be listed in China. As a result, China is not closely connected to the international securities markets. This shows that China should continue its efforts to strengthen the construction of enterprises. Thus they can have the strength to compete with domestic and foreign listed companies. In addition to the bonds issued by the central bank, all levels of governments and listed companies, there're a number of bonds privately issued by civil society organizations. Generally, the earnings of those illegal bonds are higher than that of other regular bonds. However, their safety factor is very low due to a lack of legal provisions regarding issuance, transaction and repayment. They are the bond markets that should be banned by law, but the reason of obstacles is a lack of the legislative system as well.

3 PROPOSALS FOR FINANCIAL SECURITY AND IMPROVEMENT OF FINANCIAL LAW SYSTEM IN CHINA

3.1 Establish relevant laws and regulations and implement strong supervision on inflows and outflows of cross-border capital

The reform and opening up, an established national policy, has promoted the diverse development of economic capital in China and the pace of foreign economic exchanges. Foreign capital importing into the Chinese market and domestic capital exporting out of China indicates a close contact between the domestic and international markets. In addition, it shows a great improvement of China utilizing foreign resources. In recent years, the global financial crisis caused by the U.S. subprime mortgage crisis has not completely dissipated. The economic growth of each country is still slowing down, and some countries even have negative growth. Financial security can't be guaranteed on account of turmoil in the world financial markets. Therefore, China should further strengthen the construction of a financial law system in this instability of the world's capital. The sources and normal use of capital inflows and outflows should be ensured to reduce the risk coefficient of financial security. Actually, strengthening the construction of the financial law system isn't easy but requires long-term legislative research and data collection and consulting to banks and companies with cross-border trades. Afterwards, the approximate range of legislation and regulations are determined based on these facts. And reducing and attacking cross-border transfers of cash capital should be emphasized in particular. Thus false trade practices and credit risk can be effectively reduced. After the legislation, a hearing should be held to absorb the views of all parties. What's more, a trial law should be launched to test its strength of guaranteeing financial security. It can be officially put into use after ensuring that there's no major defect.

3.2 Strengthen construction of bank law system

The bank law system has three aspects requiring construction and improvement in particular: the first is subject law. In order to better cope with financial risks and the impact of financial crisis on the financial security of banks, the law regarding cooperative banks should be improved as soon as possible. Accordingly, the most basic guidance can be provided for the cooperative ways, the holding patterns and dividend system of cooperative banks. Different law systems of loan origination are formulated according to the discrepancies between various enterprises. Thereby the development of large enterprises and the operation of small ones can be ensured. Legislation should be launched to guarantee the security of deposits from residents and enterprises and make the charge standard of banks' deposit services clear. Second, a new currency law should be established on the basis of the existing one. In accordance

with the internationalizing trend of the RMB, the guiding role of the new currency law for RMB-sovereign currency circulation is determined to safeguard the stability of the RMB exchange rate. Therefore, it's guaranteed that the circulation of the RMB in foreign countries won't affect domestic financial security. Third, the financial law regarding bank failures should be formulated to regulate the mechanism of banking financial institutions exiting the financial markets. The mechanism includes the required conditions of exiting, the exiting process and the policy of appeasement for customers after exiting. Sequentially, it's ensured that bank failures will not affect financial security or trigger turmoil in financial markets.

3.3 Proposal for improvement of securities law system

The securities market of China was established with the marketization of state-owned and collective enterprises. China relaxed the standard of issuing securities in order to support the development of these enterprises. Thus the issuers of securities are not just the central bank and local governments. Large state-owned or collective enterprises can also be financed through the issuance of securities, obtaining sufficient funds. However, there's no need to give concessions regarding securities since the market competitiveness of these enterprises has reached the desired expectation. Therefore, the present Securities Law should be revised under the premise of protecting the financial security of markets. Stock markets on different levels are uniformly adjusted and planned to create fairer development and competition of all the securities markets. In addition, bond markets should be unified: a systematic law regarding the conditions for issuing bonds, transaction principle and ways of repayment. Various acts of illegally issuing bonds should be cracked down, thereby inhibiting the flow of illicit funds in financial markets. In addition to formulating and improving various securities laws, securities standard should be further detailed to strengthen the enforceability of the law. Thus it's ensured that every detail in financial markets has a corresponding law to obey.

3.4 Promote development of supervision law system

The reason for the frequent accidents of financial security in China is that the financial supervision law system has many problems. It can't effectively manage and supervise financial markets and activities. Financial supervision law is a restraining system based on the nature of law. It specifically guides and manages financial activities and is directly responsible for legislative institutions and law enforcement agencies. Essentially different from some government policies, it's not affected by the political activities of the government. In fact, a comprehensive financial law system achieving the supervision and management of the entire financial market is not enough. Secondary and tertiary financial supervision law systems should be built according to different circumstances of each local financial market and social and economic development. Guided by the overall financial law system, they implement specific tasks of local supervision. Financial supervision institution is the authority taking financial supervision law system as the guiding ideology. To effectively implement laws, financial supervision institutions should improve the working quality and efficiency. Thereby the behavior disrupting financial markets can be eradicated to ensure financial security.

4 CONCLUSIONS

All in all, the financial industry is one of the industries with the fastest and most active development in today's society. Financial security is inseparable from financial law system. Although there have been many problems requiring solutions, actively exploring ways to improve the financial law system can efficiently protect financial security. Thus the rapid development of the entire economy and society can be promoted.

ACKNOWLEDGEMENT

The work was supported by the project of Huanghe S&T College in 2013: funding for empirical studies of the relation between capital market and economic growth in Henan Province. Project number: KYSK-2013-04.

REFERENCES

[1] Liu Mengyang, Liu Shaojun. Status, Problems and Improvement for Financial Law in China. *Journal of The Party School of C.P.C. Qingdao Municipal Committee*, 2012(05): 110–114.

[2] Yang Chunping. Financial Security and Improvement of Financial Law System in China. *Chinese Yearbook of Commercial Law*, 2008(00): 226–232.

[3] Zhang Wenjia. Brief Talk about Financial Security and Improvement of Financial Law System in China. *Legal System and Economy (late)*, 2014(01): 69–71.

Future Communication, Information and Computer Science – Zheng (Ed.)
© *2015 Taylor & Francis Group, London, 978-1-138-02653-7*

Sample handling method research in analytical chemistry

M. Li
Yancheng Vocational and Technical Institute of Health, Yancheng, Jiangsu, China

M. Wang
Yancheng Institue for Drug Control, Yancheng, Jiangsu, China

ABSTRACT: Analytical chemistry is broadly applied to many fields, to practical analytical process. The sample extraction shall be performed firstly, certain purification treatment is usually performed on samples in order to guarantee the accuracy of the analytical result, this work mainly performs analysis on the method of sample processing, and an in-depth research is performed on the application of the common chemical reaction method and instrumental analysis combined with the factors which affect sample handling in analytical chemistry based on the concept and characteristics of analytical chemistry, thus the selection of the handling method is obtained based on practical analysis requirements and the current equipments, and the analytical results shall meet the practical requirements.

Keywords: analytical chemistry; sample; handling method

1 INTRODUCTION

With the development of modern natural science, many new subjects appear based on the requirements of practical application. Chemistry is an important one, which is extensively applied in material research and development, analysis and handling. Chemistry itself has become very complete after years of application, chemistry as a subject has been set in many universities/colleges currently, and lots of chemistry talents are trained [1]. However, the economy and science in China started late due to special historical factors, thus a certain gap exists in the application of chemistry compared with some western developed countries; even though China has become the world's second largest economy after 30 years of development by the reform and opening-up, the overall technical level has been improved greatly, and chemistry is broadly applied in practical production; through practical investigation, China mostly adopts traditional analytical methods in practical sample handling, which influences the application of analytical chemistry in China to a large extent [2].

2 BRIEF INTRODUCTION OF ANALYTICAL CHEMISTRY

2.1 *Concept of analytical chemistry*

As an analysis of chemistry, analytical chemistry is a subject which is gradually formed with the application of chemistry, because chemistry mainly aims at the handling of elementary substance and compounds, corresponding chemical reaction is used in practical application process to reach the specific purpose, but before the practical chemical treatment, the components of the material should be firstly analyzed, the targeted selection of reaction environment and formula can be performed only when the components of material are understood in detail. The appearance of analytical chemistry is mainly for confirming the components of material, a part of material will be usually extracted for an experiment in the practical work process, the advanced chemical equipment and technology is used to detect the components of the material. Analytical chemistry is applied to many fields, considering the importance of analytical chemistry, the corresponding courses are set in many universities/colleges, and lots of analytical chemistry talents are trained [3]. In current sample testing, analytical chemistry will be mostly adopted to perform treatment, in the initial stage of chemistry development; some traditional methods can only be adopted due to the limitation of the technical level at that time, for example, simple chemical reaction is applied to perform a test on the components of the material; with the development of electronic information technology, many automatic chemical testing equipments appear currently, the equipment can automatically analyze the component of the sample, which greatly improves the efficiency of sample handling [4].

2.2 Development of analytical chemistry

The application of chemistry can be traced back a long way, but was limited by the technical level at that time. The recognition of an element is limited, only simple application can be performed by means of some phenomenon of chemical reaction; the development of modern nature science and the appearance of the periodic table of elements greatly improved the development of chemistry itself, the research and development of the new materials can not only be performed by means of current chemical knowledge, and the unknown material composition can also be analyzed. In the early stage of analytical chemistry, some existing materials could only be applied to the practical analysis due to the low technical level, the reaction and product could only be observed by means of simple chemical reaction, the analysis on the composition of a sample was affected by the factors of many similar chemical reaction phenomena, the effect of this handling method was worse, and some reaction needed a long waiting time, and the handling efficiency was lower, too. With the development of electronic information technology in these years, the electronic equipment is popularly applied. People implant the intelligentized chips to the corresponding equipment, which can be operated automatically. According to the established procedure, people introduce electronic equipment to the chemical field, and lots of chemical equipments are produced, for example, in analytical chemistry, lots of analytical equipments appear on the market. The equipment can automatically perform treatment on a sample only when the corresponding sample is implanted to the equipment, and then the analytical result is obtained.

3 FACTORS OF AFFECTING SAMPLE HANDLING METHOD IN ANALYTICAL CHEMISTRY

3.1 Professional quality of the work staff

In traditional analytical chemistry, staff are mainly adopted to operate, and is required at each link to control it, therefore, the effect of sample handling will be affected by the quality of the staff. If the quality of the work staff is lower, the reaction process can't be favorably controlled, causing the inaccuracy of a sample analytical result; even though lots of automatic electronic equipments will be adopted in the sample handling of analytical chemistry after years of development and the function of staff has become weak, the electronic equipments need the staff to operate them. If the work staff has no sufficient knowledge of these equipments, the operation capacity is obviously weak, generating the worse effect of sample handling, especially the development speed of electronic information technology is fast, the upgrading speed of the corresponding chemical equipments is quick, each technical upgrade will improve the efficiency of sample handling to some

degree, the corresponding equipments must be constantly replaced to guarantee the effect and efficiency of sample handling in the practical application of analytical chemistry. The continuous replacement of equipments puts forward higher requirements on the quality of the operation staff; but through the practical survey, students can excellently master the relevant theoretical knowledge of analytical chemistry by means of learning in universities/colleges due to the limitation of our educational level, the practical ability is weak, students can hardly complete corresponding work content after graduation, which affects the sample treatment effect in analytical chemistry to a large extent.

3.2 Handling methods

In practical application of analytical chemistry, different handling methods will be adopted after the corresponding samples are extracted. Certain differences exists in the handling efficiency and effect; currently, many kinds of handling methods can be chosen after years of application, such as the traditional chemical reaction method. Some known substance is used to react with the samples, observe the reaction phenomenon, and then confirm the compositions of the samples. The same handling method will be adopted in currently many field tests. If the composition of a sample is very complicated, or specific data is required, the test needs to be performed in a lab; usually, some advanced analytical instruments will be arranged in the lab, which can be used to quickly finish tests and obtain the analytical data of the sample, but the test in a lab needs some certain cost, which will increase the cost of the sample handling. It's limited by the Chinese technical level, the traditional chemical reaction method will be mostly adopted in the practical application of analytical chemistry; in recent years, even though many professional labs have been constructed with the development of the Chinese economy, and foreign advanced analytical instruments and technology are introduced, considering the analytical cost, the handling methods application of the chemical instrument is less, the equipments in the lab will only be used to perform corresponding treatment on samples in some important fields.

3.3 Advancement of chemical instrument

In current applications of analytical chemistry, the corresponding instruments will be mostly adopted to handle samples. The handling effect can be greatly improved by using chemical instruments, which have a very important function in practical sample handling, especially with the development of the electronic information technology in recent years, many automatic detecting instruments appear, the analytical results can be quickly obtained only if the samples are placed in the corresponding equipments. However, though the practical investigation, is limited by the Chinese technical level, most of the chemical instruments in the current market are produced by the foreign companies, we must introduce them from foreign countries for use, which needs higher cost, and this phenomenon influences the effect of sample handling in Chinese analytical chemistry to a large extent, China has invested lots of human power and material resources in these years to perform research on chemical instruments in order to better solve this problem, but no obvious effects are obtained. So we can see the importance of chemical instruments on sample handling, advanced chemical instruments must be adopted to guarantee the handling efficiency, in the practical selection process, an optimum instrument can be selected pertinently according to the sample conditions and analytical requirements, thus the requirements of sample handling can be satisfied, and the cost can be controlled to some degree.

4 SAMPLE HANDLING METHODS IN ANALYTICAL CHEMISTRY

4.1 Pretreatment of samples

Samples are usually extracted from the raw materials, some impurities will usually be mixed in the practical extraction process due to the limitation of the technical level, which will affect the analytical result of samples; meanwhile the sample itself may contain some sundries, certain pretreatment can be performed before the practical sample analysis in order to better solve this problem, and the purity degree of samples can be improved by means of simple extraction and chemical reaction, thus the accuracy of sample analytical results can be guaranteed. In practical sample treatment, if the on-the-spot chemical reaction is adopted to perform a qualitative test on samples, a simple pretreatment is also needed, for example, the samples can be divided into several parts, the part with more impurities shall be disposed of by visual observation, and then the specific chemical reaction will be performed, thus the effect of sample treatment can be obviously improved. If the chemical instruments in labs are selected to perform treatment on samples, the pretreatment is more important, it has strong pertinence due to the application of chemical instruments, for example, certain instruments can only perform analysis on the same kind of chemical substance, if the sample contains excessive impurities, the analytical results will be severely affected, therefore, in practical sample treatment, corresponding pretreatment technology must be adopted to eradicate the impurities in samples to a large extent, thus the accuracy of analytical results can be guaranteed.

4.2 Chemical reaction method

In many construction sites and other fields, samples need to be analyzed, a short time is usually required

in this kind of sample treatment, and the requirements on analytical results are not high, and then the simple chemical reaction method can be adopted, some existing chemical reagents can be used to make some simple chemical experiments on the spot and confirm the composition and chemical property of the sample, because this kind of treatment method is very easy and low cost is needed, therefore, it's broadly applied to many fields. However, the obtained data precision is low by means of this rough treatment method, so the use of the chemical reaction method has great limitations; the simple chemical reaction can't achieve the corresponding target for the treatment of complicated samples, the advanced analytical instruments can only be adopted to perform in-depth analysis, it's just because of this characteristics of the chemical reaction method, the treatment result of samples in practical application process can only be used as reference. Chemical reaction method is a good choice for some substance to be qualitatively tested on the spot, which can not only save corresponding costs, but also quickly obtain results, especially in the construction site of some projects, the material property is usually to be measured roughly, because materials have been transported to the site, which need to be used quickly, therefore, the time of material testing must be short, and chemical reaction method just meet the requirements.

4.3 Chemical instruments analysis method

With the development of electronic information technology, people add an intelligentized chip on the basis of traditional equipment, the chip can control the equipment only when the corresponding procedure is put in, the content of the procedure can be automatically performed according to the content of the procedure set, lots of chemical instruments appear in the field of analytical chemistry, in practical sample treatment, if these instruments are adopted, the treatment result can be quickly obtained only by adopting these instruments to place the sample into the instrument, and the precision is high. It's just because of this characteristic of the chemical instrument, which is popularly applied quickly, but through the practical investigation, the application of precise chemical instruments require a strict environment, therefore, in the actual application process, a lab must be established to place chemical instruments in the lab; the specific sample treatment shall be performed in the lab environment, and the establishment of labs need some certain costs, the corresponding sample treatment also needs some certain expense. We can see from this, even though the chemical instrument analysis method is more precise in the result of sample treatment, the corresponding cost is high, too; in the practical sample treatment, if there are special requirements, the chemical instrument analysis method is seldom adopted due to the cost; the sample will be sent to the specific lab for corresponding treatment only when some precise sample data is required.

5 CONCLUSIONS

We can know from the analysis of this paper that analytical chemistry is broadly applied to practical production. The sample composition can be quickly known by means of analytical chemistry technology so as to confirm various natures of samples, which have a very important function in the use and authentication of materials, but limited by the technical level, some impurities are usually mixed in the practical sample extraction process. Sometimes problems exist in the sample itself, which influences the analytical result to some degree. Under this background, how to perform treatment on the extracted samples becomes a critical problem. The current sample pretreatment technology has become very mature after years of application, several kinds of pretreatment methods appear, in practical sample treatment process, the optimum treatment method can be selected pertinently to eradicate the impurities in samples according to the current conditions combined with the actual situations of the samples, thus guaranteeing the accuracy of analytical results to a large extent.

REFERENCES

[1] Shao Hongfei, Research Progress of Analytical Chemistry Sample Pretreatment Technology, Chemical Analysis and Meterage, 2007(05): 81–83.

[2] Xing Yun, Wang Guifang, Li Quanmin Solvent Extraction Technology in Analytical Chemistry, Physical Testing and Chemical Analysis Part, 2005(09): 694–696.

[3] Jiang Pengzhong, a Kind of New-type Sample Pretreatment Technology Discussion—Solid Phase Microextraction Technology, Modern Commerce and Trade, 2008(09): 375–377.

[4] Jianhuan Wei, Huafeng Sun, Shengjie Yang. The cycle measurement of sulfur blank value with the CS-444 infrared ray carbon sulfur analyzer. Advanced Materials Research, 2012: 2173–2176.

Future Communication, Information and Computer Science – Zheng (Ed.)
© *2015 Taylor & Francis Group, London, 978-1-138-02653-7*

Analysis of effective management of data calls based on database

J. Jiang

Modern Education Center, Zhengzhou Huaxin University, Zhengzhou, Henan, China

ABSTRACT: As a new data management mode, a database increases the efficiency of the data storage and data call. In the work, the effective management of data calls is analyzed based on the concept and characteristic of the database. Combined with the factors influencing the effective management of data calls, it is found that only by improving the quality of the operation staff and the performance of the hardware storage and establishing the index can it ensure the effectiveness of data calls based on database. While selecting the actual hardware storage devices, it is not necessary to select those advanced as long as they can meet the actual requirement.

Keywords: computer; database; data call; framework

1   INTRODUCTION

With the development of the application of the computer, the related hardware and software technology have been improved a lot. The handled tasks are more and more complicated, and the storage data are becoming more and more. In this context, in order to improve the efficiency of data management, database technology is developed [1]. The stored information on a computer is managed by a database system, which can not only increase the data storage, but also make data calls more simple and convenient. It is important for the application of computers. Science and technology started late in China because of special historical factors, and the application of computers falls behind compared to western developed countries. In recent years, a lot of manpower and material resources have been invested and research has been done on the technologies like database in China, improving the level of database to a certain extent [2].

2   BRIEF TO DATABASE

2.1   *Concept of database*

Database is a technology along with the application of computers. Computers, which are not as good as electronic calculators at present, can only deal with some simple tasks limited by the technology at that time early in the development of computers. The performance of the computer has been improved a lot and the stored data have become more and more with the application of transistors and integrated circuits. H-languages like the C language have been invented and a lot of software with different functions, which can be applied to different fields, has been developed based

on computer languages. The concept of database is developed on the basis of the computer software. For the convenience of the data storage and data call on the computer, the corresponding software like the operation system has been developed and all the data have been stored according to certain types and features. Therefore this is the concept of a database [3]. There are many database systems on the market, however, the actual investigation shows that the systems are mainly researched and developed by software companies from abroad, and few database systems are researched and developed in China. It is indicated that the research level of databases in China is low. The database is taken more and more seriously with the popularity of computers. Some preferential policies are proposed to encourage and support the development of software companies in China, and several data storage centers are built based on the actual condition of our country. Although it costs a lot, it extremely increases the application of the database.

2.2   *Characteristics of database*

The storage and call mode of a database have distinct characteristics compared with the traditional data management mode. Firstly, the data are stored on a hard disk with a level of TB, occupying a little space. If the data are stored by using paper, they will occupy lots of space, and they are easy to lose because of problems like fire, water, etc. in the process of storage. The storage mode of a database can solve the problems, especially, the hard disks of high performance such as the waterproof, the shockproof, etc. which occur. The data storage becomes simple by using the database management mode. The computers are used to build a LAN. If the client has enough authority, it will manage the database and store the related data without the limit of time and address. The storage of the data is very simple and convenient. The database management extremely increases the efficiency of the data storage and makes the data call simpler. In traditional data management, it takes a long time to search the needed data especially in large places like the library. If the managerial staffs lack the understanding of books, they have to search the books one by one. However, the management mode of a database can solve this problem. If the related keywords are loaded into the advanced computer, the required data will be found soon.

2.3 Development of database

According to the actual requirement, some corparate users build some data service centers. It is a large database which can manage the data storage and the data call, playing a very important role in the daily operation of the company. Early in the occurrence of the database, few data were stored and the database technology level was low limited by the hardware storage equipments. The computer was not popular with its high cost, making the application of the database little. In this context, only in some special fields can the database be used to manage the stored information. In addition, a perfect database system has not formed yet. With more and more data stored in the computer, a large number of application software put forward high requirements to the database. The operation mode of the data must be improved to satisfy the requirements of the actual data management. And the database gets a fast development out of the actual application. There are many kinds of database technologies on the market, and different database systems are developed according to the requirements of different fields. The database has been perfect for several years' application. Almost every computer can adopt the some database technology. According to the actual requirement, some corporate users build some data service center. It is a large database which can manage the data storage and the data call, playing a very important role in the daily operation of the company.

3 IMPACT FACTORS OF DATA CALLS IN DATABASE

3.1 Performance of hardware equipments

The performance of the computer hardware equipments has been improved a lot after years of use. Moore law shows that: every 18 months, the performance of the computer hardware doubles while the cost is constant. And in a sense, the equipments (e.g. the computer storage, etc.) follow this law. So hardware storage equipments are quickly replaced by the new ones. Early in the development of the computer, the storage equipment stays in level MB. However, after decades of development, the storage equipments on the market almost reach level TB. Besides the increase of

the storage space, the speed of storing and reading data has been improved a lot. In order to increase the storing speed of the traditional hard disk, etc., the cache has been added to the hard disk. Stored and called data pass through the cache, improving the efficiency of data management. In the actual building of the database, the performance of the hardware equipments can affect the working of the database to a large extent. The investigation shows that the performance of the hardware storage equipments is poor out of the cost control in the building of a database in China. As they (e.g. hard disk, etc.) cannot be produced in China, the storage equipments must be imported from foreign countries in the actual building of a database, which affects the application of database technology in China to a large extent.

3.2 Mode of data search

If the storing information needs calling, the corresponding search system will be used. For searching data conveniently, the computer nowadays has the corresponding search function. When the dependent keywords are put in, the computer will match the dependent contents automatically. In the process of searching, the computer will contrast the input keywords with all the information in the database. If both of them are the same, then the information will be the information searched for. The search function has been improved a lot over several years, and the fuzzy search has been added. When some keywords are put in, more information will be searched. In a traditional database system, the stored information is so little that the computer can finish searching quickly by contrasting the information of the database one by one. However, the stored data are getting more and more along with the application of the database. Especially, the large data storage center occurs. It takes a long time to match them one by one among plenty of information. So, in data call of databases, when some common keywords are put in, lots of data will be found. And a long time of waiting will affect the use of the database. In the context, in order to improve the condition of data call in the database, the present search mode and the efficiency of searching will be improved by using some new technology and methods.

4 EFFECTIVE MANAGEMENT OF DATA CALL OF DATABASE

4.1 Improvement of quality of the manager

The computer should be operated by the actual managerial staff. In special, both data storage and data call should be operated by the staff. And the quality of the staff can affect the working condition of the database to a large extent. The staff with a poor professional quality cannot perform the operation in data storage and data call. The investigation shows:

the education in China stays on a low level; after graduating from college, the students are good at theory but bad at practice; it takes a period of time to learn and then the work can be finished well. Considering the database technology is quickly replaced by the new, the databases have large differences in different fields. In order to operate the database, it should be understood. And to solve the problem, the quality of the managers must be improved. In the staff recruitment, the corresponding standard should be increased to employ some professional staffmembers. If allowed, the staffmembers with abundant experiences will be employed to improve the working efficiency of the database. For the present staff, the ability of practice can be improved to the largest extent by training.

4.2 Improvement of performance of the equipment

It is considered that the storage equipments are important to the performance of the database. So advanced hardware equipments can be used in database building. Considering that the hardware storage equipments cannot be produced in China, the advanced equipments must be bought abroad, costing much money. Therefore the hardware storage equipments which can satisfy the actual requirement will be selected according to the need of data management, the strength of the company, etc. The relationship between the cost and the performance of the hardware equipments can be balanced very well. The investigation shows that the database system has been applied and popularized at the present. Many science and technology companies have developed several database systems to satisfy the requirement of different fields. To improve the performance of the database, the science and technology companies are constantly upgrading. And the replacement of the software technology in the database system puts forward certain requirements to the hardware storage equipments. The performance of the hardware cannot meet the requirement of the software, resulting in some functions unrealized and affecting the application of the database. In the context, some advanced hardware equipments should be used in the data building so that the hardware can meet the requirement of the actual application even if the software of the database system is upgrading in a coming period of time. So it is very important for data storage and data call in the database.

4.3 Building index

Considering more and more stored information in the database, the traditional one-by-one retrieval method cannot meet the actual application. In order to ensure the effectiveness of the data call in the database, the present retrieval method must be improved. In this context, the data in databases are classified and one index is built in each classification. In the process of retrieval, if the keywords are put in, they will match the index to confirm the classification of the corresponding index and contrast them with the concrete contents, extremely raising the efficiency of retrieval. Nowadays the above method has been used to increase the effectiveness of data call in many data systems. However, in the actual index building, there is not an agreed classification standard at present as the index has just been applied for a short time. Different databases uses different standards while building the index, making the time for building the index and the size of it is different. The index extremely increases the effectiveness of the data call in the database, especially for fuzzy retrieval. In traditional retrieval methods, each of the information in the data should be contrasted with the found dependent data. And the building of the index is on the basis of the given classification. So in fuzzy retrieval, the contents under the index are directly listed. The above shows that the index has made a great progress in the improvement of the effectiveness of the data call in the database.

5 CONCLUSION

The analysis of the whole work indicates that the importance of the database is more and more obvious with the popular application of the computer and the internet. However, limited by the technology level of our country, the required hardware storage equipments are introduced from foreign countries in the database building. The database systems on the market are mainly developed by software companies abroad, affecting the application of the database in China to a large extent. The database is mainly used to manage the data, and the data storage is for data call to a large extent. So the efficiency of the data call in the database is an important index to measure the performance of the database. The efficiency of the data call is low, affected by the factors of the performance of the hardware storage equipments, the efficiency of retrieval, etc. in data using. The problem has been researched and improved for several years, but it has not been solved completely.

REFERENCES

[1] Huang Shiquan. Analysis and Realize based on Network Data Back-up. Journal of Yangtze University (science edition) Volume of science and technology, 2009(02): 261–262.
[2] Yang Xiaoyan, Yin Ming, Dai Xuefeng. Research on Query Optimization of Oracle Database. Computer and modernization, 2008(04): 4–7.
[3] Gao Pan, Shi Weiran. Optimization of SQL Based on Oracle. Programming technique and maintenance of computer, 2010(22): 38–39.

Future Communication, Information and Computer Science – Zheng (Ed.)
© 2015 Taylor & Francis Group, London, 978-1-138-02653-7

Analysis on risk prediction of enterprise marketing

M. Lui
Shaanxi Institute of International Trade & Commerce, Shaanxi, China

ABSTRACT: As the link of the realization of enterprise profit, marketing activity is the finishing point and foothold of the production and management process of an enterprise. In contemporary marketing and management activities, the existence and harmfulness of marketing risks are ignored categorically. Meanwhile, the predictions of various kinds of risks still lack technical methods. For this reason, according to fundamental theories of marketing, this study, combining the economic analysis method and the latest achievements of industrial organization (regarded as innovation), discusses the sources of marketing risks and corresponding prediction methods.

Keywords: marketing; risk prediction; market environment

The basic contents of marketing activities include the market fixed price, product design and promotion, building of sales channels and sales promotion influencing consuming behavior. With the gradual development of the socialist market economy, operating personnel in China have gained increasing awareness that the marketing risk of an enterprise plays an important role in enterprise risks [1]. This study explored the sources of marketing risk by analyzing several basic aspects of marketing activities and proposed some technical methods of risk prediction based on this analysis.

1 FOUNDATION OF MARKETING RISK PREDICTION OF ENTERPRISES

Marketing risk prediction first needs to define the sources of marketing risk so as to know the objects and key links of enterprise risk prediction, which is the foundation and premise of marketing risk prediction of enterprises.

Firstly, with the continuous deepening development of the market economy and the gradual advancement of national upgrade industries, the risk of the intrinsic pricing model has already been the direct source of enterprise marketing risk. The source of the pricing risk refers to the possibility that the losses brought to enterprises caused by mismatching the fixed price and marketing environment of enterprises. Along with the continuous development of the market economy, the market competition is increasingly white-hot and potential competitors are becoming the important sources of competition pressure of enterprises. Accompanied by the gradual liberalization of securities market and monetary market, it can be easier

for enterprises to get funds and upgrade their production technology. Meanwhile, foreign commercial enterprises can rapidly enter a certain market and change the intrinsic supply and demand structure of this industry, which brings more risks to relevant enterprises. As China has already established the status of a big manufacturing country, many Chinese enterprises have to encounter the pricing risk brought by the sudden entering of foreign enterprises. For the low-end product market, the sudden entering of foreign enterprises may cause great losses of clients for Chinese enterprises in a short time, which undoubtedly poses a great threat to the steady development of enterprises in the long run [2].

Secondly, in China, with the improvement of the living standard, the consumers' demands for the function of an end-product tend to be differential and heterogeneous [3]. However, during the process of product update and promotion, enterprises inevitably have to face the possible risk of money loss caused by R&D failure, which is a barrier for the development of many enterprises. For those high-end enterprises in industry chains, R&D risk brought by advanced technology can hardly be avoided. Enterprises can only develop products in the fumble because no precedent of product technology can be followed.

Thirdly, risk factors demonstrated in market distribution activities are becoming the important sources of risk in the process of producing and managing of an enterprise. Market distribution activity is an important intermediate link for an enterprise to promote its products to the consumer market. In fact, the risk of distribution activity is hard to perceive directly because it often relates to internal management and financial domain beyond the products flow of enterprises. The capital chain of downstream firms is a vital factor influencing the smooth products sales and capital turnover of upstream enterprises. However, it is hard for enterprises to capture relevant information, which can bring some managing risks and possibility of potential losses.

Finally, sales promotion, under the economic condition of the modern market, is an important method for enterprises to broaden product marketing and open up the market. However, due to the great enhancement of information dissemination brought by the development of mass media, sales promotion environment of enterprises may tend to be complicated. If sales promotion methods like advertisement are improperly used, product marketing of enterprises may shrink, which is related to not only the customers' subjective sensation towards advertisement propaganda, but also some factors like facticity and acceptability of information dissemination.

In a word, marketing risk widely exists among different links of marketing. The analysis of the sources of marketing risk recounts the objects of risk prediction, which can make the marketing work shoot the arrow at the target and thus lay a foundation for the risk prediction of marketing.

2 STRENGTHENING THE PREDICTION ANALYSIS ON THE CHANGES OF CONSUMPTION PREFERENCE

The main risk of R&D and promotion of products is that products may not satisfy the needs of the market, thus leading to capital waste. It is the vital link of marketing risk to strengthen the prediction evaluation on consumption preference and the judgment of a possible new segment market. The judgment of the changes of consumption preference can rely on the prediction through statistical measures like trend extension and seasonal adjustment of existing sales. Meanwhile, the judgment of a possible new segment market can be conducted by consumption investigation. Moreover, with the development of modern electronic and information technology, electronic business system can offer a convenient wayside for enterprises to judge the changes of consumption information.

Firstly, in order to judge the changing situation of consumers' demand information, enterprises can build a seasonal adjustment model or trend extension model based on their sales data to analyze the historical data. The influence of seasonal changes on consumption preference shows more about the impact of the changing natural environment on the variation of consumption preference while trend extension model reflects the influence of human factors like social economic development on the changing trend of consumption preference. The combination of the trend extension model and seasonal adjustment model can demonstrate both the seasonal factors and the changing situation of the consumption demand.

Secondly, in order to judge the changes of the segment market, enterprises can investigate relevant market information by connecting the electronic business system to the demand management system. As more and more consumers have begun to use an online payment platform for their purchasing, judging whether the consumer demands have similarity or not equals to judging whether the consumption choices have certain concentration distribution or not. It is an important way to judge relevant consumer demands by giving out questionnaires in the purchasing process to let consumers express their willingness with the simple choice that they hope to get relevant products.

Finally, the judgment of changes of consumption preference and consumer market should be transformed into product strategy of enterprises immediately, which can avoid the R&D deviating from market demand, wasting R&D resources and capitals. To share and integrate the management information system and electronic business system is an important technological support to solve this problem, which can also be beneficial to flatting management of enterprises.

3 ANTICIPATING THE ENTERING TREND OF OUTER-RING ENTERPRISES

The price fluctuation caused by the sudden entry of outer-ring enterprises can change the structure of supply and demand in the market and the price anticipation of supply and demand in a short time. Therefore, there can be a market pricing risk, and to avoid such risk, enterprises should have appropriate anticipation on the entry trend of outer-ring counterparts and capitals.

Firstly, enterprises should have careful analysis and judgment on the flowing trend of intermediate products in China. Intermediate products often serve as element functions in the production process. The changes of intermediate products indicate the upcoming change of the structure of supply and demand in a certain industry. If outer-ring enterprises want to enter this industry, the demand for intermediate products will largely increase, leading to the larger demand and higher price for upstream enterprises of intermediate products, which brings about more production of intermediate products. The changing flow scale of intermediate products reflects the potential trend of competitors' entering the market. Paying close attention to such information can offer an important channel for the enterprises' conduct risk prediction. With the gradual development of the socialist market economy, the restriction upon the flow of production among different regions is gradually reducing. Enterprises can hardly detect their potential competitors throughout the country. However, through the flow direction changes of intermediate products of upstream industries, enterprises can timely know the development trend of potential competitors, and thus prepare coping measures against price fluctuation.

Secondly, enterprises should conduct effective analysis on their structures of cost – income to judge the influence of the market entry of outer-ring enterprises. Meanwhile, enterprises should not adopt the strategy of limiting price to resist the entry of outer-ring counterparts, which can cause damages to their long-term benefits. It is especially important for those enterprises that want to change their main business. If the duration of existence of an enterprise in a certain industry can be predicted, then according to the Chain of Reverse Reasoning, the preventing strategy can not bring benefits to enterprises. Likewise, to judge whether the counterpart has cost advantage over the basis of its technology is also an important anticipation work for coping measures. If the counterpart enterprise has no cost advantage, the strategy of limiting price can cause a cycle of price competition, which brings about the Edgeworth cycle. This is very important for those enterprises whose production capability can not cover the whole industry.

4 STRENGTHENING RISK EVALUATION BY COOPERATING WITH BANKS

One important part of enterprise channel management is to select retail traders or suppliers. For enterprise managers, one of the biggest risks in the distribution process are unmarketable products and delayed payment caused by bad business conditions of inferior retail traders, leading to the loss of capital time value or the risk loss of capital turnover failure. As inferior retail traders are not the subsections of an enterprise, it is hard to get their business information, which is a hinder for the risk prediction.

For enterprises, one tool that can be used to conduct risk evaluation on retail traders is the credit information service of banks. Compared with other information survey, the credit information service is featured by short-term, high efficiency and rare interference. According to the law of China, the cash deal among different enterprises often needs to be achieved through the transfer of bank accounts. Getting some information about the status of enterprise capital is very easy for banks. Moreover, the banks' judgment and knowledge about customer information have the character of timeliness; so compared with common enterprises, banks are more sensitive and accurate to learn about the financial risks of their customers. This is because the credit behaviors are related to the analysis and judgment of the customers' credit status and the judgment of the customers' risk is an important premise to guarantee the security of loans. Therefore, compared with other investigation agencies, banks can get more detailed and accurate information. At the same time, it is necessary for listed companies to disclose their account and financial information; with the support and approval of relevant laws, knowing about such information, will not arouse legal disputes.

By prediction, one important method to reduce the risks brought by retail traders is to change the model of information dissemination between suppliers and retail traders. There was often some information blockade between suppliers and retail traders because the information dissemination of supply and demand would influence the market power of enterprises and even lead to mutual competition among suppliers and retail traders, which goes against the stability of supply chains and the long-term business of enterprises. To achieve information sharing and reduce the risk of information leakage, it is a good strategy to establish an information sharing platform between suppliers and retail traders. The platform itself is an information carrier to conduct risk prediction; it also shows the internal requirements of cost factors in the risk prediction process. Meanwhile, the downstream enterprises can offer some suggestions on promotion strategies. For those large or medium-sized enterprises with a wider market coverage, the market response to promotion methods like advertisement can be hardly gotten in a short time, so downstream retail traders should shoulder the responsibility to feedback information on the information sharing platform.

Banks can also play a role of information calibration in the establishment of the information sharing platform between suppliers and retail traders. The financial information offered by banks can serve as the inspection mechanism on the accuracy of information disseminated by the platform, which in fact is a direct reflection that financial management of enterprises smoothly connects to the marketing activities.

5 CONCLUSIONS

The marketing risk is an important factor that needs to be considered for the enterprises in making the strategies of price, products and distribution. Marketing risk derives from market environment of enterprises, which is the reflection of possible changes of the structure of supply and demand. To lower the difficulty of management and reduce marketing risk, enterprises, according to the Theory of Industrial Organization, should get relevant information about supply and demand as well as the development trend of potential competitors through modern economic business management; they also need to get financial information of downstream retail traders through modern financial statistical resources to select agencies and retail traders, reducing the sales risk of enterprises. Thus it can be seen that the prediction of marketing risk is an organic integration of internal and external resources of enterprises, which needs the cooperation of all managing personnel of enterprises.

REFERENCES

[1] Chen Shuifen, Wu Shiqi, *Current Network Marketing Risks and Avoidance Measures.* Modern Marketing, 2011, (6):51–53.
[2] Zhang MaoLei, Zheng Yingqun, *How to Strengthen the Control of Enterprise Marketing Risks.* Shandong Coal Science and Technology, 2012, (4):303–304.
[3] Wang Hongchuan, *Strategic Analysis of Marketing Risk Transfer control of Enterprise.* New Finance Economics, 2012, (2):115.

Future Communication, Information and Computer Science – Zheng (Ed.)
© 2015 Taylor & Francis Group, London, 978-1-138-02653-7

Influence of mental health education on students' values

F. Fang
School of Marxism, Wuhan Textile University, Wuhan, Hubei, China

ABSTRACT: Mental health and values formation of students have long been the focuses of quality education. However, it requires families and parents to reexamine the traditional education of mental health for recent accidents on students, caused by the social background and the general misunderstanding of mental health. With multi-angle cooperation and civil participation in this field, this study proposed a series of solutions and improvements for the new challenges of the present students' physical and psychological health, as well as the related future development of society.

Keywords: mental health; values; student

1 INTRODUCTION

With the popularization of compulsory education, growth problems of student groups are always attracting close attention of the society. From any perspective of parents, teachers or common members of society, students' mental health and values orientation quite reflect social morality and future development of China. However, recent hot news reported a poor performance of students in this field, indicating the demand of more attention and constructive improvements from the public [1].

2 PROBLEMS OF STUDENTS' MENTAL HEALTH AND VALUES ORIENTATION

2.1 More negative factors affecting mental health in the new era

Physical and mental environments around students now have greatly changed, compared with those of several decades ago. While adults born in the 70s or 80s are usually behind the frequent changes and update of social topics and new products, students born in the 90s or even 00s easily absorb these products of times and social development and fairly reflect the influence in their behavior and values [2].

Instead of marbles and sandbags, current students are more interested in convenient social software based on the Internet and online games with peculiar experience. In the special growth period, both psychology and physiology of students are facing great changes, while the plural era, with the indispensable existence of internet, usually brings unavoidable influence to students [3]. The situation, negative for physical and mental development, leads to students'

numerous tragic examples, such as addiction to online games, violence in reality and being housebound to the Internet with lack of actual communication ability.

Additionally, stress from parents, schools and society has made students breathless. Students' mental health is overshadowed by the greatly enlarged public pressure of social development on the students' immature mind. High quality of life and advanced material guarantee are misrepresented by the society as heavy loads of students. The entire situation leads to present mournful and shocking consequences. Recently, a sixth-grade student in Chuzhou jumped down from the sixteenth floor after leaving a suicide note. And what was the exact reason for the painful choice of such a teenage boy? His note said that "I hate school" and "I hope dad and mum to live strong without me". With a feeble explanation for the lost life, the ending is so terrible that it is urgent to find the source for the prevention of such tragedies.

2.2 Increasing elusion and fear of mental health problems

While these shocking and mournful incidents highlight the students' mental problems, some changes of figures also bring the sense of crisis and urgency to the public. In the middle of the 80s, the number of students with mental problems accounted for 15.7%, coming up to 20% in the 90s. The recent resurvey by the China Youth Daily showed that about 40% of the students had short-term or long-term mental problems during the period of school, and without any effective consultation or assistance from teachers and parents, most of them chose silence. The survey indicated the fact of increasing student groups with mental problems, the elusive preference of mental problems and even the fear and exclusion of analyzing one's own problems.

Mental problems, with specific concealment and latency, reduce the alertness of the public. However, the unimpressive symptoms and thoughts usually become the source of mental problems in a students' period of rebellion and stress. In a survey of one middle school in Langfang of Herbei, for the survey question – do you think you have mental problems, only 12% of grade three students indirectly or passively admitted their mental problems, while they refused to speak out or seek assistance from people around them. And more students, in consideration of face, just said to be in "perfect" condition without any mental problems. With limited awareness of mental problems and ignorance of small details, students usually don't realize the distortion of values and the danger in the future brought by the influence of these negative mentality and emotions. Elusion and untimely solution of mental problems, preventing the full development of the students' physical and mental quality, lead to the failure of the aim of "emphasizing mental health and building correct values", and completely obliterate any expected achievement in the future.

2.3 *Increasingly serious consequences caused by individual mental problems*

Up to a certain level, mental health will become a group problem, even a social problem. As an organized and related special group, students usually have a larger uncertainty in their behaviors. Their group psychology and curiosity will magnify and radiate one student's interesting mentality and behaviors. Without regulation and control, the influence will bring unexpected shocks to student groups and social consensus environment.

With the rapid speed of information dissemination and communication, and people's increasing attention to social hotspots, students' mental health problems have come into the public light. On March 13, 2014, a Grade 3 student of Xiaozhen High School suddenly jumped down from the fifth floor. Described by his classmates, the student, who was from an impoverished family in the rural area, laughed weirdly and jumped down because of the disallowance of his teacher for the English exams that night. Additionally, Lin, a student in Fudan University, poisoned the water dispenser just to vent his resentment, leading to the death of his roommate; Zhang in Ankang Middle School, due to personal grudges, gathered the idle loafers to slash students in Guanmiao Middle School on his way home, resulting in one death and two critical injuries. Besides ineffective regulation from schools and parents, the reason for these tragedies ultimately was the parties' lack of deserved mental heath education, which then caused the distortion of mentality and values, the lack of reason and these final colossal mistakes. The lessons of blood hurt individual victims and brought devastating repercussions to involved families

as well. These children, carrying the hopes of the family, buried their bright future and family happiness by themselves, which caused a public panic of students' mentality and security.

3 IMPROVEMENT AND ATTEMPT OF MENTAL HEALTH EDUCATION

3.1 *Cultivation of self-judgment from the perspective of students themselves*

As the receptors of mental health education, students have to make changes on the level of themselves. It is not difficult for each individual student to achieve the goal of improving his own quality and forming a healthy mentality if they attempt to build correct values. The fact that students always identify and judge ideas from teachers and parents based on their formed values indicates that students' self-judgment ability determines the effect of the mental health education. And students need to emphasize the cultivation of self-judgment and learn about autocriticism and empathy from details in life. Meanwhile, proactive self-education and clear distinction between right and wrong are also required for the formation of a correct values system in students' growth.

3.2 *Strengthened guidance of correct values from the perspective of students*

In the early state of quality education, most schools responded to the call of the Ministry of Education, setting up psychological consulting rooms, requiring teachers and counselors with the qualification of psychological consultant and encouraging regular mental health education activities. Despite the good intentions of these policies, the practical implementation and feedback showed counterproductive results that psychological consulting rooms gradually became useless decorations, the requirement of teachers was just pro forma and even mental health education activities of students were just perfunctorily scheduled. Obviously, traditional policies have arrived at a fatigue period with unworthy achievements, urgently needing reform and transformation in the new situation.

First of all, status surveys of the students' mental health are required for the arrangement and implement of next targeted and personalized education schemes. The collection of basic information such as interpersonal sensitivity of boys and psychological fluctuation of girls caused by physical discomfort, determines the effect of schemes for the guidance of students' values. And then schools can humanizedly increase the interestingness and flexibility of mental health activities for the change of the mental health education situation and the correction and building of students' values. For instance, the activity of "role play for a day", in the cooperation of relevant departments, gives students an opportunity for playing other social roles, which develops the students' responsibility, gratitude and kindness. And with the feedback of parents and the conclusion of teachers to measure the results, this kind of education activities greatly increases the students' passion for participation, and follows the trend of universal education, providing an example for all kinds of educational cause including compulsory education.

3.3 Creation of positive and harmonious atmosphere from the perspective of society

As the future successors of construction cause, students shoulder the meaningful responsibility of the times. The correctness and advancement of the students' values attract more attention of the public and obtain privileges in relevant legislation. At the Second Session of the Twelfth Committee, CPPCC members underlined the physical and psychological health and the values cultivation of teenagers. With *Some Suggestions on Implement of Compulsory Education* enacted by the Ministry of Education, quality education in the student stage was hoped to be emphasized. Additionally, TV stations in China, as available mass media, have launched some meaningful and warm television broadcasts. *Dad, Where Are We Going*, a successful example of Hunan Satellite Television, showed the true contradictions and disagreements in celebrity families in the program. With the communication between parents and children throughout the viewing, it helps develop children's correct values, self-judgment ability and their personal characters of endurance, solidarity and friendship. Finally, children unconsciously realize their self-improvement in entertainment and the public is influenced by virtues around.

With visible input and effective schemes, the complete formation of student's correct values usually has an unexpectedly significant effect. Most successful people, active in different fields, have a healthy psychology and correct values as their solid foundations for every achievement. For example, Liu Jian, a 28-year-old little village official, gained the municipal exemplary construction project for his Xianfeng Village in Wudu Town, and guaranteed the completeness of this project on time with multi-coordination and collaboration. With the hope of doing some good for villagers, he has stayed in this seemingly ordinary and humble position for three years. And the motivation will support him to continue his striving until realizing the original promise to the village. From the insistence of Liu Jian for his position and promise, it can be inferred that the cohesion in the values of diligence and responsibility can become a vast immaterial wealth of the public and highly improve the development of society.

4 CONCLUSIONS

An all-round development in the student stage needs the learning and mastery of knowledge and skill as well as the mental health education and the establishment of correct values. Despite enough knowledge culture, the lack of correct values orientation and positive attitude still prevents the full realization of men's life value, even leading to crimes and dangers in society. Therefore, the successful approach to talent cultivation and sustainable development exactly starts from every right step of students and the positive influence of a harmonious environment in schools and society. For each student, healthy psychology and correct values fairly are the most powerful capital for their creation of life value and realization of dreams.

REFERENCES

[1] Chen Michen. Thoughts on Improvement of College Students' Quality. Popular Science. 2009 (9): 13–17.
[2] Zhang Yanyan. Reform of Moral Education In Institutes of Higher Learning under the Influence of Internet. East China Normal University. 2010 (1): 23–24.
[3] Zeng Zhaopeng. Research on Theory and Measurement of Environmental Literacy. Nanjing Normal University. 2011 (4): 45–49.

Future Communication, Information and Computer Science – Zheng (Ed.)
© *2015 Taylor & Francis Group, London, 978-1-138-02653-7*

Analysis and judgment of the unreliability of computer software technology

G. Li
Xuzhou Vocational Technology Academy of Finance & Economics, Jiangsu Union Technical Institute, Xuzhou, Jiangsu, China

ABSTRACT: Computer software is a form of achieving the function of logic language. Currently, computer software technology still needs maintenance and updates because of its unreliability. Therefore, in order to improve the use efficiency of software, the reliability of software should be analyzed, and some measures should be taken. Results show that factors affecting unreliability can be reduced properly by selecting appropriate programming language and strengthening software testing. Furthermore, with the development of technology, the unreliability of computer software will be solved.

Keywords: computer; software technology; unreliability; analysis

1   INTRODUCTION

Gradually, computer technology becomes the most widely used technique, especially in the field of software technology. Furthermore, the computer plays a very important role in the development of various fields. However, certain unreliability still exists in the development of the software development process. For the unreliability of the current software development process, analyzing the principles of software technology development seems important [1]. The work analyzed the unreliability factors of computer software technology, studied and judged the reasons of its occurrence.

2   ANALYSIS OF COMPUTER SOFTWARE TECHNOLOGY

Software technology, the mainstream technology in computer technology, has led the development of computer technology. Computer software technology is one of the computer technologies based on logic language. Using the advanced language of the computer, software technology can achieve corresponding functions. In various fields, computer technology is the most widely used technique and has become the mainstream of computer technology [2].

As a device based on hardware, the computer was originally designed to solve the high-rate calculation of aerospace data. Subsequently, machine language appears in computer technology. Machine language is a very simple visual language, achieving a man-machine interactive function according to the demands set by users. Then, high-level computer language emerged in the late period. As a landmark language,

C language has improved the efficiency of software programming to a certain extent, and improved the efficiency of computer programming greatly. Based on C language, many advanced programming languages emerged subsequently. Among them, the object-oriented programming language is a modular programming language, which is suitable for modern modular development. Certainly, java language and some others are also included. These programming languages have greatly improved the development of computer software technology and laid the foundation for the future development of software. So far, computer software technology has been widely used in various fields, and the automatic intelligent control system is achieved in the industry. In healthcare, amounts of medical equipment based on computer software technology are applied constantly to improve the medical standards [3].

3   UNRELIABILITY OF COMPUTER SOFTWARE TECHNOLOGY

At present, the unreliability in computer software technology still cannot be solved effectively but only be reduced to some extent without being eliminated exhaustively. So why should computer software technology exist the unreliability?

3.1   *Reason analysis of the unreliability of computer software technology*

Based on the hardware, computer software technology can achieve logic operation and certain functional requirements by compiling the logic language of computer. So, the unreliable reasons of computer software can be generally analyzed from the following aspects.

Firstly, computer software is compiled through logic language, because computer language is a logical language, regardless of machine language or advanced language. During the compiling of logic language, the occurrence of varying degrees of logical loopholes will undoubtedly result in software loopholes. That is, software vulnerabilities are caused by logical errors during compiling. Moreover, absolutely no loophole in the logical relationship is impossible. In other words, computer software technology has some loopholes inevitably, forming one of the unreliable reasons.

Secondly, there are some errors during the compiling process in computer software technology. Error is different from a mistake that can be avoided. For example, the mistakes during the compiling of software can be detected and corrected easily by revising the procedure. However, the logical error cannot be revised completely at present. For example, during the minimization segmentation process, it is possible to get close to minimization segmentation, but it is never possible to really achieve it. And this is the biggest difference between error and mistake. During the development of software, compiling the program is necessary. While compiling will undoubtedly cause some logic errors, resulting in the unreliability of software technology. Therefore, the compiling error is an inevitable problem. That is, software technology will be more and more accurate, while the errors still exist.

Thirdly, the negligence during the testing process of computer software will result in failure of detecting all software bugs. Usually, the development of software technology is equipped with a professional testing team. In fact, the task of the software testing team is to detect whether there is a bug or mistake in the software in the process of compiling. Testing the program can minimize errors and bugs. However, testing a software program still cannot fully detect the software problems. Therefore, a professional testing team is essential for a software development team to increase the bug correction of software. Their purpose is to ensure the software with few problems before using it.

Finally, the users' logic cannot be predicted during the using process of software. While using the computer software, the majority a users just discover and explore the rules by feeling instead of caring about the users' guide. In addition, few users follow the precautions of software. Thus, many uncertainties of users will result in the unreliability of the software. Take the C language program that detects whether it is a leap year for example. General programmers just compile a program in their limited logic scope. Usually, when setting the input port of the process, the programmers habitually think software can be detected by entering the year first and click the query later. But in fact, the users are likely to enter some characters that cannot be recognized, causing the paralysis of the software. Therefore, the unpredictable habits of users

will inevitably result in the unreliability of computer software.

3.2 Solutions for the unreliability of computer software technology

Analyzing the unreliable factors in computer software technology helps to clearly understand the potential problems in spite of the rapid development of computer software technology. Then, during the using and development process of computer software, how does the unreliability show itself? What is the effective ways to reduce the unreliability of computer software technology?

Firstly, one of the solutions for the unreliability of computer software technology is to select appropriate computer language before developing software. At present, with the development of the computer, computer software technology has achieved a technological leap. The languages related to computer software also have achieved great development, especially the advanced computer language. The emergence of advanced language – C language – makes computer software programming enter a fairly advanced stage. C language is the foundation of all advanced languages. However, C language, a process-oriented language, will result in a lower efficiency during the compiling process. Besides, because C language is a process-oriented language, one process problem will cause problems of software compilation or even affect the overall software development process during the compilation process. Therefore, the object-oriented language was designed on the basis of C language. Object-oriented language can avoid the problems of process-oriented language, and create a different package module by inheriting to modular management. This will greatly improve the efficiency and accuracy of the software compilation. Therefore, one of the solutions is to choose an appropriate software compiling language for the computer, thereby reducing the unreliability of computer software.

Secondly, another solution for the unreliability of computer software technology is to strengthen software testing and extend software testing time during the software development process. Because in the process of compiling, the technical staff with thinking formulary cannot be aware of the error and bug in the software development process. Therefore, it is required to strengthen the role of the test team. In the phase of software testing, plenty of professional testing should be done to find the bug in the software development process. Then software user experience will be improved, also the reliability of software will be reduced directly to a certain extent.

Thirdly, use the software correctly as well as maintenance. As mentioned above, the cause of the unreliability of software is clear. While using software, the user's habitual and arbitrary operation is likely to cause problems. Thus, it is necessary to take these causes into consideration during the software development process. In order to avoid the user's irresistible erroneous operation, some measures should be taken to prevent the software from crashing. So in the development process of software, a degree of crash protection measures is necessary. Therefore, in order to reduce the unreliability of software, this protection should be enhanced, thereby reducing the number of crashes of software.

Fourthly, to make use of technology to update software and fix bugs. During the using process of software, relevant issues can be found in the management log, which can be summarized as the latter maintenance basis of software. In the using process of software, most software will upgrade and update according to get a better user experience. This operation can reduce the probability of software unreliability to some extent. In order to ensure the real needs of software, it is necessary to regularly update software and fix bugs. These measures will ensure software runs with low probability of crashes or bugs, thus improving the user experience of the software.

4 UNRELIABILITY OF COMPUTER SOFTWARE TECHNOLOGY

As for developing software based on advanced computer language, its unreliability can be studied from the fundamental problem, and some useful conclusions can be acquired during the language selection. For example, programming with C language will cause problems in the process-oriented method. While an object-oriented solution applies modular thinking to address the existing problems. Furthermore, the specific judgments of unreliability are as follows.

Firstly, the selection of computer programming language: selecting advanced languages can reduce the unreliability of computer software technology. So, if certain software adopts a relatively low class language, then the appearance of unreliability increases correspondingly.

Secondly, frequent updates and maintenance means some problems exist in the software. Maintenance can make software faster and more convenient and fix the unreliability in a timely manner, thus reducing the unreliability of the software. So, if the software often collapses, then it really needs updates and maintenance timely.

Thirdly, the use experience of software: use experience generally includes the use efficiency, the convenience of use and the consequences emerged during the using process. While using a piece of software, the system collapses or restarts, showing the software really has an obvious unreliability. Unreliability of software technology will be increasingly low, but software technology without error cannot be achieved in a short term.

In a word, perfecting the software technology can reduce the unreliability of the software to some extent and improve its use experience. Therefore, developers of computer software technology should try to find the reasons of unreliability from a technical point of view. Thus, the user experience of the software as well as the technical quality will be improved.

5 CONCLUSIONS

The work outlined the development of computer software technology and the existing problems. After analyzing the unreliability of computer software, results show that reasons are diverse. During the maintenance, several aspects should be taken into consideration: the computer language of the programming software, software maintenance, later updates and bug fixes, etc. However, something should be clear that unreliability of software technology will be increasingly low, but the software technology without error cannot be achieved in a short term. Therefore, developers of computer software technology should try to find the reasons of unreliability from a technical point of view. Thus, the user experience of the software as well as the technical quality will be improved.

REFERENCES

[1] Zheying Li, Li Luo, Development of the Application Technology of SOC and SCM, Microcontroller and Embedded Systems, 2003 (4):15–16.
[2] Tingxiao Lu, Reliability Design and Analysis, National Defense Industry Press, Beijing, 2005 (10):22–22.
[3] Yungang Zhang, Chunmao Liu, Analysis of Software Testing Technology, Technology & Market, 2011 (02):18–19.

Future Communication, Information and Computer Science – Zheng (Ed.)
© *2015 Taylor & Francis Group, London, 978-1-138-02653-7*

Analysis on technological core of refrigeration technology of air conditioner

T. Pan
Architectural Engineering College, North China Institution of Science and Technology, Sanhe, Hebei, China

ABSTRACT: Air conditioner, acronym of AC, has been widely applied, because AC controls the ambient temperature. The work analyzed core refrigeration technology of AC, on the basis of concept and characteristics of AC refrigeration technology, combining factors affecting the refrigeration technology of AC. It shows that only when adopting the new type refrigerant can high performance compressor and reasonable control procedure assure good refrigeration effect of AC. Application of AC can be improved if humanized settings such as the sleeping mode can be added into the control procedure.

Keywords: AC; refrigerant technology; technological core; refrigerant

1 INTRODUCTION

With the development of science and technology, for solving the problem of hot whether, AC as an advanced device has appeared. Using specified refrigerant and devices such as compressor and condenser, air temperature can be lowed well. AC is popularly used in a hot summer. China has become the 2nd largest world economic unit with more than thirty years of reform and opening, whose economic ability has been greatly improved. This lays a good economic foundation for the using of AC. The operation of AC refrigeration technology has been used for many years. The technology has already been perfect. So far, there are many AC devices. According to practical using requirements, different AC devices have certain performance differences, nevertheless, Chinese economic and science started late having been affected by special historical factors [1]. Compared to some western Countries, there is still a certain distance. In recent years, considering the importance of AC refrigeration technology, China has invested a lot of human and material resources to research the AC refrigeration technology, while a remarkable effect has been acquired.

2 BRIEF ANALYSIS ON AC REFRIGERATION TECHNOLOGY

2.1 *Concept of AC refrigeration technology*

The reason why AC can refrigerate is dependent on the circulation of the refrigerant. Compressor firstly compresses the refrigerant to a liquid state. Some heat is released in this process. Generally, AC is divided into two parts, an inner part and outer part. Released heat will be discharged to the outside, and the compressed

refrigerant can be turned to a gaseous state after flowing through the evaporator. Also the heat around the area can be absorbed in the process which is done within the inner room. Inner temperature can be lowed well through the above process [2]. Inner room temperature can be controlled as a cooling state with such continuous circulation. With the development of electronic information technology, AC already has a certain automation. Intellectual chip as control core of AC can automatically control the operation of AC. Nowadays most of the ACs on the market have such a function. Inner room temperature can be continuously kept at a certain range by AC once the corresponding temperature is set. Refrigeration technology has been greatly improved through years of operation, devices like refrigerant and compressor have also been developed. For example, a formerly used refrigerant like Freon can discharge certain harmful gas into the air and cause certain damage to the ozone layer although it has well refrigeration effect [3]. For protecting mankind's living environment, the present used refrigerant will not cause damage to the environment.

2.2 *Development of AC refrigeration technology*

Great attention was paid to AC refrigeration technology from its beginning, anyway, AC was only used as a luxury, but it can't be used widely because of the high production cost which was limited by the technical level at that time. Also because of the bad refrigeration effect, the compressor made quite a loud noise in the practical operation process. With the global climate changing to become warmer and the development of modern industry, AC refrigeration technology has been greatly developed which is forced by the above two issues. In the beginning of the development of AC refrigeration technology, China couldn't produce

AC devices on its own because of the limited technical level, however, China has a great population and vast area, AC has quite a big market. Under this background, for boosting China's AC industry development, the governor department issued many preferential policies to encourage and help the development of local AC enterprises. At present, many famous AC manufactures such as Gree, Haier etc. their's AC refrigeration technologies have reached world advanced level. The present refrigeration devices like AC have been applied widely through actual investigation. The AC production cost was reduced to a certain extent with the development of industry production technology. More and more individuals can afford to buy AC. Different using environments of AC have different performance requirements, such as a shopping mall. Central AC devices are generally used because of a vast area.

3 THE PRESENT SITUATION OF CHINA'S AC REFRIGERATION TECHNOLOGY

3.1 Factors of affecting AC refrigeration

There are many factors which affect the refrigeration in the AC operation process. First of all the outside environment, AC requires larger power to reach a certain refrigeration effect if the outer weather is hotter. AC performance itself can directly affect refrigeration besides external issues, i.e. the present AC devices on sale have two types, a vertical type and hanging type. The vertical type AC has a better refrigeration performance, while the hanging type has a worse effect. Compressor is the core device to an AC device. The performance of the compressor can affect the refrigeration to a large extent. It's known that China can produce high performance compressors limited by a domestic technical ability through practical investigation. Core parts like the compressor need to be imported from abroad although China can produce AC devices. The local development of AC refrigeration technology is affected by such production modes to a large extent. Moreover, the use of a refrigerant can have an affected on practical refrigeration. Nowadays, the refrigerant in AC can be the important mark of measuring the performances of AC. There are more than eighty refrigerants. Environment issues will be taken into consideration during the present usage of a refrigerant, because some refrigerants like Freon will cause certain damage to the environment.

3.2 Problems during usage of AC refrigeration technology in China

With many years of development, AC has been popularly used. However, there are still some problems influencing the refrigeration in the practical using process. Closure property of the inner room must be assured to maximize the AC refrigeration effect. Worse refrigeration will occur if AC is used in an environment with quite good ventilation. Only large scale AC devices can meet the practical refrigeration requirement of a large scale location like a shopping mall with a larger population flow and bigger air ventilation. Although some AC manufacturing enterprises in China possess AC refrigeration technology and reach the world advance level, the overall performance of AC is badly affected by the production process. The reason why China's AC can cover a large marketing share is the low price of AC. Compared to foreign AC, there is a certain distance in energy-saving and practical refrigeration effect. Especially to ordinary clients, they have limited knowledge of the compressor and refrigerant. Clients could not tell between good and bad even some make use of a lower performance compressor. That's why the makers use a lower performance compressor for the sake of cost-saving. These compressors can also reach a certain refrigeration effect. anyhow, there are problems of large energy-consumption and loud noise etc during practical operation.

3.3 Application of AC refrigeration technology in China

Considering the importance of AC refrigeration technology, attention is paid to the research of related technology. A lot of technical patents are applied. Each AC manufacturer has developed AC with different characteristics according to the practical market requirement, for example, central AC are used in a shopping mall, the hanging type and vertical type AC are used for individual clients. Each model has powerful features. Some AC have a good soundproof effect, while some AC have a better energy-saving performance. A client can select the best matched AC according to the practical refrigeration requirement during the AC selection. AC makers will adopt plenty of patent technologies during the production to improve AC refrigeration effect. However, based on the working theory of AC, the refrigeration mainly depends on the compressor and refrigerant. AC performance can only be assured by a high quality compressor and matched refrigerant to a large extent. China's AC refrigeration technology level has reached the world advanced level through years of development.

AC manufactured in China have been exported to many countries and are widely accepted by users. One of the most important reasons is the low cost, anyway, it proofs that AC can meet the refrigeration requirement, and this is the most important reason. It is believed that China's AC industry can be further developed if AC refrigeration technology can be improved by learning from the features of foreign AC makers.

4 CORE TECHNOLOGY OF AC REFRIGERATION

4.1 Selection of refrigerant

The selection of the refrigerant can affect AC refrigeration effect to a large extent, so it can be taken as core technology of refrigeration. The refrigerant is mainly using Freon in the traditional AC refrigeration. Because of features as non toxic, non combustion, non explosion, and safe operation, many AC refrigeration devices use Freon as refrigerant. Freon can be well dissolved in organics to form a refrigerant like R12 and R22. It can meet the requirement in different environments, however, after long-term operation, a refrigerant can be discharged to air. Freon can make a chemical reaction with ozone in the ozone layer, this causes the ozone layer to get thicker, although Freon itself has no toxic and harm. The reason why human beings can survive under sunshine is because ozone can absorb the ultraviolet. The human' living environment will be damaged if the ozone layer disappears. Under this background, the using of Freon was starting to get limited. Hydrocarbon refrigerants are used because these new type refrigerants can not only solve the environmental issue, but also improve the energy-saving. They play an important role in AC refrigeration technology. A traditional refrigerant like Freon needs to be added at a fixed period because it will spread into air, or the refrigeration effect can't be reached, anyway the new type refrigerant has no need to fulfill and has few diffusion.

4.2 Selection of compressor

Compressor is the key part of AC. The practical operation process mainly depends on the compressor. Noise from AC is caused by the compressor. The performance of the compressor decides the compressing situation of the refrigerant and finally affects the practical refrigeration. The selection of the compressor is very important, in China a superior compressor can't be produced because of the limitation of technical ability. Compressors are imported from foreign countries in the AC assembling process. The production cost of the China AC is increased by this situation, and this situation can not be solved in a short time, the only way is to choose the best compressor to balance the relationship between cost and performance. It is found that foreign companies have produced multitype compressors according to the practical requirement through the practical investigation. Different types have great differences. The standard balancing the quality of one compressor is mainly depending on the power of the compressor, energy-consumption, noise etc. An advanced compressor makes a low sound and keeps energy-consumption at a certain range, also possesses high compressing power in the working process. A compressor with above three points can be sold at a higher price, only some high level AC will be equipped with such compressors in the practical AC production process.

4.3 Control procedure of AC

With the development of electronic information, there are many control chips arising. The chip will control the whole operation of the circuit according to the procedure written into the chip. Control chips are assembled in the AC during the production to improve the automatic level of AC. For different AC, corresponding control procedure can be quite different to satisfy different using conditions. AC is belonging to an electric device from a certain definition. Electric can be separated in to hardware and software, that's why the control procedure of AC is the technical core. The refrigeration situation can be influenced if problems happen in the control procedure, even the AC can't work. Especially, the present AC is equipped with a remoter, the operation of the remoter is built on the basis of procedure, and therefore the importance of control procedure can be seen. In the practical procedure designing, the designer can add corresponding functions according to the performance of AC and practical requirement of markets. For example, there are functions such as control of temperature, wind strength, movement of fan, some AC are added with a humanized setting such as sleeping mode, all these functions are realized on the basis of control procedure.

5 CONCLUSIONS

The analysis in this work shows that AC has been widely applied because it can well control the inner room temperature. No matter the hot summer or cold winter, AC can cool or heat according to practical requirement, however, AC is mainly used to refrigerate in the practical usage. Attention is paid because refrigeration technology can directly affect the working condition of AC. Selection of refrigerant and compressor must be focused on to improve the refrigeration technology of AC. Combing with advanced AC control procedure, the performance of AC can be finally confirmed.

REFERENCES

[1] Nima Pingcuo, *Briefly Discussion on the Influence of Refrigerant R12 to Environment and Selection of Its Substitute*, JOUNAL OF TIBET UNIVERSITY (natural science), 2008(02): 115–119.
[2] Zhang Xianming, Mu Ying, Wang Licun, Chen Guoqiang, *Research and Prospect on Technology Development of AC Refrigeration Compressor*, Compressor Technology, 2009(06): 58–60.
[3] Ma Yitai, Tian hua, *Present Situation and Progress of Key Energy-saving & Pollution –reducing technology of refrigeration AC* JOURNAL OF MECHANICAL ENGINEERING, 2009(03): 49–56.

Future Communication, Information and Computer Science – Zheng (Ed.)
© 2015 Taylor & Francis Group, London, 978-1-138-02653-7

Technology core and application management of remote automatic control technology

J. Wang
Ezhou Polytechnic, Ezhou, Hubei, China

ABSTRACT: Considering the wide application of automation in the modern industrial field, the work studied the technology core and application management of remote automatic control technology. Based on the idea and concept of automatic control technology and combining the current application status of remote automatic control technology, it is found that smart chip selecting, circuit designing and programming are the technological core of remote automatic control technology. Specific management about hardware, software and its function in actual application management is suggested to improve the performance of the remote automatic control system.

Keywords: remote; automatic control technology; technology core; application management

1 INTRODUCTION

With the popularity of the computer, automatic control technology has been paid more and more attention. Particularly, the use of a smart chip in recent years has provided a profound foundation for automatic control technology. After the two industrial revolutions, western countries first entered the industrial age and production efficiency was greatly improved. However, with the rapid growth of the economy, relying on people to operate mechanical equipment can no longer meet the needs of the market. In this context, the computer was developed to realize the automatic control of industrial production. Due to the limitation of the size of the computer, it cannot be directly used as the control core in actual industrial control. Therefore, a single-chip microcomputer was developed to achieve the control function of the computer by sacrificing some performance and with referencing the structure of the traditional computer. In the actual application, the microcomputer can be directly welded on a circuit board, and the whole system can run automatically after being planted in the corresponding control program [1]. Moreover, remote automatic control can also be achieved with the help of modern network technology.

2 BRIEF INTRODUCTION TO REMOTE AUTOMATIC CONTROL TECHNOLOGY

2.1 *Concept of automatic control technology*

Automatic control technology is based on the actual needs of industrial production. Traditional mechanical equipment needs personnel to operate. But employment produces certain cost and employees need break time. Therefore, if mechanical equipment can run automatically, the production cost can be reduced. Besides, mechanical equipment can realize 24 hours of uninterrupted production without any rest, which is of great significance for the actual industrial production. The development of electronic information technology, especially the emergence of the logic controller such as PLC, provides profound foundation for achieving automatic control. Once designed with a specific control circuit and equipped with corresponding programs according to the needs of actual control, those smart chips can successfully control the operation of the whole system. Nowadays, many western developed countries have built automated production lines, greatly improving the efficiency of industrial production [2]. China started late on the development of economy and science, resulting in a large gap on the developing level of automatic control technology between western developed countries and China. Although some large enterprises in China have built their own automated production lines according to the needs of actual industrial production, most technologies and equipment are introduced from abroad, which to a large extent constrains the development of automatic control technology in China. In addition, the middle and small-sized enterprises in China still adopt personnel to operate mechanical equipment, resulting in a low actual production efficiency [3].

2.2 *Idea of remote automatic control technology*

The application of automatic control technology greatly improves the efficiency of industrial production. But in the actual production process, electronic components are prone to break down. Any problem of any component will have a big influence on production lines or even will result in the shutdown of the production line. As a result, some technical personnel are needed to regularly exam the condition of equipment. Once any problem appears at a certain point, they can fix it in time. But considering the quality of those technical personnel, some serious problems cannot be fixed in time. Thus with a remote automatic control system, remote control can be achieved and a local area network can be constructed by using the existing network technology. Once connected to the network and certain privileges, it is possible to control the actual production status. Therefore, remote automatic control technology is of great importance to the actual production. Especially, the use of a large number of sensors helps to achieve the remote monitor of the production situation in real time. Moreover, specific adjustment of production lines according to market demands equips enterprises with flexible production and enables enterprises to gain profound interest with a head start in the fierce market competition.

3 TECHNOLOGY CORE OF REMOTE AUTOMATIC CONTROL TECHNOLOGY

3.1 Use of smart chip

The achievement of remote automatic control technology is mainly established on the basis of smart chip, which means the selection of smart chip is very important. At present, there are various smart chips, and different companies launch a variety of smart chips such as different models of single-chip microcomputer and programmable logic controller according to the needs of different fields. In the process of actual selection, it is very important to understand the characteristics of different types of signal chips first and then select the specific one according to the actual needs. Actual survey finds that at present most smart chips in the market are produced by foreign companies. China imports those smart chips, causing a high construction cost of the remote control system in China. In addition, constrained by the quality of technical personnel themselves, the technicians in China feel hard to maintain the equipment when some problems appear, which to a certain extent impacts the production efficiency. Moreover, hiring technical personnel from abroad greatly increases the cost. Therefore, the above phenomenon constrains the development of remote automatic control technology in China. Each type of smart chip has different characteristics. Thus, in the actual process of smart chip application, it is important to compare different types of smart chips first and then consider their costs and performances. The original intent of remote automatic control technology addresses the effort to improve the performance of the remote automatic control system when controlling its cost. Therefore, the selection of smart chips is of great importance in the construction of the remote automatic control system.

3.2 Control circuit design

The design of the control circuit is very important to realize remote automatic control technology. At the beginning of developing remote control technology, only few control functions could be realized due to the limitation of the technology level. A small amount of electronic components and a simple control circuit can satisfy the actual control needs. But with the development of the industry, there is a higher requirement for remote automatic control technology. At the same time, the development of electronic information technology provides profound foundation for realizing more control functions of the remote automatic control system. Under this condition, the control circuit becomes more and more complex, using more and more electronic components. In the early stage of developing remote automatic control technology, the design of hardware circuit and the compilation of control programs were usually accomplished by one person. However, the modern remote automatic control system needs several staff to corporate. The design of the complex control circuit needs even more staff to complete together. As the foundation to achieve remote automatic control technology, the control circuit can directly determine the effect of remote control. Only a scientific, reasonable and stable circuit can ensure a smooth progression of industrial production. Practical survey finds that with the development of remote automatic control technology, its application scope is becoming wider and wider, and it also plays a very important role in the manufacturing intelligent robot besides the application in actual industrial control fields.

3.3 Control program compilation

In a sense, remote automatic control technology belongs to the category of electronic information field, and the application of electronic information technology can be divided into two aspects: hardware and software. That is similar to remote automatic control technology, which requires a corresponding program to be compiled in the designed control circuit. In actual application, technical personnel mainly deal with control programs so as to control the production through the human-computer interaction interface of programs.

4 APPLICATION MANAGEMENT OF REMOTE AUTOMATIC CONTROL TECHNOLOGY

4.1 Hardware management

Because remote automatic control technology can be divided into two aspects: hardware and software, actual application management should also start respectively from hardware and software. The management of hardware mainly monitors the work status of the circuit and electronic components and makes specific adjustment according to the actual needs. Generally, no big changes can be made after the hardware circuit is completely designed. Therefore, actual hardware management mainly focuses on maintenance and repair. Now a large number of sensors are adopted in the construction of the remote automatic control system, and those sensors can monitor every segment of the system and observe the work status of each part of the central server. If any problem happens to the system, the sensors can quickly define the problem and take specific measures in time. It proves that hardware management plays a vital role in the normal operation of remote automatic control system. When some small electronic components get damaged, they need to be replaced timely. But if some large equipment fails to function, it takes a certain time to fix it, which will affect the actual production efficiency. As a result, it is essential to maintain hardware equipment regularly to ensure the effectiveness of hardware management.

4.2 Software management

Software is considered as the foundation to achieve its control function, but some faults often happen due to the instability of the software itself. Usually, those faults can be fixed well, but they can still impact the work condition of the remote automatic control system if the problems are not fixed in time. In the actual work process, some wrong operations may happen due to the low quality of the operation personnel, leading to failures of the software system. All those indicate the importance of software management. Some bugs inevitably exist in programming because of the characteristics of the software itself. In daily application, it is not easy to locate those bugs because they are only triggered under certain conditions. Thus, software companies update their software regularly since it is difficult to deal with those bug problems only by relying on technical staff once the problems appear. In addition to add some new functions, software companies also repair existing bugs. As for software management, here are some suggestions for software companies. On one hand, they should develop a standard process to regulate the operation of the staff, which can at the most extent help reduce software failures resulted from wrong operations. On the other hand, they should install firewalls and anti-virus software. In this modern society with developed Internet, virus and Trojan are widespread much. If remote automatic control system is infected with a Trojan or virus, it will have a serious influence on the normal work of the system. As a result, it is required to maintain software on a regular basis and repair the known bugs as far as possible.

4.3 Function management

According to the needs of practical application, current remote automatic control systems are equipped with many control functions. Thus, management of those functions also becomes an important part of the application management of the remote automatic control system. Function management in actual work is to control the operation of functions. Only with accurate sequence and appropriate function execution, can function management play its full control role in remote control systems. Actual surveys find that people haven't realized the importance of function management in the actual application of the remote automatic control system. The system directly performs the corresponding function when needed, and there are no relevant post-process measures afterwards. If certain functions are not closed timely, it is likely to cause some problems to the production and even result in serious economic loss. All those indicate the importance of function management in remote automatic control technology. In the actual work process, technical personnel should try their best to understand each function and operate according to certain criteria. After executing the corresponding function, they should also pay attention to the status after execution so as to prevent problems caused by the continuous work of a certain function. Only by attaching great importance to function management, can the normal operation of the remote automatic control system be guaranteed to the maximum extent.

5 CONCLUSIONS

The work comprehensively explained that remote automatic control technology is widely applied in actual industrial control. What's more, the use of the remote automatic control system can help realize the automation of industrial production and greatly improve production efficiency. At present, some western developed countries have built automated production lines based on remote automatic control technology. However, the automation technology in China is still at a relatively low level due to the limitation of the technology level. Consequently, in order to solve this problem well, it is of high priority to carry out researches on the technology core of remote automatic control technology and its application management. Combing with the practical situation, remote automatic control technology in China can be further improved with lessons from foreign advanced technologies and experience. Only in this way, can the development of remote automatic control technology in China be enhanced to the largest extent. It is firmly believed that the application of remote automatic control technology in China will be better and better with the development of Chinese information technology.

REFERENCES

[1] Lu Gengben, Xu Guoben, Pan Chaoyang, Bao Haoda. Achievement of Automatic Control for Windows and Ventilator of Large Warehouse by the Use of Power Line Carrier Technology, Grain and Oil Storage Science and Technology, 2007(04): 43–44.

[2] Sun Xiaofang, Pan Haitian, Xia Luyue, Cai Yijun, Fan Hong, Computer Control System Design Based on Gas-Phase Copolymerization Reaction of PLC Olefin, Experimental Technology and Management, 2007(09): 60–62.

[3] Zhang Zuying, Research on the Application of UnPBX DCS Control system in Small and Medium-Size Production Process, Science & Technology Information, 2006(10): 68–68.

Future Communication, Information and Computer Science – Zheng (Ed.)
© *2015 Taylor & Francis Group, London, 978-1-138-02653-7*

Analysis on optimization-concept-based mode of multimedia information processing

L. Sun
Lanzhou University of Arts and Science, Lanzhou, Gansu, China

ABSTRACT: Multimedia information is an important way of storage on the computer. Since original multimedia information occupied a large space, the process was conducted in practical storage. This work analyzed a multimedia information processing mode based on the optimization concept. In terms of the concept and development of multimedia information processing, the application of the optimization concept, combined with the current situation, keeps the space occupancy and quality of multimedia information in balance. In practical processing, the latest computer is not indispensable, but a computer with performance meeting the practical demands is an option.

Keywords: optimization concept; multimedia; information processing

1 INTRODUCTION

With the computer's popularization and application, multimedia replaced the traditional carrying mode of information, playing a crucial role in the area of television and the Internet. However actual survey found that the computer is able to process multimedia information and control the clarity and storage occupancy of multimedia information. With years of development, multimedia information processing has grown into an independent discipline offered in many colleges and universities, with lots of relevant personnel cultivated. Affected by special historical factors, China's economy and science and technology advanced late, resulting in a gap between China and western countries in terms of using multimedia information [1]. Through thirty years of reforming and opening up to the outside world, China has developed into the second largest economic entity in the world, enjoying a great improvement of overall economic strength. In this context, developments are also seen in the multimedia and other hi-tech industries. However, multimedia information processing still has some problems.

2 BRIEF INTRODUCTION OF MULTIMEDIA INFORMATION PROCESSING

2.1 *Concept of multimedia information processing*

Multimedia information, compared with the traditional carrying mode of information, is regarded as a huge progress. Transforming information into digital form stored on the computer both saves storage space and facilitates the use of data. Owing to differences in storage architecture of different computers, processing is necessary for multimedia information during storing and using [2]. Different processing approaches will lead to different storage occupancy and different call efficiency of multimedia information. In order to save storage space, a main approach to process multimedia information is compressing. According to practical processing demands, some kinds of processing software are developed specifically, equipped with ability to process multimedia information in various aspects. In daily use, multimedia information exists mainly in the form of images, video, audio and so on. And at present, many kinds of processing software for images, video, audio are available on the market. From above, it can be seen that multimedia information processing is a general concept though contains much specific content. With years of development, multimedia information processing, now, has spawned a number of disciplines. For instance, colleges and universities are offering corresponding subjects particularly about images, video, audio and other multimedia information. However, in a general way, few students will take all subjects [3].

2.2 *The development of multimedia information processing*

In the early period of computer development, the technical limitations limited the use of multimedia information. With the invention of integrated circuit and transistor, computer performance earned much improvement. Thereafter, the internet covering the whole world was established to meet the people's practical needs, laying a good foundation for use of multimedia information. Benefiting from this, images, video and other multimedia information began to spread in the field of the computer. Poor storage technologies and limited storage space such as the hard-disk facilitated the adoption of compression technique to reduce occupancy by multimedia information with the aim to release more space for more multimedia information. For example, MP3, an audio compressed format, is produced by a compression technique through deleting the unheard sound in the original sound fragment. With little influence on the quality, compression technique reduces a great deal of occupancy by the audio file, which is of great significance for computer storage. As information technology develops, multimedia information processing makes huge progress. According to the practical needs, many kinds of software, used to process images, video and other multimedia information, have been developed, such as Photoshop, Corel Videostudio Pro Multilingual. In practical work, all these kinds of software are popularized and widely used. However, surveys indicate that most current software available on the market is developed by foreign software companies, greatly affecting the development of multimedia information processing in China.

3 STATUS QUO OF MULTIMEDIA INFORMATION PROCESSING

3.1 Influencing factors of multimedia information processing

With the popularization of the computer, due to the large influence on the applied effect of the computer, multimedia information processing technology attracts much attention, as every country attaches importance to the research of it. But China, because of a late start in science and technology, interacts with multimedia information processing for a relatively shorter time. In consideration of the significance of multimedia information processing, dozens of colleges and universities are offering relevant courses. However, limited by the overall educational level, students are only able to master the theoretical knowledge well but operate worse. Many factors may influence the processed effect, when practically processing. Among the hurdles, the operator's quality serves as the major factor. As we all know, no processing needs no operator. The operator's quality if too poor to proficiently operate the software will certainly result in an unsatisfied processed effect. Therefore, only with enough professional quality, the processed effect of multimedia information will earn a guarantee. In addition, the performance of the computer's hardware device will also greatly affect multimedia information processing, especially when massive calculation is demanded by the processing of large videos and high-definition images. Owning a low performance, the computer will spend more time waiting or even fail to satisfy the needs of processing. The Moore's Law says hardware performance will be doubled every eighteen months,

while cost remain the same. Thus considering the rapid replacement and change of the computer, it is essential to change computers frequently with the aim to keep the hardware device advanced. However, cost of multimedia information processing is likely to rise to some extent.

3.2 Problems in China's multimedia information processing

Surveys show that China's multimedia information processing still falls behind, compared with western countries. In terms of television and film production, China's product lacks high picture quality, mainly resulting from the operator's low professional quality. With the development of economic globalization, effect of China's multimedia information processing fails to reach foreign processed effect, even with the advanced computer and applied software adopted by some enterprises. Some enterprises begin to employ foreign technicians to find a way out, which consequently brings lots of influence on the development China's multimedia information processing. For some medium or small production team, low performance computers will be a priority due to their cost budget during their processing work. This kind of computer has difficulty meeting demands in the real processing situation, especially after the recent emergence of high definition video which requires large computation ability of the computer. The reason is that much software has minimum requirements for the hardware's performance and also requires the best configuration. Though meeting the minimum requirements, many computers are not equipped with the best configurations, thus contributing to the failure to achieve a number of functions. From the above, we could conclude that China's multimedia information processing has two main problems: the operator's low professional quality or low staff quality and low performance of hardware device. As for the first one, the problem of the operator's own quality belongs to the educational matter, which is hard to improve in a short time but a long-term educational reform oriented to increase the educational level may help. By contrast, low performance of hardware, the second problem, is easy to solve. Considering the cost, it is not indispensable to choose the most advanced computer, but a computer meeting practical the demands is an option, while selecting the computer.

4 OPTIMIZATION-CONCEPT-BASED MODE OF MULTIMEDIA INFORMATION PROCESSING

4.1 Brief analysis of optimization concept

Compression technique is taken by traditional multimedia information processing as a main approach to control space occupied by multimedia information, consequently reducing the quality of multimedia information in most cases. For example, MP3, lossless audio compression mode, is supposed to sound like the original sample theoretically, after removal of the unheard fragment. However, more advanced players reveal relatively bigger differences between the processed and the original. In order to solve this problem, several compression formats are created to meet the processing demands of multimedia information in various fields. From above, it can be seen that former multimedia information processing, limited by storage, attaches more attention to compression efficiency but less importance to the quality. As a consequence, the use of multimedia was seriously influenced. With years of development, the current computer storage has already reached a high level where a large amount of multimedia information could be stored. With this achievement, some original formats are used again in the practical processing, such as images format DWG and sound format WAV. All these original formats occupy a large storage space, for instance, DWG and WAV occupy tens of megabytes, while video file demands larger space. For this reason, even when the capacity of a hard-disk is expanded into TB, the personal computer still has no capacity for a large amount of multimedia information. Therefore, to find the balance point between storage space and multimedia information quality, forms the basis for generating the optimization concept. The optimization concept can be expressed in this way that it is aimed to maximally increase the quality of multimedia information, based on the ability to store, after specific processing of multimedia information according to real storage space.

4.2 Application of optimization concept

Present multimedia information processing makes use of processing software which has, for years, been updated into multiple versions with improved functions of each version. Owing to a high intelligence degree, the current processing software which contains many plug-ins allows users to make adjustment according to their own demands. For example, users are allowed to set the compression ratio and the quality of multimedia information, and to preview the processed effect and space occupancy before processing. Therefore, the use of processing software offers a good foundation for the application of the optimization concept. During practical processing work, some matters are supposed to be considered before using the optimization concept. At first, estimation of the capacity for bearing space occupancy by multimedia information shall be conducted to get an overall knowledge of the storage condition clearly. Secondly, while setting parameters, best efforts will be made to increase the quality of multimedia information on the basis of storage space satisfaction, confirming to the optimization concept. Understanding of the optimization concept ought to come from two aspects: space occupancy and quality, both of which must be optimized at the same time. At present, optimization concept is widely applied into current multimedia information processing, especially after internet facilitates the emergence of various websites of high definition

video. In order to survive in the fierce competition, websites have to increase picture clarity and reduce internet transmission bandwidth occupancy so that they could attract more users. However, the goal of more users could be achieved only with the optimization concept which helps specifically optimize multimedia information through analyzing our average bandwidth and estimating bandwidth occupancy by video stream.

4.3 Optimization-concept-based approach of multimedia information processing

With the popularization of the Internet, multimedia information processing becomes increasingly important. Since a large number of multimedia information is available on the internet, lots of users visit websites mainly to get access to the information. Unfortunately, limited by our internet transmission bandwidth, transmission efficiency of multimedia information is relatively low. For this reason, the best way for websites to attract more users is to devote their best effort to increase the quality of multimedia information, based on the reduction of their own internet transmission bandwidth. Owning the very advantage of satisfying the websites' need, the optimization concept is adopted in all multimedia information processing of websites. However, taking advantage of the optimization concept asks the operator to learn about the content of this idea at first. Thereafter, a specific processing plan ought to be made according to real transmission bandwidth, storage condition, computer performance as well as processing software. At last, in practical work, multimedia information will be processed step by step in accordance with the plan. Only in this way are the processed effect and balance between storage space and quality be likely to be maximally guaranteed.

5 CONCLUSIONS

From this work, it can be seen that multimedia information processing grows increasingly important, with the popularization of the computer and the Internet. However, our multimedia information processing still performs at a low level, due to the limitations of technologies. In addition, the processed effect of multimedia information is largely affected by the low professional quality of the operator himself and low performance of the computer used by the staff, when practically processing. Hoping to solve these problems may be achieved by the adoption of the optimization concept into practical processing. With the adoption, it is of great significance for websites to complete the following three steps. The primary step is to confirm the space occupancy by multimedia information according to real internet transmission bandwidth and storage condition. Adjusting parameters through advanced processing software is the second step. And the final step falls on increasing the quality of multimedia information with best efforts on the basis of storage space satisfaction. China's multimedia information processing is well improved due to the increase of the operator's professional quality resulted from the continuing educational reform of recent years. And the use of advanced computer also contributes to the improvement in consideration of the significance of multimedia information. As a conclusion, there is every reason to believe that our multimedia information processing is sure to grow better along with the development of technology.

REFERENCES

[1] Cao Ying, The Application Of The Streaming Techniques In Digital Library's Multimedia Information Processing, *Sci-Tech Information Development & Economy,* 2006(07): 17–18.

[2] Wu Linjun & Li Yanwen, The Application Of Support Vector Machines In Intelligent Multimedia Information Processing, *Science and Technology Consulting Herald,* 2007(07): 11–12.

[3] Guo Jun, Multimedia Information And Multimedia Information Process, *Journal of Liaocheng University (nat. sci.),* 2005(01): 83–86.

Future Communication, Information and Computer Science – Zheng (Ed.)
© 2015 Taylor & Francis Group, London, 978-1-138-02653-7

Modern embodiment of socialist core values

H. Li
Liuzhou Teachers College, Liuzhou, Guangxi, China

ABSTRACT: Socialist core values are an important part of China's socialist ideology system, representing the soul of the socialist advanced culture and playing an important role in guiding the construction and the direction of the development of China. The innovation of this work lies in deploying the analysis on "the significance of socialist core values' modernization" and "what is the modern embodiment of socialist core values". Meanwhile, it provides an agglomerating theory strength for national development and reform based on the actual situation of the new era of China.

Keywords: Socialism; core values; modernization

1 OUTLINE OF SOCIALIST CORE VALUES

1.1 Concept of socialist core values

Socialism is a system of values aggregation, which can be classified into two kinds: socialist core values and socialist general values. Socialist core values, on the core position in the socialist value system as its name implies, functions as a general guidance, which could give the most scientific, extensive and deepest answers to a series of questions, such as "what is socialism?", "what is the essential attribute of socialism?" etc. The concept of the socialist value system meets two necessary conditions: one is that the national ideology must be socialism; the other is that the content of values must be the ideals that both country and people strive to achieve [1]. Socialist society, compared with the capitalist society, has the characteristic of public ownership of the means of production, advocating the community or government to manage and allocate the means of production according to the interests of the public. The common ideal principle of the nation and people demands that socialist core values take people's interests as the most fundamental guiding thought and that it cannot ignore the public's ideals instead of satisfying the special needs of a particular class. Socialist core values are a relatively stable concept, changing within a small range with the development of socialism compared with socialist general values. Yet, every time socialism enters a new stage of development, socialist core values will be endowed with a brand new zeitgeist [2]. Therefore, the modern embodiment of socialist core values is of great significance to the development of contemporary China.

1.2 Connotation of socialist core values

In China, the connotation of socialist core values refers to the connotation of socialist core values with Chinese characteristics, mainly focusing on the following five aspects, our party's guiding thought Marxism, socialist common ideal with Chinese characteristics, the spirit of times with reform and innovation as the core, national spirits centering on patriotism and Socialist Concept of Honor and Disgrace, including three aspects – nation, society and people. From the perspective of the nation, the proposal of "prosperity, democracy, civilization and harmony" shows the goal of the current stage of China, the primary stage of socialist development [3]. Meanwhile, these four points are also the overall layout program of China in economy, politics, culture etc. From the perspective of society, "freedom, equality, justice, law" are the guiding thoughts for the legal construction in China. As a socialist country, China, under the guidance of Marxist basic theory, should constantly promote and strengthen the construction of politics, rule of law and human rights, thus serving the interests of the public better. From the perspective of the citizens, the socialist core values of "patriotism, dedication, honesty, friendliness" contain the connotation of related moral standards, playing a guiding role in every aspect of the citizens' social life and having an indelible impact on the formation of the citizens' personality and moral trait.

2 MODERN SIGNIFICANCE OF SOCIALIST CORE VALUES

2.1 Adapting to the development of the times and social transformation

The 21st century is an important stage of development for China's social transformation, as well as an opportunity period for promoting political, legal and cultural examples. With the ever changing world scientific, economic, political, and cultural environment, China's domestic situation is also progressing with each passing day. New characteristics and challenges of the times have emerged in front of people. The development of globalization brings the communication and cooperation between nation and nation closer. As a socialist country, China's contact with Western capitalist countries in economy, culture and science and technology has also become closer. With the introduction of advanced science and technology and management philosophy, it is inevitable to suffer some culture and thought shock. Due to the differences in national ideologies and historical development, the value-orientation of China has a very big difference from that of Western capitalist countries. Shocked by the trend of Western thought, some bad values, such as money worship, egoism, weakening of ideals etc., have appeared, seriously declining the social moral climate in China. In spite of this, it is still worthy for us to learn and accept some thoughts related with freedom and democracy in Western culture earnestly. As a result, the persistence and affirmation of socialist core values under the background of the development of the times and social transformation could play a leading and guiding role in the formation of mainstream social consciousness and avoid the occurrence of thought deviation.

2.2 *Promoting the construction of a well-off society*

China is and will remain in the primary stage of socialism for a long time. Establishing an all-around affluent society is the goal that contemporary China should strive to achieve ceaselessly. The modernization significance of socialist core values has an obvious promotional function for the construction of a well-off society. In the first place, in terms of ideology, socialist core values could agglomerate the power from the public and enhance the public's sense of identity for national decision making. In social life, only people with the same value-orientation could form the same social cognition and interest appeals more easily. Therefore, socialist core values unify the public mind, thus forming a common standard in the personal activity principle, ethical codes and social behavior. With a common standard, the public will strive for the same direction or goal, with which national cohesion and integration will increase and the ultimate goal of promoting the development of socialist modernization will be achieved. In the next place, the modern embodiment of socialist core values also has an important differential significance, which could put the values that are the most abreast of the times to the dominant position, spontaneously reject the expansion and flooding of other ideological systems and reduce or eliminate the impact of old-fashioned values. In this way, socialism won't grow complacent but develop and progress better under the lead of new values.

2.3 *Leading the construction of socialist harmonious society*

Social harmony is of great importance to insure the national stability and public safety. Therefore, building a harmonious society is a reform that current China must pay attention to and carry out. The implication of the harmony of socialist core values could unify the public's basic value orientation which could be converged into a mode of thinking and code of conduct so as to lay a solid foundation for the formation of a good social moral climate. Socialist core values are an ideological system formed on the basis of Marxism. Therefore, the construction of a harmonious society should always adhere to the guidance of unified thought and avoid disturbing the formation of the public's consciousness by multiple guidance thoughts. A harmonious society also asks the public to have common ideals and goals so that it can agglomerate the strength in one direction, thereby displaying infinite power and driving force which is a mere coincidence with the formation condition for socialist core values. A harmonious society is an inevitable outcome of the development and progress of national spirit and time spirit. The modern embodiment of socialist core values could not only promote the inheritance and carrying forward of national spirit, but also amalgamate the spirit of the times with national spirit, hence cultivating Chinese traditional virtues in the whole society and spreading the good moral quality concepts such as helpful, doing boldly what is righteous, honesty and trustworthiness etc., to the public, which would lay an ideological foundation for the development of a harmonious society.

3 MODERN EMBODIMENT OF SOCIALIST CORE VALUES

3.1 *Policy regulation aspect*

A policy with solid theory basis and long-term implementation possibility is of great importance to national development, just like the reform and opening up policy, which not only enlarges the range of foreign trade in China, but also stimulates the rapid growth of national GDP. The modernization of socialist core values plays a dominant and leading role in policy regulation for the reason that socialist core values maintain the fundamental interests of the masses while people's vital interests are closely connected to national policies. The development policies formulated by the state have somewhat regulating effects on the country's political, economic, and cultural activities. If the orientation of the policies is unanimous with that of socialist core values, it will be much easier for people to accept them and have a deep understanding, thereby working consciously in accordance with policies and regulations in social life. However, if the modern embodiment of socialist core values can't be shown in national policies, it will be very difficult for the public to accept and carry out the policies. For example, in the face of a rapid accumulation of wealth and a growing gap between rich and poor, the government should pay more attention to people's livelihood when formulating policies. If the government pursues the accumulation of national wealth slavishly instead of paying attention to the basic living of citizens, it is contrary to the core value of "people-oriented". Therefore, the formulation of our current national policies should be carried out under the guidance of the modern embodiment of socialist core values so that they can further strengthen the masses' confidence in the state and maintain the safety and stability of the society effectively.

3.2 National spirit guidance aspect

If a nation wants to stand in the advanced national forest of the world and doesn't want to be beaten and oppressed, in addition to a certain economic strength, it still demands a strong spiritual support, which reflects a kind of common struggling and enterprising goal. As one of the oldest nations in the world, China has a long history and a wide range of national culture and connotation. The national spirit of the Chinese nation is an important safeguard for China's independent development. During the Anti-Japanese War, it is because of the national spirit of patriotism that the majority of the Chinese people could fight against the enemy bravely and indomitably and finally won the national liberation. Though China has realized the independent construction of the nation currently, national spirit is still the foundation of a nation and the cohesive force of the country's people. The modern embodiment of socialist core values in national spirit can usually be divided into two aspects: one is to inherit the historical spirit of the Chinese nation and the other is to carry forward the spirit of the times of the Chinese nation. Inheriting the historical spirit of the Chinese nation could enable people to learn lessons and gain experiences from historical events, thereby further strengthening the understanding of patriotism spirit and the inheritance of Chinese traditional virtues. Carrying forward the spirit of the times of the Chinese nation is proposed and advocated according to the current situation of China. Enlarging the impact of national spirits and strengthening the construction of a Socialist Concept of Honor and Disgrace could form a good atmosphere for social development so that the public could do various production activities harmoniously.

3.3 Ideological education aspect

Socialist core values have some social and class characteristics, on the basis of the ideological perception form of the public under the current society and class. The socialist core values need to be inherited and carried forward unceasingly so that they could play a guiding role throughout the construction of socialism in China. Therefore, it is worth for people to think seriously and dig in deeply its influence in ideological education. Ideological Education is not only for the modern students, but also

for the adults who have reached the legal age. The force of ideological education can't be reduced. Since the students' ideology is still immature, they don't understand the deep connotation and core idea of socialism and only know little about the construction of socialism in the country. Therefore, China draws up related education plans to enforce ideological education to students so that they can devote themselves to the career of socialist construction after becoming a capable person. Meanwhile, in ideological education, theoretical knowledge should be used to make the students' ideological perception form meet the needs of the times, which is an important manifestation of the modernization of socialist core values. The ideological education for adults still further embodies the modernization significance of socialist core values. It is because adults have experienced a systematical learning of socialist ideology and have a general understanding of its development history already. Therefore, they put more emphasis on the interpretation of the spirit of the times and the thinking of new ideas.

3.4 Legal construction aspect

Policy and ideology only have a "mild" guidance and restriction to people's behavior, which advocates people to be self-disciplined under the guidance of core values. As a result, once people's behavior appears to be in error, it will be very difficult to achieve the purpose of warning them by punishing these wrong behaviors. The restriction of the law to the public's social activities is compulsory, which defines the most basic and normal public behavior criterion and formulates the corresponding penalties with strong authority and justice for a possible violation phenomenon. Besides, the restriction of laws and regulations to people's behaviors needs to be unanimous with the spirit advocated by core values, otherwise it would reduce the fair and impartial image of law in people's hearts. At the same time, people should also be aware of the imperfections of China's legal system construction and a lack of effective prevention from some ugly behavior at the edge of the law. Therefore, national relevant departments must adhere to socialist core values and stand firm, and take them as one of the basis for formulating laws and regulations. It requires us to start from the connotation of socialist core values and fix the content advocated by socialist core values in the form of legal provisions on the national, social and civic level. For example, in order to promote justice and encourage people to do what is right boldly, it is necessary to identify and reward the voluntary actions against injustice in a legal way besides a moral praise. A posthumous awarding as a martyr or other status is also necessary, especially to those unfortunate sacrificed people.

4 CONCLUSIONS

With the constant development of the socialist core values theoretical system and its important guidance on nation-building, people begin to pay more and more attention to the modern embodiment of core values, especially in economy, culture, science and technology, and law. The modern embodiment of core values is pervasive and is playing a tremendous leading role as well. Therefore, the modern embodiment of socialist core values need to be further researched and analyzed so that ideas that are conducive to our socialist construction could be extracted from it.

REFERENCES

[1] Hong Xiaonan & He Meizi. *The Interpretation of the Connotation of Socialist Core Values with Chinese Characteristics*. Journal of College Counselors, 2013(01): 1–6.

[2] Bao Zhendong. *Sticking to Socialist Core Values and Inheriting the Chinese Spirit*. Culture Journal, 2010(5): 9–12.

[3] Ke Tizu. *The Research on Socialist Core Values*. Red Flag Manuscript, 2012(02): 4–7.

Future Communication, Information and Computer Science – Zheng (Ed.)
© *2015 Taylor & Francis Group, London, 978-1-138-02653-7*

Study on construction technology of green rubber concrete

Z. Hu
Yibin Vocational and Technical College, Yibin, Sichuan, China

ABSTRACT: Cracks, deformation and other issues often exist in the process of using concrete, a commonly used building material. Adding a certain amount of powder or rubber particles in concrete can be a good solution to these problems. This work studied construction technology of green rubber concrete. It is found that only by emphasizing the selection of tire and rubber particles as well as fabrication and application of rubber concrete can the quality of construction engineering be improved to the highest extent, based on the concept and characteristics of rubber concrete and combining them with the status of their applications in building construction.

Keywords: environmental protection; rubber; concrete; construction technology

1 INTRODUCTION

With the development of building technology, demands for high-performance materials are growing. In traditional architecture, mud and stones and other materials are mainly used, resulting in poor quality of construction and impossibility of constructing high-rise buildings. Yet with the development of modern natural science and technology, the discipline of material science emerges, playing a significant role in the construction sector [1]. For example, the use of steel, concrete and other materials can greatly improve the quality of construction work. Subject to historical factors, China started late in science and technology and there are still gaps between China and some western countries in the study of concrete and other materials. Especially with the emergence of skyscrapers in recent years, the requirements for the performance of concrete are higher. In this context, different materials are added to the concrete according to practical needs, creating some new features in concrete. For example, adding powder or rubber particles can produce high flexibility in the concrete [2].

2 ANALYSIS OF RUBBER CONCRETE

2.1 *Concept of rubber concrete*

Concrete plays an important role in construction. The use of concrete has greatly improved the quality of construction, so it is highly valued. But with the development of the construction sector itself and the emergence of complex engineering works, concrete of superior performance must be used in order to complete these projects and demands for properties of concrete are even various [3]. In this context,

certain materials are added to concrete to promote better performance of the concrete. Naturally, tire and rubber particles are added to concrete to make rubber concrete, because rubber has good elasticity while concrete often causes cracks, deformation and other issues and even accidents in the process of using and after completion. Adding powder or rubber particles into concrete can ensure certain flexibility of the concrete, thereby avoiding cracks. In traditional concrete construction, water is easy to penetrate the interior, corrosion steel and other materials, while rubber concrete can increase tightness. Thus, the use of rubber concrete can improve the performance of concrete from different aspects.

2.2 *Characteristics of rubber concrete*

Rubber concrete has distinctive features compared with ordinary concrete. Above all, rubber concrete has good flexibility, effectively preventing the building deformation, increasing the building airtight, keeping water from penetrating into the concrete and protecting the steel and so on. Out of practical applications, people can add other materials to concrete to achieve the same goals but at a higher price than rubber. Consequently, the use of rubber is common now, producing a lot of waste and large amounts of polluting gases when it is being burned due to its corrosion resistance. Especially with the shorter lifespan of rubber products, such as three years for car tires, and the popularity of automotive applications, a large number of rubber garbage is produced annually. Therefore disposal of these rubber wastes has become an important issue. This shows that adding powder and rubber particles to concrete can not only control the costs well, but can also solve environmental problems to a certain extent. Due to rubber concrete's lower requirements to rubber, it is essential to make waste rubber products into the form of powder or granules after simple treatment. It is precisely these characteristics of rubber concrete that make it widely used in modern building construction [4].

2.3 Development of rubber concrete

With the limitation of technology in China, the study of new materials such as rubber concrete started late. While in the late eighties, the United States had already began to conduct researches on rubber concrete, adding a powder and granular form of rubber to concrete to improve the performance of concrete. Still, actual survey found that adding different components of powder and granules differentiated the performance of the concrete. However, due to the low level of technology at the time, the performance of rubber concrete could not be controlled well. With the development of material science and construction engineering, rubber concrete has been highly valued and widely used in the actual construction due to its superior performance. Through practical researches, people have gotten a deeper understanding of the impact of different amounts of tire and rubber particles to the concrete's properties. In the actual construction, people were targeted to select powder or rubber particles as well as the specific components according to the actual needs. Due to the low level of technology, rubber concrete in China has a relatively short-time development. The real estate market in China has been hot in recent years, so out of the actual need for building construction and China's actual situation, some foreign advanced technology has been learned and the country's rubber concrete technology has been improved. Also, to maximize the performance of rubber concrete, some simple experiments will be carried out in the field in the actual construction process.

3 CURRENT SITUATION ON THE USE OF GREEN RUBBER CONCRETE

3.1 Factors affecting the use of green rubber concrete

Rubber concrete has been widely applied after years of development and improvement. People continue to improve rubber concrete in the process of use, making rubber concrete relatively perfect. According to the actual needs, people are targeted to match proportions of rubber concrete. As basic materials in construction, the rubber concrete's performance can directly affect the quality of the project, so factors that affect the use of it must be paid attention to. There are lots of such factors in the actual building construction. The first one is the operator's professional quality. The operator works to mix and stir rubber concrete, only sufficient technical skills ensuring the properties of rubber concrete. The second factor is rubber concrete material itself. In making rubber concrete, people get sand and mud around the construction site to facilitate the construction project. Due to different geological

conditions in different regions, sand stone and mud vary in performance. As a result, the performance of rubber concrete will be greatly impacted by materials in poor geological conditions. Besides, environmental factors should be considered because rubber emits toxic gases. Correspondingly, rubber concrete should be handled properly in order not to influence the construction project and the personnel's health.

3.2 Problems in the use of green rubber concrete

Green rubber concrete has been very widely used in building construction because of its characteristics. And it is applied in China's construction sector but the limited technical level causes many problems in the actual applications. To start with, it is the lower quality of the construction workers. They cannot get a good grasp of the ratio of rubber concrete as well as the amount of powder and rubber particles to be added according to the actual needs. Next comes the low level of environmental protection. Most of the rubber used in rubber concrete is junk, the majority being automobile tires. These rubber products have little to do with environmental protection and emit harmful gases in the production process. Therefore, these rubber materials will be exposed in indoor environment and greatly impact human health if not properly disposed. Actual survey finds that currently less rubber concrete has been used in building construction in China despite of people's realization of its importance. The main cause of this phenomenon are fewer manufacturers of powder and rubber granules meanwhile with a relatively backward production process. They are unable to meet the needs of the actual use of rubber concrete. Thus, it is necessary to intensify research on rubber concrete to improve its application in China.

4 GREEN RUBBER CONCRETE CONSTRUCTION TECHNOLOGY

4.1 Selection of tire and rubber particles

For rubber concrete, the selection of tire and rubber particles is important. It not only directly affects the performance of rubber concrete, but also impacts costs. There are currently lots of manufacturers producing powder and rubber particles on the market, offering a variety of powder products to meet the needs of different areas. In the actual selection, optimal and targeted powder or rubber granules should be selected according to the actual needs as well as the using effect of powder. Actual survey finds that many of currently available powder and rubber particles are produced by foreign companies. While some production companies in China represent a large gap with foreign companies in the powder properties. For cost considerations, some construction units will choose products from China's companies in actual building construction due to the relatively low price of tire and rubber granules. It often results in quality problems after completion. This shows the importance of the selection of tire and rubber particles. In the actual construction process, certain experiments will be conducted after the selection to ensure the performance of the powder and rubber particles. And if conditions permit it, they can be taken to the laboratory to detect with professional testing equipment to make a comprehensive understanding of it.

4.2 *Fabrication of rubber concrete*

The proportion and mixing are very important for ordinary concrete, the same being true of rubber concrete. Adding material to concrete increases the difficulty of the proportion and mixing. Therefore the operator must understand the specific properties of concrete in different proportions in order to take good control of its performance. It is the only way to grasp the added amount of powder or rubber particles to make rubber concrete meet the needs of the actual construction. In the actual making process, the operator should pay attention to the ratio of ordinary concrete apart from powder and rubber particles. Because adding powder will affect the performance of normal concrete, mechanical experiments should be initiated in the field to ensure the performance of rubber concrete and meet the needs of actual application. The rubber concrete can only be used after the test of practice. In China's production of rubber concrete, the staff merely adds some powder or rubber particles to ordinary concrete based on their work experience due to their low qualities. Although in this approach, properties of rubber concrete can be improved, the targeted ratio cannot be accomplished according to the actual situation of the construction project. For example, some of the concrete possesses a certain degree of flexibility but still cannot meet the needs of practical use since some parts require higher elasticity.

4.3 *Use of rubber concrete*

According to the previous analysis, different adding amounts of powder and rubber particles will differentiate the performance of rubber concrete. Taking the complexity of the current project into account, different parts of the construction require various materials in the actual building construction. So the added amount of powder and rubber particles should be controlled to meet the needs of different parts of the project. Thus the use of rubber concrete should be targeted. The operator should analyze the project program and get specific understanding of different

parts' needing components of concrete, to produce the corresponding amounts of rubber concrete. In practical construction, different parts of the construction using rubber concrete with corresponding properties can improve the quality of construction works in large parts. This plays an important role in the construction of the project. Given the complexity of building construction, the rubber concrete usually requires site fabrication. After fabrication, simple experiments should be conducted in the field to verify some important properties of rubber concrete in order to ensure the performance of concrete. Only the authenticated materials can be used in the actual construction, so that the quality of construction projects can be well controlled.

5 CONCLUSIONS

By analysis of the full text, it can be seen that rubber concrete has been highly valued because of its distinctive features of improving features of ordinary concrete including its elasticity and tightness and preventing problems such as cracks in the concrete. In addition, powder and rubber particles are mostly produced from discarded automobile tires, promoting the disposal of waste rubber. As a consequence, the use of rubber concrete is very common in current building construction. In conclusion, the selection of powder and rubber particles as well as actual matching proportions must be highly valued in order to well control the performance of rubber concrete. Plus, after fabrication of rubber concrete, simple experiments can be conducted in the field to ensure that its performance meets the needs of actual construction.

REFERENCES

[1] Chen Guixuan, Liu Feng, Li Lijuan, Zeng Guangshang. Research Progress of Shock and Fatigue Properties of Rubber Concrete, Building Technology, 2011(02): 169–171.
[2] Pan Dongping, Liu Feng, Li Lijuan, Chen Yingqin. Survey of Application and Research of Rubber Concrete, Rubber Industry, 2007(03):182–185.
[3] Yu Ligang, Liu Lan, Yu Qijun. Impact of Waste Rubber Powder on Properties of Concrete Mortar, Guangdong Building Materials, 2006(02):9–11.
[4] Jianhuan Wei, Huafeng Sun, Shengjie Yang. The cycle measurement of sulfur blank value with the CS-444 infrared ray carbon sulfur analyzer. Advanced Materials Research, 2012:2173–2176.

Future Communication, Information and Computer Science – Zheng (Ed.)
© *2015 Taylor & Francis Group, London, 978-1-138-02653-7*

AHP model applied to study on the education mode of ideological and political course in universities

L. Jiang
Dalian University of Technology, Dalian, Liaoning, China

ABSTRACT: In recent years, the focus of the teachers' work has been shifted to the improvement of the ideological and political education model, due to its own characteristics – a decline in the quality of education, slow development of mode, and lack of a practical value of the course. The AHP hierarchical model, just meeting the needs of such development, brings new inspiration to the education mode of the ideological and political course through rigorous scientific simplicity. This advanced model provides some constructive ideas for the stereotype, traditional mode of ideological and political education in teaching content and form as well as external factors.

Keywords: AHP model; hierarchical model; ideological and political course in universities

1 ISSUES OF EDUCATION ON IDEOLOGICAL AND POLITICAL COURSE IN UNIVERSITIES

Nowadays, the education of the ideological and political course a assumes greater responsibility with growing emphasis on "morality-oriented education" in universities. It has been regarded as required course for students at the education stage since the ideological and political course was set up in universities, which makes requirements for teachers in the ideological and political class – continuously introducing new methods and models in striving to gain new breakthroughs on teaching of traditional subjects, so as to prevent the ideological and political course from being out of line with other advanced disciplines as well as the social and cultural development [1].

The ideological and political course has been in no relation with other subjects in university because of its long-standing academic independence. The content of the ideological and political course has not been changed for a long time, because of its characteristics – preciseness and being no doubt, the stereotype mode of teachers to a great extent. For its teaching, most teachers in universities are more inclined to indoctrinate students according to the prescribed order, although they are of a higher education background and professional quality. Some teachers have gradually developed a habit of self-inertia on teaching, in addition to the abstruse and boring content of its basic education, which makes the class as dull as the printing type in textbooks [2]. Those abstract ideas cannot be effectively transformed into the specific thoughts of being fresh and vivid as well as

easy to absorb and understand. In this way, its purpose cannot be realized, and not to mention playing a positive role of educational effect and practical significance in the students' education. In addition to this, compared with other subjects, it is difficult for students to pay attention to the ideological and political learning in universities because of the disadvantages, negative treatment and overstaffing for a long time. The students' interest can be aroused by their major courses rather than textbooks dotted with ideology and concept. The ranking of ideological and political education for students continuously falls with the value of qualifying. Based on this situation, redefinition and more attempts on the mode of ideological and political education are ready to be undertaken, while its own value and status can be promoted by long-term efforts of teachers in universities [3].

2 NEW-TYPE CONCEPT EMBODIED IN AHP MODEL

A lot of new-type teaching concepts have been proposed closer to the reality and keeping pace with the times through untiring efforts of educators and applied in practical teaching, in addition to the application of many more advanced models to the teaching field crossing boundaries. Despite the subject and object not beglonging to the same range and a lack of strong support of experience, integration between them has been achieved through artificial effects, which opens up a new working field and creates more available teaching modes.

The Analytical Hierarchy process called AHP model for short, a decision-making approach with multi-level standards pioneered by A.L. Sattys, a professor at the University of Pittsburgh, is aimed at analyzing complex issues by combining qualitative methods with quantitative ones, so as to gain more reasonable results and find optimal solutions to practical problems through calculations and comparisons. This model made in the early 1970s, which has been developing up to now through continuous exploration and practice, is better in assisting in solving practical issues in social life. For ideological and political education with special and great historical significance, the AHP model undoubtedly gives support and drive to more systematic and hierarchical attempts of it on theoretical and scientific aspects. Ideological and political education in the long-term practical teaching has shown a tiled-type teaching mode characterized by neglecting the purpose and key points. The AHP model can suit the remedy and bear the brunt of the situation, conducive to the presentation of advanced ideas hidden in the textbooks by three-dimensional means again. The constructive and specific planning can be gained in the universities' ideological and political course through the establishment of the hierarchical model, effective measure and calculation, which is conducive to freeing course education from the chaotic situation like "a pot of porridge", making the content and mode of teaching a rare sort and provides a reference for a targeted teaching scheme. Then in the next step, data collection and calculation, professional support is most needed, including absolute cooperation of teachers – offering actual reference data teaching, and strong support of students in teaching feedback, thus laying a solid and reliable foundation for complicated calculation and budget.

Related factors, generally, are divided into different levels, three layers, by experts according to their weight function – target, criterion and solution layers. The first one occupies the highest position of the model, just containing the only one factor, target. The second one, criterion layer, is in the middle, playing a connecting role in the subject, including various principles and measurements in the practice and the last one at the bottom with a strong operation and strict practice requirements. The optimal hierarchical model can be gained through the study of scientifically complex overall sorting of these hierarchical factors, finding the most satisfactory consistency, thus judging the primary and secondary status of all kinds of elements in the actual work on the scientific basis. Attempts on the ideological and political course can be made through the educational research method in support of scientific calculation and advanced model, contributing to the promotion on macroscopic and emotional teaching idea of educators in the qualitative and quantitative process. Teachers receive accurate and detailed visual feedback in the later calculation and comparison of data. In addition, they can get a better

distinction in importance, including specific course content, classroom atmosphere, course form and assessment method, thus methodically and purposely achieving the goal of ideological and political course in universities.

3 ENLIGHTENMENT ON THE EDUCATIONAL MODE OF UNIVERSITIES IDEOLOGICAL AND POLITICAL COURSE FROM AHP MODEL

3.1 Focus on practical key points of content in the ideological and political education

The content of the ideological and political education in universities is characterized by large capacity, strong stability and the vast majority of text is covered with advanced ideological and political ideas – the basic principles of Marxism, Mao Zedong Thought, Deng Xiaoping Theory, the important thought of Three Represents and the Scientific Outlook on Development. It is difficult for students to establish the reasonable ideological and political structure by just attending a lecture in their mind and grasp the ideal key points of learning in its practical teaching, although the arrangement of the course content in the ideological and political textbooks is the most reasonable according to the time order and logical sequence.

CPC Central Committee and State Council recently issued a document "opinion on further strengthening and improving ideological and political education", making increasingly clear requirements for enhancing the ideological and political quality of contemporary universities' students, the practical significance of ideological and political education so that the ideology of students is able to effectively serve the cause of socialist construction. It is imperative that emphasis on the important theory learning in the traditional teaching is preserved; in addition, pay more attention to combing the practical ideological and political ideas in current social backgrounds. The AHP hierarchical model is aimed at the ideological and political education, in great efforts to expand the practical significance of the ideological and political course, increase the weight of this factor, and pay much more attention to it, which can bring new life and hope to this situation – lacking in development and breakthrough of ideological and political course for a long time. There are a lot of attemptable approaches to meeting such demands. For example, in the establishment of the course content, it is necessary to cover the teaching content of regular teaching syllabus and some sections about the hot-button topics of current politics in the National People's Congress and the CPPCC National Committee meeting, thus achieving the purpose of learning for practice in the way to provide the opportunity for students of practical application on ideological and political knowledge, combining theory with practice. In addition, the human-oriented education of moral quality and historically Chinese traditional virtues, as flavoring agent, should be interspersed in the ideological and political class, which infuses the boring class atmosphere into fun and creative fresh elements. In this way, inner motivation can be gained for reform and development of ideological and political education itself, and teaching effect can present a qualitative leap.

3.2 Hierarchical mode formation of ideological and political education

Ideological and political education itself includes the syllabus content, activities of ideological and political education and some supplementary education forms – organizing forums on ideological and political ideas. Nowadays, it is far from enough to absolutely meet demands of the teaching syllabus of ideological and political education as well as the students' learning of the new-type education mode, only by grasping the "three aspects" as mentioned. With ideological and political education paid much more attention to by state education authorities, it is imperative that ideological and political teachers no longer follow the old routine but try to apply good ideas in practical teaching, gaining much more highlights for the traditional mode of ideological and political teaching, so as to meet the requirements of the teaching effect.

In recent years, for example, the methods reform of ideological and political teaching has been promoted in many colleges and universities, Fujian province, which not only enriches the learning experience of students in class but also makes ideological and political education more interesting and practical as well as gains the echo between course and students, thus bringing new hope to the ideological and political education. The survey, by the end of 2013, shows that 26 colleges and universities participating in the teaching quality of ideological and political education in Fujian Province, the course quality is good and above average in 25 colleges and universities. The reason why such gratifying achievements have been made in the field of ideological and political education is that these teachers conducted some hierarchical attempts of teaching to a great degree. The specific forms include some platforms with education sectors as the main body, monitoring platform of teaching quality, communication demonstration platform, regular feedback platform, teacher-student interaction platform, which can make the quality of ideological and political teaching meet the strict requirements and gain a good guarantee from the start. In addition, some teachers make the teaching plan with a good reference by cooperating and discussing in the specific teaching activities. For example, there is a rolling mode of teaching free from the textbook structure, that students can distinguish the universality and individuality of the learning content on their own by continuously instilling thematic knowledge in themselves several times, thus forming a system of ideological and political knowledge in line with their own logic mode. Also there are new multi-level and diversified education forms strengthening the teaching effect of ideological and political education, participation teaching, case teaching, and exploratory teaching.

3.3 Serving major contradictions by coordinating minor factors

The major contradictions, for teachers, need to be overcome to improve the teaching quality and effect in ideological and political education; the minor factors in close contact with the major contradictions, moreover, can provide an indispensable force by solving them. These minor factors include necessary policy support, teaching equipment, and teaching reform. Schools should carry out extensive cooperation with corporate and community groups in the teaching reform to gain a series of support. According to the requirements of dialectical materialism – grasping major contradictions of matter but not ignoring minor contradictions, it is necessary that people pay much attention to minor factors after finishing the key work, thus making their task add value and individuality. For example, recently it has been prepared by the Ministry of Education compiling a series of ideological and political textbooks. There are seven working teams adopting the system of editor-in-chief responsibility, consisted of many famous domestic writers and experts, thus providing substantial support for further development of the ideological and political course. Schools and teachers needs to seek specific supports from different aspects in order to lay a solid foundation for the whole of the AHP model. For example, some schools with a good reputation can take the initiative to some specific sectors of the government of an olive branch, looking forward to gaining the opportunity for students to experience a combination of political theory and actual work in the form of practice. In addition, classroom teaching materials keeping pace with the times and vivid teaching tools should be introduced in the class communication between students and teachers, conducive to making the theory and concept of emotional weakness vividly three-dimensional. Thus students can develop a new understanding and larger interest in the connotation of traditional ideological and political education, so that students take the initiative to join the discussion and analysis of theory again. In this way, it is easier to establish a virtuous cycle of teaching and learning in the classroom, so as to lead the entire ideological and political education gradually into an autonomous mode of action and gaining a qualitative leap in the quality of teaching.

4 CONCLUSIONS

The AHP hierarchical model is applied to the study on the teaching mode of ideological and political education in colleges and universities so as to achieve a leap in the ideological and political education. This application, its intuitive and simple description is that various factors affect the teaching – some of classroom education itself and other secondary factors existing in the periphery of the main body, are sorted and prioritized through rigorously scientific calculation and comparison, thus solving all the problems of each item in this subject on whose main contradictions are laid stress, in accord with planned sequence. In its practical application, moreover, it is the only way that must be passed to achieve development and innovation of the ideological and political course, including expanding the practical teaching content and utility range of the teaching form as well as coordinating all aspects of teaching. These human-oriented and advanced teaching experiments will bring new hope to the course of ideological and political education in colleges and universities.

REFERENCES

[1] Jin liangyuan, *Current situation of university students' ideological and political perspective*. Science and Technology Innovation Herald. 2013 (7):34–36.

[2] Wang xiaoyun, *Reflections on strengthening the construction of ideological and political education team*. Study of ideological education. 2013 (7):34–36.

[3] Research Group of the League Central School, China Youth Research Center; *Research report of University students' ideological and political education*. China Youth Research. 2010 (4):70–74.

Future Communication, Information and Computer Science – Zheng (Ed.)

Folk-toy craft culture and contemporary industry informatization

H. Liu
College of Art, Jiangxi University of Finance and Economics, Nanchang, Jiangxi, China

ABSTRACT: Folk toys are not only a kind of playthings for entertainment, but also an important embodiment of folk spirit in China. However, with the development and progress of time, traditional folk-toy culture has suffered an unprecedented impact. Many reasons, such as less attention from the public, the severe aging of the products and narrow range of being understood, have constrained the development of the folk-toy culture. The work innovatively analyzed how to promote folk-toy culture and improve the market competitiveness of the folk-toy contemporary industry from the perspective of the contemporary industry informatization so as to provide references for the protection and inheritance of folk-toy culture in the future.

Keywords: folk toy; culture; industrial informatization

1 BRIEF INTRODUCTION TO FOLK-TOY CRAFT CULTURE

1.1 *Technology characteristics of folk toys*

1.1.1 *Model of folk toys*

With simple models, traditional folk toys in China share typical characteristics of Chinese ethnics. Those toys are created with the inspiration of the comprehension of life and nature as well as the expression of emotion. Therefore, different from other ceramic and painting process, folk toys show more features of folk custom and popularity, and they are more easily accepted by the broad masses. In addition, the models of Chinese folk toys are very authentic, such as some historical or mythological figures and some real animals. With rich and practical significances, those toys show vivid and incisive images, explaining the pursuit and yearning for a better life of people. Besides, the models of those folk toys also show the feature of developing intelligence such as the Chinese ring puzzle, the feature of entertainment such as the clayey whistle and roly-poly, the feature of good moral such as the cloth tiger and sachet and the feature of culture such as shadow play [1].

1.1.2 *Application of color in folk toys*

The application of color in Chinese folk toys is in line with the Chinese traditional aesthetic standard for color. Different from color collocation in the western world, the application of color in China doesn't have a purely scientific chromatographic analysis. Instead, it is an intuitive reflection of natural color and realistic color [2]. Of course, the application of color in folk toys also follows color collocation, emphasizing people's visual experience. The colors of folk toys in China are mostly bright, relaxing and warm due to the lively and cheerful tone of Chinese traditional folk. Therefore, those colors also bring people the experience of happiness and pleasure. Take the common children toy cloth tiger for example. Nowadays, the color of the tiger body is no longer confined to yellow; some other colors such as red and pink are also very common. Besides, the color of tiger ears and tiger tails is no longer limited to pure black [3]. Instead, they are more common in green, cyan and blue. The color collocation of a bright red tiger body with cyan pattern plus yellow tail and blue ears leaves people the impression of the vigor and vitality of a tiger.

1.1.3 *Decoration pattern of folk toys*

Folk toys have varied decoration patterns. There are both abstract patterns and specific patterns. Abstract patterns are mostly line designs with a strong symmetry. Some of them are just conjured up without any references such as the ruyi pattern, word pattern, fretwork pattern and lock pattern. Some of them are created by the imagination of some specific matters such as the moire, silk-ball pattern, leopard grain and scale brocade. Those different decoration patterns also show different morals. For example, taotie pattern symbols ample food and clothing; flowers with paradise flycatchers imply longevity and happiness. In addition to abstract patterns, there are also some specific patterns. For example, human images are often put on tiger bodies, and some images of small animals are drawn on the bodies of big beasts, showing certain enlightenment. Besides, there are other numerous physical patterns such as flower, grass and cloud applied to folk toys.

1.2 Cultural connotation and significance of folk toys

1.2.1 Developing folk-toy culture as an important part of Chinese art culture

Implying rich Chinese traditional culture and ethnic customs, folk toys are the wisdom of ancient working people in China and form an important part of Chinese art and culture. There are different types of folk toys including some season toys, life-etiquette toys and daily entertainment toys, expressing a rich connotation and special significance. For example, a female cloth monkey for weddings is a gift and wish for a good marriage; windmills, lanterns and diabolos represent the arrival of the Spring Festival. China has been strongly advocating the inheritance and development of intangible cultural heritage in recent years. As a result, folk art is rising in popularity, and people are showing more and more interest in folk toys. Moreover, more and more people begin to understand and study folk-toy culture while purchasing and enjoying folk toys. In addition, the models of folk toys are also being applied to various propaganda posters and stamps, achieving both a good social effect and the purpose of promoting folk-toy culture.

1.3 Protecting and inheriting Chinese folk-toy culture as the necessity of developing diverse culture

The globalization trend not only promotes economic prosperity but also prompts the unicom and communication of culture. However, China has been impacted by western culture in recent years, resulting in a young generation yearning to join the world and moving away from Chinese traditional culture. Nowadays Barbie Dolls, Teddy Bears, Superman and other foreign toys are very common in people's lives, while the folk toy is gradually being abandoned by the society and market. Therefore, it is of great importance to strengthen the protection and inheritance of culture so as to enhance the attraction of Chinese traditional culture to the younger generation. The development of cultural diversity has important guiding significance for the spiritual construction of modern people, and the rich cultural connotation and numerous forms of Chinese folk toys are exactly the foundation of cultural diversity.

2 RESEARCH OF FOLK-TOY CULTURE AND CONTEMPORARY INDUSTRY INFORMATIZATION

2.1 Definition of the contemporary industry informatization of folk-toy culture

Contemporary industry informatization tries to enhance the competence and increase the wealth of the modern industry with the auxiliary of modern information technology. Information technologies include information collection, information classification, information processing, information analysis and information application. However, it is a narrow definition to consider contemporary industry informatization construction of folk-toy culture as the only application of advanced information technology. Instead, it should have a more profound connotation. In terms of content, contemporary industry informatization includes three aspects, namely enterprise informatization, industry informatization and connection informatization among industries. However, contemporary culture industry bears its own particularity. In addition to improve the market competitiveness and occupancy of cultural products, it should address more the protecting and inheriting of folk-toy culture. Therefore, contemporary industry informatization of folk-toy culture should also include social informatization and government informatization so as to enhance the guidance of society and government to folk-toy culture construction. Although contemporary industry of folk-toy culture has begun to promote the informatization construction of modern industry, informatization still faces many obstacles, such as a weak consciousness of informatization management, less investment on informatization construction and shortage of informatization talents. Thus, operators and managers of modern industries should pay more attention to solve those problems.

2.2 Necessity and feasibility of the contemporary industry informatization of folk-toy culture

As an important part of traditional craft culture in China, folk-toy culture hasn't gained enough attention from people due to its own characteristics of formation and development. Nowadays Chinese culture has been more and more integrated with world culture, and the culture of different ethnics communicates frequently and contacts more now. Thus, it is essential to improve the influence of the contemporary culture industry so as to maintain the superiority of folk-toy culture itself. Therefore, contemporary culture industry informatization is an inexorable developing trend, indicating the necessity on the contemporary culture industry informatization of folk-toy culture. In addition, the contemporary culture industry informatization of folk-toy culture also requires feasibility, helping to achieve a better management effect. On one hand, folk-toy culture possesses huge space for developing resources, meaning that it is inexhaustible to some extent. According to the different meanings of folk toys, people can show different folk toys to the public in a particular period and thus promote culture connotation and its influence. On the other hand, the strong popularity of folk-toy culture has equipped itself with a certain ideological recognition in the mind of the masses. With the support of the masses to promote the progress of the contemporary culture industry, the construction of the contemporary industry faces less resistance.

2.3 Misunderstanding of contemporary industry informatization of folk-toy culture

At present, there are many misunderstandings in the construction of the contemporary industry informatization of folk-toy culture, in which the most prominent ones are misunderstanding of culture cognition of folk toys and misunderstanding of the contemporary industry informatization construction. Folk-toy culture is an important part of Chinese traditional culture. But some people consider it a completely old culture for entertainment. It doesn't have the connotation of deep exploration and research like painting and ceramics culture. Besides, folk toy is a kind of handcraft, and the manufacturing of folk toys is very simple. Consequently, folk-toy culture with infomatization is just a so-called "face project". This view is generated on the basis of indifference of Chinese folk craft culture. Although China has lots of elegant arts and culture, folk craft culture possesses a relatively extensive mass basis and is much easier to be understood and accepted. At this point, informatization construction needs to better promote this culture, so it is not just a "face project". The manufacturing industry of folk toys belongs to labor-intensive industry, and it requires simple manufacturing technologies and materials. Nowadays, human power is the key to develop a contemporary industry. However, informatization construction is unable to fully exploit this advantage, resulting in some people thinking informatization development is unnecessary. In this misunderstanding, human resource is viewed as an absolute advantage. But the rising of labor cost has become an inevitable trend for the development of all contemporary industries. Consequently, folk-toy contemporary industry in China should give up the old rough developing model and turn to make use of advanced information technology in order to improve the comprehensive developing capacity of modern industry.

2.4 Ways to implement contemporary industry informatization of folk-toy culture

2.4.1 Combining the creation of new products and promotion of old products

The spread of folk-toy culture greatly relies on its products. The model of cultural products shows its culture connotation, and the spread of culture is achieved through the purchase of products. Therefore, advanced information technologies should be applied to innovate new products and promote old products. New innovative products take advantage of new technologies and materials to present Chinese national culture and customs, endowing the products with new performances and applications. Besides, people may pay more attention to the products due to the novelty of new products, making sales much easier. Meanwhile, people try to understand the implicit culture when they are attracted to some products. Although some people show their esthetic fatigue to old folk-toy products, it can't be denied that those products have created much culture value and a wide range of recognition. When people introduce folk-toy culture to those without a good understanding of this culture, they still prefer to illustrate their explanation of the old products. Therefore, old folk-toy products also play an important role in modern industry. Currently, people's connections to folk toys are mainly from introduction or random purchase. But this kind of promotion method is relatively narrow. However, the culture contemporary industry construction of informatization promotes information in a wider way so as to widen the spread of culture. For example, build specific websites on the Internet to introduce folk toys or display promotion videos on TV or radios.

2.4.2 Strengthening contemporary industry informatization management

Although the contemporary industry of folk-toy culture has had the consciousness of the informatization construction, it is still necessary to strengthen informatization management of contemporary industry in order to achieve a better effect. Firstly, informatization management of contemporary industry should firstly get its breakthrough to avoid the influence of the traditional old management mode. Secondly, informatization management of contemporary industry should be top-town, inside-out and step-by-step, following certain orders. Thus, the gathered information can be more effectively used, and the development of the contemporary industry can be more balanced, avoiding uneven phenomena and problems. Thirdly, corresponding management plans must be made according to scientific and reasonable principles. Starting from the actual characteristics of modern industry, the plans should properly plan management procedures and details, avoiding blindly referring to other informatization management means and methods. Finally, carry out effect evaluation and testing for the effect of informatization management so as to track and correct the deficiencies of management. Thus, the construction of the scontemporary culture industry can be further perfected. In addition, it is also essential to accordingly develop information processing software to manage product design, material selection, procurement, promotion and other procedures for dynamic information management.

2.4.3 Strengthening talent team construction for the contemporary industry informatization of folk-toy culture

Different from other contemporary industries, folk-toy culture contemporary industry has a broader cultural connotation. Therefore, the management of talents for this contemporary industry requires managers to be equipped with not only appropriate management knowledge but also a certain cultural foundation. With a unique understanding of Chinese history and Chinese traditional folk, they should try their best to apply culture into products so as to fully play the cultural function of products. In recruitment, enterprises and relevant departments should evaluate the cultural literacy of candidates and select talents with high quality. Moreover, enterprises should provide trainings for their employees working in the contemporary industry to introduce them to folk-toy culture and enhance their knowledge and cognition of folk-toy culture.

3 CONCLUSIONS

In this modern materialistic society, there are various kinds of folk-toy crafts. Folk-toy culture is an important part of the Chinese traditional craft culture. Combining it with informatization not only can strengthen the market competitiveness of this industry, but also can enlarge the influence and spreading range of folk-toy culture.

ACKNOWLEDGEMENT

The work was funded by "Research on the Aesthetic Interest of Folk Toys in Jiangxi", an art and scientific planning project in Jiangxi Province in 2013, Project Number: YG2013005.

REFERENCES

[1] Ji Xianghong. Technological Characteristics of Chinese Folk Toys. Journal of Soochow University Engineering Science Edition, 2008(03): 77–79.
[2] Kong Yun. Contemporary Value Orientation of Chinese Folk Toys. Art & Design, 2005(01): 115–116.
[3] Sun Li. Protection and Inheritance of human intangible cultural heritage—Resources of Chinese Folk-Toy Culture and Art. Toys World, 2006(09): 50–52.

Future Communication, Information and Computer Science – Zheng (Ed.)
© 2015 Taylor & Francis Group, London, 978-1-138-02653-7

Analysis of image sharpening and tonal processing technology

H. Wang
Guangdong Engineering Polytechnic, Guangzhou, Guangdong, China

ABSTRACT: At present, computer image processing is the main method for web design and image design. It can be applied to many industries, enabling the image industry to achieve unprecedented development. Based on the analysis of current image processing methods, the work researched image sharpening and tone processing and principles of post processing, and forecasted the future development of image processing. Ultimately, the image-processing method on the basis of computer technologies will inevitably become the mainstream of image processing in the future.

Keywords: computer; image processing; method; analysis

1 INTRODUCTION

Image industry develops very quickly from the initial film photos to present digital images. Image processing is usually built on the basis of computer technology. The digital camera is commonly applied to modern photography to generate digital images that can be recognized by computer image processing software [1]. So-called post processing is to further process those digital images on the basis of photographing. Post processing enables images to better meet the aesthetic requirements and satisfy the demands of customers. Therefore, the work focused on the analysis of the post processing method of computer images.

2 THEORY OF COMPUTER IMAGE PROCESSING

The development of computer technology adds some new elements into the photography industry. Image processing based on computer technologies is more colorful, leading to more potential space for the development of the original photographing industry. As a result, the photography industry develops from the original film photography to the present digital photography. The basic theory of digital photography is to make use of digital identification technology of the computer to conduct post processing for images. Those images are taken by digital cameras and can be recognized by image-processing software [2]. Therefore, it is of great importance to identify the theory of computer image processing in order to get a better understanding of the image post-processing method.

2.1 *Foundation of processing computer digital images*

The development of computer technology has enabled the photography industry of film-camera technology to make great progress. The traditional photography industry generates images through light-sensed film, and pictures can be printed out after a complex film-developing process. However, this kind of method still has many deficiencies [3].

First of all, film is expensive and can only be used once. Thus, film should be properly stored due to its high cost, and early exposure declares its scrap. Consequently, the cost of film-camera photography is relatively high. Moreover, film must be developed in a specific room, meaning a large workload and complicated process.

Secondly, the photography of a film camera is constrained by many factors such as a good location and the weather. In order to take more pictures, people need to take those rigid requirements into consideration, greatly impacting the conditions of photography.

Finally, film-camera photography is very tedious and has extremely high requirements for the cameraman. In each photographing process, the cameraman needs to do much complex calibration work, causing photography to be a profession for the minority. People just starting to enter this profession cannot totally handle it at all.

However, the emergence of the digital camera easily solves those problems. Based on the theory of pixel, the digital camera doesn't need film any more, eliminating all the inconvenience caused by film. In addition, digital cameras make use of memory cards, making it easy to carry. The memory card can be directly connected to the computer, thus ensuing quick delivery of photos. In a word, digital cameras and the application of computer technology make modern photography technology more convenient and efficient.

The foundation of processing computer digital images can be analyzed through several aspects.

Firstly, the computer can be equipped with software for image post processing. Nowadays, lots of image-processing software can conduct post processing for pictures.

Secondly, digital images are generally in jpeg format and can be recognized by the computer. Thus computer can conduct post processing.

Thirdly, the transmission of digital images to the computer can be achieved by a data cable and memory card. Moreover, there are many easy and convenient ways to store photos.

Therefore, digital image is an image mode based on computer technology. Installing image-processing software can easily achieve the post processing of images.

2.2 *Theory of computer image processing*

Analyzing the theory of computer image is helpful to understand the theory of image post processing because the format of digital images is developed based on computer technology. In the process of design, the computer views pixel as the smallest unit of images. Therefore, pixel is the smallest unit of digital images. These pixels can be bright spots or dark spots of different colors. In the process of imaging, pictures are composed and imaged through those pixel spots. Therefore, a computer equipped with image-processing software can carry out post processing for images. Specific analysis of this theory is as follows:

Firstly, the design of computer image-processing software is very important. In general, the computer must first recognize images before processing them, requiring relevant image software to complete. For example, viewing pictures requires installing picture-viewing software; post processing of images requires installing image-processing software. At present, the popular image-processing software on the market includes Photoshop and MeiTu. All those are the main image-processing software of the computer.

Secondly, images should meet the recognition standard of computer software. Usually, digital images can be recognized by commonly used image-processing software. Therefore, it is necessary to save images in a common format before image processing, making it easy to view and edit images. Generally, JPG format is considered as a basic identification pattern.

Thirdly, images can be recognized by computer software because recognition happens at the interface port of software between the computer and images. What's more, pixel is the smallest unit of an image, and the recognition pattern of software also takes pixel as its unit, further proving that images can be recognized and processed by computer.

3 METHODS OF COMPUTER IMAGES SHARPENING

Multiple processing methods are usually applied to the post processing of computer images to achieve a better imaging effect. Sharpening is commonly used and also the most basic method for post processing.

3.1 *Concept of computer image sharpening*

Image sharpening in fact is opposite to image blurring. There are a variety of operations for image blurring. Contrary to image blurring, sharpening is to make unclear spots clear and bright spots brighter. In general, if focus shooting at gear P or M is not applied in the early stage of the photographing process, pictures have no sense of texture and show no quality. In addition, the main body of the image and spots needing highlight are not highlighted due to overall shooting method, leading to the failure of the desired effect. At this moment, image-sharpening process can make up for those deficiencies.

3.2 *Methods of computer image sharpening processing*

Sharpening processing is very common, especially for photos of bad photographing effect. Sharpening can make the image show a better sense of texture. There are several common methods of sharpening processing, explained as follows:

Firstly, local sharpening: to highlight part of the picture. Sharpening is to make unclear spots of the image clear. Local sharpening can achieve this purpose and make the local effect more prominent at the same time.

Secondly, smart sharpening: to sharpen the picture through intelligent operation. In the operation process, operators just need to select the image and choose smart sharpening, and then the picture will be sharpened in an intelligent way. This operation is very easy, but personal ideas cannot be applied in this processing. In addition, the sharpening effect is also different from the desired effect.

Thirdly, edge sharpening: to realize sharpening processing for the whole picture. Among multiple processing methods, edge sharpening is in line with personal standards with a relatively better sharpening effect. The concept of the edge exists in the process of imaging. In an image, relatively referring edge can be considered as one edge. Edge sharpening can realize the sharpening effect for the whole picture. Generally, edge sharpening is considered an integrated operation of smart sharpening and local sharpening.

Finally, selection sharpening: to sharpen the part needing sharpening processing. This operation is of high autonomy, because most image-processing software has the function of selection. Thus, operators just need to sharpen part of the image according to its actual condition. The operator should first select the part that needs processing, and then conduct sharpening processing for this part. This operation is more suitable for people with strong processing ability.

3.3 *Skills of computer image sharpening processing*

Sharpening operation should conform to the actual situation of the image. Four basic operations of processing were analyzed above. Operators can choose any of them according to the actual condition to finish image sharpening. If the whole image is blurred or the color of the image looks dark, sharpening processing after adjusting its color contrast and brightness can make the original image brighter. Directly sharpening the image without any previous processing requires a clear precondition: the original image has a good effect. Otherwise, direct sharpening will not achieve a good effect. In addition, sharpening processing should be based on the actual condition of the image. Therefore, the image can present a perfect effect in the processing. If not familiar with image processing, users can choose smart sharpening or overall sharpening. Those sharpening operations are much easier and can achieve an obvious sharpening effect. It's better to use edge sharpening to make the whole image present the effect of high brightness and good texture. Skilled users can choose selection sharpening, requiring certain skills of sharpening processing.

4 TONAL PROCESSING METHODS OF COMPUTER IMAGE

In addition to sharpening processing, tonal adjustment is another important and essential operation of post processing for computer images. Tone is the overall effect of a picture. Tonal adjustment is to make a picture present a certain effect to achieve the significance the image. Tonal adjustment is commonly applied in many scenery images and figure photos to express the potential meaning of the images. For example, when taking pictures of the sunset view in the evening, the scenery of the sunset will be hazy with a quiet and peaceful effect. At this moment, post tonal adjustment can perfectly manifest this effect. In addition, the collocation of color is mostly addressed in the process of character photography. At present, many image-processing software can set the scene for images to make the photography easier. However, the portrait sketch effect cannot be achieved through photography, but post tonal adjustment can make it. Post tonal adjustment presents the image with black and white effect and makes the color inclined to dark and gray, thus achieving a sketch effect easily.

5 COMPOUNDING PRACTICE OF COMPUTER IMAGE SHARPENING AND TONAL PROCESSING

In general, combining the operation of image sharpening and tonal processing can achieve better texture and quality of the image and strengthen its line feature. Sharpening is to make the image clearer so as to better show its texture. But when only sharpening processing is applied in order to achieve a better texture of images, an originally good image may be overexposed. At this moment, proper tonal processing can avoid this problem. For example, when processing a landscape image, the operator can first select the part needing processing, then adjust the tone for this part to a desired tone range, and finally conduct smart sharpening for the picture. Smart sharpening is an intelligent operation, so operators only need to select the image and then choose smart sharpening, and the picture will be sharpened through that intelligent operation. Therefore, the sensitivity and the overall texture of the image will be significantly enhanced through such two continuous steps. In addition, the combination of image sharpening and tonal processing is also very common in the design of web pages. Usually, Internet has standards for the size of pictures, so web pictures must be properly treated. But in order to keep a high fidelity of the image and also successfully compress the image, image sharpening and tonal processing should be combined to meet the above standards to satisfy the requirements of web pages.

6 CONCLUSIONS

The analysis of post processing of computer images manifests that an image is composed of data. A computer, equipped with software that can recognize digital images, is able to realize post processing of images. Generally, image sharpening processing and tonal adjustment are the main operations in image post processing. Sharpening processing can make original unclear pictures show better texture and present brightness and darkness of the image more obvious, making the image more vivid. While tonal adjustment usually is to make up for color distortion caused during early photographing and to make the integral color more realistic. As a result, these two processing methods are to make up the inadequacy of image photographing so as to present the image with a better aesthetic feeling and enhance image fidelity.

REFERENCES

[1] Li Kerong, Problems of Non-Linear Editing System in Post Production, Science & Technology Information, 2008(05): 56–56.
[2] Xu Shenghua, Analysis of the Post Editing and Special Effect Synthesis of Digital Videos, Scientific Consult (Science & Management), 2011(03): 78–79.
[3] Zhao Liang, Influence of Post Production of Digital Pictures in Photography Works, Art Research, 2010(1): 134–135.
[4] Jianhuan Wei, Huafeng Sun, Shengjie Yang. The cycle measurement of sulfur blank value with the CS-444 infrared ray carbon sulfur analyzer. Advanced Materials Research, 2012: 2173–2176.

Future Communication, Information and Computer Science – Zheng (Ed.)
© 2015 Taylor & Francis Group, London, 978-1-138-02653-7

Research and practice on dual-degree undergraduate personnel training mode under new situation

M. Wang
Foreign Languages College, Dalian Jiaotong University, Dalian, Liaoning, China

ABSTRACT: This study analyzed specific conditions and practical ideas faced by a dual-degree undergraduate personnel training mode under the current new situation. Above all, it mainly analyzed the practical conditions of the dual-degree undergraduate training mode, and then explored the positioning of talent cultivation of a dual-degree undergraduate. Finally, it made some reference to the practice of a related training mode based on the above statements. It is particularly critical to cultivate dual-degree undergraduate talents under the new situation. And the key to this mode is to optimize and adjust the course structure around enhancing comprehensive abilities of students. In addition, this study has analyzed the effort points of dual-degree undergraduate training mode and made new proposals based on it.

Keywords: new situation; dual-degree; undergraduate; personnel training

1 INTRODUCTION

The demand for high-quality compound professionals of foreign languages is growing increasingly in current society. Meanwhile, the dual-degree undergraduate talent training mode has gradually become a mainstream mode of cultivating foreign language talents under the new situation. By this mode, talents on the basis of a foreign language can further develop integrated capabilities and comprehensively enhance their ability of innovation, thinking as well as subject synthesis, to better meet the requirements of the times. Since the twenty-first century, with China's accession to WTO, China has had some connections with other countries in many ways. Specifically, there have been closer economic exchanges and cultural exchanges apart from political communication [1]. However, talents of a pure foreign language master only the skills of language knowledge and often appear inadequate for application in specific professional areas. Thus, training foreign language talents in the form of dual-degree can make them better enhance their professional skills and possess improved adaptability to the market and society.

2 REALISTIC CONDITIONS FACED BY DUAL-DEGREE UNDERGRADUATE TRAINING MODE

With China's constant evolvement in economy and society under globalization, higher education has a more demands to meet and especially has more prominent sense of a professional mission to serve the society. Nevertheless, ordinary colleges and universities often base themselves on distinctions between disciplines and majors to cultivate talents. Consequently, graduates have a relatively simple knowledge structure, and thus lack employability and competitiveness. Therefore, foreign language professionals should further strengthen their comprehensive mastery of knowledge and further enhance their adaptability. It allows students of colleges and universities to achieve more full employment, apply their knowledge in society, and then gradually realize their life values in the process of serving the society and the public [2]. Thus, it is relatively realistic and urgent to innovate foreign language professionals training mode in universities.

2.1 Changes faced by undergraduate foreign language education

Positioning of professional teaching faced by undergraduate foreign language education in universities is also constantly changing. Currently, the whole market requires more foreign language talents with professional application ability than those that master only pure language knowledge. Therefore, despite the traditional scholastic education, foreign language talents should transform into compound talents with stronger application features, wider scopes and a comprehensive knowledge structure. Thus, the positioning of foreign language professionals in professional training should be possessing relatively solid fundaments of foreign language knowledge, a broader range of knowledge, stable expertise knowledge, strong capabilities of comprehensive knowledge and a better overall quality.

2.2 International pattern facilitating dual-degree undergraduate training mode

With the constant development of Chinese economy and society, as well as deeper international exchanges, larger limitations have emerged in both methods and ideas of traditional foreign language professional personnel training. The knowledge composition of traditional foreign language professionals in training and education is often based on a single subject knowledge. Expertise unitary tendencies emerge especially in the course setting of foreign language professionals, preparation of course-related teaching materials and daily teaching processes. With such a unitary set of subject knowledge, students can still continuously form stronger abilities of building comprehensive knowledge through a repeated learning process [3]. However, they cannot form more comprehensive and integrated capabilities of analysis, particularly capabilities of interdisciplinary comprehensive analysis.

By the same token, expertise unitary tendencies also emerge in the training process of international politics or law major students. Although students are adequate in terms of expertise, they are deficient in a comprehensive range of knowledge. The negative impact of expertise unitary tendencies is serious. For example, students majoring in international politics lacking foreign languages knowledge during the learning process will affect the range of their collecting foreign literature. Therefore, limitations of external exchange and cooperation appear. Thus, students majoring in international politics and other types often possess no strong comprehensive practical ability despite the internationalization of their courses. And foreign language skill is their short board.

3 POSITIONING AND EFFORT POINT OF DUAL-DEGREE UNDERGRADUATE TRAINING MODE

As can be seen from the above discussion, undergraduate dual-degree talent cultivation is currently the urgent need of reality. The current training mode of professional talents is no longer able to meet the needs of the community for combined talents with comprehensive application ability. Therefore, the traditional personnel training philosophy should be further transformed to put the development of society and current demands for talents as important guides. It is urgent to further innovate and reform the current training mode of higher education. At the same time, ordinary comprehensive universities should further play to the advantage of available disciplines, optimize the professional structure of disciplines and integrate resources of high quality teaching.

On the basis of such a reality, undergraduate dual-degree training mode of higher education should be further developed and improved. And it should be positioned on the basis of training practical and compound talents. Thereby students can further master a

particular area of expertise on the basis of proficiency in foreign languages, so as to enhance their overall quality. And a particularly strong talent team is trained and formed. Consequently, the interdisciplinary disciplines of various professionals in universities are promoted and the practical degree of professional disciplines knowledge is enhanced. Undergraduate dual-degree training mode should focus on the following effort points.

3.1 Effort point of breaking the unitary tendency of subject knowledge

Undergraduate dual-degree training mode should break the unitary tendency of professional subject knowledge as well as optimize and integrate the subject knowledge of various majors. Universities should further promote general education and especially widen the relevant knowledge structure of students so as to make it have systematicness and horizontal thickness. Therefore, the dual-degree undergraduate talent training mode should further combine relevant knowledge as well as further strengthen the targeted and practical application of knowledge imparted in courses. Especially it ought to combine both related language knowledge and expertise knowledge to enhance the effectiveness of relevant courses.

3.2 Effort point of cultivating ability of using comprehensive culture

Undergraduate dual-degree training should focus on developing the abilities of students. The emphasis is to enhance professional creativity of students in using comprehensive culture, so that students can think independently and then gradually form innovative capacities. In the process of dual-degree training, students are supposed to raise relevant questions more independently and analyze them; they are supposed to further imitate relevant professional and language knowledge as well as integrate and analyze them on the basis of imitation. Therefore, the teaching process should not only allow the teachers to teach one-dimensionally, but also promote thinking and knowledge inspiration of the students; it increases discussion of teaching knowledge as well as the interactivity between students and teachers. Thus, students will be more active to participate in learning processes and enhance their comprehensive abilities.

3.3 Effort point of protecting the foreign language quality of dual-degree students

The effort point is to protect the foreign language quality of dual-degree students. These students should reach the level of professional foreign language knowledge required by the state. During the dual-degree learning process, they ought to ensure their abilities of foreign language knowledge in listening, reading, writing, translating, speaking and other aspects; they ought to master more excellent cross-cultural communication skills and awareness as well as proficiently organize and communicate in terms of international politics, law, culture, language and other aspects, in relation to the foreign exchange situation. As a result, exchanges and coordination will be gradually realized in collisions of cultural differences.

3.4 Effort point of enhancing the overall quality of dual-degree students

The effort point is to enhance the overall quality of dual-degree students. On the basis of upgrading the professional and language knowledge in students as well as allowing them to better grasp the knowledge, it promotes the upgrading of quantity and quality of morality, intelligence, aesthetics and other aspects. In the training process, it will gradually improve the level of theory and policy analysis capacity in students. So that students will be better able to critically accept diverse world cultures as well as carry forward and inherit the excellent culture of China. And ultimately, the demand for international exchanges under globalization will be met.

4 THOUGHTS ON PRACTICAL WAYS OF CULTIVATING DUAL-DEGREE UNDERGRADUATE TALENTS

4.1 Building targets of undergraduate dual-degree training

Undergraduate dual-degree training should focus on training innovative and compound talents, and it is also the core of this training mode. Therefore, the targets of undergraduate dual-degree training should be determined. The objective of this area is to fully strengthen the overall quality of students in universities. And its core goal is to cultivate and integrate capabilities of students in four aspects including capabilities of innovation, independent analysis, as well as gaining and applying knowledge. Thus, it contributes to the overall development of talents in morality, intelligence, capability and other aspects. Consequently, talents will possess more solid basic skills in foreign languages and more capacities in both theory and practice. Therefore, cultivating undergraduate dual-degree talents is firstly a comprehensive and innovative combination of the capacity level. It allows students to gradually form a competitive sense in dual-degree undergraduate courses. Also, students will possess more prominent foreign language proficiency as well as master the necessary professional knowledge and ability, in order to become practical and complex talents.

In terms of employment orientation for dual-degree undergraduate talents, students can make full use of their abilities working in foreign affairs after graduation, such as foreign enterprises, external contacts, trade and economic activities, education, foreign affairs and other aspects. Especially they can use their language expertise to further engage in relevant works such as translation or professional interpretation, as well as teaching work of using expertise in foreign language.

4.2 Measures to strengthen dual-degree undergraduate personnel training mode

Firstly, cultivating undergraduate dual-degree talents should be a training system with features. According to the characteristics and advantages of expertise in two different degrees, it optimizes talents nurturing especially the personnel training program and related course content. In training relevant personnel, the time proportion of language skills curriculum and expertise courses should be coordinated. And it should be rationally and scientifically configured according to the actual situation and needs of students. In the case of identified training objectives, the training objectives should be decomposed and achieved in phases. Meanwhile, it should combine the supplement of the second classroom and extra-curricular knowledge to achieve the full complement of talent cohesively.

Secondly, it is to teach bilingually. Foreign language or compound talents actually need a fluently spoken language as the basic attainment of communication skills, an important manifestation of undergraduate dual-degree talents achieving comprehensive abilities. Therefore, to better enhance abilities of the learner in using language, language tone and other aspects, bilingual education should be introduced in comprehensive training courses. Through full use of the bilingual teaching platform and original foreign language teaching materials, students will be exposed to international forefront knowledge of professional disciplines.

5 CONCLUSIONS

The dual-degree undergraduate personnel training mode can actually further scientifically combine knowledge of foreign languages and relevant professionals. It has gradually developed into an independent education system in the higher education of common universities and was granted the Innovative Dual-degree Personnel Training Mode after being certificated by the Ministry of Education. With the specific needs of Chinese economy and society as an important guide, the concept of training and educating all-round talents is a direction of practical talents cultivating. Therefore, strengthening the dual-degree undergraduate personnel training mode and optimizing this mode plays an important role in improving the social structure as well as enhancing the social adaptability of university students.

REFERENCES

[1] Wu Yanan. Implementation Strategies of Dual-Degree Undergraduate Education under the New Situation.

Shandong Youth Administrative Cadres College. 2010 (05): 91–93.

[2] Wang Guangyang, Zhang Zhihong, Zhou Yan, Cao Congzhu. Exploration and Practice on College Dual-Degree Undergraduate Education. Anhui University of Technology (Social Science Edition). 2007 (03): 110–111.

[3] Peng Li, Xia Yongmei. Suggestions on Improving Dual-Degree Undergraduate Education of China. Southwest Agricultural University (Social Science Edition). 2011 (05): 159–162.

Future Communication, Information and Computer Science – Zheng (Ed.)
© 2015 Taylor & Francis Group, London, 978-1-138-02653-7

Analysis of English audio-visual-oral instruction model based on network information technology

S. Zhao

Qiqihar Medical University, Qiqihar, Heilongjiang, China

ABSTRACT: The emergence of multimedia networking, which greatly accelerates the reform of the English instruction model, has very profound influences on English teaching in aspects of teaching method, teaching strategy, etc. Developing from task-oriented teaching to diversified and liberalized instruction, the current English instruction stimulates the active learning interest of students. With the help of networking technology, the more abundant learning resources of students further benefit the improvement of the student's English level in nature. This work discusses the development situation of English audio-visual-oral instruction model based on network information technology, current problems and solutions.

Keywords: network information technology; English audio-visual-oral; instruction model

1 INTRODUCTION

With the rapid development of the economy and technical level in our country, new requirements and task goals for the English teaching capacity and level are put forward by the national education department step by step. At present, English learning is almost everywhere in the student learning process. Therefore, the Ministry of Education proposes to improve English teaching by network information technology and comprehensively reform English teaching in aspects of the teaching method and teaching strategy. By taking full advantage of current network technology and information technology, the English teaching level is entirely improved increased in listening and speaking, and the learning environment of students is also improved to a great extent. This kind of audio-visual-oral instruction model can not only motivate the learning interest and initiative of students, but also give students more abundant learning resources [1]. Therefore, they are able to choose things interesting and suitable for them to learn and further promote the improvement of the English level.

2 DEFICIENCIES OF TRADITIONAL ENGLISH INSTRUCTION MODEL

Traditional English instruction method is based on tasks. Teachers will prepare the contents for class and exercises in advance. In class, they will instill these contents in students and let them accept and exercise passively. However, this model is single in both teaching content and method. Firstly, students have to accept the knowledge prepared by the teacher, which makes the teaching content too simple. For example, the listening content prepared by the teacher will neither cater to every student nor be suitable for everyone's level [2]. Therefore, students will be passively accept knowledge, rather than being highly interested in English learning. Secondly, learning resources for students are quite few. If students can get more resources in the same period rapidly and conveniently, their learning efficiency will be greatly improved.

2.1 Failure in stimulating students' learning interest

The traditional English instruction model is task-oriented. The task of the teacher is instilling all the contents of this class in students, while the task of the student is to understand and grasp these contents. However, this instruction model is too simple and insufficient in teaching content, and teachers cannot know the students' absorption situation with this method. In addition, being passive in learning, students can only get a limited knowledge because they can only learn what teachers give. Therefore, students gradually feel bored and lose their initial interests in English learning [3]. They don't pay attention to listening and speaking trainings because of their lack of initiative, positivity and interest of English learning. Finally, they just learn dumb English for the exam, while their level in listening and speaking is not high.

2.2 Insufficient learning resources

Traditional English instruction pays great attention to textbook with limited exercise materials for listening, so students always recite words and read texts in books. With different English levels and interests of students, the textbook-oriented instruction model greatly limits the learning resources and the view exploration. Taking the listening lesson as an example, teachers do not take different English levels and interests of students into the listening exercise arrangement in the traditional instruction model. They are not aware that listening difficulty should be in a certain gradient and diversity to fit different English levels and interests of students. With the continuous and rapid development of network information technology, students can not only obtain abundant learning materials via the internet, but also find suitable learning resources they are interested in. Therefore, the application of network information technology will greatly improve the current situation that the learning resource in English instruction is too simple.

2.3 Insufficient teaching time and lack of communication and interaction between teachers and students

In the traditional English instruction model, students need to learn many specialized courses and participate in some extracurricular practical activities every day, especially the non-English major students in colleges and universities. Therefore, the time for English learning every week is very limited, only two to three classes for comprehensive exercises, including listening, speaking, reading and writing, per week. The key of English learning, however, is persistent learning step by step. For example, there are only two classes for listening every week, which cannot satisfy the requirement of English improvement at all. It needs persistent training every day for a long term to improve the listening level, while for students it is difficult to find other learning platforms and resources for English listening except class time. In conclusion, the traditional English instruction model is adverse to improve the students' English listening.

The time of interaction between teachers and students in class is very limited in the traditional English instruction model. Teachers cannot accurately know the understanding degree of students, whose English level and comprehensive capability are quite different, after arranging learning tasks. Actually, teachers can only take the overall level of most students instead of every student's advantages and disadvantages into practical English instruction when arranging teaching tasks and formulating teaching contents and plans. Therefore, students with a poor foundation will lose their interests of English because the content is difficult for them, while students with a high level cannot benefit from it because it's simple for them.

3 INTRODUCTION OF ENGLISH AUDIO-VISUAL-ORAL INSTRUCTION MODEL BASED ON NETWORK INFORMATION TECHNOLOGY

Because of the new requirements of talents cultivation for the present English instruction put forward by the country, it is necessary to improve the quality of English instruction and change the single instruction model of teacher lectures. In this new era, network information technology is used to improve the quality and environment of English teaching, which helps students study English more diversely and independently. There is no doubt that a listening and speaking capacity is very important in English learning. Listening comprehension and fluid expression mean creative study and application of English. Therefore, English instruction based on a network environment pays attention to listening and speaking and strengthens the memory and learning interest of English learning by visual stimulation, including network video, actual role-playing dialogue in English, etc. The student's English level will be rapidly improved with this instruction model.

The instruction model based on network technology mainly consists of the following basic processes in three aspects: (1) instruction model combining watching and listening; (2) instruction model combining listening and speaking; and (3) role imitation and playing. In instruction model, combining watching and listening, teachers can download a video close to daily life from the network to watch in class. For example, by watching English movies, students will simulate oral expression in birthday party and other scenes and remember common expressions for communication in such scenes unconsciously. This model can not only stimulate the students' learning interest, but also strengthen their memory with visualization. As time passes, students will feel like they are learning in the environment of the mother language and will blurt English out after a long time of training.

The instruction combining listening and speaking emphasizes that students will grasp a great amount of common presentations and communication expressions after plenty of listening training. By continuous digestion and absorption, students have imported a great amount of communication expressions. Next, they will exercise the output capacity of the language, which is the English presentation skill. In order to connect listening and speaking training well, teachers can collect some materials for English learning in listening and speaking on internet, and classify these materials according to different dialogue scenes. Students can selectively imitate and have training according to the pronunciation and intonation of dialogue in listening and speaking materials to greatly improve their learning efficiency.

Role imitation and playing mean actual application of the accumulated knowledge after the previous two trainings for a while. For example, teachers can divide students into different groups in English class at college to prepare and simulate a scene of daily life for a topic. Sometimes, the teacher can ask a few students to play a drama. Therefore, students can apply knowledge into daily practice to improve confidence and achievability of English learning.

4 ENGLISH AUDIO-VISUAL-ORAL INSTRUCTION MODEL BASED ON NETWORK INFORMATION TECHNOLOGY IS THE NECESSARY DEVELOPMENT TENDENCY

4.1 Advantages of English instruction based on network information technology

Compared with the traditional English instruction model, the English instruction model based on network information technology, which changes the traditional teaching idea fundamentally, is much more advantageous. This new instruction model can greatly improve the quality and efficiency of English learning by stimulating the students' learning interest more greatly and giving them more activity in learning and more freedom in learning time and location. Actually, the task load of English learning is large because students have to strengthen training in listening, speaking, reading and writing for English learning, and these tasks are difficult for non-native speakers. Therefore, students always feel dull to study passively and the class effect cannot meet the expectation in the traditional instruction model. With multimedia information technology, however, students and teachers can obtain resources from the internet, and teachers can stimulate the students' learning interest and passion by simulating dialogue scent with voice, animation and video. In addition, students can feel the English environment in the English instruction model based on the internet. Everyone will join the discussion by answering in the group, and they will not be afraid of the presentation without face-to-face learning. Moreover, students will have enough time to consider, and get more interests through music and movies.

4.2 Existing problems of instruction model based on network

With limited teaching resources, one teacher will teach many students in the present English instruction, so some students cannot concentrate in class. Therefore, it is necessary for teachers to take full advantage of electronic resources, such as voice and video on the internet to attract the students' concentration and improve their learning interest.

One computer will be equipped for every student after developing the English instruction model based on network information technology. However, English teachers are always lacking knowledge related to network and computer technology, so they cannot resolve the student computer failure in time. Therefore, it is necessary to arrange relevant staff for preparation before class to ensure instruction efficiency and learning effect. Meantime, the computer and network technology level of English teachers should be improved.

Many teachers have misunderstood the English instruction based on network information technology that in relevant audio playing can completely replace the explanation in multimedia instruction. However, every teacher has an instruction style, while multimedia teaching resources and methods can only assist them rather than replace them in class. As the major leader in the class, teachers should make adequate explanations for audio materials to stimulate the students' learning activity and interest. In addition, explanation in class can accumulate affection between teachers and students. In a virtual internet environment, the teacher's care and style still have great influences on students in learning. Therefore, the role of the network instruction model can be better played upon by combining traditional instruction models.

5 CONCLUSIONS

The rapid development of network information technology greatly affects the English instruction model and drives the development from the traditional model of single instruction to the instruction model combining instruction with network education. This new instruction model pays more attention to listening and speaking training in English study. With image, audio and other information, the students' learning interest is stimulated and the boring feeling of students in the traditional instruction model is changed to improve the quality of English instruction. Therefore, this instruction model can better stimulate the students' learning initiative and activity. However, the role of the teacher in class is still important. In conclusion, the instruction model combining multimedia instruction with traditional instruction should be adopted to drive English instruction towards the stimulation of the learning initiative and activity, free learning time and location, rich learning resources and teacher leading.

ACKNOWLEDGEMENT

This work was supported by 2012–2013 foreign language teaching and research project of national basic education, *Modern Information Technology and Establishment of Audio-visual-oral Three-dimensional Education Model*, Project No.: JJWYZD2012015.

REFERENCES

[1] Wu Guangping, Duan Zhongyu, *Discussion on College English Audio-visual-oral Instruction Model*

in *Multimedia Network Environment*, China Education Innovation Herald, 2012 (09): pp. 90–91.

[2] Peng min, Wang Rui, *Application and Research of College English Audio-visual-oral Instruction Model in Network Environment*, Journal of Liaoning Educational Administration Institute, 2009 (03): pp. 18–19.

[3] Li Chili, *Comments on the Application of Multi-Media Teaching Model and College English Teaching Reform*, Journal of Shenyang Agricultural University, 2007 (04): pp. 67–67.

Future Communication, Information and Computer Science – Zheng (Ed.)
© *2015 Taylor & Francis Group, London, 978-1-138-02653-7*

Promotive function of literary works to English-language literature

C. Zhao
Huanghuai University, Zhumadian, Henan, China

ABSTRACT: Literary works have a crucial influence on the evolution of English-language literature. A literary work is the main manifestation and material carrier of language literature, also the driving force of the evolution of English-language literature itself. The development process of English-language literature may be divided into several phases by representative works, with each phase reflecting a particular era background and spiritual value. This work discusses the promotive function of literary works to English-language literature through analyzing representative English literature works from different periods.

Keywords: literary works; English literature; background of times; social environment

1 INTRODUCTION

Literary works are concentrated reflection and material carriers of the development of language literature of a specific nation. It is the specific era and social background that fosters the literature's expressing spiritual pursuits and living state of a nation and epitomizes the development of the civilization [1]. This work explores the remarkable role of English literary works in English literature history by combining the social background and English literature works of a particular historical time.

2 EVOLUTION PROCESS OF ENGLISH-LANGUAGE LITERATURE DRIVEN BY LITERARY WORKS

It may be recognized that the evolution process of English-language literature equals the course of constant emergence of English literature works. In order to analyze the promotive function of literary works to English-language literature, the process of English-language literature driven by literary works needs to be made clear at first [2].

The English-language literature originated before the Middle Ages when the dawn of civilization driven by the Roman culture gradually faded away because of the northern barbarians' strike. Therefore, England produced neither relatively mature literature forms nor relatively classical literary works for a long time. England did not brush his history of no literary masterpieces in a true sense until *Beowulf* was composed. As a long poem describing battles, *Beowulf* completely presents readers with sufferings and cruelty as well as bloodiness of battles, containing strong a feeling of sadness and melancholy. It is this strong feeling

that developed into a keynote for the development of English-language literature, influencing generations of litterateurs and their works in English.

Stepping into the Middle Ages, English literature, due to the overwhelming influence of religion throughout Europe, inevitably took themes from religious materials. During this period, prose and poetry developed well enough to replace former poems as popular literature forms, to some extent. For example, *Gawain and the Green Knight* was the representative of this period. In addition, various doctrines in religious materials developed into mainstream literature works while the expressiveness of literature grew as literary forms increased [3].

In the late Middle Ages, the Renaissance began in western literature and art. Thereafter, it developed into the mainstream of western literature to emphasize human values and dignity, to replace the religious sublimity and abstractness with felicity and a pleasure of life. Without doubt, English literature was influenced by the Renaissance at that moment, resulting in an increasing favor of Shakespeare's comedies and novels such as *The Canterbury Tales*. Spirits of renaissance and connotations of humanism both enriched the contents of English literature works and elevated the spiritual values of English literature to a definite new height.

With the industrial revolution and sweeping bourgeois revolution throughout Europe, the United Kingdom was firstly pushed into modern society. During that period, it was taken as mainstream in philosophy, culture and art to pursue freedom and equality and ponder about society and life with rational views. Within English literature, masterpieces written by litterateurs of critical realism showed up, reflecting the authors' critique to wicked reality and concern about people at the bottom of society. Examples are *A Tale of*

Two Cities and *Oliver Twist* written by Dickens. Those writers noticed not only the huge progress of social productivity but also the big gap between rich and poor as well as the painful struggle of people at the bottom of society. Therefore, English litterateurs were divided into schools such as romanticism and critical realism, with each having its own blueprints and thoughts for the future.

From the above, it could be concluded that the whole evolution process of English literature actually means the continuing creation of literature works, everlasting enrichment and diversification of thought and literature forms.

3 CONSTANT ADAPTION OF ENGLISH-LANGUAGE LITERATURE TO TIMES DRIVEN BY LITERARY WORKS

In early English literature evolution, the orally spread poem and story were the main forms of literature and art, transforming into poems favored by the court aristocracy and stories told by bards later. However, the stories that the upper class loved dominated English-language literature for a long time without touching people's life. Fortunately, works reflecting the entire society such as *Beowulf* appeared and began to sparkle in English literature. Against this background, English literature started its own history in a true sense. The description of cruelty and slaughter in *Beowulf* truly presented the course of resistance by the English people to the intrusion of the northern barbarians.

The Middle Ages, characterized by a theocracy overpowering the secular power, was an era when religious ideas widely disseminated throughout Western societies. What's more, Christian ideas, propositions and spiritual pursuits turned into common ethical standards and social code of ethics of the entire British society. Therefore, literary works in this period naturally formed a core theme of propagating Christianity. Form the above, the process of gradual adaption of English-language literary works to the social environment and social existence is, in fact, a course of the continuing creation of English literature works. Styles and contents of literature works were firmly inlaid with features of the times during its creation. Constant development of English-language literature truly equaled the process of changing themes and contents.

4 FORMS OF ENGLISH-LANGUAGE LITERATURE IN DIFFERENT PERIODS ENDOWED BY LITERARY WORKS

English-language literature, in different phases, displayed different style characteristics which were actually achieved by the evolution of forms of literary works. Forms of English-language literature had gone through stages such as poetry, prose, comedy, and fiction, each of which interrelated with the others in terms of time sequence. Besides, inner relationship,

like learning from and influencing each other to some extent, could be found between the different forms.

In the first phase, early English literature was mainly in the form of poetry and prose. *Gawain and the Green Knight* was an appropriate combination of prose and poetry in terms of narration. Spanning from poetry to prose not only created a more flexible form for English-language literature, but also presented contents more clearly. The reason was that poems or other literatures were mainly spread orally in early British society, with the popular English as the main language. However, less thought existed in literary forms, which became a problem within English-language literature at the moment. By contrast, prose, much more free, is a form of narration oriented towards readers who have a more high cultural cultivation.

The second phase, spanning from prose to fiction was, in fact, also an important historic reform for English-language literature. Compared with prose, fiction enjoyed much more space to compose, after which the possibility for greater length and more detailed description gradually showed up. Fiction itself, being able to integrate prose with poetry organically, became another peak of literary creation and a relatively mature and senior form in modern literary creation. Between prose and fiction existed comedy, a transition of literary forms. The development of comedy itself offers birth premise and earliest living space for fiction; therefore, most early fictionist had experiences of composing comedy.

Furthermore, the form evolution of fiction constantly enhanced organic docking between literature and other art forms. Fiction creation gradually presents diversified styles and composing concepts in modern English literature. With an increasing number of fiction reproduced as film and TV series, the inner characteristics of fiction, such as great vitality and gradual adaption to modern mass media, were reflected in this process itself.

At last, in Western societies, the formation of modern standard language was closely connected with the creation of some literary masterpieces. Without the guidance of relatively typical and standard English literature works, modern English would not possess such rich vocabularies and standard grammars, so to speak. However, its guidance had a remarkable significance on the development of British history and culture.

5 DOMINANT ROLE OF CRITICAL LITERARY WORKS IN THE EVOLUTION OF ENGLISH-LANGUAGE LITERATURE

Regarded as a pioneer of one literary era but a terminator of another, the critical literary work creatively transported some new elements into the kernel of literature creation occasionally. It is those important works with epoch-making significance in English literature history that guarantee consistent self-renewal and progress of English-language literature.

In fact, *Gulliver's Travels* and *Robinson Crusoe*, two books from the Renaissance and Enlightenment eras, played a crucial pioneering role. The two books changed perspective into small potatoes and highlighted pioneering and an adventurous spirit in their life, whose connotation displayed qualities of the emerging bourgeoisie when starting an undertaking. Such qualities were bold pioneering spirits and no fear of hardships and dangers. They were brave enough to pursue personal interests and existing values, taking an attitude of suspicion, criticism and negation to the feudal monarchy. In fact, these works mapped special social contradictions in a particular historical period. Protagonists in these novels generally possessed such spiritual qualities as exploring, adventuring and pursuing personal interests, which in the end, transformed inner spiritual value into a common value orientation of the whole society. So to speak, English literature works at that time played an irreplaceable role both in the development process of English-language literature and the course of social emancipation of the mind.

Characters in *Vanity Fair*, *Oliver Twist*, and A *Tale of Two Cities* significantly promoted the development of English-language literature. The births of these masterpieces lead more attention of British literature to social problems. Besides, British intellectuals were guided to change people's views about social issues with their pens, ultimately aiming to promote reform and progress of society. Most British intellectuals, coming from the upper class, lacked concern about the sufferings of ordinary people. Instead, they paid more attention to of discussions and analysis of some abstract issues or inserted energy into love novels. It is the present moment in the evolution of English-language literature that truly brings wide attention to common social problems and their solutions. These masterpieces both promoted the development of English literature and brought social issues to the litterateurs' observation, also ultimately promoted social development and progress of the United Kingdom by changing and correcting the people's spiritual world.

6 LITERARY WORKS IN HISTORY SERVING AS LIGHTHOUSE FOR THE DEVELOPMENT OF BRITISH LITERATURE

British literature works of the previous generation developed in a specific historical background and ultimately transformed into inexhaustible fortune for later writers to study and learn from. For plenty of English literary writers, learning from previous masterpieces is an important way to increase literary cultivation and a key method to change the creating direction of their literature and make innovations.

On the whole, previous masterpieces formed an inner feature of composing with melancholy in the evolution of English-language literature. Therefore, the creation inspiration and styles of subsequent British writers were largely influenced by former writers. The living environment where the British had already survived for thousands of years probably made them inclined to observe life with a slightly sad perspective. What's more, overcast London became another reminder for them to observe and think with an attitude of depression and self-introspection. British scholars are believed to adhere to former ideological tendencies and continue to leave the world literary history forms of literature with British characteristics.

Throughout British literature history, returning to tradition repeatedly is also a prominent phenomenon. For example, *Harry Potter* and other bestsellers constantly set and arranged scenes with existing Druid culture, even in modern times. British characteristics presented in this literary form resulted from the writers' returning to and learning from British traditional culture and forms of early British literature and art.

7 CONCLUSIONS

Being a flower eternally blooming in world literature history, English-language literature appoints literary works of generations of writers in British history as its inner carrier. These accomplishments promote constant development and innovation of British literature in aspects of forms and contents, et.al. Progress, from poetry to drama to fiction, is a process for English-language literature to be mature, in which transformation of content from simple story to deep meditation of society and life occurs. The meditation ultimately results in a social concern and human concern unique in British literature. Stepping into modern society, English-language literature is continuing to move forward on the path paved by its predecessors.

REFERENCES

[1] Ye Lixian, Review of Recent Studies in the Origination of English Literature Canonization. *Journal of Tianjin University (Social Sciences)*, 2013(5):468–475.
[2] Jiang Hong, A Comparative Study of English Editions of Short History of English Literature. *Journal of Sichuan International Studies University*, 2007(4):139–144.
[3] Wu Songhua, Analysis of Relationship between Pre-Raphaelite Brotherhood and English Literature. *Beauty & Times: Fine Arts Journal*.

Future Communication, Information and Computer Science – Zheng (Ed.)
© 2015 Taylor & Francis Group, London, 978-1-138-02653-7

Influencing factors on the customer's buying behavior of E-commerce

B. Xi

Faculty of Business, Wuhan Polytechnic, Wuhan, Hubei, China

ABSTRACT: In E-commerce, the customer's buying behavior will have different characteristics. Grasping the factors influencing the customer's buying behavior in E-commerce could help manufacturers understand the inherent law of consumer behavior changes. In E-commerce, group psychology, time cost and the content of the product portfolio are the main influencing factors for the customer's buying behavior. Meanwhile, traditional factors such as price, income and the like will also continue to impose great impact on the customer's buying behavior. The innovation of this work is to analyze different influencing factors systematically and put forward proper strategies for manufacturers to adopt to maximize their own profits.

Keywords: E-commerce; buying behavior; consumer psychology; influencing factor

1 INTRODUCTION

E-commerce, on the basis of network and information technology, brings enormous influence on the modern retail industry of consumer goods. In an era in which E-commerce is widely spread and accepted, it is very important for enterprises to think of how to make reasonable marketing strategies by combining their own factor endowment and making use of the influencing factors of consumer behavior and the law of function of these factors [1]. This work, based upon consumer psychology and relevant theories of industrial economics and marketing, discusses the factors that influence the consumer's purchasing psychology and probes into the ways by which these factors influence consumer behavior.

2 TRADITIONAL FACTORS' GREAT INFLUENCE ON CONSUMER PURCHASE

In an economic sense, price and income are the main influencing factors for consuming and purchasing behavior. The price reflects the consumer's consumption willingness, i.e. WTP, and contains the influence of consumer's consumption preferences. Meanwhile, the income of the consumer, as the constraint for consumer purchasing, has a restrictive effect on the customer's buying behavior. The consumer's consumption target and constraint condition haven't changed in E-commerce [2]. Though E-commerce provides a more prompt buying tool, traditional factors could still impose great impact on the customer's buying behavior.

2.1 *Price still being an important influencing factor for the customer's buying behavior*

Price advantage is one important reason why the E-commerce system can be widely spread and accepted by most consumers on a C2C level. E-commerce provides an excellent opportunity to reduce intermediate circulation links for retailers. Meanwhile, the increase of their sales speed and acceleration of information transmission could help enterprises realize the scale economy [3]. Therefore, under the dual effect of scale economy and the elimination of double marginalization, enterprises could take advantage of absolute cost to determine a price lower than that of manufacturers' direct selling, so as to realize the price advantage. It has become an important prerequisite for the E-commerce system to replace the entities store's direct selling.

At the same time, the difficulty for consumers to get information is greatly reduced in E-commerce. As a result, it will be much easier for consumers to master the differences of price and the market will be more fully competitive. Under the circumstance of very similar product quality and characters, price differences will directly lead to the change of consumer's buying choices. The higher substitutability of products is a direct result of the widely spread E-commerce.

2.2 *Income as a guiding factor for producers*

As a limiting factor for consumer demand, income has an effect on the customer's buying behavior in a form of constraint condition. In E-commerce, the income of consumers will be likely to have significant impact on the producer's production and management decisions. The income of consumers in fact determines the market orientation of related products of manufacturers. In the previous retailing mode, it is relatively difficult for manufacturers to get information of the consumer's income and the cost for survey is also very huge for enterprises. It inevitably leads to a low frequency of investigation conducted by enterprises for demand information. And data with a relatively long-time historical period might be continued to be used. However, in E-commerce, the difficulty of acquiring information will be greatly reduced, which will certainly spur manufacturers to acquaint themselves with their customer groups more frequently to determine the market orientation of their own products.

In summary, though E-commerce has been widely accepted, consumer's consumption behavior is still restricted by price and income. It is an important tool for manufacturers to make adjustments to their own product quality and target groups. Compared with the previous sales mode, the acquisition difficulty of these traditional factors is relatively lower, among which price is an important reason for E-commerce's wide spreading while income should deserve more attention from enterprises.

3 AN OUTLINE OF EMERGING INFLUENCING FACTORS IN E-COMMERCE

Through the above analysis, it's not difficult to know the significant influence of traditional factors in the customer's buying behavior. However, in E-commerce, some new factors which would influence consumer purchasing will emerge due to the changes of the transaction mode and transfer mode of information flow and logistics in the supply chain. These factors have such an important impact on manufacturer's management decisions that manufacturers should pay attention to them, trying to achieve the maximum of their own profits and long and stable development of enterprises.

Firstly, the factors influencing customer's buying preferences include the complexity and currency of online transaction, both of which are an organic component of products. The capital paid by consumers exists in a form of tangible cost expenditure while the difficulty and time of getting commodity is an intangible cost that consumers pay. The intangible cost is a component of manufacturer's services, as well as a kind of expenditure from consumers in strength and energy. If consumers buy related products for management purpose, whether these products can be delivered rapidly in fact also concerns seller's capital time cost.

Secondly, group psychology and the condition of the product portfolio are more important influencing factors for the customer's buying behavior. Group psychology could be reflected in the manufacturer's sales record, credibility and the like in E-commerce. In contrast with the traditional sales mode, the sales record is more widespread and well known to consumers more easily. This determines that consumers could know the enterprises' product quality indirectly according to previous sales records, thereby choosing their own buying decision. Meanwhile, product assortment and mix, in fact, is moreover a means to bring certain differences to the products of enterprises, which could change the consumer's understanding of product function, even bring about induced consumption.

4 COMPLEXITY OF ONLINE TRANSACTION BEING AN IMPORTANT INFLUENCING FACTOR FOR PURCHASING

Since China has a vast territory, unbalanced productivity level and a jagged popularity of information technology, the complexity of online transaction varies greatly in different regions. E-commerce mainly uses the electronic information technology platform, thus relying heavily on network transaction. However, for Chinese consumers on the Internet and even with different familiarity to online transaction, the same kind of transaction system might get different comments in different regions. Some consumers, who are unfamiliar with Internet and electronic transaction platform, might feel strange to online transaction and even have uncomfortable feelings. For these consumers, some closely related electronic transaction activities often could leave an impression of hard operation on them, even produce a repulsion for electronic commerce system because of it, even though the products provided by the electronic commerce system have an advantage in price.

Since the transaction system is relatively easy and convenient for consumers, higher requirements for the logistics management and order management system of enterprises are put forward. It is because the E-commerce system carries multiple contents of information collection and allocation of capital and goods itself. If consumers can all finish this series of links easily, relevant procedures need to be finished by the transaction personnel of enterprises. The simplification of the online transaction system, in fact, has also brought enterprises a transaction mode with high efficiency.

5 SUPPLY SPEED OF GOODS AND CONSUMER PURCHASING

The supply speed of goods, closely related with the logistics system and supply chain system of enterprises, is an influencing factor on consumer purchasing. Goods supply involves capital allocation, goods distribution and reassignment, the selection and application of transportation means and so forth. For consumers, getting related products and the information of product transmission early could increase the consumer's trust to manufacturers and satisfaction.

To realize the supply of goods as soon as possible, manufacturers could either establish their own supply system, or adopt a third-party logistics to deliver goods. For some small manufacturers, cooperating with a third-party logistics company possessing good faith could bring stronger sense of trust to consumers on its own. Meanwhile, some manufacturers, deciding to do the goods distribution by themselves, need to pay higher transportation cost. Even though, for those enterprises whose products have high technical content and need special transportation means, a transportation team of their own could guarantee the product quality to the greatest extent.

Modern E-commerce enterprises have universally taken the transportation process of goods as a kind of necessary information to provide for consumers. Yet, in fact, E-commerce enterprises could establish a necessary connection between the information feedback of payment and logistics process. Making payment after the goods have been delivered to the designated route or transport node could reduce the consumer's reserved distrust to the credibility of some medium-sized and small enterprises.

6 GROUP PSYCHOLOGY'S DIRECT INFLUENCE ON THE CUSTOMER'S BUYING BEHAVIOR

Group psychology imposes an obvious influence on most consumers. Since E-commerce purchases and sells goods relying on information technology, consumers get no chance to palpably perceive the products, thus giving rise to the following purchasing behavior. The buying decisions of other consumers are important for these consumers. Against this background, manufacturers might face a problem of dependence on old products and risk of renewing products. Consumers tend to purchase the products that sell more while having much doubt about the products that sell less, which makes related manufacturers slow down the updating of products to ensure the purchasing volume, and to obtain greater market share.

But under the circumstance that products have been promoted in the market over a long period of time while technology has no great update, substitute products will appear in a large amount, with which the homogenization of products will intensify. As a result, relevant enterprises will be inevitably perplexed by high competition pressure, constantly falling prices and shrinking profit margin, etc. Under this background, enterprises will likely face peer cut-throat competition.

Group psychology, generated mainly because of an inadequate understanding of product quality and characters, is an initiative reflection of consumers for the disadvantage of incomplete information. However, after the product's related information and manufacturer's credibility have been recognized to a certain extent, consumers would be more inclined to choose the product portfolio independently. It is an inevitable result of the long-term development of the market that the influence of the purchasing volume of other consumers will become smaller and smaller. For most manufacturers, it will lead to an objective fact that

the original advantage of brand and credibility will be gradually reduced.

7 PRODUCT ASSORTMENT SALES PROMOTING ENTERPRISES SALES BETTER

Among the factors influencing the sales status of products in E-commerce, product differentiation is a traditional and important factor. In E-commerce, product assortment selling is an important measure to achieve the enterprise's control of prices. Yet only in E-commerce could enterprises be able to carry out the product assortment selling more effectively.

The premise for product assortment sales is that different consumers have a heterogeneity of demand. For different consumers or different segment markets, the consumer's willingness-to-pay is in a reverse relationship to different products in the consumption portfolio. The existence of this relationship is an important prerequisite for making profit from assortment selling, for the reason that if consumers have homogeneous requirements to product portfolio, manufacturers would have to face such a situation that they have to win some consumers with a sacrifice of losing some other consumers. The economic profit brought by the product portfolio for enterprises, in fact, has no essential difference from that of separate selling for each product.

In the meanwhile of product assortment selling, manufacturers should maintain a certain number of assortment sales because it could enable manufacturers to make more profit from consumers with heterogeneous requirements. But it could also lead to no purchasing at all for some market portfolio which do not have strong preferences to products. To solve this problem, manufacturers should make great efforts to adjust their product assortment, enabling consumers to make their own choice from tied-in products and separate products, thereby realizing profit maximization.

Product assortment sales are an important means to realize induced demand and enhance product differences. The product itself contains the core product, augmented product, formal product and so forth. Increasing product assortments and realizing combined sales are actually an important strategy to add augmented products and formal products. In the practical process, combining complementary products for sales would have some similar effect to that of package price sales mode. When combining products that are not complementary in function to sell, whether the customer group's demand of enterprises has heterogeneity should be considered.

8 CONCLUSIONS

In E-commerce, the customer's buying behavior is affected by traditional factors such as price and income, as well as by other multiple aspects such as group psychology, the difficulty of purchasing and the delivery speed. With the constant development of E-commerce and continuous improvement of the market economic system, the influence of group psychology will be gradually weakened in an unified market. Meanwhile, the marketing strategies of product assortment and hybrid product assortment will be likely to have a more significant influence. When facing these influencing factors on consumer behavior, enterprises should change their own marketing strategies according to the characteristics of their own products and market environment, so as to enlarge their own profit constantly.

REFERENCES

[1] Zheng Ying-long. E-commerce Innovation and Consumer Behavior Changes Based on the Relationships Network. Chinese Business and Market, 2012, 26(10): 91–98.

[2] Pei Yuling. Analysis on Consumer Psychology and Behavior in E-commerce. Chinese Business & Trade, 2011, (30): 123–124.

[3] Wei Mingxia & Xiao Kaihong. Analysis on Perceptual Factors Influencing the Online Consumer Behavior in B2B E-commerce. Science and Technology Management Research, 2006, 26(6): 175–179.

Future Communication, Information and Computer Science – Zheng (Ed.)
© *2015 Taylor & Francis Group, London, 978-1-138-02653-7*

Application and reflect on modern English language and literature

J. Lei

School of Foreign Languages, Leshan Normal University, Leshan, Sichuan, China

ABSTRACT: This work studied ways to enhance the applicability of English language and literature, analyzing the characteristics of English language and literature talents and problems in the cultivating process. Based on it, ways to improve the applicability of English language and literature are introduced. English Language and Literature calls for a reinforcing applicability and professionalism. Students and graduates are supposed to raise their comprehensive ability to use the language continuously. The innovation of this work is that it stresses the comparison of the same type of subjects, finding out the limited applicability and the competitive pressure of English Language and Literature.

Keywords: English language and literature; application; teaching; reflection

1 INTRODUCTION

Having made relevant researches on the reform of English language and literature, the State Education Committee points out that China has entered the 21st century. The international exchanges and trades are blossoming with an admiring speed, forcing the domestic market and society to have more diversified requirements for foreign language talents. In the past, China trained foreign language talents by fixed patterns, their basic foreign language skills unable to gratify the requirements of the growing market. While English language and literature graduates are boosting in recent years, their competitiveness is going downward. It is impossible for China to absorb so many English Language and Literature graduates each year, but the comprehensive foreign language talents are not enough, such as legal English, financial English and English news [1]. Therefore, during the training of English language and literature students, their ability to apply should be the central issue. The training characterized by the school of economics will succeed in enhancing their social applicability. As a result, an important direction for English language and literature is the wide-scope and complex talent, in particular, business English, an important way to apply English.

2 CONNOTATION OF ENGLISH LANGUAGE AND LITERATURE TALENTS WITH ADAPTABILITY

Applied talents of English Language and Literature actually belong to complex talents with diversified functions. The main specialty of applied personnel rests in that they take English as a tool to communicate in the transnational commercial exchanges. Naturally, applied English talents not only need to arm themselves with more solid professional skills, but also have to be able to take advantage of English in cross-border situations [2]. As a consequence, those talents are required to equip themselves with more comprehensive English culture, better English listening, reading, writing and communication skills, among which listening and speaking are the comprehensive connotation of those talents. Despite that the traditional English level is largely tested by written forms, the society is still eager for the comprehensive talents with excellent listening and speaking, so that fluent communication with foreigners will be realized. Therefore, the prominent listening and speaking is an important part of the connotation of applied English language and literature talent.

3 STATUS QUO AND PROBLEMS IN APPLICABILITY OF ENGLISH LANGUAGE AND LITERATURE

3.1 Good market prospect for English language and literature talents

It being more than a decade since China has become a member of WTO, companies from different countries are pouring into China to find opportunities. Meanwhile, China's enterprises also have found more opportunities on the world market, entire China entering a foreign trade era with closer foreign exchanges. Under such a circumstance, Chinese enterprises and various departments have gained numerous opportunities to establish external foreign contacts, and China's foreign trade regime is also underway to deepen reform [3]. Hence, frequent export and import business is deepening the overall economy gradually. From the related data, it can be known that the number of foreign trade enterprises is increasing by nearly 100,000 per year, masses of hidden foreign services not included. So the talents of English language and literature have a vast job prospect. Although there being a lot of people in foreign trades, the number of the capable and high-end professionals is unsatisfying. Comprehensive talents of English language and literature are continuing to be popular and relatively favored by the market.

3.2 *Challenges towards English language and literature talents*

After China accessed the WTO, many Chinese enterprises sold their products and services on foreign markets without the help of foreign trade companies. Business communication, political communication and various types of cultural exchange are becoming a whole new orientation for national development. China's foreign exchange tends to be more purposeful, being an inexorable tendency of the market instead of a purely utilitarian. Consequently, demands from the market for talents specializing in English language and literature are lower than before, and English language and literature should be more integrated with practical application. Higher education has become relatively common in China, a large number of students having a good command of English and many graduates having passed the CET4 or CET6. Compared to them, it is obvious that students of English language and literature are less competitive. If comprehensive applicability is not based, then many graduates with professional English Certificates will completely meet the demand of the market. Thus, English Language and Literature learners are confronted with more fierce market competition and challenges. Many graduates are faced with restructuring, and the reform of English education is urgent.

3.3 *Key issue of English language and literature: professional applicability*

Current application of English mainly focuses on its professional application, especially in a more open business environment. English is required to be applied in the communication between customers and clients, particularly, to promote exchanges between them. Students majoring in English language and literature have a systematic learning of the English language. Nonetheless, they tend to concentrate on word formation, grammar and so on. Also professional issues are often overlooked, especially because the professional learning of English language and literature places relative importance on literature without an adequate understanding of the social and business functions. Then a dilemma occurs. It can be seen that

the characteristic of language as a carrier of information is obvious in the international communication. For students of English language and literature, the applicability in certain areas needs to be nourished to adapt to society, except for planning the career and cultivating ability from a professional perspective. If graduates of English Language and Literature do not have the professional capacity, they will spend a long period developing it and adapting to the job.

4 REFLECTIONS ON THE TRAINING OF APPLICABILITY OF ENGLISH LANGUAGE AND LITERATURE

4.1 *Analysis of status quo of English language and literature education*

Students majoring in English language and literature are not always exposed to specialized knowledge in the daily teaching. Although they tend to have a better knowledge of literature in their learning of professional language, they are lacking business knowledge, legal knowledge and other comprehensive knowledge related. Many English language and literature majors have been taking examinations since high school, without many social experiences and comprehensive work experience. All those have become the bottleneck of English language and literature majors when entering the job market. There being no clear teaching object in the training of English language and literature majors, the arrangement of the teaching curriculum is a bit confusing. In particular, a small amount of business knowledge is added to the cultivation. It is clear that teaching is still immature in the solving process, ignoring the practical and comprehensive upgrade. So certain effects can be seen, but the teaching dilemma will still continue.

4.2 *Making a clear position of professional disciplines, and enlhancing the applicability*

For the current position and cultivation of English Language and Literature, it is needed to focus on the application of English Language and Literature. Therefore, according to the actual situation, the routine teaching scope of English language and literature is expanded to concern the professionalism except for the literature. It is necessary to consider the scope of English as a communication tool. It may include the following categories: the first one is the combination of commercial and professional knowledge of English, being the direction of business application. Currently, it is also a very important direction for many universities to open an English major. The direction pays attention to how students expand their own expertise to grasp more knowledge in activities on economy and trade. Second is the combination of legal knowledge and the English language. The current external activities in economy, trade and cultural exchanges are increasing, thus overseas litigation or arbitration will inevitably emerge. Litigation and related legal actions, initiated by a foreign court, are often written in English without using the language of the second country to write legal instruments. Therefore, this aspect requires the combination of English and law. Third, it is essential to develop the education of teaching Chinese as a foreign language. Because the Chinese language has very special syntax and word formations, foreigners will inevitably meet difficulties in learning it. However, the education of teaching Chinese as a foreign language requires professionals mastering both Chinese and Western languages. Especially the students majoring in English language and literature, will become important talents in this area with a good command of both English and Chinese.

4.3 Changing the mode of teaching and combining practice with theory

In terms of the development of English language and literature, it is founded on the students' basic skills of international languages, particularly the basic knowledge of engaging in international communication activities and those related. Hence, it is significant to integrate issues related to external exchanges and teach the knowledge to be used in the practical situations, according to the problems students may encounter in future work.

Most talents of English language and literature are required to fully get hold of English grammar and vocabulary, and further their understanding of professional knowledge, better promoting the ability to fulfill the integrated and cross use of Chinese and English. These capabilities include some basic processes involved in foreign exchange activities, understanding of basic business trades, political and economic exchanges, as well as the command of some general rules and related policies used in international communications. Theory and practice can both be used to gain more comprehensive skills and knowledge. These skills and knowledge need to be put into practice. Under certain conditions, it is the duty of colleges and universities to create more favorable conditions to practice so that the comprehensive technology and information scheduling will be improved, using modern office hardware and software to practice language ability and skills.

Hence, under such a big background, it is required for universities to connect with related units or companies out of school, especially foreign enterprises and units, to provide more practice chances to further enhance students' skills of business activities, and thus become useful talents for the market and society.

5 CONCLUSIONS

The competition in the 21st century is very fierce, and the competition between talents is the same. As a result, English language and literature graduates are more inclined to have a sense of crisis, and they should take initiative to adopt the habit of raising a vocational capacity in the daily learning process. What English language learners in colleges should do is deal with challenges of employment, rather than to accept passively. They also ought to depend on their own talent and strength of the English language and accumulate English language skills, gradually training themselves to be comprehensive talents. The future being an era with not only competition, but also opportunities, English language and literature ought to bring out its new characteristics of the times to seize opportunities and brave challenges. For those graduated English language and literature majors, they had better update and optimize their language structures and knowledge structures to respond to the demanding market, create and achieve their personal values.

REFERENCES

[1] Wang Sha, *Relationships between English Knowledge and Business Knowledge in the Teaching of Business English*. Journal of Nanjing Institute of Industry Technology, 2004(9):61.

[2] Huang Weixin, *From Teaching Business English to Teaching Business in English: An approach to develop graduates with English proficiency and business expertise*. International Economics and Trade Research, 2005(6):12.

[3] Xian Xiubin & Tang Wenlong, *New Reflect on Disciplinary Orientation of Business English*. Higher Education Exploration, 2005(2):61.

Future Communication, Information and Computer Science – Zheng (Ed.)
© *2015 Taylor & Francis Group, London, 978-1-138-02653-7*

Analysis on underlying factors of mental health

J. Qi

College of Mental Health, Qiqihar Medical University, Qiqihar, Heilongjiang, China

ABSTRACT: The work analyzed influencing factors of mental health of all groups in the current society, and investigated the problems contained in the analysis of these factors. Then, underlying influencing factors from various respects were analyzed. Finally, it was concluded that positive reinforcement and guidance should be paid attention to, and a positive psychological subconscious be developed. The innovation lies in the application of Freud's psychoanalytic theory and Skinner's behavioral analysis theory, as well as a conjoint analysis on underlying factors of mental health.

Keywords: mental health; influencing factors; underlying factors; behaviourism; psychoanalysis

1 INTRODUCTION

In the current fast-paced society, the people's enterprise, toughness and endurance are becoming stronger and stronger. Meanwhile, more and more pressure begins to emerge from the external and inner world. Various groups in society are facing different types of psychological stress. For example, primary and secondary school students have to finish their homework, thus facing pressure from promotion to elite schools; while college students and fresh graduates need to face more and more employment problems; workers inevitably have to confront larger and larger work pressure [1]. Thereby, various problems occur to people of all ranks, such as conflict between ideality and reality and that of interpersonal relationship. However, the mental health of people should be attached importance to, so as to achieve sustainable development in China. The masses will indeed contribute themselves to the whole society with nothing but a guarantee of their mental health. Hence, the underlying factors of mental health should be taken into account.

2 CURRENT ISSUES EXISTED IN ANALYSIS OF INFLUENCING FACTORS OF MENTAL HEALTH

The research and development of psychology were relatively slow in China, especially when new China was founded. Psychological research achievements in the former Soviet Union had been paid much attention to during a period of time. Though certain lessons emerged, China was different from other countries and nations due to its unique social formation [2].

Thus, after China entered the era of reform and opening up, psychologists of new generations started to make investigations on personality formation and mental health problems of Chinese people. Despite some achievements, there are still some disadvantages in current methods and contents. In particular, the quality of psychological research should be promoted.

2.1 *Lack of the humanistic spirit in the current study and intervention of psychological experiments to influencing factors of mental health*

Currently, the lack of investigations of psychological experiment on different social groups exists in psychological research. The psychology surveys, including that of influencing factors of mental health, were conducted mainly through the distribution of questionnaires, fill out of psychology scales and statistical analysis. This research approach was similar to that of medical science owing to the importance attached to data integration and analysis [3]. Nevertheless, in fact, too much emphasis on data integration and analysis tended to bring about the lack of the humanistic features of psychology and of investigations on relevant data. Based on data statistics of survey forms, many experiments and findings were just statistical reports with no reference value. Because no theories of psychology were formed in combination with the humanistic spirit in these experiments. On the other hand, current researches were summaries of historical theories with no new theoretical perspectives, thus the depth of articles was not enough. Furthermore, most of the reports only focused on the building of a related theoretical system of psychology. Therefore, without exploratory experiments on related psychological issues, the theories made few contributions to the solution of existing issues.

More studies on individual psychological phenomenon and less explorations to the influencing factors and mechanisms

The achievements in contemporary psychological research, especially journals and articles laid emphasis on psychological issues with the application of related tools, such as psychology scales. The main research was still conducted on neuropathy, limb sensation and interpersonal communication. However, these studies were aimed at the exploration of the superficial phenomenon but no internal mechanisms. Thus, the internal mechanisms of negative factors of the crowds should be explored, as well as environmental factors. Actually, influencing factors behind the mental health problems should be analyzed among Chinese people, so as to settle the problems of various groups essentially.

2.3 *Lack of comprehensive analysis on mental health problems of social groups*

Although the mental health problems of social groups have been paid attention to by psychologists, little analysis is made from a comprehensive and holistic perspective. Mental health education and psychological counseling are current solutions with limited systemic and comprehensive researches on the internal factors. So these stopgap studies have the one-sidedness without rounded analysis on psychological abnormalities, and no timely measures are adopted against key mechanisms. Therefore, the correct analysis and understanding on the main factors are not only the key to get rid of various psychological problems, but important prerequisites to improve the mental health of social groups.

3 ANALYSIS ON INFLUENCING FACTORS OF MENTAL HEALTH WITH PSYCHOANALYSIS THEORY

The psychoanalysis theory was put forward by the Austrian psychologist Freud. In clinical practice, he found psychiatric disorders were caused by a trauma psychic occurred in human beings. Thereby, psychanalysis methods should be taken to make targeted analysis on these unhealthy conditions, thus relieving or removing the repressed psychological conflicts.

3.1 *Effects of subconscious on mental health*

The subconscious creates a profound effect on the mental health. Psychological conscious activity that everyone could perceive is only a part of the overall human mentations; however, most of the conscious activities are hidden, which is the so-called human subconscious. Freud believed that the human subconscious could create prodigious effects on human consciousness and specific behaviors. Actually, it acted as the internal motivation for people's daily

activities. Meanwhile, it was the deep-seated cause of psychological problems and related psychological illness. Many patients suffering from a mental illness, especially phobia and obsessive-compulsive disorder, would show some extrinsic behaviors out of control owing to the potential psychological conflicts. These were the reasons for the emergence of various absurd or complicated behaviors occuring in patients with psychological illness. Thus, effects of the subconscious on mental health should be explored and analyzed in the psychological study.

3.2 *Effects of complex factors generated at tender ages*

Freud believed that the unhealthy conditions of human psychology were actually due to the pain, experience and trauma psychic occuring at tender ages. Human beings at tender ages tended to have some physiological impulses which could not be controlled by rationality, especially instinctive desires. These ideas out of rationality, including the expression of instinctive desires, could not be recognized by the guardians. Meanwhile, the guardians would repress these behaviors of desire deliberately. Ultimately, the desires would transform into the human subconscious and became a unique complex. Actually, this complex equalled to a series of ideas with emotional strength in the subconscious, such as the "Electra complex" or "Oedipus complex" mentioned in Freud's research. From these ideas, individuals were susceptible to environmental impacts. Furthermore, the accumulation of psychological factors, especially the accumulation in the subconscious, would eventually become an important element of psychological disorders. However, the effects of the spiritual subconscious on mental distortions were just a kind of inference of psychology theories and logicality. Therefore, without scientific inferences, the detailed analysis could not be made on the importance of complex factors generated at tender ages with this speculative view.

3.3 *Effects of internal conflict factors appeared in personality*

Human personality was divided into three distinctive parts by Freud from perspectives of the psychoanalytic school, including Id, Ego and Super-ego. As the most vague and most primitive spiritual personality, Id referred to one hedonistic view followed by humans. In fact, it was explained by the accumulation of deep-seated desires and impulses. Because Id could not distinguish the good and the evil, it would attempt to search for the part where the desires could be relieved and satisfied whether it should or not. So, behavioral transgressing and disorders would emerge without moral and legal restraint.

However, Ego represented the part composed of reality and rationality in humans. This personality could not only adapt to various external environments, but affect the demand of Id and accept the supervision of Super-ego. Thus, it was capable of intervening in Id and making investigations and reflections on the thoughts of Id.

Moreover, Super-ego actually referred to compliance to morality and pursuit to the highest specifications and perfect concepts, and was beyond the human desire for survival. This was one personality carried out in accordance with moral laws.

From the above, intrapsychic conflicts and collisions appearing in the personality actually represented three forces: the pursuit of Id to happiness, the confronting of Ego with reality and the longing of Super-ego for perfection. So the contradictions and conflicts were inevitable which were generated between personalities. Thus, human psychology was healthy with their personalities maintaining stable. On the contrary, without control, harmony and balance among the three respects, serious damages would occur in the personality. Especially when the human mind and behavior were dominated by Id or pursuing to perfection too strict, there would be an imbalance in the personality. In this way, the adaption to the requirements and development of society could not be made. And even mental disorders might come up.

4 ANALYSIS ON FACTORS AFFECTING MENTAL HEALTH BY APPLICATION OF BEHAVIORISM THEORY

Behaviorism theories of psychology were proposed by Watson, an American psychologist. From the viewpoint of behaviorism, the mental health problems were generated by impacts of various mental conditions in the spiritual world.

Pavlov and other scholars found that abnormal behaviors or morbid personality states were generated by repeated learning and strengthening of some conditions. Actually they were a fixed state gradually formed in the past. At early times, personal behaviors were laid much emphasis on by behavioristic psychology. In fact, this passive learning effect was equal to behavioral abnormal fixation, which was gradually formed under the conditioned reflex. Later, Skinner et al suggested that behavioral abnormalities were not only due to passive learning behaviors,

but strengthening of operational conditions. Thus, behaviors themselves could be influenced by the character of strengthening behaviors. Harmful strengthening meant harmful behaviors. Therefore, abnormal strengthening brought about abnormal behaviors. However, one positive and effective psychological strengthening mode was needed to maintain mental health. As a result, psychological treatment and perfect psychology could be promoted better.

Nevertheless, social imitation was another important factor affecting mental health. Most of the behaviors were imitated or learned from social communication behaviors. Thereby, social imitation was also considered as learning in a wide range of society. As to some antisocial behaviors, they were imitation processes of the advanced neural cortex in the brain, such as murder or drug abuse. Maybe they were outcomes of conscious or unconscious learning. Therefore, the positive strengthening and imitation guidance of psychological behaviors should be paid more attention to, so as to maintain mental health.

ACKNOWLEDGEMENTS

This study was supported by Projects of Heilongjiang Provincial Education Department for humanities and social sciences in 2013 with project number 12532437 – *Research on System Programs of Mental Health Maintenance in Students of Heilongjiang*.

REFERENCES

[1] Ou Minhong, Huang Zhouzhong, Huang Yunkun. *Investigation of Patient with Severe Mental Disorder to Cause Trouble in Rural Areas*. Journal of Clinical Psychological Medicine. 2010(04): 254–255.

[2] Wang Yuhuan, Liu Yanhui, Huang Fangchao. *Relationship of Social Support and Life Quatity of the Disabled Elderly People*. Journal of Nursing Science. 2010(10): 80–82.

[3] Zhang Ruikai, Dai Jun, Li Hongwu. *Status and Development Dilemma of Community Mental Health Services— The Research Based on Beijing's 164 Communities*. Social Work (second half). 2010(05): 42–45.

Future Communication, Information and Computer Science – Zheng (Ed.)
© 2015 Taylor & Francis Group, London, 978-1-138-02653-7

Technology application based on computer programming software

P. Wang
Chengdu Aeronautic Polytechnic, Chengdu, Sichuan, China

ABSTRACT: A computer can be divided into two parts, software and hardware. Software mainly directs the operation of hardware, which demonstrates the importance of software. This study mainly focused on the technology application of computer programming software. Based on the concept and characteristics of computer programming software, this study, combining the current situation and some problems of the application of computer programming software, proposed that only by improving the quality of programmers and adopting advanced programming software as well as appropriate methods can the development of the software industry in China be promoted to the highest degree.

Keywords: computer; programming software; C language

1 INTRODUCTION

With the popularization of computers, people have developed application software with corresponding functions according to the practical needs of different industries. However, along with the improvement of the performance of computer hardware, the software is becoming more and more complex and the programming is getting harder and harder. Against this background, the way to better use computer programming software is on the study agenda of many experts and scholars. In order to simplify the programming, a lot of computer programming software emerged. By using such software, especially some visual programming software like VB and VFP, the programmers can also take the job of simple software programming without good command of programming knowledge [1]. It is the application of such software that greatly promotes the development of computer software technology.

2 BRIEF ANALYSIS ON THE COMPUTER PROGRAMMING SOFTWARE

2.1 Concept of computer programming software

The implementation of the computer order mainly relies on the software. In the early stage of computer development, restrained by their volume and performance, computers could only process some simple mathematical computation. The performance of early computers, with high manufacturing cost, even could were not as good as current electronic counters and thus were not used popularly. With the emergence of transistors and integrated circuits, the performance of computers has been largely promoted and their size has been controlled to some extent. Although computers in this stage could fulfill some simple tasks, their performance was still low compared with modern computers since the accomplishment of a specific task must rely on corresponding programs [2]. As the application of computers became wider, the number of programming had been larger and larger. In order to simplify programming, people developed high level languages like C language on the basis of the original computer languages. Meanwhile, according to the characteristics of high level languages, some programming software was developed, which could easily help fulfill the programming work. Nowadays, there are various kinds of high level languages and each of them has its corresponding programming software. In view of the different characteristics of different computer languages, relevant programming software can be chosen according to the needs of practical use [3].

2.2 Characteristics of computer programming software

Compared with traditional programming methods, the appearance of computer programming software is a revolutionary progress. The traditional programming could only adopt low level languages like assembly language, so it had low efficiency. With the help of programming software, much work can be done by software; for instance, programmers only need to design functions when debugging and compiling programs. Thus it can be seen that the application of programming software has greatly simplified the programming process, which is of vital importance for programming. With the enhanced performance of computer hardware, the software is becoming more and more complex. Modern computer software has reached GB level, so the programming of such huge software must rely on groups consisting of many individual programmers. On the contrary, traditional programming can not be achieved by group work. With the help of programming software, this problem can be easily solved. Especially, some high level languages like C language can divide complex software into several modules. Therefore, if software is complex, each function can be designed as a form of module; then a scanning function should be put into the main program to invoke a corresponding module when needed. The realization of this software technology mainly depends on computer programming software.

2.3 Development of computer programming software

Since their emergence, computers have been paid much attention to because they can automatically process tasks. As software is the basis of computers, the development and design of computer software can exert a considerable influence on the application of computers. Therefore, people have studied on the enhancement of the efficiency of software development for a long period, which stimulated the birth of computer programming software. In the early stage of computer programming software, constrained by the technology of that time, programming software had few functions and its debugging function and compilation facility were not perfect; there were a lot of bugs that could not be checked out in programs, which influenced the application of programming software to some extent. Considering the characteristics of programming software, the practical programming always can not be inseparable from relevant programming software, correspondingly, software could be improved during their application process. After many years of application, current computer programming software has many functions. The software companies have developed different kinds of programming software according to the practical needs. However, during the concrete programming process, the best programming software instead of a random one must be chosen on the basis of flow chart and application area after the comparison of the characteristics of several programming software.

3 CURRENT SITUATION OF THE APPLICATION OF COMPUTER PROGRAMMING SOFTWARE

3.1 Factors influencing the application of computer programming software

After many years of application, the computer programming software has become much welcomed by programmers because it can largely enhance the efficiency of programming. Now, programming software has been popularized in the current programming

work. However, influenced by its special historical background, the economic and technological progress started late in China. Compared with some western developed countries, China lags behind in the level of computer programming software. In order to narrow the huge gap between China and western countries in this field, factors influencing the application of computer programming software must be studied. Research shows that there are many factors that can influence the application of computer programming software. The first factor is the quality of programmers. If programmers do not have a good professional quality, then they can not have a good command of programming software and its various kinds of functions, which is bound to influence the efficiency of programming. In addition, the influencing factors include the performance of the computer and the quality of computer software. The operating of computer software involves a large amount of calculation and the implementation of different software puts different demands on the hardware performance of computers. Therefore, the development of large software places high demands on the performance of computers. If poor performing computers are used to develop large software, then the debugging and compiling work will take a long time, which can cause no-response or system halted and even lead to the failure of programming software.

3.2 Problems in the application of computer programming software

The software industry in China is still on a low-level of development though China has paid much attention to the industry and input of a large amount of human resources and material resources in order to build China's own software industry. However, influenced by some problems like piracy, the development of the Chinese software industry was very slow, and even other developing countries like India have surpassed China in this industry. The main reason is that China has a lot of problems in its application of computer programming software. Firstly, the programmers in China do not have good professional qualities. The development of modern software, especially large-scaled software, is a heavy workload which usually needs a team to accomplish accordingly to the characteristics of programming; the job like analyzing demand and designing programming flow chart should be divided into several links. After many years of development, the functions of the current programming software are perfect and all the work can be done with the help of software. For example, the design of the flow chart can invoke some function plug-ins of programming software. Almost all the programming software on the market have this function, so the development of a whole program can be done by a single programming software. However, research shows that different software on the market has different characteristics and the widely used programming software in China is C++. Obviously; it is not scientific and appropriate to use only C++ instead of choosing specific programming software on the basis of the practical needs of programs.

4 APPLICATION NOTICE OF COMPUTER PROGRAMMING SOFTWARE

4.1 Enhancing the professional qualities of programmers

Considering the importance of software development, China attaches much importance to the cultivation of corresponding talents. Many universities in China have set up curriculums of software development and cultivated large numbers of programmers. However, constrained by the education level of China, students have poor practical abilities of software development although they can command theoretical knowledge through relevant education. This is because students have little chance to learn advanced programming software through the education of universities; classroom teaching in China exerts much pressure of the examination-oriented education system on students, which compels students to command relevant theoretical knowledge so as to get good grades. Though China has been reforming its education system over the recent years, the reform can hardly get striking achievements in a short time. Under this background, enhancing the professional qualities of programmers should be firstly demonstrated in recruitments. A strict assessment system should be established in recruitments to offer chances for programmers to have a practical operation, selecting high-quality talents based on assessment results. If the performance of an enterprise is excellent, the enterprise should recruit some programmers with rich experiences. For existing programmers, enterprises can adopt periodic training, letting them know advanced ideas of program design and direction for use of programming software through learning and practicing. Only through these efforts can the comprehensive qualities of programmers be enhanced so as to ensure that programmers can finish their jobs in program development.

4.2 Adopting advanced programming software

After many years of application, programming software has been very perfect. At present, there are many kinds of computer software and technology companies that develop programming software with different characteristics according to their understanding. Each kind of computer language has corresponding programming software; if the programming software has been inappropriately used, the development of programs will be greatly influenced. Research shows that having been influenced by some factors like cost, the software development usually adopts some outdated versions. Considering the cost of software

development, the application of programming software needs some expenses and the software industry is booming so fast that a new version is released at regular intervals. If enterprises want to use the new version, they need to buy software once again. Therefore, in order to ensure the progressiveness of programming software, frequently changing software is unavoidable, which is bound to increase the cost to a large extent. To better solve this problem, enterprises need to compare the different software on the market, analyze their characteristics and choose the optimum one to enhance the efficiency of programming software based on the practical needs of software development.

4.3 Adopting appropriate methods of application

The computer programming software is changeless, but different methods of its application can bring different effects. Meanwhile, the summary of methods of application needs the accumulation of rich experiences. Only through a long time engagement in this industry, continuous experience accumulation and summarized appropriate skills can the efficiency of programming software be enhanced. Therefore, in software development, programmers with rich experiences are more efficient than new programmers. Thus it can be seen that for the current application of programming software, it is a problem being studied by many experts and scholarship on how to make programmers learn the characteristics and application skills of software in a short time. Some scholars thought the rich experiences of some programmers could be demonstrated on paper to let other programmers learn from them. However, study shows that many programming skills can not be transformed into the written language and they can only be commanded through the repeated use of this software. To make programmers learn about the characteristics and skill of programming software rapidly, the apprenticeship system can be adopted. Specifically, new programmers can know the attentive problems in programming by learning from older programmers rapidly. Then, new programmers can try some simple software development and gradually heighten the complexity, thus they can finally summarize the suitable methods of using the programming software for themselves.

5 CONCLUSIONS

Through the analysis above, we can know that programming software has been used widely in current software development. The application of programming software simplifies the process of software development and enhances the efficiency of software development to a large extent, which is important for the work of software development. However, restrained by the influence of the economic level, there are still many problems in the software development of China, which seriously affected the development of the software industry in China. Under this background, many experts and scholars have been studying the enhancement of the application efficiency of computer programming software. Large practical researches show that the quality of programmers and the technological level of software and hardware must be improved; on this basis, according to the current situation of the software industry, China can mirror the experiences of western developed countries and import advanced computer software and hardware technology, promoting the application of computer programming software to a large degree.

REFERENCES

[1] Jia Jinlan. Discussion about the Computer Programming Languages Courses in Mechanic Colleges, Occupation, 2010(17):120–121.

[2] Chen Yanli, Zhang Chenhui, Hong long. Research and Implementation of Automatically Construction of assembler Programs, Computer Engineering and Design, 2006(20):3887–3889.

[3] Zhao Meiyun, Xu Heguo, Qu Kui. Simulation Research on Improving Production System of an Enterprise Base on Witness, Journal of China Three Gorges University (Natural Sciences), 2011(02):88–91.

Future Communication, Information and Computer Science – Zheng (Ed.)
© 2015 Taylor & Francis Group, London, 978-1-138-02653-7

Analysis on strategy of computer foreign language interactive teaching enhancement

Y. Wu

Qiqihar Medical University, Qiqihar, Heilongjiang, China

ABSTRACT: In practical international communication, foreign language plays an important role. After years of development of reform and open-up, China has become the second largest economy in the world. The increasing communication with foreign countries results in a high demand of foreign language experts. Based on the method and development of computer foreign language teaching and combined with the problems in computer foreign language interactive teaching, this work mainly studied the strategy of computer foreign language interactive teaching. The conclusion is that only by improving the computer skills of the teachers and ensuring the application of advance hardware and software can the interactive teaching effect of a foreign language be well improved.

Keywords: computer, foreign language, interactive teaching

1 INTRODUCTION

As an international general language, English is highly valued around the world. It is always chosen as the means of communication between different countries. However, with the diversification of the economy, if English is not the first language for both of the countries, the communication in English between them may cause trouble because of their cultural differences. For this reason, China starts to develop experts in various foreign languages. Limited by the economy and technology, the education level in China is relatively low compared with some developed western countries. During the foreign language teaching, the poor infrastructure and backward teaching method brings up students with strong theory but weak practical abilities such as listening and speaking, which denies the original purpose of teaching [1]. In order to solve this problem, new teaching strategies have to be applied.

2 PRESENT SITUATION OF COMPUTER FOREIGN LANGUAGE TEACHING

2.1 *Method of computer foreign language teaching*

The computer has been highly valued for its advanced performance since it appeared. However, in the early stage of computer development, limited by its volume and performance, the computer could only deal with simple tasks, which was even not as good as the electronic calculator nowadays. With the application of the transistor and integrated circuit, the performance of the computer has been highly improved. Based on the

computer language, people developed a lot of application software to reach the demand in different fields. In this context, some scholars introduced the computer into education, and took good advantage of it via the application of multi-media, which highly improved the efficiency of teaching [2]. Compared with other theoretical subjects, foreign language teaching emphasizes more the training of listening, speaking, reading and writing because of its specific characteristics. The application of the computer can meet the needs for this kind of teaching where apparently the traditional methods cannot. After over 30 years of development of reform and open-up, China has already become the second largest economy in the world, with a significant improvement on the overall economic level. In order to improve China's education, the government invests a lot to establish the educational infrastructure every year. Nowadays, most of the schools in China are equipped with multi-media facilities, which lay a good foundation for the application of computer. In practical education, teachers are enabled to use courseware to conduct a flexible teaching method, and to improve the teaching effect [3].

2.2 *Factors that influence computer foreign language teaching*

In practical foreign language education, there are a lot of factors that influence the teaching effect such as the teachers' professional quality. According to a survey, older teachers in Chinese schools gained rich experience in teaching after years of working, but limited by their knowledge about advanced facilities such as computers, many of them cannot even type, which obviously disables them to apply computers to education, leading to a waste of resources. As for the use of computers, it is already very popular in foreign language teaching. However, according to surveys, the application of computers is relatively simple. Most teachers only make simple courseware or even just download some directly from the internet, making the form of courseware monotonous. During the early stage of this introduction, students will be attracted by this novel method for sure, with their study interests inspired. But they are prone to get visually fatigued after a period of time, which leads to interest loss. Besides those factors mentioned above, there are still many such as the establishment of infrastructure, students' situation, etc. In the foreign language teaching of western countries, computers have been popularized. A complete system has formed after years of application, and improves itself continuously during practical application. China's foreign language teaching has to reference the advanced experience in western countries in order to optimize its application.

2.3 Development of computer foreign language teaching

As an important product in the information industry, computers upgrade very fast. With the improvement of the computer performance, its application is developing all the time. In the early stage of computer application to the foreign language teaching, the performance of the computer was low. People developed some simple study software based on their needs to train the practical ability of the students. But the effect was not satisfying because of the high price of computers, low popularity in application and the absence of a phonetic room at school. With the rapid development of the information industry, the price of computers becomes lower and lower. Many schools purchased a lot of computers and built phonetic rooms based on their teaching plan. Computers had been popularized during this period. However, because of the limited understanding of advanced computer language, large scaled application software was absent. Thus only simple listening and speaking exercises could be conducted with the help of an earphone. Now the computer application has reached the stage of diversification. In traditional courses, the teaching content has been made into courseware, represented more lively to the students. In practical courses such as listening, the advanced multi-media technology and the learning software of listening and speaking are highly utilized to improve the students' practical abilities. The computer simulation technology can also build up a pleasant atmosphere of listening and speaking.

3 PROBLEMS EXISTED IN COMPUTER FOREIGN LANGUAGE INTERACTIVE TEACHING

3.1 Teachers' low professional quality

According to a practical survey, senior teachers take a large part in the staff. Although efforts are made on faculty reformation by introducing lots of junior teachers, the share of senior teachers is still considerable, especially in the north western area with a backward economy. These senior teachers have rich teaching experience that enables them to explain theoretical knowledge meticulously in traditional class. Some schools introduced computers and multi-media facilities to meet their practical needs in teaching. But there is only a small amount of teachers who know how to manipulate these facilities in China's foreign language teaching. The major teachers can only present their knowledge to their students via simple courseware, which apparently fails to maximize the deserved performance of these facilities. Nowadays many schools established their own phonetic rooms. Plenty of practice of foreign language teaching in advanced western countries shows that the application of computers in a phonetic room can achieve good interactive teaching, which improves the teaching effect. This method is of significant importance in improving the students' practical abilities in the foreign language they study.

3.2 Backward hardware

The realization of foreign language interactive teaching is based on advanced computer hardware, especially after the appearance of large scaled application software for foreign language teaching, which requires a higher performance of the computer. However, limited by the economic level, currently the hardware in China's schools is relatively backward. There's hardly any school that is willing to purchase an advanced computer, but some eliminated computers instead, in consideration of the cost, which apparently fails to achieve the requirement of foreign language interactive teaching. A similar problem also appears in choosing teaching software. The cost of developing software results in a corresponding charge for its application. In China's computer foreign language interactive teaching, some software with simple functions, or even uncompleted pirated software is preferred, which cannot maximize the effect of foreign language interactive teaching. Thus, the backward hardware is a serious problem in China's foreign language interactive teaching. Although with the development of China's economy, increased investment in education improved this situation to a certain extent, it still takes a long time for this problem to be completely solved.

4 STRATEGY OF COMPUTER FOREIGN LANGUAGE INTERACTIVE TEACHING ENHANCEMENT

4.1 Improving teachers' computer skills

In order to improve the effect of computer foreign language interactive the teaching, improving teachers' computer skills should be the priority. Only by deeply understanding the computers, can teachers conduct a novel teaching method with computers and inspire the students' interests to improve the effect of interactive teaching. Considering China's low education level, teachers graduated from university only have the knowledge in theory with a low practical ability. The improvement of teaching skills cannot be achieved within a short period of time. But the computer skills can be trained accordingly based on the situation of computer foreign language interactive teaching. Teachers graduated under this training will be able to accomplish computer foreign language interactive teaching, which is of significant importance in foreign language teaching. There are two ways to improving the teachers' computer skills. Firstly, some computer operation tests should be set during the recruitment to force the teachers to deeply understand computer foreign language interactive teaching. Secondly, for the senior teachers, trainings can be introduced to improve their computer abilities. During the training, senior and junior teachers should team up so that the senior teachers will be led by teachers with the newest teaching concept.

4.2 Introducing hardware with high performance

According to the above discussion, low performance hardware is a serious problem in computer foreign language interactive teaching. The best solution to this problem is introducing a newer type of facilities. However, the survey shows that newer facilities also have a higher price, which results in high costs to introduce these facilities. Considering the current education fund, these new type of facilities are not affordable. Instead of blindly choosing facilities by their performance, facility purchasing should be based on the needs of foreign language interactive teaching. The requirement for computer performance varies with the languages. Thus according to the actual requirement, half of the average performance may be sufficient enough to use, which can save considerable cost. Therefore, the selection of computer hardware should balance between price and performance, which means ensuring the performance of function and minimizing the price at the same time. In practical purchasing, the factor of upgrading should also be considered. For example the upgrading of the teaching software and other auxiliary facilities will require a higher performance of the computer. Thus computers with higher performance can last longer without replacement.

4.3 Installing advanced teaching software

The application of computers is mainly based on application software. People developed software with various functions according to the requirements in different fields, which highly promotes the popularization of computers. In education, the application of computers is quite popular. Software companies developed a lot of teaching software according to the teaching condition. There is plenty of teaching software of foreign language interactive teaching on the market. Customers have to pay for the software to cover the cost of its development. The more comprehensive the function is, the more they have to pay. Currently, in China's computer foreign language interactive teaching, some software with simple functions are chosen, which can only undertake basic teaching without fully achieving interactive teaching. In order to solve this problem, advanced software has to be introduced. Especially for the rapid upgrading speed of the software, the software companies will release new versions every other period of time, with some new functions as well as bug fixing to improve the customers' experience. In the software selection, customers can compare software on the market by its function, and choose the most suitable one as the interactive teaching software.

5 CONCLUSIONS

According to the discussion in this work, after years of development, computers have already been popularized. In foreign language teaching, computers are very popular. However, limited by China's technology and the teachers' quality, problems still exist in China's foreign language interactive teaching. One of them is the cost of hardware, which cannot be solved within a short period of time. Under this condition, in order to improve to effect of interactive teaching, it is suggested to reference the advanced experience of western countries according to the actual situation in China. Advance hardware and software have to be purchased. Last but not least, the teachers' quality has to be improved to acquire good computer operation ability. Therefore flexible interactive teaching can be conducted according to the course content.

REFERENCES

[1] Luo Zhongmin, Liu Jian, *Research on P2P Network Technology and Foreign Language Teaching Mode*, Journal of Kunming University of Science and Technology (Social Sciences), 2009(08):94–98.

[2] Zhang Fake, Zhao Ting, *Design of University English Listening Teaching Mode Based on English Audio-visual Learning Website*, Computer-assisted Foreign Language Education, 2007(01):36–40.

[3] Yi Su, *Analysis of English Interactive Teaching Mode in University Under the Environment of Information Technology*, Journal of Xinxiang Teachers College (Social Sciences), 2009(01):171–175.

Development and application of database based on SQL

D. Li
Guizhou Institute of Technology, Guiyang, Guizhou, China

ABSTRACT: Currently in the market, there are various databases with its own characteristics. The work mainly focuses on the research of database development and application based on SQL, analyzing the application of SQL in China's database and problems according to its concept and characteristics. It can be seen that only by introducing advanced development concepts can developers with sufficient professional quality and advanced storage devices improve the performance of a database thus gaining acceptance and adoption for more users in the competitive market.

Keywords: SQL (Structured query language); database; application software

1 INTRODUCTION

With the popularization of computers, targeted database systems have been developed, because of a lot of information stored on the computer in digital form for better management. Compared with traditional data storage, the new database with distinctive characteristics is not subject to natural factors – fire, water and time. It can be preserved permanently as long as the data is stored on the appropriate hard disk. The level of database technology is poorer in China than that of some western developed countries influenced by special historical factors, so that most of the database systems are developed by foreign software companies [1]. It is necessary to spend a lot on them, thus largely restricting the development of database technology in China, while their programming and functions have some differences according to the needs of various fields.

2 PROFILE OF SQL

2.1 *Concept of SQL*

With the development of computer software technology, computer language has been highly developed, thus the emergence of a variety of high-level languages on the basis of traditional computer language, such as C language and JAVA. The efficiency of software development has been largely improved by the application of these languages, which plays an important role in the development of software industry. SQL, an acronym for structured query language, is a language derived, targeting the system of database. Much more attention has been paid to the database system along with the development of the computer in recent years

for which courses of SQL language have been set up in many colleges and universities in order to improve the database technology in China, cultivating a large number of related technicians. The survey shows that students are skilled in theoretical knowledge but not in practical ability after graduation, subject to the level of education in China. It is simple to master theoretical knowledge of structured query language; moreover, there is the key point of database development and its application through use of the knowledge, which is the major function of SQL language. Apparently without being aware of the importance of SQL, developers can only achieve simple optimization of database application according to some existing databases and idea of simple structured query language [2].

2.2 *Characteristics of SQL*

The purpose of data and information is mainly for future access, thus taking into consideration the efficiency of data using regardless of any method of storage. In the storage of traditional paper data, a lot of libraries storing multiple books were built for facilitating the information query, where people could find the materials they wanted if necessary [3]. The larger libraries are, the longer it takes to find what people want, along with the increasing number of books. The emergence of a database system provides a good solution to the problem that related information in the computer can be searched automatically by the function of retrieval due to the data and information stored on the computer. After years of development, the current structured query language has been gradually formed, whose emergence, in a sense, is mainly to improve the efficiency of data calls. There are multiple database systems among which the competition is very intense in the current market so as to meet the need of a large amount of information after entering the information age. It is imperative that one database is with advanced performance, especially the efficiency of data calls, in order to play a much more important role on the market. Many developers of databases, therefore, believe that the importance of query language has exceeded that of programming language.

2.3 *Development of SQL*

Early in the development of the computer, there was no concept of database, with only some simple mathematical calculation achieving subject to its performance and size, even its performance was not as good as that of a current electronic calculator. With the advent of transistors and integrated circuits, computer performance and hardware storage devices have been greatly improved while a lot of complicated tasks can be fulfilled and much more data stored. In this context, there is an issue analyzed by many experts and scholars – how to refer to the data in the computer quickly; meanwhile the targeted database system has been developed according to the actual situation of data storage in order to manage all the data and information. Data query can be retrieved only item by item, and they are matched with the information automatically in the database when corresponding key words are put in, thus filtering out what people need. The data query in such way can be accomplished right now if less information is stored in the database. With the development of the information technology and internet industry, the scale of database storage itself is increasingly larger, while the level of hard disk has reached TB. Despite of the superior performance of the computer, it takes longer to match these data one by one due to the huge storage space of the database.

3 DATA DEVELOPMENT AND APPLICATION BASED ON SQL

3.1 *Development and application of SQL data in China*

Compared with some western developed countries, there is still a large gap in the level of database technology in China since economy, science and technology developed relatively late. In recent years, despite being aware of the importance of a database, a lot of manpower and material resources have been put into the study on database technology, but the development of the software industry is slow influenced by the pirated software market in China. The developing country, India has become a developed country of software industry, exceeding the level of China. The survey shows that the course of SQL database has been set up in colleges and universities so as to cultivate its high-quality developers, but it is imperative that the theoretical knowledge is mastered if students want to graduate with honors subject to examination-oriented

education system. In the actual teaching, the theoretical mode is adopted in class, thus lacking operating practice, while they are not skilled in programming software at work after graduation. Structured query language as the retrieval system in a database, is a new query concept with characteristics more distinctive than those of the traditional mode, retrieved one by one. Currently many of the available databases on the market are based on structured query language, thus greatly improving the efficiency of data calls, which plays an important role in the application of the database.

3.2 *Issues in the development and application of SQL database*

The performance of a database depends on the efficiency of data calls and storage, especially the former, because of not much difference in the storage space of existing database systems. It is the main purpose of data storage for the future use, but the situation cannot be accepted – it must take a long time to wait for the query results after entering relevant keywords if needed. Thus only by adopting some new query concepts, like structured query language, can the efficiency of data query and the performance of the database be improved. The survey shows that cost and storage efficiency has been paid much more attention to than the data calls in the use of China's database. A lot of people hold the view that the efficiency of database storage and calls can be improved as long as advanced storage devices of hardware are used, which is obviously not scientific. Even using advanced devices of hardware, some outdated software systems of the database will be adopted subject to reducing the cost of database construction, while most have traditional query methods, it still takes a long time to wait after retrieving the mass data information, provided that he reaction time of the computer is doubled. Thus the structured query language plays an important role in the application of database.

4 STRATEGIES OF DATABASE DEVELOPMENT AND APPLICATION BASED ON SQL

4.1 *Advanced concept of database development*

The advanced concept plays an important role in the software development. Given the characteristics of the program, multiple statements and logic relations can be chosen in order to achieve some specific functions while the efficiency of program execution depends on its programming mode. Nowadays, it is focused on using resources effectively so that all the developers seek the optimization of program development. Generally, the birth of a new concept will serve its improvement of practical application, such as the structured query language, a new one compared with the traditional mode of data query. For the database development work, the demands of database application cannot be met only by the traditional retrieval methods of data calls, while the emergence of structured query language provides a solution to the problem. The efficiency of data calls can be largely improved through some new targeted methods of the query developed, according to its characteristics. In the actual operation, programming personnel with the limited understanding of new concepts usually design databases in the traditional way, influenced by various factors. It needs a period of time, for them, to adapt the new concept and find the importance of structured query language. They can, meanwhile, be skilled in the structured query language to apply it to development and application of the database, thus improving the performance of the database.

4.2 Improving the professional quality of developers

In the actual development of software, the personnel, as the main body of the whole development process have to control all the steps. The accuracy-needed results of analysis will not be gained if the quality of developers is poor, thus not drawing the appropriate program flow chart according to the analysis, and the actual program cannot be written by using the existing programming software on this basis. Thus the application of structured query language will be affected unavoidably to a great extent. Consequently, the importance of the developers' quality should be realized. In the recruitment of personnel, they need to meet much higher standards in order to solve the problem better while it is necessary that certain tests set up to grasp whether the knowledge of structured query language and database has been mastered well by them. Then it can be guaranteed that hiring developers has professional quality through the actual operation procedures and even the experienced are supposed to be employed if conditions allow it. For the existing developers, they need training in order to have a good knowledge of advanced concepts and some new computer languages like structured query language, in which it is advisable that the focus is on the cultivation of their practical ability. In the actual application, moreover, it can improve the professional quality of them by learning from foreign advanced database systems and development concepts as much as possible.

4.3 Adoption of advanced storage devices

The computer consists of two parts – software and hardware. There is a close connection between the two in that hardware lays a foundation for software operations while the former guides the work of the latter. It is necessary for the database system to be equipped with hardware as a storage device, a basis for the system and software as operating system. Specific requirements of the performance of a storage device depends on different database systems while advanced storage devices are supposed to be adopted for ensuring the application of database systems. The hard disk in the current market has reached the capacity of TB to store a lot of information after the storage devices developed for many years. For the same capacity of the hard disk, moreover, the characteristics differ according to the needs of actual use. The part of a cache has been added to some of the hard disks so as to improve the efficiency of actual storage. In the construction of a database, efficiency will be theoretically enhanced if the retrieval system based on SQL is used on the condition that the performance of storage devices can meet the requirements of software operation. In the selection of actual hardware devices, it is targeted to select an optimal storage device according to the characteristics of the database system to ensure its good operation.

5 CONCLUSIONS

A conclusion is drawn from the analysis of this work that data information has been stored in the computer with the development of the computer itself, and some targeted database systems have been developed in order to facilitate the data management. In the early application of the database system, only some simple management functions could be performed subject to technical conditions at that time, while these functions could fully meet the actual needs due to a relatively small amount of information stored. More and more information has been stored on the computer with the development of computer performance and internet, thus building multiple data storage centers in China to store large amounts of data. It is necessary to realize plenty of complicated functions for the management of such huge data. The purpose of data storage is for its retrieval and usage later so that the data query function is important. The application of a database, therefore, can be improved by using advanced language of structured query and enhancing the retrieval function of data.

ACKNOWLEDGEMENT

This work was supported by the school-level project *"Teaching Reform of Database System"* of Guizhou Institute of Technology, with Project Number: Educational Reform 20132020.

REFERENCES

[1] Li Cuimei, *Security Research of Database Access Control Based on SQL*, Modern Computer (Professional), 2011(04):31–34.

[2] Wei Chenyan, Yang Jianming, Yao Sili, *Application of Stored Procedure and Triggers in SQL Database*, China Information Times, 2011(06):59–60.

[3] Yu Yang, Lai Dongyin, Ye Yutang, Liu Ping, *Design and Development of Database Online Marking Software Based on SQL*, Software Guide, 2011(09):142–144.

Future Communication, Information and Computer Science – Zheng (Ed.)
© 2015 Taylor & Francis Group, London, 978-1-138-02653-7

Deficiencies and suggestions of physical training teaching in universities

J. Dai
Suzhou Polytechnic Institute of Agriculture, Suzhou, Jiangsu, China

ABSTRACT: Physical training teaching in universities, based on training as its teaching method, is a new type of physical teaching mode with the ultimate goal of improving the students' physical quality and sports skills. The work studied the deficiencies of physical training teaching in universities, and proposed some suggestions about improving its teaching effect, hoping to be helpful for the future development of physical training teaching in universities.

Keywords: physical training; deficiency; suggestion

1 IMPORTANCE OF PHYSICAL TRAINING TEACHING IN UNIVERSITIES

Physical training teaching in universities is a weak aspect of university education all the time. Students show little interest, the teaching method is outdated, and teaching facilities are defective, greatly impacting the physical teaching effect. Effectively improving the teaching quality of physical training teaching is of positive significance not only for students but also for the development of the school and social physical career. In modern society, college students should be equipped with not only extensive knowledge and excellent working skills but also with a good physical quality, thereby satisfying the requirements of comprehensively constructing talent mechanism in China. Physical training teaching is not simply a teaching method. Combining training with teaching organically, it improves the students' skills, techniques and tactics of physical sports, and also enables students to practice and get familiar with sports skills and tactics [1]. This combination of theory and practice greatly improves the teaching qualities, achieving double effect with half work. In addition, physical training teaching also helps to achieve a closer relationship between students and teachers, because they have more communication opportunities in training teaching than in the traditional teaching mode. Students gradually eliminate their fear and prejudices to teachers in constant talking and communications and gain a better understanding of physical teaching, thus finishing training tasks with high quality and efficiency.

2 PROBLEMS IN PHYSICAL TRAINING TEACHING IN UNIVERSITIES

2.1 Little interest to physical training from students

Different from the management in high school and middle school, the university respects more the interest and choices of students, giving them more spare time to develop their interest. In universities, students usually only have courses of two or four units in one day; sometimes they even have no class for the whole day. As to courses like physical sports, universities attach less importance. Some teachers won't force students to attend the class if they don't want to. Some universities even don't include a PE grade into the final score of a student, resulting in less and less enthusiasm of students for physical training even though this kind of management mode gives students more freedom. After stepping into free and easy universities from busy and hard high schools, some students involuntarily develop a loose and undisciplined habit, and some just stay in the dormitory surfing the Internet or playing Internet games, lacking interest in study and physical exercises [2]. In addition, some students insist on high school thoughts and cognitions, considering study as the whole of student life and physical training as an entertainment. Those wrong thoughts lead to the students' shallow understanding of physical training. Interest is the source of students to take physical exercise actively. Therefore, lack of interest is the most obvious problem of physical training teaching and also the most fundamental reason for a low efficiency of training teaching.

2.2 Indifference to physical training teaching from teachers

A common phenomenon in society is the universities' indifference to physical teaching. Neither good teacher resources nor comprehensive hardware facilities are invested in for physical teaching. In this context, the enthusiasm of the teacher for physical teaching also gradually decreases, resulting in the indifference of physical training teaching. Moreover, even though some universities pay more attention to physical courses, students take physical training just in order to improve their physical quality, and they don't systematically study sports skills and tactics [3]. As a result, teachers won't spend much energy on physical training teaching, and seldom summarize and analyze the efficiency of training teaching. Furthermore, compared to teachers of other courses, physical teachers are disadvantaged in curriculum theory and progress schedule. They make schedules and progress control of physical training courses according to their own feelings. For example, if they can't complete the teaching task of this class, they just leave it to next class, showing no consciousness of completing a task goal timely and making a reasonable arrangement for each class. As a result, the teaching quality of physical training is reduced. The teachers' thought affects the students' attitude. Therefore, if students have a certain interest in physical training but the school and teachers show indifference, physical training teaching will be just perfunctory. Consequently, students will gradually generate negative emotions and their original interest will finally be worn off.

2.3 Backward teaching system of physical training in universities

The teaching system of physical training in universities is the framework of teaching activities; the development and accomplishment of all teaching tasks should abide the principles and constraints of its teaching system. Therefore, the teaching system is the fundamental basis to ensure the quality of physical training teaching. However, the current teaching system of physical training in universities is still affected by traditional thoughts, meaning its administrative purpose is greater than the teaching purpose. At present, physical training teaching in universities is just to satisfy the requirements of schools for its administrative management, forming a kind of formalism of physical training teaching. Universities only care that they have physical training teaching, not considering the teaching quality and effect. There is a certain risk for physical training teaching. Wrong training methods may lead to accident harms. Therefore, some universities cancel the trainings programs with a large amount of exercise and high risk and set up some easy and safe ones to avoid severe accidents during physical training. Although it helps to reduce the risk of injuries, it also reduces the effect of training to a certain extent because students are not able to take a more authentic training and to feel the significance of training programs. Take long-distance running for example. Students can not only exercise their vital capacity and endurance but also strengthen their willpower through the real exercise.

2.4 Outdated teaching methods and facilities of physical training teaching

School is blocked from society to a certain extent. After a long time of working in universities, some physical teachers gradually have less contact with the society, resulting in incomprehensive understanding of some new teaching methods and failure of applying those effective methods properly and skillfully. In addition, under the influence of little emphasis of physical training teaching from universities, physical teachers lack innovation spirit and prefer traditional teaching methods. Those teaching methods are summarized from long-term practices. On one hand, it is easy to use; on the other hand, it has a low risk of errors. However, those traditional teaching methods have largely failed to keep pace with the development and progress of times, no longer suitable for the current teaching environment and student demands. In addition, those traditional methods are not designed specifically for physical training teaching. Therefore, physical teachers should actively innovate new teaching methods and make full use of those methods in training teaching.

In addition to outdated teaching methods, some facilities in universities also can't keep up with the requirement of physical training. For example, if a school is not equipped with tennis courts, students only have to do stimulated training in basketball or football courts when physical teachers want to arrange trainings about tennis. In this way, students cannot understand rules of tennis and have no chance to practice serve according to the size of the courts. Therefore, facilities are important auxiliary tools for physical training teaching, requiring schools to intensify efforts and investment.

3 SUGGESTIONS ABOUT ENHANCING PHYSICAL TRAINING TEACHING IN UNIVERSITIES

3.1 Strengthening the leading role of models

The leading role of models has meanings of two aspects: one is the model influence of some excellent athletes, and the other is the exemplary role of teachers. College students are still in the developing stage of their life, and their mind is not so mature. So in trainings, they doubt their abilities and feel unconfident about their potentials, inevitably bearing some negative emotions due to difficulties. At this very moment, they urgently need comfort and encouragement to inspire their enthusiasm for physical training.

3.2 Innovating teaching methods of physical training

Physical training bears certain repeatability, requiring students to repeat one action over and over again to improve the skill proficiency. At the same time, students gradually feel dull and bored of repeating a single action. Therefore, physical teachers should innovate teaching methods and improve the interestingness of physical training to change the dull and boring atmosphere of physical training teaching. Moreover, innovative teaching methods should meet the demands of both the physical and mental development of students, enabling students to enjoy the training and relieving their pressure from life and study at the same time.

3.3 *Reinforcing the construction of faculty team*

Physical teachers are designers and implementers of training plans and schedules, and their teaching levels directly affects teaching qualities and training effects. Their teaching enthusiasm also indirectly decides the students' attitude towards physical training. Reinforcing the construction of a faculty team firstly requires strict control of recruitment. Universities should develop a standard for recruiting teachers according to the physical qualities of their students and the desired effect of their physical teaching. As for applicants that don't meet the standards, schools cannot hire them. Instead, as for applicants well conforming to the standards, schools should choose the outstanding ones, thus guaranteeing the quality of teachers from a fundamental source. Then, schools should provide vocational skill trainings for those new teachers after they begin their work of physical teaching in order to help them quickly adapt to the requirements of that position. The vocational skill trainings include an introduction to the teaching system, teaching objectives, teaching methods and students' conditions. In addition, because schools are relatively blocked from the society, they should provide further-study opportunities for teachers to open their mind and get to know more innovative teaching methods. Thus, teachers can get some practice in better schools or education institutions, and they will be able to return to their schools to make a further contribution to their physical training teaching system.

3.4 *Applying scientific testing methods in physical teaching*

The 21st century is a time of informatization and scientization, and the development of modern physical training needs the guidance of the computer and scientific methods. Applying scientific testing methods to the physical training teaching system achieves a more scientific analysis of training results, and thus teachers can develop and adjust future training plans. Scientific testing methods can be applied in two aspects: one is to test the external training effects of students; the other is to test the internal physiological and biochemical indexes of students. Current testing equipment for the students' training effect is relatively outdated. For example, at present, stopwatches are used to test the scores of sprinters, and teachers can only see the current scores. If they want to compare the current score with previous ones, they need to manually look up previous records. However, the rapid development of informatization provides convenience for testing. Teachers can make use of advanced computer software to input the students' basic information and training information to set up a small archive for each student. Thus the teacher can better test the students' training results through automatic analysis and various statistical charts from the software. Testing the internal physiological and biochemical indexes of students, such as blood test and pulse test, helps to better understand the students' physical condition, indirectly controlling the effect of training.

3.5 *Giving full play to the linkage effects of in and out of class*

In recent years, researches from relevant state departments for the physical quality of contemporary university students show that the comprehensive physical quality of them presents a downward trend year by year, arousing great attention of physical teachers. Physical training can't only rely on the limited time of classes; instead, it is a long-term process of constant persistence. Only in this way, can students gain better training results and achieve the goal of strong physical fitness and health. Therefore, teachers should expand teaching space as well as teaching time and positively guide students to do some extracurricular sports activities besides the tasks in class. Extracurricular sports requires students to complete it by themselves without on-side guidance of teachers, so physical teachers need to make extracurricular training plans for students and teach students some countermeasures for accidents that might occur.

4 CONCLUSIONS

In a word, physical teachers should actively find out and analyze the deficiencies existing in current physical teaching, and then search for solutions in the physical training teaching in universities to improve the teaching effect as well as the students' physical qualities and sports skills.

REFERENCES

[1] Kong Jie, Ways to Cultivate Interest in Physical Training in Universities. Contemporary Sports Technology, 2013(02): 65–67.
[2] Tan Zhuoyue, Analysis of Physical Training and Teacher. Science and Technology Information, 2012 (12): 297–298.
[3] Wei Zhenqiang, Present Situation of Extracurricular Physical Training and Its Reform Measures. Course Education Research, 2013(25): 220.

Future Communication, Information and Computer Science – Zheng (Ed.)
© *2015 Taylor & Francis Group, London, 978-1-138-02653-7*

Exploration of network marketing under micro-blogging environment

D. Zhang
School of Economics and Management, Hulunbuir College, Hulunbuir, Inner Mongolia, China

ABSTRACT: This work mainly analyzed methods of network marketing under a micro-blogging environment such as micro-blog marketing. Through contrast analysis of network marketing methods in the network era, the advantages and disadvantages in domestic and foreign companies, small businesses and large enterprises were explored, demonstrating that network marketing under a micro-blogging environment is a good network marketing channel. The innovation of this work lies in analytical methods of the advantages and disadvantages, dialectically analyzing the network marketing channel – micro-blog.

Keywords: micro-blogging; network marketing; enterprise; market promotion; publicity

1 INTRODUCTION

Micro-blog in China came into being relatively late. In May of 2007, the first web site providing a network micro-blog in Mainland China, Fanfou.com, opened to the public. In 2009, the beta version of Sina Weibo in Mainland China appeared, becoming the most massive scale users among micro-blogging products in Mainland China. Other micro-blogging products are also sprouting out gradually. To some extent, micro-blog, namely micro blog, is an extension of the traditional blog. Its main feature is the certain restriction of text published, such as Sina Weibo, which is limited to one hundred and forty Chinese characters. And so it does, every micro-blog can be read fast, adapting to the characteristics of contemporary readers living in the fragmentation of time [1]. Facing the rapid development of these micro-blogging network products, and particularly realizing micro-blog has spread beyond the propagation breadth and velocity of other network media products, many companies are beginning to rethink how to realize their promotion and marketing through such a new network media – micro-blog.

2 BASIC CONCEPTS AND FEATURES OF MICRO-BLOG MARKETING

Based on the relationship between micro-blogging users, micro-blog is a product of network communicating, and also a comprehensive network platform for spreading, sharing and acquiring relevant information. Micro-blogging users can construct their own speech community via a variety of Internet ports, organizing the information they want to release through about a hundred Chinese characters [2]. This sharing mode has advantages of immediacy, meaning the texts can be quickly read. Besides, it is highly interactive with a fast and broad spreading range. Also, it possesses a greater influence.

In 2010, micro-blogging products in Mainland China began to flourish. Based on WEB2.0, micro-blog marketing is a marketing return of new media. The general steps of micro-blog marketing are as follows. Firstly, the company needs to create a micro-blog, and then publishes relevant cultural information, product information, some news and information related to its business areas on the micro-blogging platform. Moreover, through the micro-blogging platform, company can communicate with relevant audiences interactively, thus attracting potential target customers, further building a better brand and product image and achieving the purpose of marketing. The forms that enterprises realize marketing on the basis of micro-blog mainly include customer service platform, advertising model, brand promotion and other related marketing activities. Moreover, according to the current development of micro-blogging products, micro-blog marketing has the following several different characteristics [3].

2.1 *Spreading faster and broader*

The first characteristic of micro-blog marketing is its faster and wider spread. Through the micro-blogging mode, the information that an enterprise needs to propagate can spread faster. This faster propagation velocity is based on a simpler and more efficient operation of micro-blog, and it also takes related more powerful functions such as comment and retransmission of micro-blog as the condition. Dissemination of corporate information is mainly achieved through the micro-blogging users focusing on the enterprise. Namely, fans of the enterprise will forward and pass the information of the enterprise they are interested in. Further, information of the company's products or services can be conveyed to the potential customers, so that the traditional point-to-point propagation model gains a new breakthrough. However, other traditional media are not available to such a propagation speed and amount.

2.2 Strong interaction of micro-blog marketing pattern

Secondly, micro-blog marketing has a lot of interactive forms, and the level of interaction is far beyond other marketing forms. The communication patterns on micro-blog platforms are diverse, such as one-to-many and one-to-one. It also allows consumers to focus on the company, the products or the services and to make a comment. Whether supporting or giving advice, as long as the number of people involved in the topic is large, the spreading influence of the information will be on the play. Besides, enterprises can get a part of target customer's trust and attention through some activities and even allow the target customers to participate in related marketing activities, eventually making them become the company's loyal customers.

2.3 High accuracy of micro-blog marketing behavior

The third feature of micro-blog marketing is its high accuracy. Corporate marketing behavior is mainly aimed at specific target groups and customers, causing the attention of target groups. Most groups and customers of micro-blog marketing have a certain degree of consuming ability, and most micro-blogging users are young people. Although not necessarily having a lot of wealth or savings, they definitely have a certain spending power. Besides, most micro-blogging users have a college education background. Therefore, highly educated and younger consumer groups inevitably become a significant customer base. In addition, these consumers are familiar with the micro-blogging platform, and they can freely use and switch the various related functions. Hence, promoted on the micro-blogging platform, the commercial information can be read more easily by young micro-blog users. This fan base that enterprises build up through a micro-blogging platform are indeed potential buyers. In this way, contacting the potential market in the first line through micro-blog, enterprises can communicate with target consumers online, allowing enterprises to better collect the real situation of the market, so that the marketing activities such as enterprise interaction and brand promotion can be more accurate.

2.4 Lower cost of micro-blog marketing

The fourth feature of micro-blog marketing is a relatively low marketing cost and high return rate. Through micro-blog marketing, enterprises can better reduce unnecessary administrative examination and approval procedures. But for setting out advertisements in the traditional media, relevant managers need to check the advertisement, and the required marketing cost is relatively high. In the process of micro-blog marketing, as long as enterprises edit the micro-blog content, relevant information can be delivered, and a related effect will be produced. For the enterprise, these aspects don't need a lot of marketing cost. Thus, the marketing cost is relatively lower, and companies can obtain the results of marketing through relatively less expensive marketing cost in a short period of time. Accordingly, micro-blog marketing is certainly an enterprise marketing behavior of high efficiency and low cost.

3 INFLUENCE OF MICRO-BLOGGING ENVIRONMENT FOR MOST CHINESE ENTREPRENEURIAL MARKETING BEHAVIOR

On the current Chinese market, more than ninety-nine percent of Chinese enterprises are of small or medium type, actually forming an important force in the whole national economy and having a very great influence on the development of the national economy in China. China has a large number of small and medium enterprises, and the difference of their products in the same industries is not very obvious. Especially, the current price of products and services are gradually transparent and are almost the same.

Hence, if the majority of small and medium enterprises in China truly want to expand their market and break the ice of their own products and services, they need to build their brands and carry out marketing behavior as planned. However, most of small and medium-sized companies have a limited financial strength, not being able to actually win more market share. Therefore, small and medium enterprises in China are often trying to specialize their business areas in the market operation process, so as to getting a greater degree of income. Actually, small and medium enterprises need to consider the following advantages and disadvantages in developing marketing under micro-blogging environment.

3.1 Low cost of marketing

In the gradual socialization process of current media channels, the cost of micro-blog marketing is relatively low, which is more suitable for current marketing needs of small and medium scale enterprises in China. Current Chinese micro-blogging products are generally carried through free registration. After free registration, companies can enter the marketplace and conduct marketing activities. Besides, micro-blog operation is very simple. Whether having a smart phone or fixed desktop computer, the enterprises can update client-related business information at anytime and anywhere as long as they get a client terminal of micro-blog and start it. Thus, actually the cost in the process of micro-blogging marketing is relatively low for the entire enterprise, narrowing the gap between small-scale enterprises and fast developing large-scale enterprises, further bringing convenience of network marketing for small and medium enterprises.

3.2 Reducing the level of information asymmetry

Information propagating through micro-blog goes fast, and the penetrating power of information is relatively strong. Also, there are many patterns of publishing information. Therefore, companies can update the information more timely. Through this marketing channel – micro-blog, companies can communicate with potential consumers more quickly and interact with them on the micro-blog network platform, leading to a higher time effect of information flow and feedback. Therefore, owing to the timely transmission of information, enterprises can face more accurate consumer groups and control the situation of information asymmetry between consumers and businesses. Thus, the majority of small and medium enterprises get the same contacting opportunities with potential target consumer groups as large-scale enterprises can, further offering an information window to potential target consumers. Consequently, the majority of small and medium enterprises have a very great future value in this micro-blog marketing platform.

3.3 Micro-blog marketing influenced by the amount of information and the number of fans

Whether enterprises can use micro-blogging marketing successfully or not largely depends on the popularity of their micro-blog account. To use the micro-blog network marketing under a micro-blogging environment, a micro-blog account of enterprise needs more followers, namely enough micro-blogging fans, so as to give better publicity on the micro-blogging platform. Otherwise, enterprises will encounter more obstacles and challenges in marketing promotion. Yet this would not happen to the large-scale enterprises. Relying on the personal charm of the general leader or the high popularity built up off line, large-scale enterprises can accumulate more fans for their micro-blog accounts on the Internet within a very short period of time. However, lacking influence or popularity, the majority of small and medium enterprises will not be able to attract the attention of micro-blogging fans more quickly. Therefore during the initial period, once the company fails to accumulate more micro-blogging fans, there will be little effect in micro-blogging promotion process for small and medium enterprises.

Furthermore, the speed of micro-blogging, information publishing and updating are all very fast, bringing a certain negative impact to small and medium enterprises during the process of micro-blog marketing. Although the relevant information edited in micro-blog by enterprises is very abundant, if it is released at the wrong time and if much micro-blogging information appears simultaneously, the fans will not actually read it carefully. Meanwhile, if wrong messages are released in the micro-blogging content, micro-blogging fans will criticize it quickly. Accordingly, the correct time and right content of a micro-blog needs to be further discussed.

3.4 Existing risks in the process of micro-blog marketing

The information of micro-blog propagates very fast and broad. If mishandled, there may be some risks in the process of micro-blog marketing. Thus, micro-blogging content needs to be carefully considered. If the content does not comply with some national provisions and requirements, then most likely it will not be issued and even may be banned. Moreover, if the company operates the official micro-blog account incautiously, thus occurring errors and leading to a negative effect, maybe the propagation of viral negative news will be triggered, becoming a joke among neighborhoods and marketplaces easily. All of these above risks should be taken into consideration during the course of the micro-blog marketing. Of course, controlling these risks and making them play a positive role for micro-blog marketing is significant for the enterprises.

4 CONCLUSIONS

Foreign micro-blogging networking products have formed a new network marketing method for companies, such as Twitter, which carries out network marketing through a wealth of data and a more powerful search engine, and also creates a variety of senior business certified accounts. Furthermore, as an information exchange platform between companies and consumers, Twitter also promotes the development of network marketing. In terms of current network marketing in China, especially the initial stage of development of network marketing under a micro-blog environment, further exploration is needed for entrepreneurial network marketing on the new platform. At present, the majority of China's small and medium enterprises have a relatively small audience. Therefore, if the niche manufacturers or merchants want to be found by the potential target audience, a particular target audience platform is certainly needed. Consequently, according to the features of micro-blog products, micro-blog marketing is a better platform for small and medium enterprises. However, in the process of micro-blog marketing, enterprises should take possible risks into account, adjusting their network marketing strategy under the controllable risks, so as to promoting the development of the enterprises.

REFERENCES

[1] Zheng Lei, Li Shenghong. Information Propagation Model Based on Micro-blog Network. Communications Technology, 2012(02): 39–41.
[2] Ge Jingping, Zou Liqing. Investigation of Micro-blog Stereo Marketing Strategy on Film. Contemporary Film, 2012(02): 112–115.
[3] Ning Jinpeng, Wang Xiaoyu. Design and Implementation of Users Analysis on Micro-blog Marketing System. Computer Applications, 2011(02): 233–236.

Research on trans-boundary mode of EGP and ESP vocabulary

J. Wei

Applied Technology College of Dalian Ocean University, Dalian, Liaoning, China

ABSTRACT: With the development of economics and society, the professionals with a comprehensive English knowledge are more and more popular. So, it becomes one of the main streams of the development of society to learn English well. Combined with ESP vocabulary, EGP vocabulary plays an effective role in English learning. It is an important researching topic at present. The innovation of the work is to analyze the vocabulary feature, relativity and trans-boundary mode of EGP and ESP and find a more effective research method for the vocabulary trans-boundary mode.

Keywords: EGP (English for General Purposes); ESP (English for Special Purposes); Trans-boundary Mode

1 INTRODUCTION

The economics in China have been developing at a striking speed, building an inseparable connection with world economics after taking part in the WTO. Meanwhile, the English talents are needed urgently along with the development of economics and the deep cooperation with world economics. As the most widely used language in the world, English has played a very important role in the communication and development of world economics. China is the biggest developing country in the world, and it is necessary to strengthen the English education and especially develop the English professionals in accord with the development of market economics in China, laying the foundation for the development of market economics in the future [1].

2 EGP AND ESP VOCABULARY

2.1 *EGP vocabulary*

EGP vocabulary, including all aspects of the social life, has a wider range than ESP vocabulary. EGP vocabulary is the basis of ESP vocabulary. E.g., CET-4, CET-6, IELTS, TOEFL and GRE vocabulary are all EGP vocabulary. The process of learning EGP vocabulary is to help the learner understand and master the skills of English learning, and understand application skills of different English words under different environments, laying a foundation for the learning in the future.

2.2 *ESP vocabulary*

ESP vocabulary is a very practical language branch. It has been used by many rising industries with the rapid development of society, as the extension and development of EGP vocabulary. The range of ESP vocabulary is larger and larger, including business English, foreign trade English, movie English, English for science and technology, legal English, agricultural English, restaurant English, hotel English, etc [2]. With strong practicability, ESP vocabulary is studied and applied to some certain industry especially.

3 RELATIONSHIP BETWEEN EGP AND ESP VOCABULARY

3.1 *Relationship*

EGP vocabulary, as the basic stage of ESP vocabulary, has a larger range of vocabulary than ESP vocabulary. In addition, ESP vocabulary, as a practical branch of EGP vocabulary, is the extension and development of EGP vocabulary [3].

3.2 *Necessity of the combination and trans-boundary development*

After taking part in the WTO, China has made contact with the world economy more and more closely and cooperated on economy with other countries more and more frequently. The frequent foreign communication has made people realize that the pure English vocabulary cannot meet the requirements of economic trade. However, EGP vocabulary has many kinds in content and is complicated to learn. So the vocabulary with different varieties have been summerized and concluded based on EGP to form the specific ESP vocabulary.

Compared with EGP vocabulary, ESP vocabulary is more professional. One kind of ESP can be selected for English learning according to the requirement. It can save time, increase the learning efficiency;

besides, it is beneficial to make a deeper communication of economy, improve the economy trade level and develop the society. F.e., hotel English is a kind of ESP, and the English vocabulary involved is set according to the hotel's business, including the vocabulary used for receiving the guests from the airport, asking the guests about the housing needs, all the hotel facilities, the service items, etc. Besides, EST, as a kind of ESP vocabulary, is professional and popular in modern society. Development of science and technology depends on that of EST; the innovation and development of ESP is driven by that of science and technology. ESP can be divided to more levels in science and technology because science and technology have a large range. The development of ESP above has advanced the communication of science and technology among the countries in the world and the increase of the total level of world science and technology. It has been one of the important elements, affecting the 3rd world science and technology revolution nowadays and getting more and more powerful.

EGP vocabulary has great relationship with ESP vocabulary. Furthermore, if there is no EGP vocabulary, there will be no ESP vocabulary. However, with the development of economics and the intensity of competition, the pure learning of ESP vocabulary cannot meet the requirement for persons of ability. The one mastering EGP vocabulary and using ESP vocabulary skillfully will be paid more and more attention to. To adapt to the keener market competition, the combination mode of EGP and ESP vocabulary is searched. In this mode, people can learn the knowledge of EGP vocabulary and master that of ESP vocabulary. The trans-boundary mode of English vocabulary can improve competition of the individuals, society and countries and make work and life in society more stable. So, the trans-boundary mode of EGP and ESP vocabulary has been the research focus.

4 TRANS-BOUNDARY OF EGP AND ESP VOCABULARY

Trans-boundary mode of EGP and ESP vocabulary mainly reflects in the teaching mode, and the teaching mode has a great influence on the transfer of EGP to ESP vocabulary. So research on the teaching mode has been kept going among the discussion and study of the trans-boundary mode of EGP and ESP vocabulary.

4.1 Difference of EGP and ESP vocabulary

4.1.1 Difference of the main teaching content
In order to make the students deeply understand and master EGP vocabulary, the historical accounts of past events are selected as the teaching material and the above English vocabulary belongs to the public language system, without being directed the special industry and purpose. The purpose of learning EGP is to make the students understand skills of daily spoken English communication and English learning, laying the foundation for English learning in the future. However, ESP is an extension of EGP. The main language content focuses not on the general public language system but some certain industry. The teaching content is not general content about literature and history any longer, but the professional vocabulary. It refers to lots of professional vocabulary, phrases, sentences and even the long difficult professional sentences, including the theories and terms of some specialty.

4.1.2 Different teaching mode of EGP and ESP vocabulary
The teaching mode of EGP is more or less the same as that of the traditional teaching mode. That is to deepen the impression of some words by learning and reading some work. However, the teaching mode of ESP vocabulary is different. Because of the special teaching purpose, ESP vocabulary needs more practice compared with the vocabulary in the book. The purpose of ESP learning is to perform timely application, so only the actual training and experiment can make the ESP learning more efficient. ESP vocabulary has a special teaching theory called GSP teaching theory. The core of GSP teaching theory is to achieve one certain target by using the language. Furthermore, the teaching content is not to purely remember vocabulary, but to learn some vocabulary, understand some theory and master some technique by unscrambling some professional work. This is different from the teaching purpose and mode of EGP.

4.2 Procedure of trans-boundary teaching of EGP and ESP vocabulary

4.2.1 Selecting the teaching material
The teaching material is the base and direction of teaching. It decides the knowledge level of teaching in the future. Firstly, the teaching material should be improved. The selected teaching material only accords with the stipulation of the teaching outline for EGP vocabulary, without a certain requirement. In general, the teaching outline of EGP vocabulary consists of the history and culture of English speaking countries, Chinese economy, society customs, etc. Through it the students can understand the history and culture of China and foreign countries. However, it is professional to select the material aiming at ESP vocabulary. The teaching material is selected according to the major of the students, and the content is almost the work and vocabulary related to professional knowledge. The purpose is to make the students learn professional theory and basic knowledge on the basis of the work. Besides, the teaching material of ESP vocabulary is professional. Some vocabulary emerges to explain some professional terms, and it cannot be found in the dictionary. This is the biggest difference of ESP and EGP vocabulary.

In order to combine EGP vocabulary with ESP vocabulary, different teaching materials should be selected and used by combination. EGP vocabulary is put in the teaching process of ESP vocabulary. So the students can master not only ESP but also EGP vocabulary.

4.2.2 *Selection of the teacher*

The subjects of the teaching activity are the teacher and the students. The teacher plays a very important role in teaching the students English vocabulary, imparting knowledge and solving problems of the students. In the trans-boundary process of EGP and ESP vocabulary, the teacher should have a powerful ability of controlling, teaching and professional quality. For EGP vocabulary, the teacher is selected based on English language ability, skills of language use, ability to teach and to communicate with the students, etc. However, the teacher is more important in the teaching of ESP vocabulary. ESP vocabulary teaching is a kind of vocabulary teaching around some specialty. To solve all kinds of problems in the process of English learning and perform teaching successfully, the teacher is selected on basis of professional quality, professional English quality, teaching quality, etc. It is more rigorous than that of EGP vocabulary. So it demands that the basic English teaching is performed by the teacher that can teach EGP vocabulary and the further teaching is done by the teacher that can teach ESP vocabulary in the trans-boundary mode of EGP and ESP vocabulary. In addition, the contradictions of EGP and ESP vocabulary are being communicated and studied constantly, making both of the vocabulary integrate and the students more efficiently mater ESP vocabulary.

4.2.3 *Teaching mode combining theory with practice*

EGP vocabulary is different from ESP vocabulary in teaching mode. According to the character of EGP vocabulary, the teaching mode of it focuses on not practice but theory. However, ESP vocabulary is very professional, with a teaching mode of combining theory with practice. E.g., in the teaching of EST vocabulary, the teacher should not only perform in classes but also practise in getting touch with every single word, making the students deepen the memory of professional vocabulary in scientific experiments.

So, research is done on improvement of the teaching mode in the trans-boundary mode of EGP and ESP vocabulary, and the teaching mode of combining theory with practice is applied to the teaching of both vocabularies. That makes EGP vocabulary not separated from practice and ESP vocabulary get a higher level based on EGP vocabulary. And EGP vocabulary can easily be transferred to ESP vocabulary for the purpose of trans-bound.

5 CONCLUSIONS

EGP and ESP vocabulary are connected with each other and not separated. The former is the basis of the latter, and the latter is the extension of the former. In the 21st century, the comprehensive English talents, doing well in EGP and ESP vocabulary, are needed with the improvement of the economy and society in China. So the talents with excellent professional knowledge, who are good at listening, speaking, reading and writing, are developed. It is another huge task of education in China and also an important guarantee for China's economy. It is necessary that EGP vocabulary will be transferred to ESP vocabulary as the economy raising.

Although lots of researches have been done based on the trans-boundary mode of the two, many problems have not been solved and it takes much more time. It is believed that there will be a comprehensive trans-boundary mode, combining both of them in the future.

REFERENCES

[1] Wang Yin, Zhou Ping. Comparision of EGP and ESP. Shandong Foreign Language Teaching Journal, 2011(6): 39–42.
[2] Liu Fagong. Teaching Relationship of EGP and ESP. Foreign Languages and Their Teaching. 2010(1): 20–23.
[3] Xie Xuehua. Relationship of EGP and ESP. Journal of Shandong Education Institute, 2011(4): 45–48.

Future Communication, Information and Computer Science – Zheng (Ed.)
© *2015 Taylor & Francis Group, London, 978-1-138-02653-7*

Strategies on cultivating the ability of English-majored students in normal universities based on current employment situation

F. Huang
School of Foreign Languages, Ningxia Normal University, Guyuan, Ningxia, China

ABSTRACT: At present, English-majored students in normal universities in China should make an effort to improve their employability and employment level. The work analyzed the root causes of employability problems of English-majored students in normal universities from several aspects: professional qualities, knowledge structures and ideologies. It is concluded that students in normal universities should focus on improving their practical abilities and broadening their knowledge. Addressing the analysis of employment problems on the modern market as an innovative point, the work also studied the strategies to improve the abilities of English-majored students in normal universities.

Keywords: normal university; English-majored normal university student; teaching reform

1   INTRODUCTION

Normal universities form an important component for the teacher training system in China. It is imperative to improve the teaching abilities and levels of normal universities to enhance the level of basic education in China. But at the moment, some changes have happened to the structure and scale of college and university graduates; thus the abilities that normal university students should learn and obtain in schools become more varied. According to the current employment situation and training model of English-majored students in normal universities, the work explored some problems about the ability and attainment of those students, and proposed some strategies and measures for improvement [1].

2   SCARCE CAPACITY OF ENGLISH-MAJORED STUDENTS IN MODERN NORMAL UNIVERSITIES IN CHINA

English-majored students in normal universities are the main source of the English teaching force in secondary schools in China. At present, the qualities and abilities of English-majored students in normal universities still need to be improved due to the defects of the teaching content and methods of universities [2]. Currently, English-majored students in normal universities have some defects in quality and ability including: relatively single structure, low professional quality and weak consciousness of employment.

2.1   *Professional abilities of English-majored students in normal universities to be improved*

Professional ability is a basic quality for English-majored students in normal universities to be capable of working in front-line teaching. The level of higher education in China still needs to be improved, and the cultivation mode and methods are still relatively outdated. Therefore, the professional ability of English-majored students in normal universities cannot completely adapt to the requirements of the front-line teaching staff in all levels of schools.

A certain gap exists between the English knowledge accomplishment of students and the training objectives of colleges and universities, manifesting that the ability of English-majored students in normal universities needs to be improved. In order to improve the English level of secondary school students, education departments in China require secondary schools at all levels to improve the students' abilities of listening, speaking, reading and writing. Besides, skillfully using English to study and work has also become an important goal of English teaching [3]. However, normal universities generally emphasize more the teaching and accumulation of written knowledge and theories. As a result, the improvement of the English knowledge accomplishment and teaching level can hardly be manifested in the requirement of practical curriculum systems, resulting in English-majored students in normal universities paying more attention to knowledge accumulation than practical teaching ability. Lots of English-majored students in normal universities express that their application ability of practical knowledge lags behind the course system they learned. At present, failure of converting the knowledge they learned into professional teaching ability is a common problem that a large number of students from normal universities are facing.

The finiteness of the professional ability of English-majored students in normal universities directly determines whether they can work smoothly. It is generally believed that English-majored graduates from normal universities cannot convert the knowledge they've learned into professional ability in a short period of time. Usually, it takes a long time and relatively high cost to cultivate the teaching staff, to a certain extent affecting the successful transition of English-majored normal students from being a student to being a teacher.

2.2 *Relatively narrow range of knowledge of English-majored students in normal universities*

The narrow knowledge of English-majored students in normal universities is an important factor affecting the improvement of the teachers' comprehensive capabilities. Under the current education system of colleges and universities, basic cultivation of the English language and literature is the mainstream among various cultivating models of English-majored students in normal universities. The relative neglect of the English teaching practice results in lots of those students lacking teaching practice experience and failing to be qualified as a teacher right after graduation.

Modern higher education increasingly develops towards integration; thus improving the comprehensive quality of students and cultivating compound talents becomes an important internal goal of the present higher education in China. However, in addition to an English professional knowledge, English-majored students in normal universities rarely learn knowledge of other majors as an auxiliary, failing to meet the requirement of passing on comprehensive and abundant knowledge to students in the modern education of middle schools.

Highlighting humanistic concern and literacy in higher and secondary education in China has become an important goal of liberal arts teaching. However, numerous of English-majored students in normal universities feel difficult to carry our practical teaching and be qualified of corresponding teaching work due to a narrow knowledge, leading to difficulties in the employment.

2.3 *Employment consciousness of English-majored students in normal universities to be enhanced*

Employment consciousness refers to the consciousness to get a job through integrating into society via various ways. Enhancing employment consciousness does not only require students to take the initiative to learn professional knowledge. According to a relevant national course setting and cultivating goal, employment consciousness to be engaged in related work also requires students to comprehensively improve their personal abilities and accomplishments and broaden their knowledge, thus to achieve employment through multiple ways.

On one hand, English-majored students in normal universities currently show a weak employment consciousness, actually as a result of lacking a general education in higher education. General education is the precondition of higher education in the professional stage in the cultivation models of western universities. Before the study in the professional stage, students have already learned and broadened their basic knowledge and developed basic humanistic qualities. Thus they can get a better and full understanding of their interests, so that they can choose their own major orientation and life path. But English-majored students in normal universities come from passive acceptance, and they don't have a full understanding of other professional content. Therefore, they are easy to generate learning-weariness, resulting in a lack of employment consciousness and active enthusiasm of broadening their knowledge. Some students are frustrated by changing majors and thus give up study, fiddling away the four years in university.

On the other hand, being engaged in teaching is a cultivation goal and even the life goal of most English-majored students in normal universities. But at present, many of those students have not considered or considered little about changing the single channel of employment and achieving employment in multiple ways. In order to change this situation, employment ideas about adapting to the market economy condition should be stressed in the ideological and political education of those students to encourage them to realize employment through various forms.

3 STRATEGIES TO CULTIVATE THE ABILITY OF ENGLISH-MAJORED STUDENTS IN NORMAL UNIVERSITIES IN CHINA

The above analysis manifests that three major problems hinder the ability cultivation of English-majored students in normal universities, namely: narrow knowledge, gap between theory and practice and weak consciousness of employment. The work proposed corresponding strategies based on the reform of higher education in China in order to solve the problems.

3.1 *Improving the cultivation model of English-majored students in normal universities*

The problems in the process of cultivating English-majored students in normal universities are mainly reflected in two basic aspects: knowledge and practical ability. In order to solve these two problems, normal colleges and universities at all levels strive to adjust the cultivation model of their students, and carry forward the traditional model of cultivating college and university students: stressing major knowledge and emphasizing the teaching and understanding of basic theories. At the same time, they should also strive to set up a wide-scope cultivation model and incorporate more content of history, philosophy and social science into the cultivation planning of normal university students, thus to broaden the students' knowledge scope and solve related problems caused by narrow knowledge.

Practice should be more emphasized in the process of cultivating English-majored students in normal universities. Because practice itself does not belong to classroom teaching, it is still emphasized by the traditional education force in Kairov's pedagogy. In China, it is also applied to the teaching process in colleges and universities, but not so emphasized as knowledge teaching. As a result, students lack practical abilities because they have no opportunities to practice teaching. Besides, the evaluation mechanism of students does not cover their practice, so students are not directly encouraged to take part in practicing activities, resulting in the neglect of practice. Therefore, universities should encourage students to take part in practicing activities to avoid such problems. Furthermore, practice should be designed as an essential section for students in the process of education.

3.2 Adjusting the cultivation goal of normal university students

At present, the cultivation goal of English-majored students in normal universities is relatively narrow, resulted from a long-time copy of the Soviet education model in the higher education process in China. The Soviet education model advocated carrying out teaching activities specialized in small and specific disciplines, stressing the cultivating speed and degree of specialization. The Chinese education model is deeply influenced by an important basic idea from the Soviet education model that great effort should be made to cultivate specialized talents in a short time to serve the socialist construction. But this kind of education idea constrains the flow direction of talents within the profession, resulting in students not willing and not being able to transfer from the existing employment direction so as to adapt to the employment environment under socialist market economic system. As for the teaching staff in colleges and universities in China, adjusting the cultivation goal of normal university students is not to change the professional course catalog only. Instead, it is more to change the professional cultivation model from single cultivation to multi-targets cultivation.

The basic requirement of the cultivation goal of normal university students is to provide them with more choices. Under the market economic environment, prominent position and role of individual decision making is a basic premise of guaranteeing employment marketization. In the process of teacher education and cultivation, some students are not so interested in the professional knowledge they are learning. At the moment, they are encouraged to change their majors and their study content based on their rights

of independent choices, providing important help for them.

Cultivating teachers for institutions of higher education or secondary schools is of course the top priority for the cultivation of normal university students. However, it is also indispensable to improve the comprehensive ability of students in order to improve the comprehensive quality of the whole nation, and provide the society with workers of high quality. The employment of students under the market economic environment becomes more diversified, and the ways of employment become more liberalized. Therefore, too much limitation on the cultivation goal of normal university students in fact largely impacts the update and development of the teaching and cultivating of normal university students in China.

Adjusting the cultivation goal of normal university students indicates that normal universities should take a more solid step in the form of education and teaching among comprehensive universities. Comprehensive university is the transit hub of modern social talents, having students equipped with comprehensive knowledge and independent abilities of thinking and analyzing. Thus, normal universities should set up more comprehensive courses to broaden the knowledge range for students in related majors.

3.3 Intensifying efforts to change students' ideas

One prominent problem in current Chinese higher education is the relatively poor employment ability of students. Some key universities are even facing employment problems and pressures, as a result from the cultivation process and methods of students failing to adapt to the requirements of the modern market economical environment. In addition, the students' unclear employment consciousness and unscientific understanding of employment problems are also root causes of current employment problems in universities in China.

On one hand, an important premise to solve the problems of student employment is to make them clearly understand their professional basis and expertise. Some college students even do not have a clear understanding of their professional content and characteristics, blindly studying professional knowledge and taking part in activities. As a result, when meeting employment problems, those students are totally at a loss and lack preparation. What's worse, some students even cannot get the recognition and approval from employers when some employment opportunities come to them. Therefore, relevant departments in normal universities should strive to apply ideology education into employment education, helping students to discover their own expertise and employment environment, and encouraging them to study with purpose and transfer their knowledge and abilities into the basic qualities needed for employment.

On the other hand, some students take the teaching profession as their life goal and ideal pursuit, and they confine their developing direction and space into being an English teacher in secondary schools, resulting in students failing to combine their ideals with social needs. Therefore, normal universities should adjust the students' subjective cognizance, and stop them defining their life and developing direction and path according to the cultivation mode and directional distribution in planned economy time. Instead, they should strive to improve their comprehensive abilities to meet future challenges with more flexible and sufficient preparation.

4 CONCLUSIONS

In conclusion, the main problems of cultivating university students in China include: a narrow range of knowledge, lack of general education and weak practice ability. Those problems are also very obvious in English-majored students in normal universities. Therefore, in order to adjust the education and teaching model of English-majored students in normal universities and improve their comprehensive abilities, normal universities should start from teaching as well as a cultivation mode, and change their outdated teaching methods. In addition, they should also strive to strengthen the students' consciousness of independent and flexible employment, encourage them to bravely take part in social practices to enrich their life knowledge and experience, and thus verify the correctness of their learned knowledge.

REFERENCES

[1] Deng Fangjiao, Current Situation and Strategies of Learning Burnout among Senior Students in Normal Universities. Education Exploration, 2013(12): 146–148.

[2] Luo Mingli, Discussion about Pre-Service Training and Cultivation of English-Majored Students in Normal Universities from the Perspective of Teacher Professional Development. Education and Vocation, 2011(8): 178–180.

[3] Zhang Zhenhua, Research on Applying Task-Based Teaching to Cultivate Competence-Based Teaching Skills for English-Majored Students in Normal Universities. Journal of Xinjiang University (Philosophy, Humanities & Social Science), 2009(05): 151–156.

Author index